Genetics, Genomics and Breeding of Plants

Genetics, Genomics and Breeding of Plants

Edited by Kiara Woods

SYRAWOOD
PUBLISHING HOUSE

New York

Published by Syrawood Publishing House,
750 Third Avenue, 9th Floor,
New York, NY 10017, USA
www.syrawoodpublishinghouse.com

Genetics, Genomics and Breeding of Plants
Edited by Kiara Woods

International Standard Book Number: 978-1-68286-581-1 (Hardback)

Cataloging-in-Publication Data

Genetics, genomics and breeding of plants / edited by Kiara Woods.
    p. cm.
Includes bibliographical references and index.
ISBN 978-1-68286-581-1
1. Plant genetics. 2. Genomics. 3. Plant breeding. I. Woods, Kiara.
SB123 .G46 2018
631.53--dc23

TABLE OF CONTENTS

PREFACE

I am honored to present to you this unique book which encompasses the most up-to-date data in the field. I was extremely pleased to get this opportunity of editing the work of experts from across the globe. I have also written papers in this field and researched the various aspects revolving around the progress of the discipline. I have tried to unify my knowledge along with that of stalwarts from every corner of the world, to produce a text which not only benefits the readers but also facilitates the growth of the field.

Plant genetics and breeding is a vast field of study. In modern times, the manipulation of plant genes in order to alter specific traits of plants has become possible, due to the advancement of science. Plant hybrids are also being created for commercial purposes. This book presents the complex subject of plant genetics and breeding in the most comprehensible and easy to understand language. From theories to research to practical applications, case studies related to all contemporary topics of relevance to this field have been included herein. This book is a vital tool for all researching and studying this field.

Finally, I would like to thank all the contributing authors for their valuable time and contributions. This book would not have been possible without their efforts. I would also like to thank my friends and family for their constant support.

Editor

Variation in Seed Germination of 134 Common Species on the Eastern Tibetan Plateau: Phylogenetic, Life History and Environmental Correlates

Jing Xu[1], Wenlong Li[2], Chunhui Zhang[1], Wei Liu[1], Guozhen Du[1]*

1 State Key Laboratory of Grassland and Agroecosystems, School of Life Science, Lanzhou University, Lanzhou, P.R. China, 2 State Key Laboratory of Grassland and Agroecosystems, School of Pastoral Agriculture Science and Technology, Lanzhou University, Lanzhou, P.R. China

Abstract

Seed germination is a crucial stage in the life history of a species because it represents the pathway from adult to offspring, and it can affect the distribution and abundance of species in communities. In this study, we examined the effects of phylogenetic, life history and environmental factors on seed germination of 134 common species from an alpine/subalpine meadow on the eastern Tibetan Plateau. In one-way ANOVAs, phylogenetic groups (at or above order) explained 13.0% and 25.9% of the variance in germination percentage and mean germination time, respectively; life history attributes, such as seed size, dispersal mode, explained 3.7%, 2.1% of the variance in germination percentage and 6.3%, 8.7% of the variance in mean germination time, respectively; the environmental factors temperature and habitat explained 4.7%, 1.0% of the variance in germination percentage and 13.5%, 1.7% of the variance in mean germination time, respectively. Our results demonstrated that elevated temperature would lead to a significant increase in germination percentage and an accelerated germination. Multi-factorial ANOVAs showed that the three major factors contributing to differences in germination percentage and mean germination time in this alpine/subalpine meadow were phylogenetic attributes, temperature and seed size (explained 10.5%, 4.7% and 1.4% of the variance in germination percentage independently, respectively; and explained 14.9%, 13.5% and 2.7% of the variance in mean germination time independently, respectively). In addition, there were strong associations between phylogenetic group and life history attributes, and between life history attributes and environmental factors. Therefore, germination variation are constrained mainly by phylogenetic inertia in a community, and seed germination variation correlated with phylogeny is also associated with life history attributes, suggesting a role of niche adaptation in the conservation of germination variation within lineages. Meanwhile, selection can maintain the association between germination behavior and the environmental conditions within a lineage.

Editor: Keping Ma, Institute of Botany, Chinese Academy of Sciences, China

Funding: This study was supported by the Key Program of National Natural Science Foundation of China (Grant No. 40930533) and the National Natural Science Foundation of China (Grant No. 41171214 and No. 31170430). Funder's website: http://www.nsfc.gov.cn/Portal0/default152.htm. The funders had no role in study design, data collection and analysis, decision to publish, or preparation of the manuscript.

Competing Interests: The authors have declared that no competing interests exist.

* E-mail: guozhendu2013@gmail.com

Introduction

Seed germination is one of the most extensively researched areas in plant biology [1]. The timing and level of germination strongly affect a plant's recruitment success and may consequently have implications for species migration [2]. According to recent studies, it is reasonable to expect that, seed germination could be affected by phylogeny [3], [4], life history attributes such as seed size [5], [6], seed dispersal [7], life form [8], [9] and environmental signals [10], [11]. Among the many environmental factors, temperature is perhaps more important in determining suitable conditions for seedling establishment, while other factors are germination triggers or cues [12].

However, there are three problems in these studies. Firstly, most studies have measured the effects of one variable at a time, ignoring the possibility that correlations among several phylogenetic and life history variables may confound the effects of any single variable on seed germination. For example, seed size is related with seed dispersal [13], habitat [14] and growth form

[15], [16]. Consequently, to assess the role of natural selection on seed germination at the community level, we should take into account as many variables as possible when measuring the effect of any single variable. Nevertheless, only a few studies focus on the effect of phylogenetic, life history and environmental correlates on seed germination. For example, Wang et al. (2009) investigated seed germination of 69 arid/semi-arid zone species [17]. Secondly, although it is important to predict future distributions of species [18] and the germination study of seeds collected from one community at the same time may provide important information to understand the dynamics of a community, very few studies have been addressed to test seed germination in an alpine/subalpine community by combining phylogenetic analysis. Thirdly, temperature is predicted to increase with climate change [19] and the warming is much more intense in mountainous and high-elevation regions than at low altitude [20], [21]. To alpine/subalpine plants, which are expected to be affected more by climate warming [22], the effects of temperature on germination have hardly been studied from a community perspective. Therefore, we expect to

advance our understanding on how phylogenetic, life history and environmental factors to regulate seed germination in an alpine/subalpine meadow community.

In this study, we chose 134 common species collected from the alpine/subalpine meadow on the eastern Tibetan Plateau, and the following questions were addressed (a) Whether differences in seed germination among species from the same community are related to phylogeny, life history traits, and/or environmental factors? (b) What proportion of germination variation among species could be attributed to the species' phylogenetic background, life history attributes, and environmental conditions? (c) Does a higher germination temperature affect germination of alpine/subalpine plants, and if so, how?

Materials and Methods

Study site

The study area is located on the eastern Tibetan Plateau (101°-103°E, 34°-35.70°N). The altitude ranges from 2800 to 4200 m, and the climate is cold Humid-Alpine with a mean annual precipitation (snow and rainfall) of 620 mm. Mean annual temperature is 2–3°C, and mean January and July temperatures are −10.7°C and 11.7°C, respectively. There is an average of 270 frost days per year. The grassland is dominated by native monocotyledons such as species of Poaceae and Cyperaceae and by native dicotyledons such as species of Ranunculaceae, Polygonaceae, Saxifragaceae, Asteraceae, Scrophulariaceae, Gentianaceae, and Fabaceae.

Seed collecting and germination tests

In this study, seeds of 134 common species (Table S1) were collected from private grasslands of the study site from July to October in 2009 after we obtained permission from the grasslands owners, our studies did not involve endangered or protected species, and all the germination experiments were carried out in our own laboratory. For each species, seeds were collected at the beginning of their dispersal period to ensure they were mature. For a single species, seeds were collected from one site but from more than 20 plants. Seeds were air dried after collection, cleaned and stored at room temperature (approximately 15°C). For every species, three replicates of 100 air-dried seeds were randomly selected and weighed, and average mass per seed was calculated.

Before the germination experiments, the viability of seeds of each species was tested with tetrazolium chloride [23]. Only species with a seed viability of ≥99% were used in the germination experiments. The germination experiments were started on the middle of March (starting season of germination in the study area) in 2010. Seeds were placed in Petri dishes (9 cm diameter) on double layers of moistened filter paper, and then placed in growth chambers (Conviron E15 Growth Chamber, Controlled Environments Ltd., Winnipeg, Canada) under five different incubation treatments: 5/15°C (12:12 h) was simulated natural conditions prevailing in the soil at 5 cm depth in April and May in the study area (control treatment). 5/20°C (12:12 h), 5/25°C (12:12 h), 10/20°C (12:12 h) and 10/25°C (12:12 h) were simulated temperature increase. This experiment used three replications of 50 seeds of each species per temperature treatment. Seeds were kept saturated with distilled water. Dishes were randomized, stacked, and placed in temperature chamber. Seeds were incubated under darkness and a relative humidity of about 70% for 60 days. The seeds were checked for germination daily, at which time they were exposed to light for a few minutes. Thus, any light requirement for seed germination was fulfilled during these exposures [10]. The visible protrusion of the radicle was the criterion for germination.

Germinated seeds were removed from the Petri dishes at each counting. At the end of each experiment, the remaining ungerminated seeds were tested for viability by staining with tetrazolium chloride, and the proportion of unviable seeds was calculated.

Statistical analysis

Germination percentage (GP) was defined as the proportion of seed germinated, and seed mortality was defined as the proportion of unviable seeds tested with tetrazolium chloride after germination experiments. Mean germination time (GT) was estimated as follows: $GT = \sum(G_i \times i)/\sum(G_i)$, where i is the day of germination, counted since the day of sowing, and G_i is the number of seeds germinated on day i [24]. Three species that did not germinate at the end of the experiments were not included in this calculation, i.e., 131 species were used in GT analysis (Table S1).

First, a composite phylogeny of 134 species was constructed with Phylomatic version 3 (http://phylodiversity.net/phylomatic/) based on the angiosperm megatree (R20120829) [25]. Branch lengths were made proportional to time using the 'bladj' function in the program Phylocom 4.0 [26] and divergence time was estimated based on fossil data [27], [28]. To test the robustness of our results to uncertainties associated with branch length estimates, we also ran our analyses on the same composite tree, but with branch lengths set to 1. The resulting phylogenetic tree was used for subsequent analyses. We tested for the existence of phylogenetic signal by estimating Pagel's λ for GP and GT using "fitContinuous" functions in the R package "geiger" version 1.99-3 [29], using a maximum likelihood framework to estimate the parameter λ, which can vary from 0 (no influence of phylogeny) to 1 (strong phylogenetic influence) [30].

Then, one-way, two-way and multi-factorial ANOVAs were used to determine the effects of phylogeny and various life history (i.e., seed size, dispersal mode, life form, onset of flowering, duration of flowering) and environmental attributes (i.e., temperature and habitat) on GP and GT. One-way ANOVAs measured the effects of each factor on the variance of GP and GT across all other variables; two-way ANOVAs were conducted to detect significant interactions and associations between factors; multi-factorial ANOVAs tested the effect of each class variable independent of the others. We conducted a series of ANOVAs which include all variables but one (incomplete model). When each of these ANOVAs was compared to the ANOVA including all variables (complete model), the difference between the proportion of the total sum of squares (ss) explained by the complete model (its R^2) and the R^2 of the incomplete model represented the proportion of the total ss explained by the deleted class variable [3]. Besides, multi-factorial ANOVAs corroborate associations between factors suggested by the two-way ANOVAs. If in the complete ANOVA, a given class variable had a lower R^2 value than in the incomplete ANOVA from which a different variable had been deleted, the increase in the R^2 value of the first variable would be due to an association (or correlation) or strong interaction with the second variable [3]. To carry out the statistical analysis, we grouped 134 species according to the following categories:

1. Phylogenetic group. Each of the 134 species was assigned to a family and an order according to Angiosperm Phylogeny Group III [31] (Table S1). When comparing the GP and GT between families, families containing more than seven species were chosen.

2. Life form. Species were grouped into two classes: annual and perennial.

3. Dispersal mode. Species were classified into four groups according to the morphological features of their seeds [4]: unassisted, ant-dispersed, adhesion-dispersed and wind-dispersed.

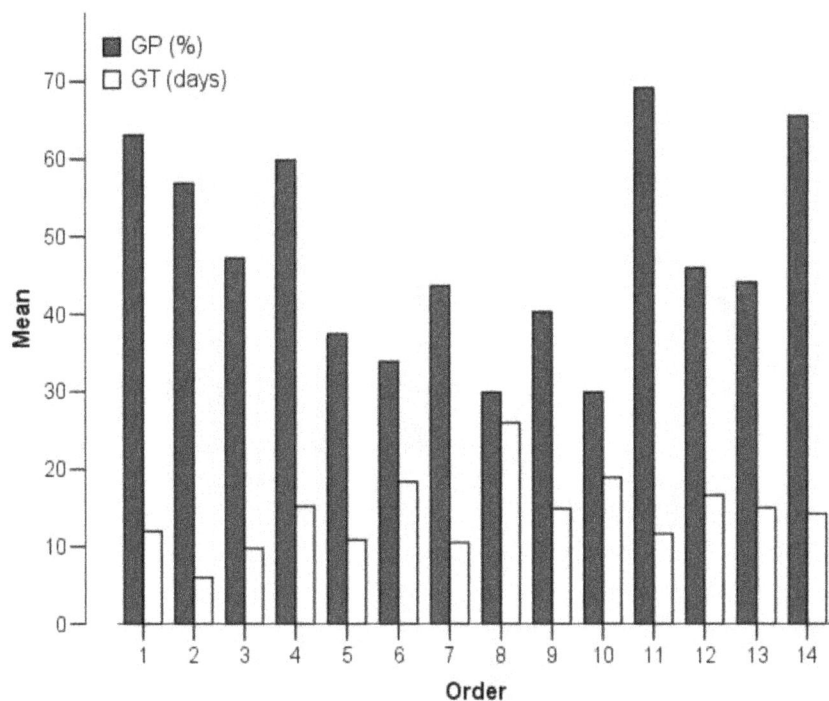

Figure 1. Germination percentage (GP) and mean germination time (GT) of seeds from 14 orders. 1 = Asterales, 2 = Brassicales, 3 = Caryophyllales, 4 = Ericales, 5 = Fabales, 6 = Gentianales, 7 = Lamiales, 8 = Liliales, 9 = Malpighiales, 10 = Myrtales, 11 = Poales, 12 = Ranunculales, 13 = Rosales, 14 = Saxifragales.

4. Seed size. The mean seed size of each species was assigned to 1 of 5 seed size classes according to Baker [32]: 0.032–0.099 mg, 0.100–0.315 mg, 0.316–0.999 mg, 1.000–3.161 mg, 3.162–9.999 mg.

5. Onset of flowering. Each species was grouped based on the Flora of China [33] and field observation records: early, flowering begins in May; middle, flowering begins in June; or late, flowering begins in July and August.

6. Duration of flowering. Each species was grouped based on the Flora of China [33] and field observation records: short, flowering duration of 1 month; median, flowering duration of 2–3 months; or long, flowering duration of ≧4 months.

7. Temperature. Based on temperatures occurring in the species' habitats, rising trend and the optimum alternating temperature regime of the most species [10], 5/15°C (control treatment), 5/20°C, 5/25°C, 10/20°C and 10/25°C were chosen.

8. Habitat. The habitats were classified into three categories: bottomland, north slope and south slope.

Because the data were unbalanced, all ANOVAs were conducted using GLM procedure of SPSS 13.0. The type III sum of squares was used to establish the significance level of each effect. In addition, both GP and mortality were arcsine square root transformed, and GT were log-transformed to improve normality and stabilize variances.

Results

Phylogenetic correlates

The results indicated that phylogenetic signals (λ) for GP and GT were 0.53 and 0.75, respectively. Both λ values were significantly different from 0 (χ^2 tests, both $P<0.001$).

One-way ANOVAs indicated that both GP and GT were significantly different among taxa (Figure 1, 2A), and order

membership could account for 13.0% of the variance in GP and 25.9% of the variance in GT (Table 1). Thus, the majority of seed germination variation took the form of variation within orders.

Two-way ANOVAs yielded significant interaction terms for phylogenetic relatedness when analyzed in combination with life history attributes, such as seed size, life form (Table 2).

The multi-factorial ANOVAs suggested the variance in GP and GT explained by order independently were 10.5% and 14.9%, respectively (Table 3, 4). In the multi-factorial ANOVAs of GT, the R^2 of life form and dispersal mode increased from 4.6% to 6.8% and from 2.8% to 5.7% respectively when phylogenetic group was deleted from the model, which suggested there were associations between phylogenetic group and life form and between phylogenetic group and dispersal mode (Table 4).

Life history correlates

Life form. One-way ANOVAs indicated that the impact of life form on GP was not statistically significant, whereas the effect of life form on GT was notable (Table 1). The GT of annuals (11.03±0.50 days, mean±SE, hereafter) showed earlier than that of perennials (13.76±0.33 days). Two-way ANOVAs showed significant interaction terms for life form when analyzed in combination with phylogeny, seed size, dispersal mode and habitat (Table 2). The multi-factorial ANOVAs suggested life form accounted for 2.5% of the variance in GT independently (Table 4).

Seed size. Generally, species with heavier seeds (seed size ranged from 3.162 mg to 9.999 mg) had the highest GP (63.58±4.34%) and the most delayed GT (17.50±1.35 days); small-seeded species (seed size ranged from 0.032 mg to 0.099 mg) had the lowest GP (43.63±3.68%, Figure 2B).

One-way ANOVAs indicated that seed size had significant effects on germination, which could account for 3.7% and 6.3% of the variance in GP and GT, respectively (Table 1). The linear

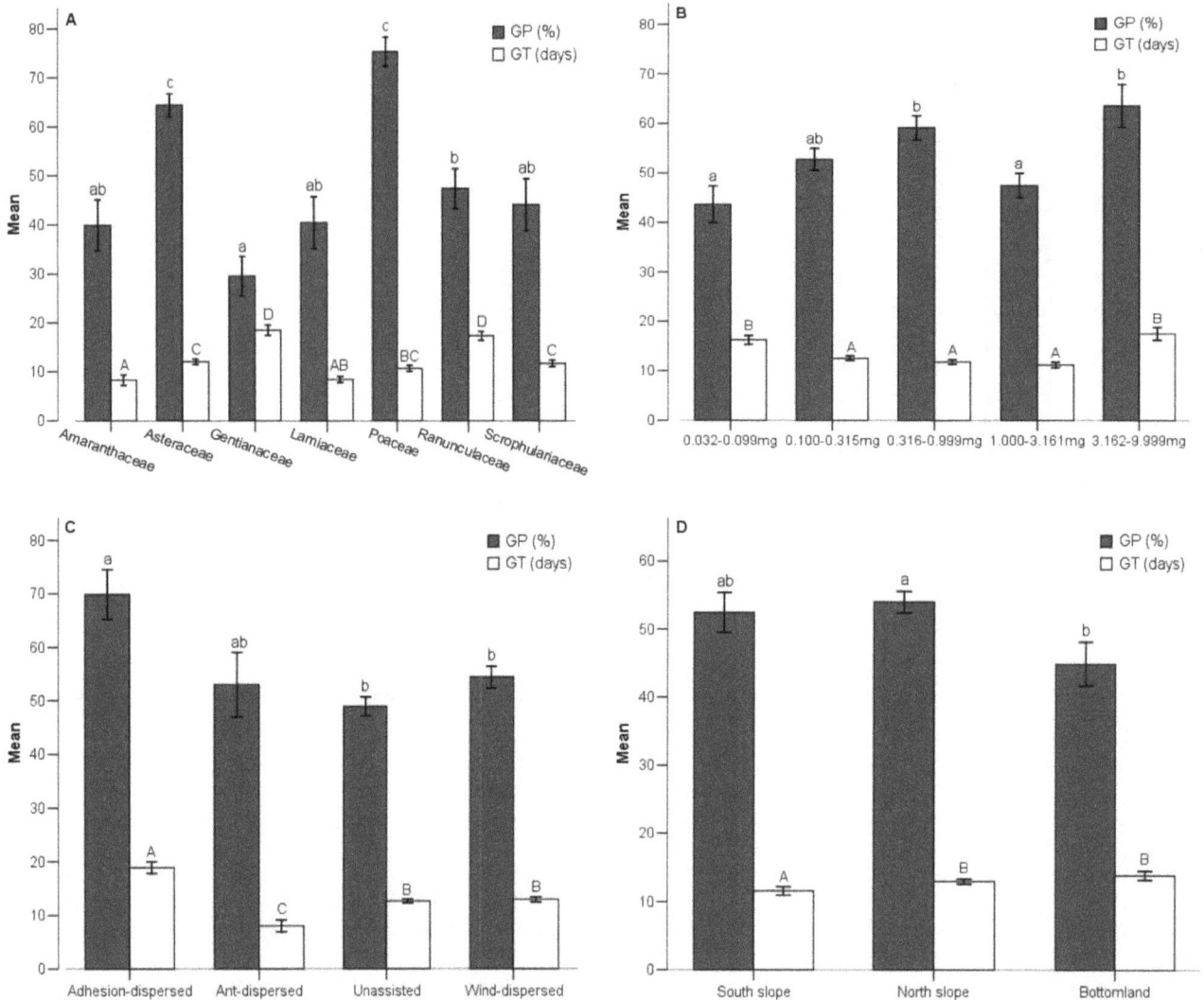

Figure 2. Germination percentage (GP) and mean germination time (GT) in different taxonomic groups. (A) GP and GT of seeds from seven families; (B) GP and GT of seeds from five seed size groups; (C) GP and GT of seeds from four dispersal mode groups; (D) GP and GT of seeds from three habitats. Bars (mean±SE) that do not share a letter represent significantly different values at $P < 0.05$ level (Turkey multiple comparison test). Different lowercase letters and capital letters indicate significant difference of GP and GT, respectively.

regression analysis showed there was either no significant correlation between seed size and GP ($R^2 = 0.005$, $P = 0.397$), or no GT ($R^2 = 0.016$, $P = 0.154$) (Figure S1A, S1B). Further, two-way ANOVAs detected significant interactions between seed size and habitat, and between seed size and other life history attributes, such as dispersal mode (Table 2).

In the incomplete models, the multi-factorial ANOVAs suggested seed size accounted for 1.4% and 2.7% of the variance in GP and GT independently, respectively (Table 3, 4).

Dispersal mode. Seeds of adhesion-dispersed species showed the highest GP ($69.84 \pm 4.60\%$) and the most delayed GT (18.89 ± 1.10 days), seeds of ant-dispersed species presented the earliest GT (8.04 ± 1.10 days), and seeds of unassisted species had the lowest GP ($48.92 \pm 1.74\%$, Figure 2C).

One-way ANOVAs suggested dispersal mode had significant effects on both GP and GT, which could account for 2.1% and 8.7% of the variance in GP and GT, respectively (Table 1). The multi-factorial ANOVAs suggested dispersal mode explained

0.7% and 1.5% of the variance in GP and GT independently, respectively (Table 3, 4).

Flowering time. One-way ANOVAs suggested that the impact of onset of flowering on GP, and the effect of duration of flowering on GT were statistically significant (Table 1). The GP of late-flowering species ($59.85 \pm 2.94\%$) was higher than that of others. The GT of medium duration of flowering species (13.11 ± 0.30 days) was later than that of others. Results also revealed that no linear correlations were observed, neither between GP and flowering time ($R^2 = 0.012$, $P = 0.212$) nor between GT and flowering time ($R^2 < 0.001$, $P = 0.927$) (Figure S1C, S1D).

Two-way ANOVAs showed significant interaction terms for flowering time when analyzed in combination with phylogeny, life history attributes and habitat (Table 2). The multi-factorial ANOVAs suggested onset of flowering explained 0.7% of the variance in GP (Table 3), and duration of flowering explained 0.5% of the variance in GT independently (Table 4).

Table 1. Results of one-way ANOVAs showing effect of phylogeny, life form, seed size, dispersal mode, onset of flowering, duration of flowering, temperature and habitat on final germination percentage (GP) and mean germination time (GT) among 134 species, R^2 is the proportion of variance explained by each factor.

Source of variation	Germination percentage (GP)				Mean germination time (GT)		
	df	F	P	R^2	F	P	R^2
Phylogenetic group	13	7.560	<0.001	0.130	17.219	<0.001	0.259
Life form	1	0.061	0.804	<0.001	52.508	<0.001	0.074
Seed size	4	6.318	<0.001	0.037	10.900	<0.001	0.063
Dispersal mode	3	4.821	0.003	0.021	20.579	<0.001	0.087
Onset of flowering	2	4.147	0.016	0.012	0.528	0.590	0.002
Duration of flowering	2	2.042	0.131	0.006	6.872	0.001	0.021
Temperature	4	8.123	<0.001	0.047	25.258	<0.001	0.135
Habitat	2	3.392	0.034	0.010	5.778	0.003	0.017

Environmental correlates

Temperature. Generally, the earliest GT occurred in 10/20°C (10.15±0.47 days), and the highest GP occurred in 5/25°C (59.50±2.81%), whereas the most delayed (17.75±0.68 days) and the poorest germination (38.16±2.82%) occurred in 5/15°C (Figure 3). Tetrazolium tests revealed that most ungerminated seeds were still alive at the end of the experiments. The percentage of ungerminated but viable seeds was 51.32%, 32.43%, 26.33%, 24.46% and 22.63% at 5/15°C, 5/20°C, 5/25°C, 10/20°C and 10/25°C respectively, and the temperature treatments significantly affected the mortality of the seeds (Figure 4).

One-way ANOVAs suggested that temperature had statistically significant effects on seed germination, and explained 4.7% and 13.5% of the total variance in GP and GT, respectively (Table 1). The multi-factorial ANOVAs suggested the variance in GP and GT explained independently by temperature was 4.7% and 13.5%, respectively (Table 3, 4).

Habitat. Generally, seeds from bottomland displayed lower GP (44.87±3.23%) than seeds from north slope (53.99±1.55%), and seeds from south slope presented earlier GT (11.56±0.65 days) than that from other habitats (Figure 2D). One-way ANOVAs suggested that habitat had a marked effect on germination, and contributed 1.0% and 1.7% of total variance in GP and GT, respectively (Table 1). Two-way ANOVAs yielded significant interaction terms for habitat when analyzed in combination with life history attributes (Table 2). The multi-factorial ANOVAs suggested the variance in GP and GT explained independently by habitat was 0.3% and 1.0%, respectively (Table 3, 4).

Discussion

Phylogenetic correlates

There is a growing concern that the optimization of organisms by natural selection may be influenced or prevented by life history or phylogenetic constraints [34], and they can be used to explain variations in ecological or other traits among taxa [5], [35]. This study has confirmed that both GP and GT are phylogenetically conserved traits in an alpine/subalpine meadow community, which suggests that, despite large interspecific variation, the range of variation in these traits is limited by phylogenetic affiliation. Seed germination, like any other trait, is shaped both by the natural history of the species and by the evolutionary history of the lineage, and a large proportion of interspecific variation in germination is correlated with taxon membership, representing lineage history. Similarly, Norden et al. (2009) reported that germination delay was a phylogenetically conserved trait [6]. Thus, our findings are consistent with Norden, and further show that germination percentage is also a phylogenetically conserved trait. This is mainly due to that closely related species tend to share similar values for a given trait, typically more similar than distantly related species [36], and seed germination has coevolved with other plant traits that are directly involved in regeneration success [6].

Life history correlates

Life form. In this study, we found that life form had an insignificant effect on GP, but annuals germinated significantly earlier than perennials. Similarity, Wang et al. indicated that there was no significant effect of life form on GP of 69 arid/semi-arid zone species [17], and Schippers's simulations indicated that being a non-dormant annual could be a viable strategy [37]. On the contrary, Rees suggested that in a variable environment, annuals tended to have more dormancy than perennials based on a large

Table 2. Results of two-way ANOVAs showing the independent effects of one of two main factors that have significant effects in one-way ANOVAs and interaction effects on germination percentage (GP) and mean germination time (GT) due to phylogenetic group (P), life form (LF), seed size (SS), dispersal mode (DM), onset of flowering (OF), duration of flowering (DF), temperature (T) and habitat (H). R^2 is the proportion of variance explained by each factor (Only significant interaction terms are shown).

Source of variation A/B	Effect of A				Effect of B				Effect of A×B			
	df	F	P	R^2	df	F	P	R^2	df	F	P	R^2
Germination percentage (GP)												
P/SS	13	6.441	<0.001	0.117	4	5.989	<0.001	0.037	23	1.637	0.031	0.056
P/OF	13	5.523	<0.001	0.101	2	1.021	0.361	0.003	12	3.853	<0.001	0.067
SS/H	4	8.663	<0.001	0.050	2	2.457	0.086	0.007	7	3.330	0.002	0.034
DM/OF	3	6.616	<0.001	0.029	2	6.390	0.002	0.019	5	5.530	<0.001	0.040
OF/H	2	1.313	0.270	0.004	2	7.443	0.001	0.022	4	4.218	0.002	0.025
Germination time (GT)												
P/LF	13	12.561	<0.001	0.205	1	19.899	<0.001	0.030	7	3.073	0.003	0.033
P/SS	13	13.267	<0.001	0.219	4	8.064	<0.001	0.050	21	2.624	<0.001	0.082
P/DM	13	9.412	<0.001	0.163	3	8.222	<0.001	0.038	9	4.867	<0.001	0.065
P/DF	13	14.506	<0.001	0.231	2	3.452	0.032	0.011	10	2.748	0.003	0.042
LF/SS	1	46.552	<0.001	0.067	4	12.122	<0.001	0.070	4	3.406	0.009	0.021
LF/DM	1	21.975	<0.001	0.033	3	5.963	0.001	0.027	3	7.007	<0.001	0.031
LF/H	1	56.872	<0.001	0.081	2	7.038	0.001	0.021	2	4.679	0.010	0.014
SS/DM	4	4.475	0.001	0.027	3	12.428	<0.001	0.055	6	6.015	<0.001	0.053
SS/DF	4	10.883	<0.001	0.063	2	0.391	0.676	0.001	5	4.258	0.001	0.032
SS/H	4	8.120	<0.001	0.048	2	6.110	0.002	0.019	7	3.119	0.003	0.033
DM/DF	3	21.719	<0.001	0.092	2	1.875	0.154	0.006	4	2.670	0.031	0.016
DM/H	3	17.752	<0.001	0.076	2	0.444	0.641	0.001	5	5.343	<0.001	0.040
DF/H	2	16.325	<0.001	0.048	2	5.107	0.006	0.016	4	6.262	<0.001	0.037

Table 3. Multi-factorial ANOVAs for the independent effects of each main factor and their associations.

Source of variation	df	F	P	R²	df	F	P	R²
Complete model					**Phylogenetic group removed**			
Phylogenetic group	13	6.559	<0.001	0.118				
Life form	1	0.130	0.719	<0.001	1	0.444	0.506	0.001
Seed size	4	2.869	0.022	0.018	4	4.127	0.003	0.025
Dispersal mode	3	1.707	0.164	0.008	3	2.037	0.107	0.009
Onset of flowering	2	2.797	0.062	0.009	2	3.727	0.025	0.011
Duration of flowering	2	0.756	0.470	0.002	2	1.065	0.345	0.003
Temperature	4	9.520	<0.001	0.056	4	8.569	<0.001	0.050
Habitat	2	1.049	0.351	0.003	2	2.132	0.119	0.007
Model	31	5.789	<0.001	0.220	18	4.710	<0.001	0.115
Seed size removed					**Dispersal mode removed**			
Phylogenetic group	13	7.006	<0.001	0.124	13	6.666	<0.001	0.119
Life form	1	0.043	0.835	<0.001	1	0.207	0.649	<0.001
Seed size					4	3.565	0.007	0.022
Dispersal mode	3	2.619	0.050	0.012				
Onset of flowering	2	2.792	0.062	0.009	2	3.822	0.022	0.012
Duration of flowering	2	1.895	0.151	0.006	2	0.669	0.513	0.002
Temperature	4	9.411	<0.001	0.055	4	9.489	<0.001	0.056
Habitat	2	1.870	0.155	0.006	2	0.889	0.412	0.003
Model	27	6.150	<0.001	0.206	28	6.206	<0.001	0.213
Onset of flowering removed					**Temperature removed**			
Phylogenetic group	13	6.728	<0.001	0.120	13	6.229	<0.001	0.112
Life form	1	0.083	0.773	<0.001	1	0.123	0.726	<0.001
Seed size	4	2.867	0.023	0.018	4	2.725	0.029	0.017
Dispersal mode	3	2.384	0.068	0.011	3	1.621	0.183	0.008
Onset of flowering					2	2.656	0.071	0.008
Duration of flowering	2	0.970	0.380	0.003	2	0.718	0.488	0.002
Temperature	4	9.467	<0.001	0.056				
Habitat	2	1.506	0.223	0.005	2	0.996	0.370	0.003
Model	29	5.962	<0.001	0.213	27	4.973	<0.001	0.173
Habitat removed								
Phylogenetic group	13	6.761	<0.001	0.121				
Life form	1	0.220	0.639	<0.001				
Seed size	4	3.290	0.011	0.020				
Dispersal mode	3	1.602	0.188	0.007				

Table 3. Cont.

Source of variation	df	F	P	R²	df	F	P	R²
Onset of flowering	2	3.263	0.039	0.010				
Duration of flowering	2	0.431	0.650	0.001				
Temperature	4	9.519	<0.001	0.056				
Habitat								
Model	29	6.115	<0.001	0.217				

Dependent variable is germination percentage (GP). For each main factor, R^2 is the proportion of the Type III sum of squares attributed to the main effect. The proportion of the variance explained by each class variable independent of others examined by the difference between the R^2 of the complete model and the R^2 of the model from which this class variable has been deleted.

grass data set [9]. These different conclusions could be partially explained by the effects of life form on seed germination, which may vary over habitats or floras. In other words, species composition and life form category are different in distinct habitats or floras, and the classification principles of life form are not identical, all of these factors will affect the final results.

Moreover, there are two major reasons why annuals germinated earlier than perennials on the eastern Tibetan Plateau. For one thing, annuals are more dependent on seeds than perennials in order to be able to persist in the environment in reproduction process [14], and early-germinating species preempt biological space and gain competitive advantage over late-germinating species. For another, the short growing season is a major barrier for the survival of seedlings in alpine/subalpine meadow on the Tibetan Plateau. Therefore, in suitable conditions, rapid germination is critical for successful establishment of annuals.

Seed size. Seed size is an important parameter of plant fitness as it may highly influence the regeneration process of a population [38]. In our study, seeds range over three orders of magnitude in size. This likely represents multiple solutions to the same problem, for example, some plants choose to make many small seeds, and some make a few large seeds. Neither of these strategies is "better", both may work equally well. This may explain why most of our factors (other than phylogeny) explained a very small amount of the overall variance: size does matter, but there is more than one right choice.

On the other hand, there is no accordant relationship between seed size and germination strategies in the previous studies. For example, many studies have indicated that there is a significant negative relationship between seed size and dormancy [11]. However, Wang et al. reported that germination percentages among species had a significant negative correlation with seed size [17]. Some other authors reported that seed size did not have a general effect on germination [39]. In this study, we proved that seed size had significant effects on both GP and GT (but not linear relation between germination and seed size). These different results may stem from the following reasons. Firstly, the important factors co-varying with seed size and/or seed dormancy may have been left out of consideration [11]. For example, Rees (1996) found a significant positive relationship between seed size and germination in species with specialization for dispersal but no such relationship for unspecialized seeds [15]. Secondly, seed size varies greatly among different floras and the distribution of seed size in alpine/subalpine meadows on the Tibetan Plateau is skewed to small size compared with other communities. For example, seed size of Wang's study ranged from 0.06 mg to 63.50 mg, with a mean of 10.54 mg, and 70% of the seeds were heavier than 1 mg, whereas seed size of our study ranged from 0.03 mg to 6.61 mg, with a mean of 0.98 mg, only 33% of the seeds were heavier than 1 mg. Thirdly, the environment of plants to be a better predictor of dormancy than are plant longevity and seed size combined [11]. Therefore, the differences among habitats should be considered, especially the special condition of alpine/subalpine meadow in the Tibetan Plateau.

Besides, our results revealed that species with heavier seeds had the highest GP and the most delayed GT. This is mainly because heavier seeds have larger embryos and more endosperm nutrients, which is associated with increased germination percentage [40]. Nevertheless, larger seeds would not germinate fast due to their usual opacity of thick and hard seed coats [41], and small seeds are expected to have a competitive advantage over larger seeds by having faster emergence, since small seeds have proportionally greater surface area for water absorption [42].

Table 4. Multi-factorial ANOVAs for the independent effects of each main factor and their associations.

Source of variation	df	F	P	R²	df	F	P	R²
Complete model								
Phylogenetic group	13	14.066	<0.001	0.227				
Life form	1	29.940	<0.001	0.046				
Seed size	4	8.135	<0.001	0.050				
Dispersal mode	3	6.078	<0.001	0.028				
Onset of flowering	2	4.495	0.012	0.014				
Duration of flowering	2	2.885	0.057	0.009				
Temperature	4	41.292	<0.001	0.210				
Habitat	2	5.897	0.003	0.019				
Model	31	19.510	<0.001	0.493				
Phylogenetic group removed								
Phylogenetic group								
Life form					1	46.681	<0.001	0.068
Seed size					4	8.694	<0.001	0.052
Dispersal mode					3	12.851	<0.001	0.057
Onset of flowering					2	1.452	0.235	0.005
Duration of flowering					2	4.616	0.010	0.014
Temperature					4	32.589	<0.001	0.170
Habitat					2	6.077	0.002	0.019
Model					18	18.501	<0.001	0.344
Life form removed								
Phylogenetic group	13	15.590	<0.001	0.245				
Life form								
Seed size	4	6.645	<0.001	0.041				
Dispersal mode	3	5.971	0.001	0.028				
Onset of flowering	2	3.841	0.022	0.012				
Duration of flowering	2	5.336	0.005	0.017				
Temperature	4	39.462	<0.001	0.202				
Habitat	2	9.046	<0.001	0.028				
Model	30	18.313	<0.001	0.468				
Seed size removed								
Phylogenetic group					13	14.302	<0.001	0.229
Life form					1	24.116	<0.001	0.037
Seed size								
Dispersal mode					3	11.765	<0.001	0.053
Onset of flowering					2	3.693	0.025	0.012
Duration of flowering					2	3.348	0.036	0.011
Temperature					4	39.494	<0.001	0.201
Habitat					2	6.247	0.002	0.020
Model					27	20.272	<0.001	0.466
Dispersal mode removed								
Phylogenetic group	13	16.031	<0.001	0.250				
Life form	1	29.728	<0.001	0.045				
Seed size	4	12.499	<0.001	0.074				
Dispersal mode								
Onset of flowering	2	2.914	0.055	0.009				
Duration of flowering	2	2.553	0.079	0.008				
Temperature	4	40.311	<0.001	0.205				
Habitat	2	6.457	0.002	0.020				
Model	28	20.452	<0.001	0.478				
Duration of flowering removed								
Phylogenetic group					13	14.435	<0.001	0.231
Life form					1	35.127	<0.001	0.053
Seed size					4	8.393	<0.001	0.051
Dispersal mode					3	5.862	0.001	0.027
Onset of flowering					2	5.171	0.006	0.016
Duration of flowering								
Temperature					4	41.045	<0.001	0.208
Habitat					2	4.917	0.008	0.015
Model					29	20.533	<0.001	0.488
Temperature removed								
Phylogenetic group	13	11.190	<0.001	0.188				
Life form	1	23.818	<0.001	0.037				
Seed size	4	6.472	<0.001	0.040				
Dispersal mode	3	4.835	0.002	0.023				
Habitat removed								
Phylogenetic group					13	14.122	<0.001	0.227
Life form					1	36.505	<0.001	0.055
Seed size					4	8.322	<0.001	0.051
Dispersal mode					3	6.454	<0.001	0.030

Table 4. Cont.

Source of variation	df	F	P	R^2	df	F	P	R^2
Onset of flowering	2	3.576	0.029	0.011	2	5.893	0.003	0.019
Duration of flowering	2	2.295	0.102	0.007	2	1.904	0.150	0.006
Temperature					4	40.655	<0.001	0.206
Habitat	2	4.691	0.009	0.015				
Model	27	12.953	<0.001	0.358	29	20.133	<0.001	0.483

Dependent variable is mean germination time (GT). For each main factor, R^2 is the proportion of the Type III sum of squares attributed to the main effect. The proportion of the variance explained by each class variable independent of others examined by the difference between the R^2 of the complete model and the R^2 of the model from which this class variable has been deleted.

Dispersal mode. The principal models propose that presence and duration of seed germination would be correlated with seed dispersal mechanism [8], [43]. In this study the effects of dispersal mode on seed germination have been demonstrated, with adhesion-dispersed seeds germinating to higher GP than unassisted seeds (Figure 2C), which is largely for the following three reasons. Firstly, some natural enemies of seeds and seedlings (such as seed predators, parasites, herbivores and pathogens) respond to density and/or distance from the parent. Secondly, sib competition may often be more severe than competition with non-sib competition, because their patterns of resource use are probably more similar. Thirdly, some species have special microhabitat for germination and establishment. Seeds of unassisted species are most likely to experience problems noted above, thus, decrease in GP can spread these risks encountered by unassisted seeds. Our results also revealed that ant-dispersed species displayed the earliest GT. This is mainly due to avoiding being buried too deeply by ants, which may result in seeds failing to germinate.

Moreover, we found that there was strong association between dispersal mode and seed size (Table 2, 4). This correlation is frequently interpreted in terms of adaption to different lifestyles [3]. Generally speaking, small-seeded species should disperse better than large-seeded species, trading off seed size with dispersal capacity [44].

Flowering time. In plants, vegetative growth, flowering, seed development, dispersal, and germination typically follow in sequence with more or less overlap between the phases. So it is necessary to have a complex perspective when assessing the impact of a single phenological trait like flowering time [45]. However, there are very few studies examining the interspecific effects of flowering time on germination. Our results indicated that no linear relation existed between germination characteristics and flowering time (Figure S1C, S1D). Nevertheless, onset of flowering had a significant effect on GP and duration of flowering had a significant influence on GT, although the percentage of variance in germination explained independently by flowering time was very small (both were less than 1%, Table 3, 4). As opposed to our results, Wang et al. (2009) indicated flowering time had no marked impact on seed germination [17]. Some possible explanations for the contradiction are conceivable. Firstly, flowering phenology changes along elevation gradients, with plants at higher elevations typically flowering later than plants of the same species that grow at lower elevations [46]. Secondly, differences in flowering time are often attributable to the degree to which flowering is related by the timing of other phenophases such as seed dispersal and seed germination [47]. Thirdly, early flowering time may imply early dispersal, germination, and thus a longer period of growth available to the juvenile. But a long juvenile period, including the unfavorable season, increases the risk of mortality before reproduction. This selection pressure will involve tradeoff [48], which may lead to difficulties to determine the impact of flowering time on germination directly and simply.

Moreover, flowering periods patterns are constrained mainly by phylogenetic inertia at the family level [49]. Our results revealed that more than 50% of the late-flowering species belong to Asteraceae, and species of Asteraceae presented higher GP in alpine/subalpine meadow community (Figure 2A), which can partly explain why late-flowering species displayed higher GP.

Environmental correlates

Temperature. We believe it is justified to pay close attention to temperature because it has been proven to be the most important environmental variable regulating seed dormancy and germination [10], [11]. Our results indicated that temperature had

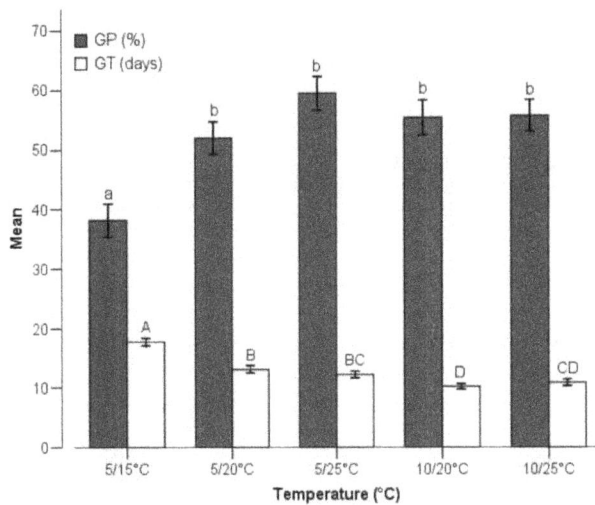

Figure 3. Effects of temperature on germination percentage (GP) and mean germination time (GT). Bars (mean±SE) that do not share a letter represent significantly different values at P<0.05 level (Turkey multiple comparison test).

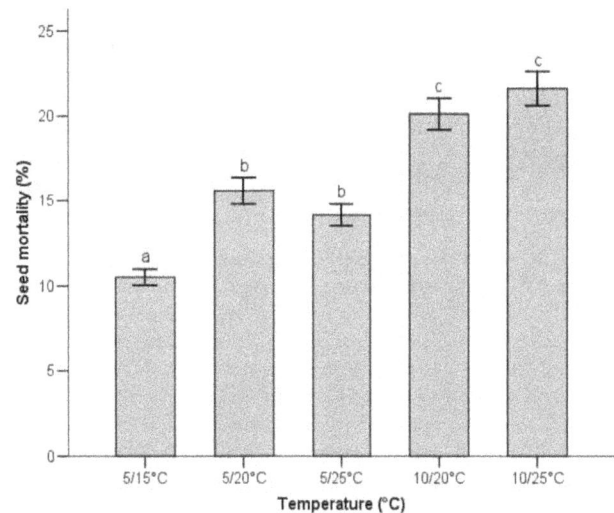

Figure 4. Effects of temperature treatments on seed mortality during germination. Bars (mean±SE) that do not share a letter represent significantly different values at P<0.05 level (Turkey multiple comparison test).

a marked effect on germination and elevated temperature would lead to a significant increase in GP and an accelerated germination compared with control (Figure 3). This is consistent to the widely accepted view. For example, Baskin and Baskin have suggested that alpine species require relatively high temperatures for germination [10]. Milbau has suggested that the germination temperature in alpine plants is relatively high in comparison with ambient temperatures [2]. On the other hand, quite a lot of seeds were ungerminated but viable in our experiments, the mean percentage of ungerminated but viable seeds was 31%. We consider that may be an adaptation to the local harsh environment. Because of the germinated seeds unable to come back formerly static status, and unable to ensure seedlings could adapt to the multivariate conditions of alpine/subalpine meadow, germination completely may eventually cause the population extinction. Spreading germination in time to disperse risk plays a key role in reproductive success [50].

More importantly, there was a significant increase in mortality rate of seeds because of temperature rise (Figure 4), and it seems that fungal attack can interpret these results. Thus, it can be inferred that the high seed mortality is likely to produce selection pressures on germination, i.e. increased germination should be selected if there is high seed mortality in high temperature, which could be one reason of good germination under a relatively high temperature environment. Furthermore, high temperature can improve and accelerate germination directly by activating enzymatic reactions occurring in the process of germination and by regulating the synthesis of hormones that affect the status of seed dormancy [10].

Habitat. In this study, we found that seeds from south slope presented earlier GT than that from other habitats, and seeds from bottomland displayed lower GP than seeds from north slope. There are some reasons responsible for this result. First of all, the abiotic conditions are different among north slope, south slope and bottomland due to different temperature, irradiation and water stress levels. Generally, north slope have better moisture relations, less variation in temperature, and generally less harsh conditions than south slope, so this condition appears to favor the establishment of perennial species [51]. Our results also confirmed

this view. In details, the percentage of perennial species is 71% in north slope and 57% in south slope. Meanwhile, perennials germinated significantly later than annuals, which can explain why seeds from south slope displayed earlier GT than seeds from north slope. Secondly, lower GP and more delayed GT of seeds from bottomland is usually related to higher dormancy levels of seeds, because adequate moisture during seed formation is expected to result in the production of more dormant seeds than in drier conditions [52].

On the other hand, our results showed habitat had significant effects on germination (Table 1) and there were strong associations between seed size, life form, flowering time and habitat (Table 2, 3, 4), which means inherent characteristics of species may play a prominent role in evolution of germination strategies, but stochastic factors such as environmental conditions are also important selective pressures. In other words, seed germination is not only constrained by phylogenetic effects but also other factors such as environmental cues.

In our study, even the most complete ANOVAs accounted for only 22.0% of the variance in GP (Table 3) and 49.3% of the variance in GT (Table 4). Thus, we can not point out the direct cause of variation in seed germination. However, we confirm that a large proportion of interspecific variation in seed germination is correlated with taxon membership, representing lineage history. Meanwhile, selection can maintain the association between germination behavior and the environmental conditions within a lineage.

In conclusion, our results indicate that germination variation is largely dependent on phylogenetic inertia in a community. Life history factors and selection in the local environment also account for the patterns of germination in plant communities. Our results demonstrate that elevated temperature will lead to a significant increase in germination percentage and an accelerated germination. Moreover, there is a significant increase in seed mortality because of temperature rise. We infer that high seed mortality is likely to produce selection pressures on germination, which could be one reason of good germination under a relatively high temperature environment. Additionally, a significant proportion of variance in germination remains unexplained in our results, suggesting that other factors are responsible for the interspecific variation in germination displayed by alpine/subalpine species. Comprehensive studies combining community level as well as multivariate approaches are needed to enhance our understanding of the evolutionary and ecological forces shaping germination strategy.

Supporting Information

Figure S1 Linear relations between seed germination and life history attributes. (A) between germination percentage (GP) and seed size; (B) between mean germination time (GT) and seed size; (C) between germination percentage (GP) and flowering time; (D) between mean germination time (GT) and flowering time. For each species, the midpoint of the flowering period is as an estimate of flowering time. (ie the midpoint of the extreme dates of a species' flowering period, given in calendar days, 1-365, starting from January 1).

Table S1 The 134 alpine/subalpine species we used in the research. The Angiosperm Phylogeny Group III(2009) was used to assign the affiliation of each species to higher levels.

Acknowledgments

We would like to thank Wei Qi, Haiyan Bu, Xuelin Chen, Yifeng Wang, Kun Liu and Shiting Zhang for help in collecting seeds and identifying species. We particularly thank Professor Carol C. Baskin for reviewing the initial draft of this manuscript and for many constructive comments.

Author Contributions

Conceived and designed the experiments: JX GD. Performed the experiments: JX WL GD. Analyzed the data: JX CZ. Contributed reagents/materials/analysis tools: JX CZ WL. Wrote the paper: JX WL CZ.

References

1. Bewley JD (1997) Seed germination and dormancy. Plant Cell 9: 1055–1066.
2. Milbau A, Graae BJ, Shevtsova A, Nijs I (2009) Effects of a warmer climate on seed germination in the subarctic. Ann Bot 104: 287–296.
3. Mazer SJ (1989) Ecological, taxonomic, and life history correlates of seed mass among Indiana dune angiosperms. Ecol Monogr 59: 153–175.
4. Leishman MR, Westoby M, Jurado E (1995) Correlates of seed size variation: a comparison among five temperate floras. J Ecol 83: 517–530.
5. Zhang ST, Du GZ, Chen JK (2004) Seed size in relation to phylogeny, growth form and longevity in a subalpine meadow on the east of the Tibetan Plateau. Folia Geobot 39: 129–142.
6. Norden N, Daws MI, Antoine C, Gonzalez MA, Garwood NC, et al. (2009) The relationship between seed mass and mean time to germination for 1037 tree species across five tropical forests. Funct Ecol 23: 203–210.
7. Levey DJ, Bolker BM, Tewksbury JJ, Sargent S, Haddad NM (2005) Effects of landscape corridors on seed dispersal by birds. Science 309: 146–148.
8. Grime JP, Mason G, Curtis AV, Rodman J, Band SR, et al. (1981) A comparative study of germination characteristics in a local flora. J Ecol 69: 1017–1059.
9. Rees M (1994) Delayed germination of seeds: a look at the effects of adult longevity, the timing of reproduction, and population age/stage structure. Am Nat 43–64.
10. Baskin CC, Baskin JM (1998) Seeds: ecology, biogeography, and evolution of dormancy and germination. San Diego, California, USA: Academic Press.
11. Jurado E, Flores J (2005) Is seed dormancy under environmental control or bound to plant traits? J Veg Sci 16: 559–564.
12. Moles AT, Westoby M (2004) What do seedlings die from and what are the implications for evolution of seed size? Oikos 106: 193–199.
13. Armstrong DP, Westoby M (1993) Seedlings from large seeds tolerated defoliation better: a test using phylogeneticaly independent contrasts. Ecology 74: 1092–1100.
14. Leishman MR, Westoby M (1994) The role of seed size in seedling establishment in dry soil conditions-experimental evidence from semi-arid species. J Ecol 82: 249–258.
15. Rees M (1996) Evolutionary ecology of seed dormancy and seed size. Philos T R Soc B 351: 1299–1308.
16. Moles AT, Ackerly DD, Webb CO, Tweddle JC, Dickie JB, et al. (2005) A brief history of seed size. Science 307: 576–580.
17. Wang JH, Baskin CC, Cui XL, Du GZ (2009) Effect of phylogeny, life history and habitat correlates on seed germination of 69 arid and semi-arid zone species from northwest China. Evol Ecol 23: 827–846.
18. Pearson RG (2006) Climate change and the migration capacity of species. Trends Ecol Evol 21: 111–113.
19. IPCC (2007) Fourth assessment report. Climate change 2007: synthesis report. Cambridge, UK: Cambridge University Press.
20. Liu XD, Chen BD (2000) Climate warming in the Tibetan plateau during recent decades. Int J Climatol 20: 1729–1742.
21. Beniston M (2003) Climatic change in mountain regions: A review of possible impacts. In Climate Variability and Change in High Elevation Regions: Past, Present & Future. Springer, Netherlands. pp. 5–31.
22. ACIA (2004) Impacts of a warming Arctic: Arctic climate impact assessment overview report. Cambridge, UK: Cambridge University Press.
23. Ruf M, Brunner I (2003) Vitality of tree fine roots: reevaluation of the tetrazolium test. Tree Physiol 23: 257–263.
24. Ellis RH, Roberts EH (1978) Towards a rational basis for testing seed quality. In: Seed Production (ed: P.D. Hebblethwaite). Butterworths, London. pp. 605–635.
25. Webb CO, Donoghue MJ (2005) Phylomatic: tree assembly for applied phylogenetics. Mol Ecol Notes 5: 181–183.
26. Webb CO, Ackerly DD, Kembel SW (2008) Phylocom: software for the analysis of phylogenetic community structure and trait evolution. Bioinformatics 24: 2098–2100.
27. Bell CD, Soltis DE, Soltis PS (2010) The age and diversification of the angiosperms re-revisited. Am J Bot 97: 1296–1303.
28. Smith SA, Beaulieu JM, Donoghue MJ (2010) An uncorrelated relaxed-clock analysis suggests an earlier origin for flowering plants. P Natl Acad Sci USA 107: 5897–5902.
29. Harmon LJ, Weir JT, Brock CD, Glor RE, Challenger W (2008) GEIGER: investigating evolutionary radiations. Bioinformatics 24: 129–131.
30. Pagel M (1999) The maximum likelihood approach to reconstructing ancestral character states of discrete characters on phylogenies. Systematic Bio 48: 612–622.
31. Angiosperm Phylogeny Group III (2009) An update of the Angiosperm Phylogeny Group classification for the orders and families of flowering plants: APG III. Bot J Linn Soc 161: 105–121.
32. Baker HG (1972) Seed weight in relation to environmental conditions in California. Ecology 53: 997–1010.
33. Flora China Editing Group (2004) Flora reipublicae popularis sinicae. Beijing, China: Science Press. (in Chinese)
34. Kochmer JP, Handel SN (1986) Constraints and competition in the evolution of flowering phenology. Ecol Monogr 56: 303–325.
35. Miles DB, Dunham AE (1993) Historical perspectives in ecology and evolutionary biology: the use of phylogenetic comparative analyses. Annu Rev Ecol Syst 24: 587–619.
36. Wiens JJ, Ackerly DD, Allen AP, Anacker BL, Buckley LB, et al. (2010). Niche conservatism as an emerging principle in ecology and conservation biology. Ecol Lett 13: 1310–1324.
37. Schippers P, Van Groenendael JM, Vleeshouwers LM, Hunt R (2001) Herbaceous plant strategies in disturbed habitats. Oikos 95: 198–210.
38. Leishman MR (2001) Does the seed size/number trade-off model determine plant community structure? An assessment of the model mechanisms and their generality. Oikos 93: 294–302.
39. Chen ZH, Peng JF, Zhang DM, Zhao JG (2002) Seed germination and storage of woody species in the lower subtropical forest. Acta Bot Sin 44: 1469–1476.
40. López-Castañeda C, Richards RA, Farquhar GD, Williamson RE (1996) Seed and seedling characteristics contributing to variation in early vigor among temperate cereals. Crop Sci 36: 1257–1266.
41. Pearson TRH, Burslem DF, Mullins CE, Dalling JW (2002) Germination ecology of neotropical pioneers: interacting effects of environmental conditions and seed size. Ecology 83: 2798–2807.
42. Sadeghi H, Khazaei F, Sheidaei S, Yari L (2011) Effect of seed size on seed germination behavior of safflower (Carthamus tinctorius L.). J Agr Biol Sci 6: 5–8.
43. Venable DL, Lawlor L (1980) Delayed germination and dispersal in desert annuals escape in space and time. Oecologia 46: 272–282.
44. Venable DL, Brown JS (1988) The selective interactions of dispersal, dormancy, and seed size as adaptations for reducing risk in variable environments. Am Nat 131:360–384.
45. Lacey EP, Roach DA, Herr D, Kincaid S, Perrott R (2003) Multigenerational effects of flowering and fruiting phenology in Plantago lanceolata. Ecology 84: 2462–2475.
46. Ziello C, Estrella N, Kostova M, Koch E, Menzel A (2009) Influence of altitude on phenology of selected plant species in the Alpine region (1971-2000). Clim Res 39: 227–234.

47. Johnson SD (1993) Climatic and phylogenetic determinants of flowering seasonality in the Cape flora. J Ecol 81: 567–572.

48. Verdu M, Traveset A (2005) Early emergence enhances plant fitness: a phylogenetically controlled meta-analysis. Ecology 86: 1385–1394.

49. Smith-Ramírez C, Armesto JJ, Figueroa J (1998) Flowering, fruiting and seed germination in Chilean rain forest myrtaceae: ecological and phylogenetic constraints. Plant Ecol 136: 119–131.

50. Venable DL (2007) Bet hedging in a guild of desert annuals. Ecology 88: 1086–1090.

51. Koniak S (1985) Succession in pinyon-juniper woodlands following wildfire in the Great Basin. West N Am Naturalist 45: 556–566.

52. Luzuriaga AL, Escudero A, Pérez-García F (2006) Environmental maternal effects on seed morphology and germination in *Sinapis arvensis* (Cruciferae). Weed Res 46:163–174.

2

Phylogenetic Responses of Forest Trees to Global Change

John K. Senior[1], Jennifer A. Schweitzer[1,2], Julianne O'Reilly-Wapstra[3], Samantha K. Chapman[4], Dorothy Steane[1,5], Adam Langley[4], Joseph K. Bailey[1,2]*

1 School of Plant Science, University of Tasmania, Hobart, TAS, Australia, 2 Department of Ecology and Evolutionary Biology, University of Tennessee, Knoxville, Tennessee, United States of America, 3 School of Plant Science and National Centre for Future Forest Industries, University of Tasmania, Hobart, TAS, Australia, 4 Department of Biology, Villanova University, Villanova, Pennsylvania, United States of America, 5 University of the Sunshine Coast, Sippy Downs, Queensland, Australia

Abstract

In a rapidly changing biosphere, approaches to understanding the ecology and evolution of forest species will be critical to predict and mitigate the effects of anthropogenic global change on forest ecosystems. Utilizing 26 forest species in a factorial experiment with two levels each of atmospheric CO_2 and soil nitrogen, we examined the hypothesis that phylogeny would influence plant performance in response to elevated CO_2 and nitrogen fertilization. We found highly idiosyncratic responses at the species level. However, significant, among-genetic lineage responses were present across a molecularly determined phylogeny, indicating that past evolutionary history may have an important role in the response of whole genetic lineages to future global change. These data imply that some genetic lineages will perform well and that others will not, depending upon the environmental context.

Editor: Han Y.H. Chen, Lakehead University, Canada

Funding: These authors have no support or funding to report.

Competing Interests: The authors have declared that no competing interests exist.

* E-mail: Joe.Bailey@utk.edu

Introduction

Elevated carbon dioxide (CO_2) and nitrogen (N) availability are expected to have important consequences for forest ecosystem dynamics [1–4], where soil N scarcity may limit the CO_2 fertilisation effect [5,6]. Atmospheric CO_2 concentrations are expected to double by the turn of the century [3], while anthropogenic N fixation rates have already doubled preindustrial rates [7,8]. Tree species have vital roles in carbon (C) and N cycling and are expected to act as important sinks for anthropogenic CO_2 emissions [2]. As the dominant constituents of forest ecosystems, tree species also have important extended community effects such as plant-plant and plant-herbivore interactions [9,10]. It is, therefore, vital that we understand the consequences of anthropogenic CO_2 and N fertilisation on tree species to predict future impacts of these important global change factors on forest ecosystems.

A recent direction in global climate change research is the utilization of phylogenetics to better understand and predict the impacts of global change [11–15]. Closely related taxa have the potential to respond in a similar manner to global environmental changes, due to shared evolutionary histories, genetic background, and phenotypic traits. Thus, taking phylogeny into account may provide generality that is more appropriate for modelling the impacts of large-scale global climate change than generalising across species that share fundamental niches. For example, Davis et al. [14] assessed the flowering time of plant clades occurring in both the United States and the United Kingdom and found that phenological responses to global climate change were shared within clades. Similar trends are likely to occur in the responses of

other plant traits to other large-scale perturbation. For example, the magnitude of CO_2-induced increases in biomass may vary much more within functional types (e.g. herbaceous vs woody species) than among them [16] though consistent differences in response may arise when functional groups align with major phylogenetic differences such as gymnosperms vs. angiosperms [17]. However, few studies utilise an explicit phylogenetic framework (see [11,13–15]) to assess the importance of phylogeny. Such an approach is important as the differences amongst plant functional groups that are currently being explored in global change studies, likely represent the evolutionary consequences of phylogenetic divergence [18–20]. Examining phylogenetic responses to climate change may capture a broader range of variation [21] among taxa for better understanding how plants can respond to increasing CO_2, N or other environmental factors.

If a phylogenetic approach is useful to understanding the consequences of global change, *Eucalyptus* represents a model genus in which to test it. *Eucalyptus* is a globally important forest plantation species that is planted worldwide and is the dominant genus in many Australian ecosystems, occurring in subalpine woodlands, cool and warm temperate forests, rainforests and tropical savannahs [22]. Having evolved under a large range of climatic and edaphic conditions the genus is also highly diverse with over 700 species displaying a wide range of growth forms, from giant forest to dwarf coastal trees and stunted, multi-stemmed, "mallee" forms in semi-arid areas [23]. This genus is of great ecological and economic importance, yet relatively few studies have been concerned with the effects of climate change on eucalypt forests in native or non-native habitats around the world. Studies to date have found that eucalypt species are responsive to elevated CO_2 and N fertilization but responses differ in direction and magnitude, with negative, neutral and positive growth responses documented [24–30]. Consequently, the response of eucalypt species to elevated CO_2 and N fertilization may not be as general as expected and a phylogenetic approach may be useful to better inform responses to global change.

On the island of Tasmania (Australia), there are 29 species of *Eucalyptus* that occur in a range of habitats from coastal wet and dry forests to alpine environments. The species belong to the two main subgenera of *Eucalyptus* (i.e., subgenus *Symphyomyrtus* and subgenus *Eucalyptus*; [23]), but lower taxonomic classifications are unresolved, with a number of authors grouping species in different ways [23,31,32]. In this paper we follow the classification of Brooker [23] in which the classification of eucalypt species is based predominantly on morphological traits, including bark, leaf, floral and fruit morphologies, and is largely supported by the available molecular data [33,34].

Using 26 of the 29 species of *Eucalyptus* found in Tasmania, we used a phylogenetic approach to better understand plant performance in response to the global change factors of elevated CO_2 and soil N fertilization. All species were exposed to factorial treatments of ambient and elevated CO_2 and low and high soil N concentrations, where it was hypothesised that plant performance of closely related species to these two global change factors would be similar. Specifically, we hypothesized that eucalypt species would respond differentially to elevated CO_2 and N fertilization based on past, shared evolutionary history. We found that; 1) individual species' growth responses are largely idiosyncratic, however, 2) when species were nested within genetic lineages, species within a particular genetic lineage shared similar responses to elevated CO_2 and high soil N concentrations, significantly differing from other genetic lineages. Utilising a phylogenetic approach may, therefore, provide a potential framework by which the responses of individual forest species to global climate change may be generalized across groups of closely related species. These

Table 1. Phylogenetic classification of all Tasmanian eucalypt species based on Brooker [23] and DArT data from this study, informed with recent genetic data from McKinnon *et al.* [33] and Steane *et al.* [34].

Subgenus	Genetic lineage[1]	Species	Section/Subsection[2], Series Brooker [21]	Code[3]
		E. obliqua	Eucalyptus, Eucalyptus	EEE
		E. regnans	Eucalyptus, Regnantes	EER
		E. sieberi	Cineracea, Psathyroxyla	ECPS
		E. delegatensis	Cineracea, Fraxinales	ECF
		E. pauciflora	Cineracea, Pauciflorae	ECP
Eucalyptus	Genetic lineage 1	E. radiata	Aromatica, Radiatae	EAR
		E. amygdalina	Aromatica, Insulanae	EAI
		E. nitida	Aromatica, Insulanae	EAI
		E. pulchella	Aromatica, Insulanae	EAI
		E. risdonii	Aromatica, Insulanae	EAI
		E. tenuiramis	Aromatica, Insulanae	EAI
		E. brookeriana	Triangulares, Foveolatae	SMTF
	Genetic lineage 2	E. ovata	Triangulares, Foveolatae	SMTF
		E. rodwayi	Triangulares, Foveolatae	SMTF
		E. barberi	Triangulresa, Foveolatae	SMTF
		E. archeri[4]	Euryota, Orbiculares	SMEO
		E. cordata	Euryota, Orbiculares	SMEO
		E. gunnii	Euryota, Orbiculares	SMEO
	Genetic lineage 3	E. morrisbyi[4]	Euryota, Orbiculares	SMEO
Symphyomyrtus		E. urnigera	Euryota, Orbiculares	SMEO
		E. johnstonii	Euryota, Semiunicolores	SMES
		E. subcrenulata	Euryota, Semiunicolores	SMES
		E. vernicosa	Euryota, Semiunicolores	SMES
		E. globulus	Euryota, Globulares	SMEG
		E. perriniana	Euryota, Orbiculares	SMEO
	Genetic lineage 4	E. dalrympleana	Euryota, Viminales	SMEV
		E. rubida	Euryota, Viminales	SMEV
		E. viminalis	Euryota, Viminales	SMEV

[1]"Genetic lineage" designation is based on AFLP [33] and/or DArT analyses [34,35].
[2]Sectional classification is given for species belonging to subgenus *Eucalyptus* (no sub-sectional classification available); all Tasmanian species from subgenus *Symphyomyrtus* belong to section *Maidenaria*, so only subsections are shown here.
[3]Taxonomic code representing subgenus, section, subsection (for subgenus *Symphyomyrtus* only) and series.
[4]No DArT data available for these species.

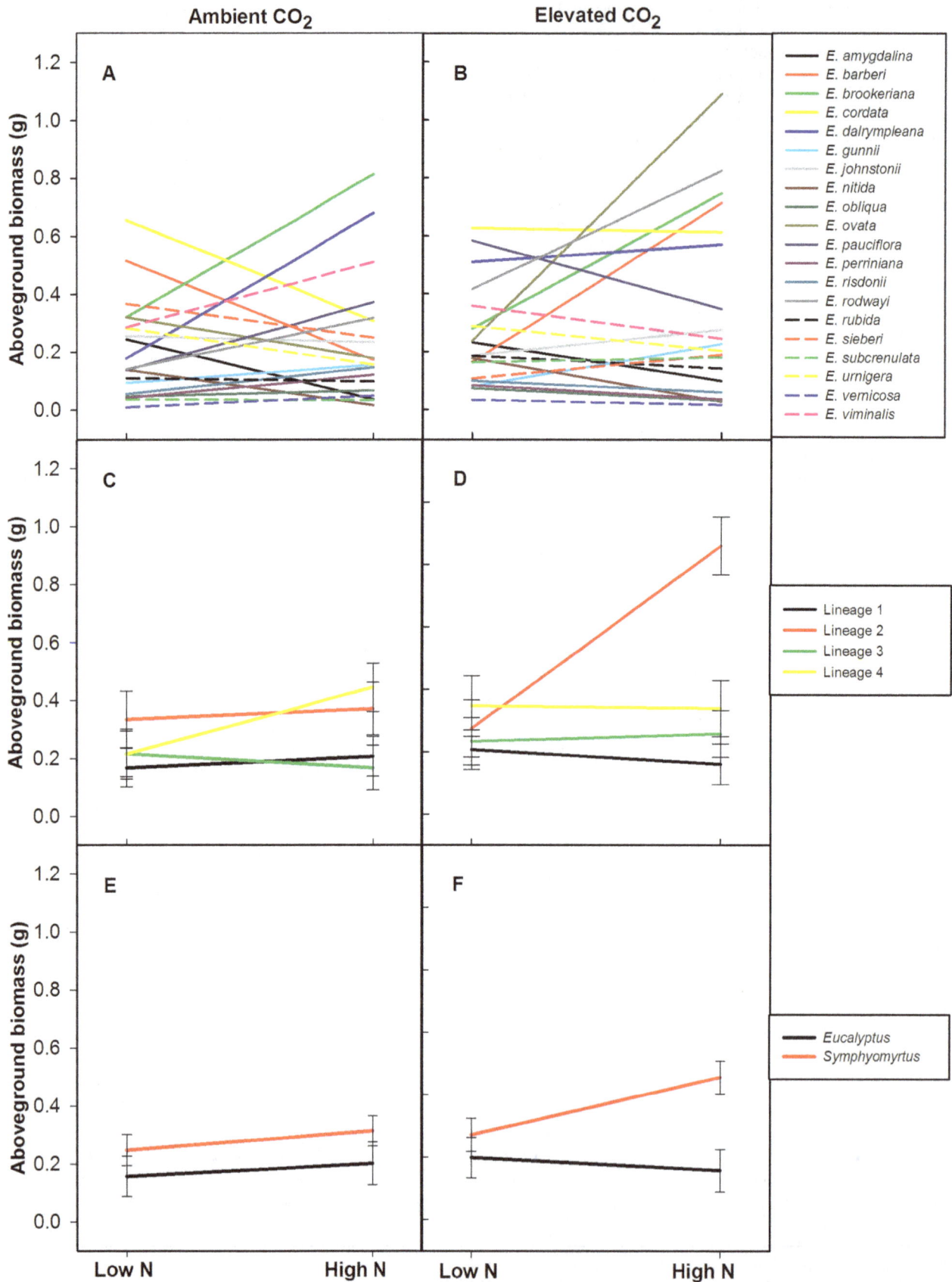

Figure 1. Aboveground biomass responses to elevated CO₂ and N. Interaction plots of least squares means ($\pm$ standard error for C–F) of the aboveground biomass of species (A and B), genetic lineages (C and D) and subgenera (E and F) with control soil N and high soil N (30 kg ha^{-1} of N

added) under both ambient (left panels) and elevated CO_2 (right panels). Standard errors are not presented on the species panels (A–B) due to space constraints. Each colour and line (solid/dashed) combination represents each species, genetic lineage or subgenus analysed, which are represented in the respective legends.

results overall suggest that there may be phylogenetic "winners" and "losers" in response to global change factors based on past evolutionary dynamics and/or history.

Methods

Plant Material

Seed of 26 native Tasmanian eucalypt species was purchased from Forestry Tasmania (http://www.forestrytas.com.au/); *E. archeri*, *E. morrisbyi* and *E. coccifera*, were not included in the study because seed was not available. Seed of each species was obtained from one to six individual maternal trees from a single population. To enhance germination, the seed of each species was vernalised before sowing. Seed of each species was folded in paper towel, wrapped in cheese cloth and soaked in a solution of water containing a drop of dishwashing detergent overnight. Each seed bundle was then squeezed the following morning to remove excess water and refrigerated for 30 days at $4°C$. After this period, the seed of each species was dispersed on the surface of 26 separate trays filled with commercial potting mix, which consisted of eight parts composted fine pine bark and three parts coarse river sand with added macro- and micro-nutrients from Nutricote Grey (Langley Australia Pty Ltd., Welshpool WA), which included N, phosphorus (P) and potassium (K) in the weight ratio of 19:2.6:10, at a rate of 3 kg/m^{-3}. The surface of the potting mix was then disturbed gently and covered with vermiculite for water retention. The trays received a daily soaking of water, in equal volumes, in a greenhouse while seeds germinated. Germinants were grown for three weeks until the majority of seedlings of each species had developed the first pair of true leaves and were uniform in size.

Phylogenetic Framework for Tasmanian Eucalypts

A phylogenetic framework for this study was devised using the classification of Brooker [23] with adjustments using recent molecular genetic information on the subgenera *Symphyomyrtus* [33] and *Eucalyptus* [34,35]. Hence, we have devised a recent, genetically informed taxonomical classification. The Tasmanian species belonging to subgenus *Eucalyptus* were placed in Genetic Lineage 1 (GL1). The four Tasmanian species of series *Foveolatae* (*E. barberi*, *E. brookeriana*, *E. ovata*, and *E. rodwayi*) formed Genetic Lineage 2 (GL2); Tasmanian endemics belonging to series *Orbiculares* and *Semiunicolores* formed GL3. The remaining (non-

endemic) species of subgenus *Symphyomyrtus* (*E. globulus*, series *Globulares*; *E. perriniana*, series *Orbiculares*; *E. viminalis*, *E. rubida* and *E. dalrympleana*, series *Viminales*) were place in GL4. These genetic lineages were used to conduct nested analyses to test for the effects of phylogenetic group and subgenus (Table 1).

Further support for these groupings was found in broad analyses of all eucalypt subgenera (that did not include all the Tasmanian species that were included in this study; [34,35]) that used a relatively new molecular marker called Diversity Arrays Technology (DArT; [36]). In this study, we used DArT to check the genetic integrity of the groupings that we defined and found that genotyping supported the genetic lineages (see [34]).

Elevated CO_2 and Soil N Fertilization Study

To determine responses to the global change factors of elevated CO_2 and soil N fertilization, eucalypt seedlings were grown under all factorial combinations of ambient and elevated CO_2 and high and low soil N fertilization. Twelve seedlings of each species were transplanted into forestry tubes filled with the same commercial potting mix (described above). These twelve seedlings of each species were divided randomly into three replicates for each of the four factorial treatments of CO_2 (ambient or elevated) and soil N (low and high). The seedlings for each level of CO_2 were then placed randomly, via random number generation, into forestry tube racks.

Seedling racks for each CO_2 treatment were placed randomly into separate, air-tight, controlled greenhouse chambers maintained at $23°C$ with a natural photoperiod; half of which were then fertilized with N at the soil surface (details below). The CO_2 treatments and their respective seedlings were exchanged between two greenhouse chambers each week; greenhouse chamber effects were avoided further by moving the CO_2 tanks and regulator as well as monitoring CO_2 concentrations with an infra-red gas analyser (IRGA) device (LiCor 6200, LiCor Inc., Lincoln, NE, USA). The forestry tube racks were also repositioned randomly during these periods to avoid seedling positional effects in the greenhouse. In the elevated CO_2 treatment carbon dioxide was elevated to 720 ppm using a CO_2 control unit (Thermoline Scientific Equipment, Smithfield, Australia) and compressed CO_2. The low CO_2 treatment was maintained at ambient CO_2 (~400 ppm) and monitored frequently for leakage of CO_2 from the neighbouring high CO_2 chamber with the LiCor (none

Table 2. General linear model results of variation in aboveground biomass (AGB), belowground biomass (BGB), total biomass (TB) and root:shoot (R:S) between the factors of species, carbon dioxide (CO_2), nitrogen (N) and all interactions.

Variable	Species		CO_2		N		Species*CO_2		Species*N		CO_2*N		Species*CO_2*N	
	F	P	F	P	F	P	F	P	F	P	F	P	F	P
AGB	7.83$_{(19,193)}$	**<0.001**	2.87$_{(1,193)}$	0.093	1.31$_{(1,193)}$	0.254	0.97$_{(19,193)}$	0.502	1.35$_{(19,193)}$	0.167	0.34$_{(1,193)}$	0.560	1.96$_{(19,193)}$	**0.016**
BGB	7.14$_{(20,198)}$	**<0.001**	6.25$_{(1,198)}$	**0.014**	0.33$_{(1,198)}$	0.565	1.01$_{(20,198)}$	0.458	1.12$_{(20,198)}$	0.336	0.11$_{(1,198)}$	0.740	1.85$_{(20,198)}$	**0.023**
TB	7.28$_{(19,190)}$	**<0.001**	4.34$_{(1,190)}$	**0.040**	0.57$_{(1,190)}$	0.452	0.89$_{(19,190)}$	0.593	1.18$_{(19,190)}$	0.288	0.34$_{(1,190)}$	0.563	1.92$_{(19,190)}$	**0.019**
R:S	3.33$_{(19,188)}$	**<0.001**	8.87$_{(1,188)}$	**0.004**	3.27$_{(1,188)}$	0.073	1.73$_{(19,188)}$	**0.042**	0.46$_{(19,188)}$	0.973	0.12$_{(1,188)}$	0.732	1.91$_{(19,188)}$	0.570

Bold, underlined values indicate statistical significance ($\alpha = 0.05$); degrees of freedom are denoted as subscript of each F value.

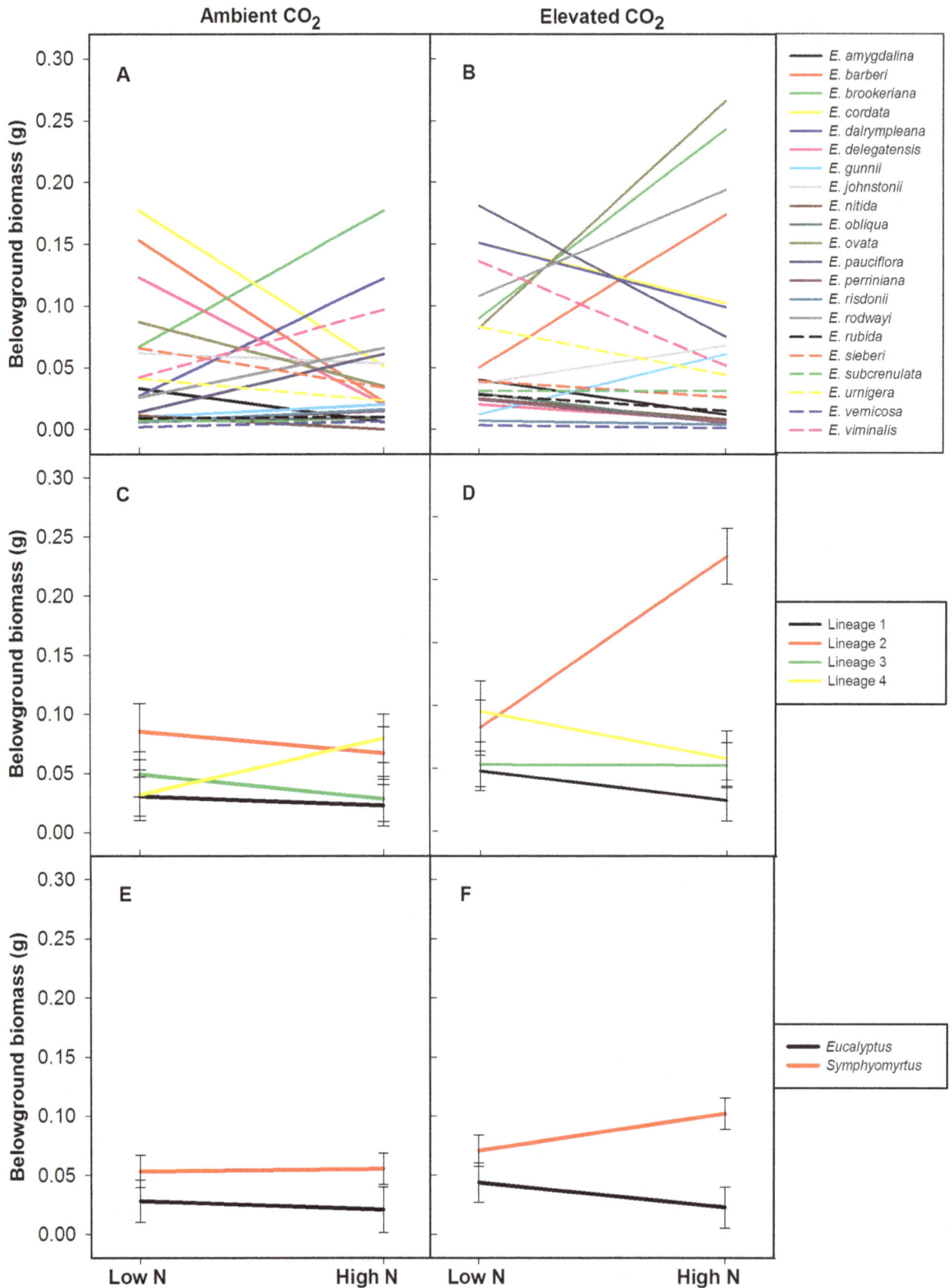

Figure 2. Belowground biomass responses to elevated CO_2 and N. Interaction plots of least squares means ($\pm$ standard error for C–F) of the belowground biomass of species (A and B), genetic lineages (C and D) and subgenera (E and F) with control N and 30 kg ha^{-1} of N added (High N)

under both ambient (left panels) and elevated CO_2 (right panels). Standard errors are not presented on the species panels (A–B) due to space constraints. Each colour and line (solid/dashed) combination represents each species, genetic lineage or subgenus.

occurred). Seedlings in each treatment were watered on a daily basis until the eighth week of the study, at which time water was applied as needed. At the same time that the CO_2 treatments were initiated, pellets of urea at an approximate concentration of 30 kg N ha^{-1}, were applied each month to the high N treatment seedlings to replicate the approximate N addition of forestry practices that alleviates N limitation [37].

At the end of the experiment (approximately 5 months later; before seedlings became root bound) each seedling was harvested destructively by carefully removing each seedling, gently shaking off as much soil as possible, and severing the aboveground from the belowground biomass at the root collar. The belowground biomass of each individual of each species was sealed in separate plastic bags after collection and the aboveground biomass was placed in separate brown paper bags. The belowground biomass was refrigerated (6°C) and the aboveground biomass was stored at ambient conditions until oven-drying (at the end of the harvest). The aboveground biomass samples were oven-dried for 48 hours at 60°C and then weighed (g). The belowground biomass of each seedling was carefully rinsed, separately, over 2 and 0.5 mm sieves to remove as much soil as possible whilst retaining as much of the fine root biomass as possible. The washed belowground biomass samples were then oven-dried for 48 hours at 60°C and weighed (g). Oven-dried belowground biomass was divided by aboveground biomass to yield root:shoot and above- and belowground biomass was summed to determine total biomass.

Statistical Analysis

All statistical analyses were conducted using the statistical package SAS (version 9.2, SAS Institute Inc., Cary USA). Other traits were measured (height, leaf biomass, leaf area and relative growth rate) but were not included in the analysis due to strong inter-correlations with all traits ($R^2 > 0.7$; data not presented). Separate analyses were run for each phylogenetic level to maintain statistical power. The responses of seedling aboveground biomass, belowground biomass, total biomass and root:shoot to elevated CO_2 and soil N fertilization were analysed to determine how each eucalypt species responded to these environmental variables. Due to mortalities in the greenhouse, whole treatment groups within particular species were absent, making analysis of these species impossible. Consequently, these species were removed from the

analysis. There was a highly significant effect of subgenus on species mortality ($P<0.001$), where 54 seedlings were lost from subgenus *Eucalyptus* and 31 from subgenus *Symphyomyrtus*. Five species were removed from subgenus *Eucalyptus* (GL1) and one species was removed from subgenus *Symphyomyrtus* (GL4) due to the death of whole treatments (n = 3). However, no significant interactions between mortality and environmental treatments were present. General linear models were used to analyse species, CO_2 (2 levels of treatment, ambient and elevated) and N (2 levels of treatment, low and high) effects for each morphological variable (total biomass, aboveground biomass, belowground biomass, and root:shoot; PROC GLM). All interaction terms were included; species $\times$ CO_2, species $\times$ N, CO_2 $\times$ N and species $\times$ CO_2 $\times$ N, where all main effects and interaction terms were treated as fixed effects. Data were tested for the assumptions of normality and homoscedasticity and appropriate transformations were applied to meet the Shapiro-Wilk test when required. Diagnostic graphical representations were also checked for normality and homoscedasticity. Aboveground, belowground and total biomass data were square root transformed while root:shoot data were power transformed (0.3). A power transformation was applied when log or square-root transformed data did not satisfy the assumptions of normality and homoscedasticity.

To test for phylogenetic patterns in the responses of the eucalypt species, mixed models were conducted in SAS (PROC MIXED). Models tested genetic lineage (four genetic lineages), CO_2 (two levels of treatment, ambient and elevated) and N fertilization (two levels of treatment, low and high) effects for each variable (total biomass, aboveground biomass, belowground biomass and root:shoot) These main effects were fixed effects while species(genetic lineage) was used as a random term to test the genetic lineage effect. All interaction terms were also included as fixed effects; genetic lineage $\times$ CO_2, genetic lineage $\times$ N, CO_2 $\times$ N and genetic lineage $\times$ CO_2 $\times$ N.

To test for the effects of subgenus, CO_2 and N fertilization, mixed models were used to test for subgenus (two subgenera; *Symphyomyrtus* and *Eucalyptus*), CO_2 (two levels of treatment, ambient and elevated) and soil N fertilization (two levels of treatment, low and high) effects for each variable (aboveground biomass, belowground biomass, total biomass and root:shoot). These main effects were treated as fixed effects while species(subgenus) was used as a random term to test the subgenus effect.

Table 3. Mixed model results of variation in aboveground biomass (AGB), belowground biomass (BGB), total biomass (TB) and root:shoot (R:S) between the factors genetic lineage, carbon dioxide (CO_2), nitrogen (N) and all interactions, using the random term species(genetic lineage) to test the genetic lineage effect (random effect not shown).

Variable	Genetic lineage		CO_2		N		Genetic lineage*CO_2		Genetic lineage*N		CO_2*N		Genetic lineage*CO_2*N	
	F	P	F	P	F	P	F	P	F	P	F	P	F	P
AGB	$2.79_{(3,21)}$	0.066	$3.70_{(1,186)}$	0.056	$7.51_{(1,186)}$	**0.007**	$1.19_{(3,186)}$	0.316	$3.68_{(3,186)}$	**0.013**	$0.65_{(1,186)}$	0.421	$6.93_{(3,186)}$	**<0.001**
BGB	$3.62_{(3,21)}$	**0.030**	$9.17_{(1,183)}$	**0.003**	$0.22_{(1,183)}$	0.642	$1.12_{(3,183)}$	0.342	$1.49_{(3,183)}$	0.219	$0.14_{(1,183)}$	0.710	$6.50_{(3,183)}$	**<0.001**
TB	$2.68_{(3,21)}$	**0.037**	$4.87_{(1,180)}$	**0.029**	$3.17_{(1,180)}$	0.077	$0.95_{(3,180)}$	0.418	$2.10_{(3,180)}$	0.102	$0.57_{(1,180)}$	0.451	$7.10_{(3,180)}$	**<0.001**
R:S	$7.28_{(3,21)}$	**0.002**	$12.28_{(1,178)}$	**0.001**	$4.11_{(1,178)}$	**0.044**	$1.18_{(3,178)}$	0.319	$0.51_{(3,178)}$	0.678	$0.44_{(1,178)}$	0.510	$2.34_{(3,178)}$	0.075

Bold underlined values are significant and degrees of freedom are denoted as subscript of each F value.

Table 4. Mixed model results of variation in aboveground biomass (AGB), belowground biomass (BGB), total biomass (TB) and root:shoot (R:S) between the factors subgenus, carbon dioxide (CO_2), nitrogen (N) and all interactions, using the random term species(subgenus) to test the subgenus effect (random effect not shown).

Variable	Subgenus		CO_2		N		Subgenus*CO_2		Subgenus*N		CO_2*N		Subgenus*CO_2*N	
	F	P	F	P	F	P	F	P	F	P	F	P	F	P
AGB	$3.69_{(1,24)}$	0.067	$1.66_{(1,195)}$	0.199	$2.77_{(1,195)}$	0.097	$0.59_{(1,195)}$	0.442	$2.29_{(1,195)}$	0.132	$0.02_{(1,195)}$	0.881	$2.02_{(1,195)}$	0.157
BGB	$5.38_{(1,24)}$	**0.029**	$5.21_{(1,192)}$	**0.024**	$0.02_{(1,192)}$	0.881	$0.32_{(1,192)}$	0.571	$2.02_{(1,192)}$	0.157	$0.03_{(1,192)}$	0.853	$0.83_{(1,192)}$	0.363
TB	$4.09_{(1,24)}$	0.054	$3.01_{(1,189)}$	0.084	$0.92_{(1,189)}$	0.338	$0.15_{(1,189)}$	0.700	$2.08_{(1,189)}$	0.151	$0.01_{(1,189)}$	0.965	$0.98_{(1,189)}$	0.324
R:S	$7.34_{(1,24)}$	**0.012**	$5.88_{(1,187)}$	**0.016**	$0.84_{(1,187)}$	0.360	$0.02_{(1,187)}$	0.891	$0.37_{(1,187)}$	0.720	$0.720_{(1,187)}$	0.396	$0.65_{(1,187)}$	0.422

Bold underlined values are significant and degrees of freedom are denoted as subscript of each F value.

Species(subgenus) was chosen over genetic lineage(subgenus) as this random term conserves a larger proportion of variation. All interaction terms were included; subgenus×CO_2, subgenus×N, CO_2×N and subgenus×CO_2×N as fixed effects.

Results

Species Analyses

There was a three-way interaction between CO_2, N fertilization and species for aboveground biomass, indicating that the response of plant species to the combination of CO_2 and soil N fertilization is variable (**Table 2; Figure 1A, B**). For example, N addition resulted in a nearly three-fold increase in the aboveground biomass of *E. dalrympleana* under ambient CO_2, whereas under elevated CO_2, N addition had less impact, resulting only in a 12% increase in aboveground biomass. In contrast, N addition under ambient CO_2 resulted in a 78% decrease in the aboveground biomass of *E. ovata*, but a large, 3.5-fold, increase under elevated CO_2.

Similar to aboveground biomass there was also a three-way interaction for belowground biomass. Species belowground biomass displayed a variety of responses to elevated CO_2 and soil N, whereby species responded differently to soil N availability depending on the concentration of CO_2 (**Figure 2A, B**). For example, under ambient CO_2, N addition resulted in a six-fold decrease in the belowground biomass of *E. barberi* whereas under elevated CO_2, N addition resulted in a 248% increase in belowground biomass. In contrast, N addition under ambient CO_2 resulted in a 252% increase in the belowground biomass of *E. cordata* but a decrease of 48% under elevated CO_2. It is clear that individual species responses to the treatment factors were highly variable.

Phylogenetic Analyses

Similar to the species level analysis, a three-way interaction of genetic lineage, CO_2 and soil N was identified in both above- and belowground biomass, indicating the importance of evolutionary history in response to these climate change factors. Both the above- and belowground biomass of genetic lineages responded differently to N availability depending on the concentration of CO_2 (**Table 3; Figure 1C, D; Figure 2C, D**). Under ambient CO_2, above- and belowground responses by genetic lineage did not significantly differ, regardless of N availability. However, under elevated CO_2, the response of GL2 to N addition significantly differed from other genetic lineages, with large increases in above- and belowground biomass of 213% and

166%, respectively. These results indicate that the response of genetic lineages to CO_2 and N are highly variable, but they also show that closely related species respond similarly, providing significantly more predictive ability than individual species responses.

Finally, unlike the genetic lineage level comparison, at the broadest phylogenetic level (subgenus), mixed models found no significant interactive effects of subgenus, CO_2 and N in seedling morphological responses (**Figure 1E, F; Figure 2E, F**), but there were, however, main effects (**Table 4**). Surprisingly, no significant effects of subgenus, CO_2 or N were revealed in aboveground biomass. However, there was a significant effect of subgenus on belowground biomass and root:shoot, whereby species in the subgenus *Symphyomyrtus* had a significantly larger belowground biomass (143%) and root:shoot (33%) than species from subgenus *Eucalyptus*. Significant main effects of CO_2 were also found in belowground biomass and root:shoot indicating that the two subgenera responded to elevated CO_2 in a similar manner.

Discussion

Species-level Response

Independently, elevated CO_2 or soil N fertilization generally enhances plant growth (e.g. [25,26,29,38,39]) and two-way interactions between CO_2 and N availability are also commonly reported in the literature [2], where N availability is expected to constrain the CO_2-induced stimulation of plant growth [6]. These interactions have been observed across a broad range of tree species from different families and ecological contexts [25,38,39,40], where studies suggest that tree species allocate C to belowground sinks to alleviate N limitation [2]. However in the present study, species displayed considerable variation in root:-shoot in response to CO_2 concentration. Variable responses in biomass allocation to elevated CO_2 may aid in explaining the variable responses of species to both CO_2 and N fertilization. Species that display inherently higher proportions of belowground biomass may more effectively exploit soil N and, therefore, respond more strongly to elevated CO_2 [2]. Evolutionary history and local adaptation are likely to be drivers of these differential responses, indicating the presence of evolutionary trade-offs [9] in the responsiveness of species to either elevated CO_2 or soil N. A number of species did not appear to respond to elevated CO_2 and N fertilization whereas, due to variability in traits (e.g. larger root:shoot [41]), others did respond to these factors, either singularly or in combination (interactive effects).

The large variation among species responses is likely to complicate our general understanding of the impacts of global change. For example, Spinnler *et al.* [42] found differential impacts of elevated CO_2 on tree species within model spruce-beech (*Fagus sylvatica-Picea abies*) ecosystems growing on either acidic or calcareous soils with either standard or increased nutrient availability. The biomass of *P. abies* was enhanced by elevated CO_2 on both soils, whereas *F. sylvatica* only responded to elevated CO_2 when grown in acidic soil. In this case, the biomass of *F. sylvatica* decreased by 10% with added nutrients and by a further 24% with no added nutrients in response to elevated CO_2. The results of our study indicate that variable species responses to global change may in fact lead to changes in ecosystem dynamics. The highly variable responses among species within this study suggest that species shifts in eucalypt dominated communities could occur under elevated CO_2 and changes to soil N due to fertilizer runoff or deposition. These variable species responses also indicate that taking a phylogenetic perspective may provide the generality required to more efficiently predict the growth responses of species to global climate change, rather than assessing the responses of multiple species individually.

Phylogenetic Similarity in an Ecosystem Response

This study is among the first, to our knowledge, to examine the growth responses of any taxa to global change within a phylogenetic context (but see [11,13–15] for non-growth responses). We found a strong, shared response in plant performance within GL2. Species responses to elevated CO_2 and soil N were determined by genetic lineage, where phylogenetically shared traits resulted in similar species responses within this group. Similar to the species level analysis, a number of genetic lineages did not appear to respond to elevated CO_2 and N, whereas due to phylogenetically shared traits (e.g. larger root:shoot), GL2 responded strongly to elevated CO_2 and soil N fertilization.

Genetic lineage level analyses showed that GL2 responded significantly differently to elevated CO_2 and N than other genetic lineages; these species demonstrated strong, similar responses to elevated CO_2 and N. The root:shoot differed significantly among genetic lineages, where GL2 displayed, on average, a 50% greater root:shoot. This shared trait possibly allowed for greater nutrient utilisation, thus stimulating the CO_2 fertilisation effect [6]. As these species in GL2 generally inhabit nutrient poor areas [32], this trait may have evolved in response to these environmental conditions.

The shared responses of GL2 to elevated CO_2 and N, suggests that the growth responses of species to the global change factors of elevated CO_2 and N may be phylogenetically biased, and a result of shared phylogenetic traits among closely related species [13,14]. Therefore, under a competitive environment containing a number of phylogenetic lineages, there may be 'winner' lineages and 'loser' lineages. For example, under elevated atmospheric CO_2 concentrations and high soil N availability, in communities containing species from both GL1 and GL2, species from GL2 may outperform those from GL1. Willis *et al.* [43] applied similar phylogenetic methods to explain the invasiveness of non-native plant groups in Concord, MA (USA), where non-native groups displayed a greater ability to adjust flowering time in response to climate change; flowering time was postulated to be linked to fitness through ecological mismatches such as pollination. The results of our study indicate that a phylogenetic approach may provide a mechanism whereby highly idiosyncratic species-specific

responses to elevated CO_2 and N fertilization may be generalised across closely related species, perhaps providing a better understanding of the effects of global change on ecosystem dynamics.

When species responses were analysed for the effects of subgenus, the subgenera differed in belowground biomass and root:shoot but did not significantly interact with CO_2 or N. These results indicate that there is a phylogenetic effect and the two subgenera respond to CO_2 and N fertilization in the same way; thus this level of phylogeny may be too broad to differentiate effects of CO_2 and N on morphological traits. Overall, these results suggest that more recently evolved traits at the lower phylogenetic grouping (genetic lineage) level are likely to be responsible for the large proportion of variation among species in responses to elevated CO_2 and N fertilization.

Conclusions and Implications

Predictions of future plant distributions and the sustainability of the services those ecosystems provide are often made using niche-based models that implement correlative methods that relate the presence or absence of species across environmental gradients [44,45]. These models may be species-specific or generalised over species with similar fundamental niches. Using species-specific responses to both elevated CO_2 and N may not be the ideal tool to use for predicting the responses of ecosystems to global environmental change [45], for conservation or climate change mitigation. The results of this study suggest that the responses of forest trees to elevated CO_2 and N may not be random, but phylogenetically biased. A phylogenetic approach may provide a possible alternative to species-specific studies since it could be applied across many plant groups and ecological contexts. To consolidate our findings, further studies are warranted. For example, field studies would determine if phylogenetic patterns in species responses persist while subject to natural conditions. Understanding the evolutionary relationships of species may help us better understand and predict the future ecology of forest communities and as phylogenetic information continues to become exponentially more available, its utilization could complement the plant functional group approaches that are currently represented in many dynamic global vegetation models [46,47]. However, this raises the question of which phylogenetic level should be chosen to best predict species responses? The utilisation of the phylogenetic level which most adequately represents the distribution of the climatically relevant traits would be most pertinent.

Acknowledgments

Thanks to the support staff at the Central Science Laboratory at UTAS for use of molecular laboratory facilities. Thanks to Brad Potts, René Vaillancourt and Mark Hovenden for helpful discussions. Thanks also to Mark Hovenden for his expertise in climate change experiments and design as well as Tracy Winterbottom for expert assistance in the greenhouse. Finally, thanks to Camilla Bloomfield and Danny Lusk for their assistance in the glasshouse and lab.

Author Contributions

Conceived and designed the experiments: JKB JAS AL SKC JO. Performed the experiments: JKS. Analyzed the data: JKS. Wrote the paper: JKS JAS JO SKC DS AL JKB.

References

1. Reich PB, Hobbie SE, Lee T, Ellsworth DS, West JB, et al. (2006) Nitrogen limitation constrains sustainability of ecosystem response to CO_2. Nature 440: 922–925.

2. Hyvönen R, Ågren GI, Linder S, Persson T, Cotrufo MF, et al. (2007) The likely impact of elevated [CO_2], nitrogen deposition, increased temperature and

management on carbon sequestration in temperate and boreal forest ecosystems: a literature review. New Phytol 173: 463–480.

3. IPCC (2007) Climate change: The physical science basis. Contribution of Working Group I to the Fourth Assessment Report of the IPCC. Cambridge: Cambridge University Press.

4. Hovenden MJ, Williams AL (2010) The impacts of rising CO_2 concentrations on Australian terrestrial species and ecosystems. Austral Ecol 35: 665–684.

5. Luo Y, Su B, Currie WS, Dukes JS, Finzi A, et al. (2004) Progressive nitrogen limitation of ecosystem responses to rising atmospheric carbon dioxide. Bioscience 54: 731–739.

6. Norby RJ, Warren JM, Iversen CM, Medlyn BE, McMurtrie RE (2010) CO_2 enhancement of forest productivity constrained by limited nitrogen availability. P Natl Acad Sci-Biol 107: 19368–19373.

7. Vitousek PM, Aber JD, Howarth RW, Likens GE, Matson PA, et al. (1997) Human alteration of the global nitrogen cycle: sources and consequences. Ecol Appl 7: 737–750.

8. Galloway JN, Dentener FJ, Capone DG, Boyer EW, Howarth EW, et al. (2004) Nitrogen cycles: past, present, and future. Biogeochemistry 70: 153–226.

9. Langley JA, Megonigal JP (2010) Ecosystem response to elevated CO_2 levels limited by nitrogen-induced plant species shift. Nature 466: 96–99.

10. Lindroth RL (2010) Impacts of elevated atmospheric CO_2 and O_3 on forests: phytochemistry, trophic interactions and ecosystem dynamics. J Chem Ecol 36: 2–21.

11. Gillon J, Yakir D (2001) Influence of Carbonic anhydrase activity in terrestrial vegetation on the ^{18}O content of atmospheric CO_2. Science 291: 2584–2587.

12. Edwards EJ, Still CJ, Donoghue MJ (2007) The relevance of phylogeney to studies of global change. Trends Ecol Evol 22: 243–249.

13. Willis CG, Ruhfel B, Primack RB, Miller-Rushing AJ, Davis CC (2008) Phylogenetic patterns of species loss in Thoreau's woods are driven by climate change. P Natl Acad Sci-Biol 105: 17029–17033.

14. Davis CC, Willis CG, Primack RB, Miller-Rushing AJ (2010) The importance of phylogeny to the study of phenological response to global climate change. Philo TR Soc B 365: 3201–3213.

15. Molnár AV, Tökölyi J, Végvári Z, Sramkó G, Sulyok J et al. (2012) Pollination mode predicts phenological response to climate change in terrestrial orchids: a case from central Europe. J Ecol 100: 1141–1152.

16. Nowak RS, Ellsworth DS, Smith SD (2004) Functional responses of plants to elevated atmospheric CO_2 - do photosynthetic and productivity data from FACE experiments support early predictions? New Phytol 162: 253–280.

17. Curtis PS, Wang XZ (1998) A meta-analysis of elevated CO_2 effects on woody plant mass, form, and physiology. Oecologia 113: 299–313.

18. Cramer W, Bondeau A, Woodward FI, Prentice IC, Betts RA, et al. (2001) Global response of terrestrial ecosystem structure and function to CO_2 and climate change: results from six dynamic global vegetation models. Global Change Biol 7: 357–373.

19. Pearson RG, Dawson TP (2003) Predicitng the impacts of climate change on the distribution of species: are bioclimate envelope models useful? Global Ecol Biogeogr 12: 361–371.

20. Esther A, Groenveld J, Enright NJ, Miller BP, Lamont BB, et al. (2010) Sensitivity of plant functional types to climate change: classification tree analysis of a simulation model. J Veg Sci 21: 447–461.

21. Cadotte M, Dinnage R, Tilman GD (2012) Phylogenetic diversity promotes ecosystem stability. Ecology 93(8) Suppl: S223–S233.

22. Williams JE, Woinarski JCZ (1997) Eucalypt ecology: Individuals to ecosystems. Cambridge: Cambridge University Press.

23. Brooker MJH (2000) A new classification of the genus Eucalyptus L'Her. (Myrtaceae). Aust Syst Bot 13: 79–148.

24. Conroy J, Milham P, Barlow E (1992) Effect of nitrogen and phosphorus availability on the growth response of Eucalyptus grandis to high CO_2. Plant Cell Environ 15: 843–847.

25. Wong CS, Kriedemann PE, Farquhar GD (1992) CO_2 × nitrogen interaction on seedling growth of four species of eucalypt. Aust J Bot 40: 457–472.

26. Duff G, Berryman C, Eamus D (1994) Growth, biomass allocation and foliar nutrient contents of two Eucalyptus species of the wet–dry tropics of Australia grown under CO_2 enrichment. Funct Ecol 8: 502–508.

27. Lawler I, Foley W, Woodrow I, Cork S (1996) The effects of elevated CO_2 atmospheres on the nutritional quality of Eucalyptus foliage and its interaction with soil nutrient and light availability. Oecologia 109: 59–68.

28. Gleadow R, Foley W, Woodrow I (1998) Enhanced CO_2 alters the relationship between photosynthesis and defence in cyanogenic Eucalyptus cladocalyx ex F. Muell. Plant Cell Environ 21: 12–22.

29. Atwell BJ, Henery ML, Ball MC (2009) Does soil nitrogen influence growth, water transport and survival of snow gum (Eucalyptus pauciflora Sieber ex Sprengel.) under CO_2 enrichment? Plant Cell Environ 32: 553–566.

30. McKiernan AB, O'Reilly-Wapstra JM, Price C, Davies NW, Potts BM et al. (2012) Stability of plant defensive traits among populations in two Eucalyptus species under elevated carbon dioxide. J Chem Ecol 38(2): 204–212.

31. Pryor LD, Johnson LAS (1971) A classification of the Eucalypts. Canberra: Australian National University.

32. Williams K, Potts B (1996) The natural distribution of Eucalyptus species in Tasmania. Tasforests 8: 39–165.

33. McKinnon GE, Vaillancourt RE, Steane DA, Potts BM (2008) An AFLP marker approach to lower-level systematics in Eucalyptus (Myrtaceae). Am J Bot 95: 368–380.

34. Steane DA, Nicolle D, Sansaloni CP, Petroli CD, Carling J, et al. (2011) Population genetic analysis and phylogeny reconstruction in Eucalyptus (Myrtaceae) using high-through put, genome-wide genotyping. Mol Phylogenet Evol 59: 206–224.

35. Woodhams M, Steane D, Moulton V, Jones RC, Nicolle D, et al. (2012) Novel distances for Dollo data. Syst Biol 62(1): 62–77.

36. Jaccoud D, Peng K, Feinstein D, Kilian A (2001) Diversity arrays: a solid state technology for sequence information independent genotyping. Nucleic Acids Res 29: e25.

37. May B, Smethurst P, Carlyle C, Mendham D, Bruce J, et al. (2009) Review of fertiliser use in Australian forestry. Forest and wood products Australia limited project number: RC072–0708. Victoria, Australia.

38. Griffin KL, Winner WE, Strain BR (1995) Growth and dry matter partitioning in loblolly and ponderosa pine seedlings in response to carbon and nitrogen availability. New Phytol 129: 547–556.

39. Cao B, Dang QL, Yü X, Zhang S (2008) Effects of [CO_2] and nitrogen on morphological and biomass traits of white birch (Betula papyrifera) seedlings. Forest Ecol Manag 254: 217–224.

40. Zak DR, Pregitzer KS, Curtis PS, Vogel CS, Holmes WE, et al. (2000) Atmospheric CO_2, soil-N availability, and allocation of biomass and nitrogen by Populus tremuloides. Ecol Appl 10: 34–46.

41. Poorter H, Niklas K, Reich P, Oleksyn J, Poot P, et al. (2012) Biomass allocation to leaves, stems and roots: meta-analyses of interspecific variation and environmental control. New Phytol 193: 30–50.

42. Spinnler D, Egli P, Körner C (2002) Four-year growth dynamics of beech-spruce model ecosystems under CO_2 enrichment on two different forest soils. Trees-Struct Funct 16: 423–436.

43. Willis CG, Ruhfel BR, Primack RB, Miller-Rushing AJ, Losos JB, et al. (2010) Favourable climate change response explains non-native species' success in Thoreau's woods. PloS ONE 5: e8878.

44. Morin X, Thuiller W (2009) Comparing niche-and process-based models to reduce prediction uncertainty in species range shifts under climate change. Ecology 90: 1301–1313.

45. Keenan T, Maria Serra J, Lloret F, Ninyerola M, Sabate S (2010) Predicting the future of forests in the Mediterranean under climate change, with niche and process based models. Glob Change Biol 17: 565–579.

46. Sitch S, Huntingford C, Gedney N, Levy PE, Lomas M, et al. (2008) Evaluation of the terrestrial carbon cycle, future plant geography and climate-carbon cycle feedbacks using five Dynamic Global Vegetation Models (DGVMs). Glob Change Biol 14: 2015–2039.

47. Prentice IC, Harrison SP, Bartlein PJ (2011) Global vegetation and terrestrial carbon cycle changes after the last ice age. New Phytol 189: 988–998.

3

Phylogenetic and Trait-Based Assembly of Arbuscular Mycorrhizal Fungal Communities

Hafiz Maherali[1]*, John N. Klironomos[2]

1 Department of Integrative Biology, University of Guelph, Guelph, Ontario, Canada, **2** Department of Biology, I.K. Barber School of Arts and Sciences, University of British Columbia – Okanagan, Kelowna, British Columbia, Canada

Abstract

Both competition and environmental filtering are expected to influence the community structure of microbes, but there are few tests of the relative importance of these processes because trait data on these organisms is often difficult to obtain. Using phylogenetic and functional trait information, we tested whether arbuscular mycorrhizal (AM) fungal community composition in an old field was influenced by competitive exclusion and/or environmental filtering. Communities at the site were dominated by species from the most speciose family of AM fungi, the Glomeraceae, though species from two other lineages, the Acaulosporaceae and Gigasporaceae were also found. Despite the dominance of species from a single family, AM fungal species most frequently co-existed when they were distantly related and when they differed in the ability to colonize root space on host plants. The ability of AM fungal species to colonize soil did not influence co-existence. These results suggest that competition between closely related and functionally similar species for space on plant roots influences community assembly. Nevertheless, in a substantial minority of cases communities were phylogenetically clustered, indicating that closely related species could also co-occur, as would be expected if i) the environment restricted community membership to single functional type or ii) competition among functionally similar species was weak. Our results therefore also suggest that competition for niche space between closely related fungi is not the sole influence of mycorrhizal community structure in field situations, but may be of greater relative importance than other ecological mechanisms.

Editor: Alfredo Herrera-Estrella, Cinvestav, Mexico

Funding: This research was funded by Discovery grants from the Natural Sciences and Engineering Research Council of Canada (http://www.nserc-crsng.gc.ca) to both authors. The funders had no role in study design, data collection and analysis, decision to publish, or preparation of the manuscript.

Competing Interests: The authors have declared that no competing interests exist.

* E-mail: maherali@uoguelph.ca

Introduction

Functional traits have long been hypothesized to influence community assembly because organism function determines the ability to tolerate climatic conditions, acquire resources and interact with other individuals [1–4]. When functional traits are shared by closely related species (i.e., conserved), phylogenies can be used to determine whether organism function has played a role in the assembly of a given community [1,4–7]. For example, if environmental filtering influences community assembly, then co-occurring species should share characteristics that enable survival in a particular habitat. As a result, communities would be phylogenetically clustered, or more closely related than expected by chance. If competition influences community assembly then co-occurring species should not share functional characteristics, resulting in communities that are phylogenetically even, or more distantly related than expected by chance. Because traits may or may not be conserved, phylogenies may not necessarily be effective proxies for assessing similarities in the functioning of closely related species. Therefore, both trait and phylogenetic perspectives are necessary to test hypotheses about the relative effects of environmental filtering and competition on the assembly of communities [6,7].

Though phylogenetic or trait information has been used to examine community assembly [6], these perspectives have been combined in only a small number of cases [5,8–10]. Moreover, studies that combine phylogenetic and trait information have been confined to communities of macro-organisms such as plant and animals. Nevertheless, communities of micro-organisms can also be structured by processes such as environmental filtering and intense competition among closely related species [11–13]. Microbial species strongly influence ecosystem processes as well as the performance of plants and animals, but hypotheses about how functional trait evolution influences community assembly are more difficult to test than with macro-organisms because of a lack of information on the traits that define microbial niches [12].

In this study, we employ phylogenetic and trait-based approaches to test hypotheses about mechanisms of community assembly in the field for an ecologically important phylum of microbes, the arbuscular mycorrhizal (AM) fungi (Glomeromycota). A majority of species are in three distinct taxonomic families (Glomeraceae, Acaulosporaceae, and Gigasporaceae) within two orders (Glomerales and Diversisporales) [14]. AM fungi are an ancient lineage of obligate biotrophs which must form associations with plants in order to obtain energy for growth and reproduction [15]. AM fungal communities are known to respond to variation in climate, soil resources and plant host identity [16–18], as well as influence plant function [19,20] and the coexistence of plant species [21]. Despite their ecological importance, little is known about the mechanisms that regulate community assembly in AM fungi [22].

Functional traits associated with spatial niches are similar among closely related species in the Glomeromycota (i.e., they are conserved) [23,24]. For example, members of the Gigasporaceae extensively colonize soil but exhibit limited and slow colonization of roots. Conversely, species in the Glomeraceae rapidly and extensively colonize roots but produce limited hyphal biomass in soil. The Acaulosporaceae form a third distinct group that tend to be poor colonizers of both soil and roots. Because of this trait conservatism, phylogenies can be used to test hypotheses about the mechanisms of community assembly in AM fungi. By experimentally manipulating the phylogenetic relatedness of AM fungal communities under uniform host and soil conditions, we have previously shown that realized species richness was highest when the starting species were more distantly related to each other and did not share similar functional traits [20]. However, fungal communities in the field are likely to be influenced by dispersal limitations, priority effects, host variation, soil heterogeneity and stochasticity [16,25], all of which may supersede trait and phylogenetic effects. Thus, field studies are necessary to determine the relative importance of various ecological mechanisms responsible for community assembly.

To determine whether phylogenetic and trait dispersion influence AM community composition under field conditions, we examined the species composition of AM fungal communities in an old field. If closely related and functionally similar species compete, it would be expected that communities would be phylogenetically even and consist of species with dissimilar trait values. If soil conditions or plant hosts act as habitat filters, it would be expected that communities would be phylogenetically clustered and consist of species with similar trait values. We tested these predictions by sampling soil at regular intervals within a 50 m×50 m grid. We characterized AM fungal community composition at each sampling point based on the morphological identification of spores. We calculated whether species composition at each sampling point was phylogenetically even or clustered and whether trait dispersion was greater or lower than expected by chance.

Materials and Methods

Site Description and Plot Layout

We established survey plots at the Long-Term Mycorrhiza Research Site (LTMRS), an old field meadow dominated by perennial herbaceous plants, and which is located on relatively even ground in the Nature Reserve of the University of Guelph Arboretum, Guelph, ON (43°32′30″ N, 80°13′00″ W). Soils at the site are generally nutrient poor, and particularly low in phosphorus (2.1 mg P kg^{-1} dry soil) [26]. Though the site has been used for agriculture in the past, cultivation was abandoned in 1967. In 2000, we placed a single 50 m×50 m gridded plot in the centre of the site. Previous analyses suggest that because of limited spore dispersal [15], AM fungal community composition is spatially structured at scales <50 cm [27,28]. Therefore, we established sampling points on the grid at 1 m intervals (51×51 points = 2601 community samples).

Sampling Species Richness

We used trap cultures to determine the species richness of AM fungal communities. Though trap cultures can exclude species that have poor rates of colonization or specific host requirements [22,29], previous research suggests that they nevertheless capture relatively high numbers of species [18]. In addition, trap cultures allowed us to include only those species that were sporulating, avoiding bias associated with the inclusion of ecologically inactive resting spores in whole soil samples [22]. Our previous results indicate that estimates of species richness obtained from 18 s rRNA-based terminal restriction fragment length polymorphism (t-RFLP) analysis were positively correlated with known species richness in trap cultures [20]. The morphological species identifications using trap cultures were also necessary to match taxa with AM fungal species cultured from the same field site that had been used to obtain trait information [20].

Each grid point was marked as the center for a plot. To characterize the species richness of the community at each grid point, we sampled species located within a 30 cm radius of the center point. We placed four stakes 30 cm from the center in each cardinal direction. In June, we collected 4 soil subsamples using a soil corer (3 cm in diameter, 15 cm deep), which were then pooled and mixed well. Our previous research suggests that closely related AM fungal species can compete to colonize plant roots [20], raising the possibility that the trap culture technique could filter species in a way that produces phylogenetically even communities. To reduce the likelihood that competition for root space would restrict the taxa recovered from soil samples, we established three separate trap cultures for each sampling point. We note that other methods of limiting competition in trap cultures are available, such as using low amounts of inoculum to initiate cultures. However, we opted to divide soils into multiple trap cultures to increase the likelihood of root colonization and sporulation. AM fungal species lists for each sampled community consisted of species pooled across the 3 trap cultures.

To establish trap cultures, we divided the soil sample into three parts and placed it in a Cone-tainer (SC10, Stuewe & Sons, Tangent, OR, USA) which had the bottom 2/3 filled with a mix of 50% inert calcined clay (Turface, Profile Products LLC, Buffalo Grove, IL, USA) and 50% silica sand. The top 1/3 of the container was filled with field soil. The resulting 7803 Cone-tainers were randomly placed on benches in the greenhouse. To provide abundant root area for fungal colonization, five seeds of leek (*Allium porrum*), a species that is frequently used as a general host for AM fungal species [30], were added to each Cone-tainer, and thinned to three plants per pot after germination. Plants were watered daily and no fertilizer was added.

To minimize the possibility that life history differences in spore germination and growth rate would bias taxon recovery from soil samples, trap cultures were harvested after 12 weeks, providing enough time for species from different AM fungal families to colonize roots [23,24]. At harvest, plant shoots and the top 1/3 of the pot containing the field soil were removed. The bottom 2/3 of the pot, which contained the growth medium, leek roots and freshly produced spores, was used to extract and identify AM fungal species. These materials were mixed in a blender, suspended in water, and then passed through a series of sieves whose mesh ranged from 1 mm, 0.5 mm, 0.3 mm, and 0.047 mm. The fraction remaining on the smallest sieve size was placed in a beaker and decanted twice to remove heavy particles that settled to the bottom. The floating fraction was placed on a wet nitrocellulose filter and sealed in a petri dish. We mounted up to 100 AM fungal spores on slides with 1:1 (v/v) of polyvinyl-alcohol-acetic-acid-glycerol and Melzer's Reagent [30] and identified them using morphological and developmental characters as described on the International Culture Collection of Vesicular Arbuscular Mycorrhizal Fungi (INVAM) web site (http://invam.caf.wvu.edu/fungi/taxonomy/speciesID.htm). Because spore abundance depends strongly on life history [15], it was not an appropriate metric for quantifying species abundance. As a result, species were scored as either present or absent. Grid points where no species were found were eliminated from the analysis of community assembly.

To describe the spatial distribution of each species based on the presence or absence at each sampling point, we calculated the Morisita index of dispersion (I_δ) [31]. Because I_δ requires information on abundance, we carried out this analysis using species presence data aggregated over 4 adjacent sampling points (i.e., derived from 2×2 m plots). As a result, the maximum abundance each species could have in the analysis was 4 (e.g., a species was found at each of the 4 sampling points included in each aggregated plot). I_δ values <1 indicate repulsion among individuals, manifested as an even distribution; I_δ ~1 suggests a random distribution; and I_δ >1 indicate attraction among individuals, manifested as clumping. I_δ could not be calculated for the rarest taxon, *Scutellospora pellucida*, which was found at only 6 points on the sampling grid.

AM Fungal Trait Data

To determine whether AM fungal traits influenced community composition, we obtained information on the extent of root and soil colonization from previous studies of the same fungal taxa collected at the same site [20]. AM fungi were cultured by inoculating seedlings of *Plantago lanceolata* with single fungal spores of each species. These cultures were grown for one year in 20 cm diameter pots containing sterilized field soil. After cultures were established, 50 g of AM fungal inocula were added to pots containing sterilized field soil along with a germinated seedling. After 1 year of growth, root colonization (percentage of root length infected [32]) and soil hyphal length (m hyphae g^{-1} soil [33]) were measured. Though fungal traits and performance can vary with plant host [15], our previous studies suggest root colonization and soil hyphal length were similar when assessed using four old field species as hosts [24]. Therefore, we assumed that these fungal traits were representative of the performance of each species, rather than being an outcome of specific interactions between the host plant and fungal species.

Phylogenetic Tree Construction and Analyses of Trait Conservatism

To determine whether shared evolutionary history could explain patterns of species coexistence, we developed a phylogenetic tree using previously published molecular phylogenies [24,34,35]. Because these phylogenies were created with different gene sequences, we manually pruned and combined these trees to produce a topology that included only the taxa found in our old field sample plot. Because of this method of tree construction, branch lengths were not available. Therefore, we set all branch lengths to 1, a conservative assumption that minimizes type I error rate in comparative analyses [36].

To verify that fungal traits were conserved [24] using the species found at our field site, we calculated contribution indices (CIs) for each node in the phylogeny and a tree-wide phylogenetic signal using the 'aotf' function in PHYLOCOM 4.2 [37]. Contribution indices vary between 0 and 1 and estimate the degree to which individual nodal divergences along the phylogeny contribute to extant trait variation [38]. A trait was considered conserved if significant variation is explained more by relatively ancient than recent divergences in the phylogeny. Phylogenetic signal is derived from the tree-wide variance of standardized independent contrasts [39]. If closely related lineages have similar traits, then the magnitude of the independent contrasts should be low, resulting in low tree-wide variance. To determine if CIs and tree-wide phylogenetic signal were statistically significant ($P\leq0.05$), they were compared to a distribution of 1000 values calculated by

randomly swapping trait values across the tips of the phylogeny in PHYLOCOM [37]. This method also generates randomized trait values at internal nodes because character reconstruction is based on the randomized tip values.

Phylogenetic and Trait-based Analyses of Community Composition

To test whether phylogenetic relationships and functional traits influenced AM fungal species assemblages, we compared phylogenetic relatedness and observed patterns of trait variation for each of the sampled communities to randomly generated communities derived from a 'constrained' null model that assumes that the probability of a species contributing to an assemblage is determined by its overall frequency across the entire sampling grid [37,40]. The null communities were created by randomly swapping species occurrences among all sampling grid points while maintaining the species richness of the observed community at each sampling grid point [41]. Although other null models were available for comparison to sampled communities [40], we used the constrained null model for several reasons. First, this null model was developed for species presence/absence data, which made it suitable for our study design. Second, simulations suggest that in groups such as AM fungi, where both traits [24] and species frequency [42] are conserved, this null model is less prone to Type 1 error [43]. Third, the spatial distribution for a majority AM fungi at our site was clumped (see results). The constrained null model preserves some degree of this spatial autocorrelation, which reduces Type 1 error rate in tests of trait and phylogenetic-based community composition [44]. All analyses were done using PHYLOCOM 4.2 [37].

To quantify the phylogenetic relatedness of co-occuring species, we calculated the mean nearest phylogenetic taxon distance (MNTD) using the 'comstruct' function in PHYLOCOM. MNTD is defined as the average distance to the closest relative of each species in the sample [37]. Communities that are phylogenetically even have MNTDs higher than expected by chance, whereas communities that are phylogenetically clustered have MNTDs lower than expected by chance. We chose MNTD over other relatedness metrics [6] because simulations indicate that in situations where functional traits are conserved, this metric is most suitable for detecting phylogenetic evenness and clustering [45] while minimizing Type I error rates [43].

To quantify trait variation within each community, we calculated the variance (VAR) of root colonization and soil hyphal length for both observed and null communities using the 'comtrait' function in PHYLOCOM. If competition structures communities, then species with dissimilar traits are more likely to co-occur, and trait variance in observed communities should be higher than that generated in null communities. Conversely, if habitat filtering influences species assemblages, then species with similar traits are more likely to co-occur, and trait variance in observed communities should be lower than that generated in null communities.

To determine whether MNTD and trait variance in an observed community differed statistically from MNTD and trait variance in a randomly assembled community, we compared observed values to a distribution of 9999 communities generated from the null model. To determine whether the observed community MNTD or trait variance was significantly different from the null community using a two tailed test ($\alpha=0.05$) we calculated whether it was in the top or bottom 2.5% of the null distribution (i.e., 250/10000).

To examine whether the overall pattern in MNTD and trait variance across the sampling grid was consistently different from null expectations, we calculated a standardized effect size (SES) for each metric for each sampling point based on the difference between the observed metric and the mean metric of the null communities. For MNTD, we calculated the nearest taxon index (NTI) as: $NTI = -1 \times [(MNTD_{OBS} - MNTD_{RANDOM})] / sd(MNTD_{RANDOM})$. A negative NTI indicates that a community is phylogenetically even, whereas a positive NTI indicates that a community is phylogenetically clustered. For trait variation we calculated SES_{VAR} as: $SES_{VAR} = [(VAR_{OBS} - VAR_{RANDOM}) / sd(VAR_{RANDOM})]$. A positive SES_{VAR} indicates that traits are dispersed in a community whereas a negative SES_{VAR} indicates that traits are clustered within a community. We used one sample Wilcoxon signed ranks tests to determine if the site level distributions of NTI and SES_{VAR} for each trait were significantly different from a null expectation of zero.

Results

Fungal Richness and Frequency

There were sporulating AM fungal species in 2532 of the 2601 sampling grid points and species richness in these communities ranged from 1 to 8 taxa. There were 2151 communities for which mean nearest phylogenetic taxon distance (MNTD) could be calculated (i.e., where richness was ≥2). A total of 15 species spanning 3 families were identified (Figure 1, 2), 57% were Glomeraceae, 24% were Acaulosporaceae and 19% were Gigasporaceae. Our sampling protocol was sufficient to reach saturation for number of species existing at the site (Figure 1, inset). Ten of 14 species for which Morisita's index (I_δ) could be calculated had values >1, indicating a clumped distribution pattern (Figure 2). The four most abundant species (Figure 1), however, had values that were ~1 or <1, indicating either a random or even distribution pattern, respectively.

Fungal Trait Conservatism

AM fungal functional traits were conserved. We detected a significant tree wide phylogenetic signal for the extent of root colonization (contrast variance = 0.497, $P = 0.002$) and hyphal colonization of soil (contrast variance = 163.2, $P = 0.004$). The bulk of extant trait variation was accounted for by deep divergences in the phylogeny (Figure 3). For root colonization, the divergence between the Glomerales and Diversisporales had the largest contribution index (CI) accounting for 83% of extant trait variation (Node A, $P = 0.001$). For hyphal length colonization of soil, the divergence between Gigasporaceae and Acaulosporaceae within the Diversisporales had the largest CI, accounting for 84% of extant trait variation (Node B, $P = 0.001$).

Phylogenetic and Trait-based Community Assembly

AM fungal communities were most frequently phylogenetically even; 36 sampling points had MNTDs significantly higher than expected by chance, whereas 17 sampling points had MNTDs that were significantly lower than expected by chance. The distribution of standardized effect sizes for MNTD, expressed as the Nearest Taxon Index (NTI) was significantly <0 (Figure 4, Test statistic = −4.15, $P < 0.0001$) and 54.8% of sampling points had NTIs <0.

The extent of root colonization in AM fungal communities was more often dispersed than clustered; 91 sampling points had trait variances significantly higher than expected by chance, whereas 89 sampling points had trait variances significantly lower than expected by chance. The distribution of SES_{VAR} for root colonization was significantly >0 (Figure 4, Test statistic = 3.56, $P < 0.0001$) and 53.1% of sampling points were >0.

The extent of soil hyphal colonization in AM fungal communities was most frequently clustered; 102 sampling points had trait variances significantly lower than expected by chance, whereas 76 sampling points had trait variances higher than expected by chance. Though 58.6% of sampling points had values <0, the distribution of SES_{VAR} for soil hyphal colonization did not differ significantly from 0 (Figure 4, Test statistic = −0.97, $P = 0.335$).

Discussion

We found evidence for phylogenetic-based community assembly in the AM fungi of an old field. A majority of AM fungal assemblages at our study site were phylogenetically even. This result is consistent with a previous experimental study [20] where we found that AM fungi were more likely to co-exist on roots of a single plant species under uniform soils when they were drawn from different families. Other recent surveys have also shown that AM fungal community composition is non-random [46]. In the current study, fungal communities also had higher than expected variation in the intensity of root colonization. Thus, species that intensively colonize roots were more likely to co-exist with those that had relatively low root colonization. That a majority of communities in the field were made of up of distantly related and functionally dissimilar species suggests that competition for root space influences AM fungal community assembly at small spatial scales, even when other factors such as host identity, soil conditions and dispersal limitations vary in nature [4,5].

Like previous studies of AM fungal community structure [27,47], we found that the fungal species assemblage at our study site was dominated by a small number of abundant taxa and that most species had a clumped distribution. In particular, two Glomeraceae species were more frequently observed across the sampling grid than species from other families (Figure 1, 2). This family specific pattern of abundance is consistent with previous local and global surveys using both spore and sequence based sampling techniques, which show that dominant species tend to be members of the Glomeraceae, particularly Group A [18,42,48,49]. Over dominance in AM fungal communities has led to the hypothesis that stochastic processes associated with the opportu-

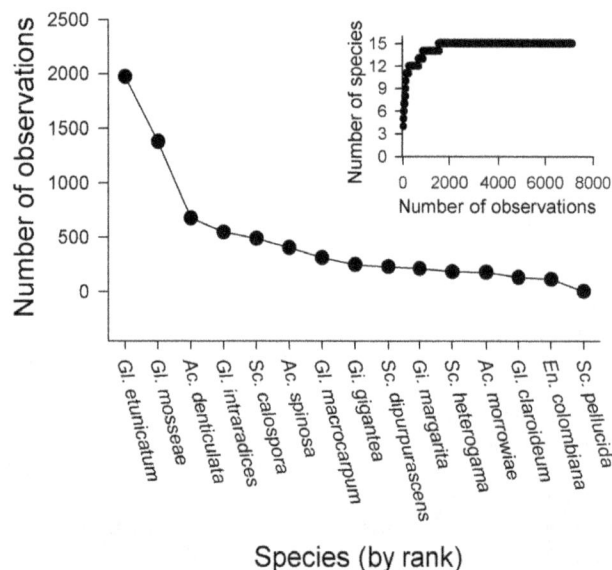

Figure 1. The frequency of AM fungal species across the 50 m×50 m sampled grid and a species rarefaction curve (inset).

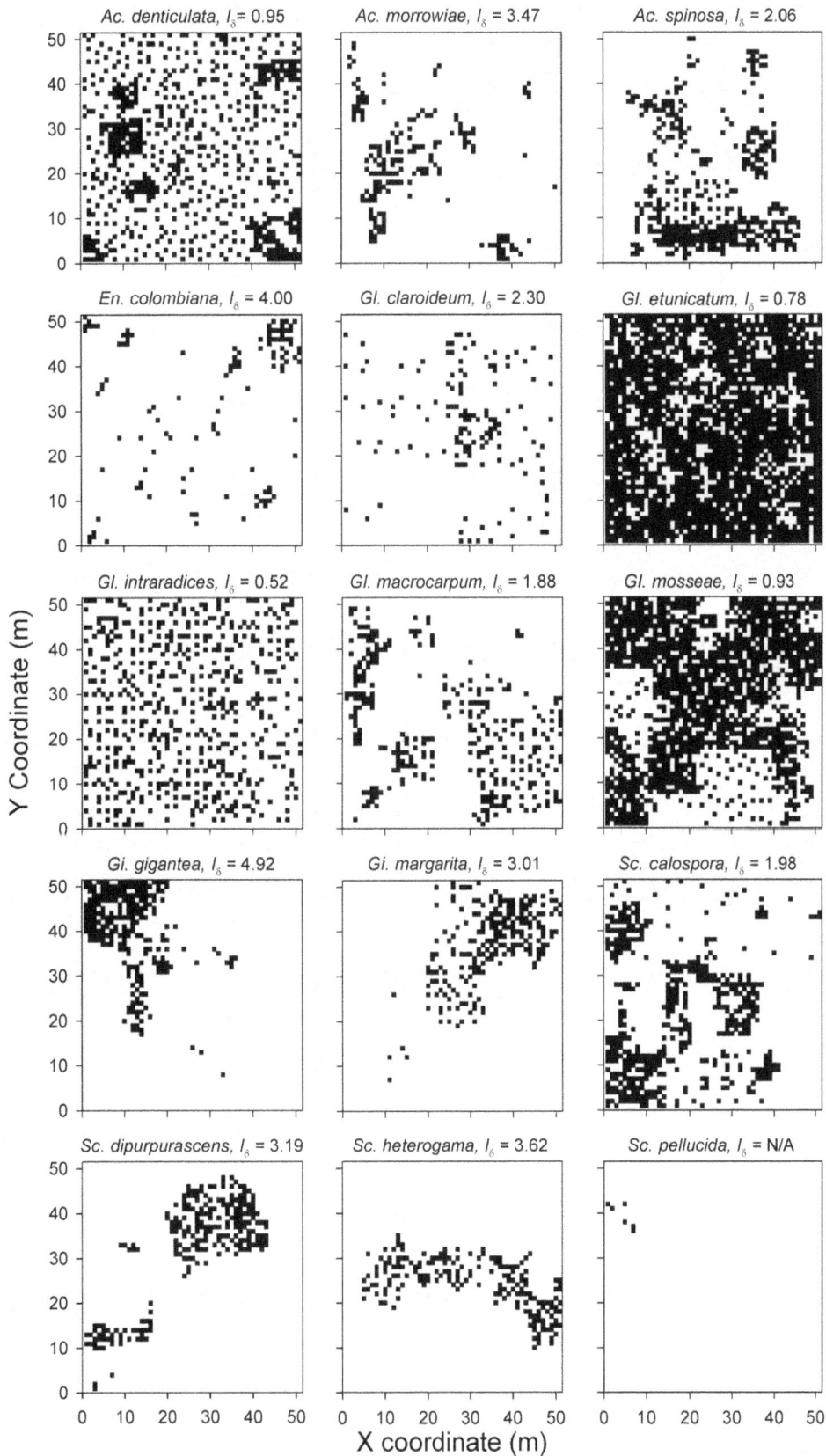

Figure 2. The distribution of each species across the 50 m ×50 m sampling grid. Species were scored as present or absent in each of the 2601 sampled communities.

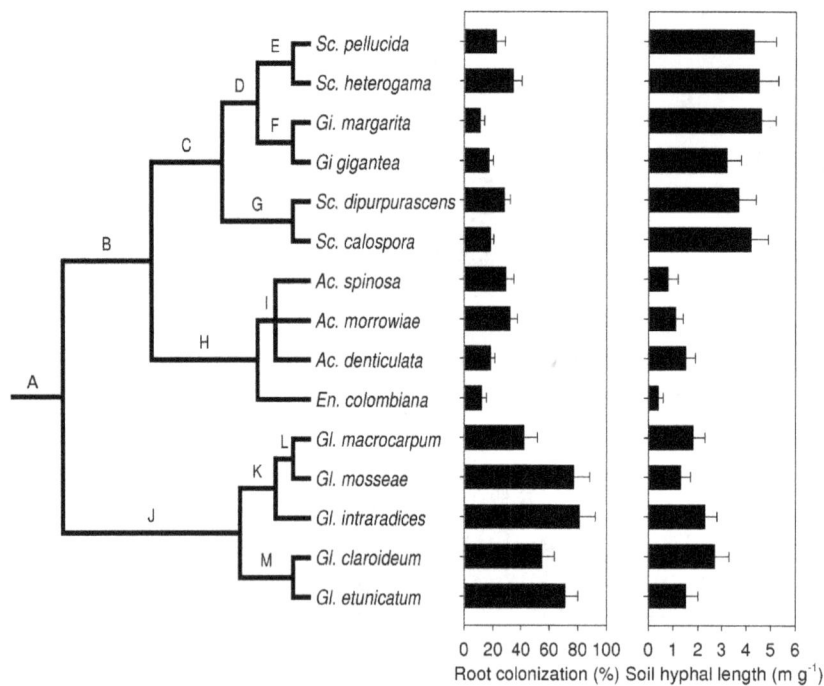

Figure 3. A phylogeny of AM fungi found in the 50 m×50 m sampling grid, along with trait values for Root Colonization and Hyphal Length mapped to each taxon. Both traits were phylogenetically conserved.

nistic colonization of roots is a primary mechanism responsible for AM fungal community structure [47]. However, our findings indicate that even when over dominance occurs (Figure 1), competition for root space among functionally similar taxa can still be a determinant of community composition.

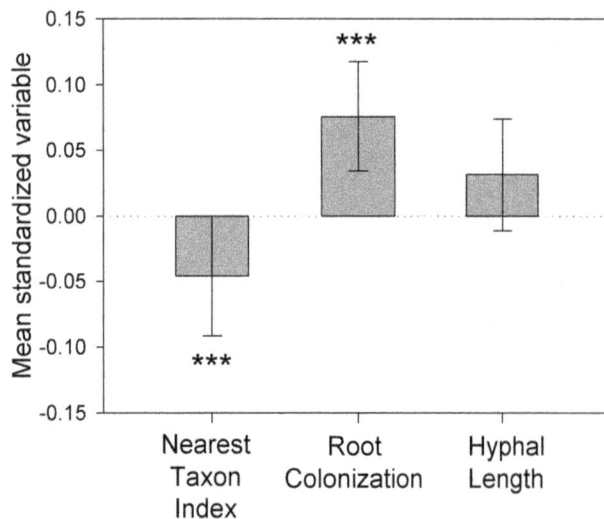

Figure 4. Mean (±95 CI) Nearest Taxon Index (NTI, A), standardized trait variance (SES$_{VAR}$) for Root Colonization (B) and SES$_{VAR}$ for Hypal Length in soil (C) in the 50 m×50 m sampled grid. Mean NTI was significantly lower than 0, indicating that communities were more phylogenetically even than expected by chance. Mean Root Colonization variance was significantly higher than 0, indicating that community trait variance was higher than expected by chance. Mean Hyphal Length in soil did not differ from 0. ***$P<0.0001$.

Our results differ from a recent global meta-analysis of AM fungal community structure [50] that found that community composition at a site is more frequently phylogenetically clustered than even. This apparent conflict may have arisen from a difference in the phylogenetic scale of the species pool used between studies [51]. To test for the ecological significance of competition at each sampling point in the grid, it was necessary for us to construct null communities assuming that only those species found within the study site were able to colonize any given sampling point, in proportion to their abundance [43,45]. By contrast, the global meta-analysis tested whether species composition within a site was clustered relative to the global diversity of AM fungi. The difference between the local versus global species pool used for tests of community assembly mechanisms suggests that our findings complement rather than conflict with those done at a global scale. For example, interactions such as competition could determine which species co-exist at small spatial scales within a site, but species composition for the whole site could be restricted by a larger scale ecological filter associated with niche requirements or climate [51].

Though the congruence between phylogenetic evenness and high variation in root colonization intensity within sampled communities suggests that this trait determines co-existence among AM fungi (Figure 4), two other hypotheses could explain the tendency for sampling points to be phylogenetically even. First, other functions that influence fungal fitness such as root colonization rate, spore production rate, frequency of hyphal network formation, uptake of P and N, and the metabolism of sugars [29] could influence co-existence if they are conserved. Testing whether these additional traits influence fungal species co-existence, however, is limited by a lack of information on them in multiple lineages [29]. Second, co-existence in AM fungi could be regulated by negative interactions with consumers, pathogens and parasites [52,53]. If closely related species share susceptibility to natural enemies [54], then negative density dependence will prevent these species from co-occurring [55]. Testing this prediction is also limited by a lack of information on the extent that AM fungi are regulated by consumers [52], and whether susceptibility to natural enemies is conserved.

Even though hyphal colonization of soil was conserved, this trait was not likely to influence community assembly. Variance in hyphal colonization of soil was lower than for root colonization, and did not differ from zero (Figure 4), suggesting that the dispersion of this trait was random among co-existing species. One explanation for the lack of an association between hyphal colonization of soil and community composition is that AM fungi in the field form mycelial networks through the fusion of hyphae with conspecifics [22] that are likely much larger than in isolated pots [56]. Thus it is possible that hyphal colonization of soil measured on species growing in previously sterilized soil in a pot was not a meaningful indicator of how this trait is expressed in the field. Alternatively, it is also possible that because the volume of soil that can be explored by hyphae is large, there is little competition among AM fungi for this aspect of the niche.

A substantial minority of sampling points contained assemblages that were phylogenetically clustered. This result suggests that environmental filtering also influenced the assembly of AM fungal communities at our study site. One potential cause of filtering is the soil environment, which could affect composition in two ways. First, high nutrient soil patches could eliminate AM fungal species that specialize on nutrient uptake because these functions would not be required by plant hosts [57,58]. If nutrient uptake capacity is conserved and is higher in specific lineages [20,24] and these species become extinct in nutrient rich patches, then communities could be phylogenetically clustered. Second, some fungal lineages may have specialized to occupy specific soil texture classes. For example, other field surveys indicate that Glomeracae can dominate on clay soils, whereas Gigasporaceae can dominate on sandy soils [25,57]. Thus, spatial variation in soil texture could have excluded specific lineages, resulting in sampling points that were phylogenetically clustered. An additional cause of environmental filtering is host identity, which has previously been shown to influence the presence or absence of AM fungal taxa [17,19,58–60]. Though we lacked information on the identity of roots that AM fungal species associated with, plant species could influence community composition of mycorrhizal fungi in different ways. For example, if the benefits fungi provide a plant species are conserved [24], it is possible that the active culturing or sanctioning of certain lineages by plant hosts in order to maximize that benefit [61,62] results in communities that contain only one lineage and are therefore phylogenetically clustered. By contrast, if the benefits fungi provide plants are not conserved [63,64], then culturing or sanctioning by plants could also result in increased phylogenetic evenness of fungal communities.

The occurrence of phylogenetically clustered assemblages could also be caused by the co-existence of closely related species that are weak competitors for niche space [4,65]. Specifically, species in the Diversisporales (Acaulosporaceae and Gigasporaceae) are closely related and share a low ability to colonize roots. If the poor ability of these species to colonize roots allows them to co-exist, then a substantial number of phylogenetically clustered assemblages should have a mean root colonization value lower than the site median of 52.5%. However, we found that only 13.5% of

sampling points had both phylogenetically clustered assemblages and low mean root colonization, a proportion that was significantly lower than expected by chance ($X^2 = 11.89$, $P = 0.0006$, df = 1). This result suggests that the co-occurrence of weak competitors for root space was relatively rare. However, we also found that a higher than expected proportion of sampling points (31.8%) had assemblages that were both phylogenetically clustered and had mean root colonization values higher than the site median. This finding suggests that closely related species in the Glomeraceae with high root colonization can co-occur frequently (Figure 3). The co-occurrence of these abundant (Figure 1) and closely related species suggests that phylogenetic clustering occurred because of dispersal by dominant species into locations where niche space on roots was not adequately filled [47,49].

A potential limitation of our study is that sampling AM fungal communities using trap cultures likely resulted in the absence of fungal species that cannot be cultured. Often, molecular methods of AM fungal identification obtain more apparent taxa (operational taxonomic units, or OTUs) than spore based methods [22], and it is therefore likely that total AM fungal species richness was underestimated in our study. Nevertheless, species richness was not outside of the range obtained in molecular-based surveys of old fields and grasslands [16,47]. In addition, we were able to obtain species from the major families of the Glomeromycota (Fig. 2), suggesting that the trap culture method did not discriminate strongly against specific lineages. Thus, even though species richness was likely underestimated, relative differences in phylogenetic community structure at each sampling point were unlikely to be biased by the trap culture method.

In conclusion, we provide evidence for phylogenetic and trait-based community assembly in AM fungi occurring in a realistic ecological setting. A majority of assemblages at the old field site we sampled were composed of species that were distantly related and differed in the extent to which they were able to colonize roots. Our results were therefore consistent with experimental evidence that competition can prevent functionally similar and closely related taxa from co-existing at small spatial scales [20]. Nevertheless, competition for habitat space on roots was not a universal determinant of the composition of AM fungal communities in the old field, and our results also suggest that habitat filtering can influence community composition at small spatial scales. Moreover, competition may also be weak in certain situations, resulting in communities shaped by the dispersal of abundant species [47,49]. More generally, our findings suggest that, as in macro-organisms, combining phylogenetic and trait-based approaches can provide insights into the mechanisms of microbial community assembly [12].

Acknowledgments

We thank E. Seifert for assistance with data management and data analysis. This manuscript was improved by comments from C.M. Caruso and two anonymous reviewers.

Author Contributions

Conceived and designed the experiments: HM JNK. Performed the experiments: JNK. Analyzed the data: HM. Wrote the paper: HM. Edited text: HM JNK. Formulated hypotheses and predictions: HM JNK. Designed the sampling protocol and collected data: JNK.

References

1. Darwin C (1859) The Origin of Species. J. Murray, London.
2. Elton C (1946) Competition and the structure of ecological communities. Journal of Animal Ecology 15: 54–68.

3. Weiher E, Keddy P (1999) Ecological Assembly Rules: Perspectives, Advances, Retreats. Cambridge University Press, Cambridge.

4. Webb CO, Ackerly DD, McPeek MA, Donoghue MJ (2002) Phylogenies and community ecology. Annual Review of Ecology and Systematics 33: 475–505.
5. Cavender-Bares J, Ackerly DD, Baum DA, Bazzaz FA (2004) Phylogenetic overdispersion in Floridian oak communities. American Naturalist 163: 823–843.
6. Vamosi SM, Heard SB, Vamosi JC, Webb CO (2009) Emerging patterns in the comparative analysis of phylogenetic community structure. Molecular Ecology 18: 572–592.
7. Cavender-Bares J, Kozak K, Fine PVA, Kembel SW (2009) The merging of community ecology and phylogenetic biology. Ecology Letters 12: 693–715.
8. Ingram T, Shurin JB (2009) Trait-based assembly and phylogenetic structure in northeast Pacific rockfish assemblages. Ecology 90: 2444–2453.
9. Swenson NG, Enquist BJ (2009) Opposing assembly mechanisms in a Neotropical dry forest: implications for phylogenetic and functional community ecology. Ecology 90: 2161–2170.
10. Kraft NJB, Ackerly DD (2010) Functional trait and phylogenetic tests of community assembly across spatial scales in an Amazonian forest. Ecological Monographs 80: 401–422.
11. Horner-Devine MC, Bohannan BJM (2006) Phylogenetic clustering and overdispersion in bacterial communities. Ecology 87: S100–S108.
12. Green JL, Bohannan BJM, Whitaker RJ (2008) Microbial biogeography: from taxonomy to traits. Science 320: 1039–1043.
13. Hibbing ME, Fuqua C, Parsek MR, Peterson SB (2010) Bacterial competition: Surviving and thriving in the microbial jungle. Nature Reviews Microbiology 8: 15–25.
14. Schüßler A, Schwarzott D, Walker C (2001) A new fungal phylum, the Glomeromycota: phylogeny and evolution. Mycological Research 105: 1413–1421.
15. Smith S, Read D (2008) Mycorrhizal symbiosis, 3rd edn. Academic Press, London.
16. Öpik M, Moora M, Liira J, Zobel M (2006) Composition of root colonizing arbuscular mycorrhizal fungal communities in different ecosystems around the globe. Journal of Ecology 94: 778–790.
17. Hausmann NT, Hawkes CV (2009) Plant neighborhood control of arbuscular mycorrhizal community composition. New Phytologist 183: 1188–1200.
18. Oehl F, Laczko E, Bogenrieder A, Stahr K, Bosch R, van der Heijden M, Sieverding E (2010) Soil type and land use intensity determine the composition of arbuscular mycorrhizal fungal communities. Soil Biology and Biochemistry 42: 724–738.
19. Helgason T, Merryweather JW, Denison J, Wilson P, Young JPW, Fitter AH (2002) Selectivity and functional diversity in arbuscular mycorrhizas of co-occurring fungi and plants from a temperate deciduous woodland. Journal of Ecology 90: 371–384.
20. Maherali H, Klironomos JN (2007) Influence of phylogeny on fungal community assembly and ecosystem functioning. Science 316: 1746–1748.
21. Bever JD, Dickie IA, Facelli E, Facelli JM, Klironomos JN, Moora M, Rillig MC, Stock WD, Tibbett M, Zobel M (2010) Rooting theories of plant ecology in microbial interactions. Trends in Ecology and Evolution 25: 468–478.
22. Rosendahl S (2008) Communities, populations and individuals of arbuscular mycorrhizal fungi. New Phytologist 178: 253–266.
23. Hart MM, Reader RJ (2002) Taxonomic basis for variation in the colonization strategy of arbuscular mycorrhizal fungi. New Phytologist 153: 335–344.
24. Powell JR, Parrent JL, Klironomos JN, Hart MM, Rillig MC, Maherali H (2009) Phylogenetic trait conservatism and the evolution of functional tradeoffs in arbuscular mycorrhizal fungi. Proceedings of the Royal Society Biological Sciences 276: 4237–4245.
25. Lekberg Y, Koide RT, Rohr JR, Aldrich-Wolfe L, Morton JB (2007) Role of niche restrictions and dispersal in the composition of arbuscular mycorrhizal fungal communities. Journal of Ecology 95: 95–105.
26. Sherrard ME, Maherali H (2010) Local adaptation across a fertility gradient is influenced by soil biota in the invasive grass, Bromus inermis. Evolutionary Ecology doi: 10.1007/s10682-011-9518-2.
27. Klironomos JN, Rillig MC, Allen MF (1999) Designing field experiments with the help of semi-variance and power analyses. Applied Soil Ecology 12: 227–238.
28. Mummey DL, Rillig MC (2008) Spatial characterization of arbuscular mycorrhizal fungal molecular diversity at the sub-metre scale in a temperate grassland. FEMS Microbial Ecology 64: 260–270.
29. van der Heijden MGA, Scheublin TR (2007) Functional traits in mycorrhizal ecology: their use for predicting the impact of arbuscular mycorrhizal fungal communities on plant growth and ecosystem functioning. New Phytologist 174: 244–250.
30. Brundrett M, Bougher N, Dell B, Grove T, Malajczuk N (1996) Working with Mycorrhizas in Forestry and Agriculture. ACIAR Monograph 32. Canberra.
31. Morisita M (1962) I_δ-index, a measure of dispersion of individuals. Researches on Population Ecology 4: 1–7.
32. McGonigle TP, Miller MH, Evans DG, Fairchild GL, Swan JA (1990) A new method which gives an objective measure of colonization of roots by vesicular arbuscular mycorrhizal fungi. New Phytologist 115: 495–501.
33. Miller RM, Jastrow JD, Reinhardt DR (1995) External hyphal production of vesicular-arbuscular mycorrhizal fungi in pasture and tallgrass prairie communities. Oecologia 103: 17–23.
34. Redecker D, Raab P (2006) Phylogeny of the Glomeromycota (arbuscular mycorrhizal fungi): recent developments and new gene markers. Mycologia 98: 885–895.
35. Krüger MKrügerC, Walker C, Stockinger H, Schüßler A (2012) Phylogenetic reference data for systematics and phylotaxonomy of arbuscular mycorrhizal fungi from phylum to species level. New Phytologist 193: 970–984.
36. Ackerly DD (2000) Taxon sampling, correlated evolution and independent contrasts. Evolution 54: 1480–1492.
37. Webb CO, Ackerly DD, Kembel SW (2008) Phylocom: software for the analysis of phylogenetic community structure and trait evolution. Bioinformatics 24: 2098–2100.
38. Moles AT, Ackerly DD, Webb CO, Tweddle JC, Dickie JB, Westoby M (2005) A brief history of seed size. Science 307: 576–580.
39. Blomberg SP, Garland T, Ives AR (2003) Testing for phylogenetic signal in comparative data: behavioral traits are more labile. Evolution 57: 717–745.
40. Kembel SW, Hubbell SP (2006) The phylogenetic structure of a neotropical forest tree community. Ecology 87: S86–S99.
41. Gotelli NJ (2000) Null model analysis of species co-occurrence patterns. Ecology 81: 2606–2621.
42. Öpik M, Vanatoa A, Vanatoa E, Moora M, Davison J, Kalwij JM, Reier Ü, Zobel M (2010) The online database MaarjAM reveals global and ecosystemic distribution patterns in arbuscular mycorrhizal fungi (Glomeromycota). New Phytologist 188: 223–241.
43. Kembel SW (2009) Disentangling niche and neutral influences on community assembly: assessing the performance of community phylogenetic structure tests. Ecology Letters 12: 949–960.
44. Hardy OJ (2008) Testing the spatial phylogenetic structure of local communities: statistical performances of different null models and test statistics on a locally neutral community. Journal of Ecology 96: 914–926.
45. Kraft NJB, Cornwell WK, Webb CO, Ackerly DD (2007) Trait evolution, community assembly, and the phylogenetic structure of ecological communities. American Naturalist 170: 271–283.
46. Davison J, Öpik M, Daniell TJ, Moora M, Zobel M (2011) Arbuscular mycorrhizal fungal communities in plant roots are not random assemblages. FEMS Microbiology Ecology 78: 103–115.
47. Dumbrell AJ, Nelson M, Helgason T, Dytham C, Fitter AH (2010) Idiosyncrasy and over dominance in the structure of natural communities of arbuscular mycorrhizal fungi: is there a role for stochastic processes? Journal of Ecology 98: 419–428.
48. Dumbrell AJ, Ashton PD, Aziz N, Feng G, Nelson M, Dytham C, Fitter AH, Helgason T (2011) Distinct seasonal assemblages of arbuscular mycorrhizal fungi revealed by massively parallel pyrosequencing. New Phytologist 190: 794–804.
49. Lekberg Y, Schnoor T, Kjøller R, Gibbons SM, Hansen LH, Al-Soud WA, Sørensen SJ, Rosendahl S (2011) 454-sequencing reveals stochastic local reassembly and high disturbance tolerance within arbuscular mycorrhizal fungal communities. Journal of Ecology doi: 10.1111/j.1365-2745.2011.01894.x.
50. Kivlin SN, Hawkes CV, Treseder KK (2011) Global diversity and distribution of arbuscular mycorrhizal fungi. Soil Biology and Biochemistry 43: 2294–2303.
51. Cavender-Bares J, Keen A, Miles B (2006) Phylogenetic structure of Floridian plant communities depends on taxonomic and spatial scale. Ecology 87: S109–S122.
52. Gange A (2000) Arbuscular mycorrhizal fungi, Collembola and plant growth. Trends in Ecology and Evolution 15: 369–372.
53. Purin S, Rillig MC (2008) Parasitism of arbuscular mycorrhizal fungi: reviewing the evidence. FEMS Microbial Letters 279: 8–14.
54. Gilbert GS, Webb CO (2007) Phylogenetic signal in plant pathogen–host range. Proceedings of the National Academy Sciences 104: 4979–4983.
55. Janzen DH (1970) Herbivores and the number of tree species in tropical forests. American Naturalist 104: 501–528.
56. van der Heijden MGA, Horton TR (2009) Socialism in soil? The importance of mycorrhizal fungal networks for facilitation in natural ecosystems. Journal of Ecology 97: 1139–1150.
57. Johnson NC (1993) Can fertilization of soil select less mutualistic mycorrhizae? Ecological Applications 3: 749–757.
58. Johnson NC, Tilman D, Wedin D (1992) Plant and soil controls on mycorrhizal fungal communities. Ecology 73: 2034–2042.
59. Vandenkoornhuyse P, Husband, R. , Daniell TJ, Watson IJ, Duck JM, Fitter AH, Young JPW (2002) Arbuscular mycorrhizal community composition associated with two plant species in a grassland ecosystem. Molecular Ecology 11: 1555–1564.
60. Hawkes CV, Belnap J, D'Antonio C, Firestone MK (2006) Arbuscular mycorrhizal assemblages in native plant roots change in the presence of invasive exotic grasses. Plant Soil 281: 369–380.
61. Bever JD, Richardson SC, Lawrence BM, Holmes J, Watson M (2009) Preferential allocation to beneficial symbiont with spatial structure maintains mycorrhizal mutualism. Ecology Letters 12: 13–21.
62. Kiers ET, Duhamel M, Beesetty Y, Mensah JA, Franken O, Verbruggen E, Fellbaum CR, Kowalchuk GA, Hart MM, Bago A, Palmer TM, West SA, Vandenkoornhuyse P, Jansa J, Bücking H (2011) Reciprocal rewards stabilize cooperation in the mycorrhizal symbiosis. Science 333: 880–882.
63. Klironomos JN (2003) Variation in plant response to native and exotic arbuscular mycorrhizal fungi. Ecology 84: 2292–2301.

64. Koch AM, Croll D, Sanders IR (2006) Genetic variability in a population of arbuscular mycorrhizal fungi causes variation in plant growth. Ecology Letters 9: 103–110.

65. Paine CET, Harms KE, Schnitzer SA, Carson, WP (2008) Weak competition among tropical tree seedlings: implications for species coexistence. Biotropica 40: 432–440.

4

Determinants of Plant Community Assembly in a Mosaic of Landscape Units in Central Amazonia: Ecological and Phylogenetic Perspectives

María Natalia Umaña[1]*, Natalia Norden[1,2], Ángela Cano[1], Pablo R. Stevenson[1]

1 Universidad de Los Andes, Laboratorio de Ecología de Bosques Tropicales y de Primatología, Centro de Investigaciones Ecológicas La Macarena, Bogotá, Colombia, **2** Pontificia Universidad Javeriana, Departamento de Ecología y Territorio, Bogotá, Colombia

Abstract

The Amazon harbours one of the richest ecosystems on Earth. Such diversity is likely to be promoted by plant specialization, associated with the occurrence of a mosaic of landscape units. Here, we integrate ecological and phylogenetic data at different spatial scales to assess the importance of habitat specialization in driving compositional and phylogenetic variation across the Amazonian forest. To do so, we evaluated patterns of floristic dissimilarity and phylogenetic turnover, habitat association and phylogenetic structure in three different landscape units occurring in *terra firme* (Hilly and Terrace) and flooded forests (Igapó). We established two 1-ha tree plots in each of these landscape units at the Caparú Biological Station, SW Colombia, and measured edaphic, topographic and light variables. At large spatial scales, *terra firme* forests exhibited higher levels of species diversity and phylodiversity than flooded forests. These two types of forests showed conspicuous differences in species and phylogenetic composition, suggesting that environmental sorting due to flood is important, and can go beyond the species level. At a local level, landscape units showed floristic divergence, driven both by geographical distance and by edaphic specialization. In terms of phylogenetic structure, Igapó forests showed phylogenetic clustering, whereas Hilly and Terrace forests showed phylogenetic evenness. Within plots, however, local communities did not show any particular trend. Overall, our findings suggest that flooded forests, characterized by stressful environments, impose limits to species occurrence, whereas *terra firme* forests, more environmentally heterogeneous, are likely to provide a wider range of ecological conditions and therefore to bear higher diversity. Thus, Amazonia should be considered as a mosaic of landscape units, where the strength of habitat association depends upon their environmental properties.

Editor: Anna-Liisa Laine, University of Helsinki, Finland

Funding: Facultad de Ciencias at Universidad de los Andes provided financial support. The funders had no role in study design, data collection and analysis, decision to publish, or preparation of the manuscript.

Competing Interests: The authors have declared that no competing interests exist.

* E-mail: maumana@gmail.com

Introduction

How tropical forests are able to harbour the Earth's richest flora is one of the most challenging questions in community ecology. One possibility to explain such diversity is that tropical regions are mosaics of landscape units, promoting plant specialization to distinct habitat conditions [1,2]. Such pattern has been reported in Western Amazonia [3], Panama [4], Borneo [5], the wet forests of Western Ghats in India [6], and in subtropical China [7]. Large-scale habitat heterogeneity is thus as an important driver of beta-diversity in tropical regions. At small spatial scales, community assembly is thought to be the result of local biotic interactions and environmental filtering [8,9]. Yet, the importance of these processes is still debated as local floristic composition is also the result of dispersal from the regional species pool [10–14]. Since beta diversity provides a direct link between diversity at local and regional scales [15], determining the drivers of floristic dissimilarity across space may yield clues into how coexistence is maintained in tropical forests.

Regions characterized by mosaics of landscape units offer an excellent framework to address this issue, as local community structure may be driven by different processes in distinct landscape units. Phylogenetic-based analyses appear to be a compelling approach because it provides valuable information to disentangle among competing hypotheses, therefore offering a conceptual framework for the development of a synthetic ecological theory [16–18]. Here, we integrate information on ecological and phylogenetic data at different spatial scales to assess the importance of habitat specialization in driving compositional and phylogenetic variation in central Amazonia. Our approach takes advantage of the occurrence of a mosaic of patches of *terra firme* and flooded forests in the Colombian Amazon. In this region, a system of nutrient poor, black water flooded plains called Igapó, where trees are subject to long periods of flooding every year [19] is embedded in a landscape dominated by *terra firme* forests shaped by historical events occurring at different moments in space [20]. These distinct landscape units exhibit differences in species composition and structure, likely to be driven by edaphic factors [21]. If so, variation in the extent of floristic dissimilarity among sample units should mirror environmental differences among sites, independently of geographic distance. The examination of this issue usually relies on the comparison of species lists from forest inventories sampled along an environmental gradient. Yet, as phylogenetic relationships among species change across space, integrating phylogenetic turnover into these analyses further provides new insight elements to evaluate the degree of habitat association beyond the species level [22].

We expect different ecological processes to shape community structure in different landscape units, depending upon their abiotic properties. The stressful conditions found in Igapó are likely to sort species out, restricting the flora to species having particular adaptations to grow and persist in these demanding conditions [23–26]. Thus, Igapó is expected to show low tree diversity and phylodiversity, as well as the occurrence of species withstanding flood. As a result of such environmental filtering, and assuming that important traits show phylogenetic signal [13], then co-occurring species should be more related than expected by chance (i.e. phylogenetic clustering) [16,17]. *Terra firme* forests, in contrast, show less physiological stress than Igapó, and a broader range of forest types underlying higher habitat heterogeneity. Defler & Defler [21] reported the occurrence of distinct physiographic units in *terra firme*, with areas of forest characterized by rolling hills dissected by brooklets (therein Hilly forests), and areas of different geomorphological history, suggesting past floodings during the Pleistocene (therein Terrace forests) [20]. Such heterogeneity at local and large scales may provide a wide range of ecological niches, allowing the coexistence of a higher number of species. Thus, we expect elevated levels of diversity and phylodiversity in *terra firme* forests, as well as an association between environmental factors and species occurrence. If local assemblages contain species with distinct ecological strategies in resource acquisition [27–29], and these strategies are phylogenetically conserved [13], then co-occurring species should be less related than expected by chance (i.e. phylogenic evenness) [16,17].

To test these predictions, we collected information of six 1-ha plots in a lowland tropical forest in Vaupés, Colombia, comprising three major landscape units: one in flooded forests Igapó and two in *terra firme* forests (Terrace and Hilly forests) [21]. Specifically, we addressed the following questions: (1) Are diversity and phylodiversity lower in habitats subject to stressful environmental conditions? (2) To which extent do environmental differences across the landscape shape species and phylogenetic composition within and among landscape units? (3) Do local plant communities in flooded forests show phylogenetic clustering whereas those in *terra firme* phylogenetic evenness, and are these patterns conserved across spatial scales?

Methods

Study Site

This study was conducted at the Mosiro-Itajura Caparú Biological Station (CBS) (01°04'12''S 069°30'55''W), in the basin of the Apaporis river, Colombian Amazonia, where the average annual rainfall is 3950 mm and the mean annual temperature is 25°C [20]. Although there is no marked dry season (month <100 mm), the study area shows an annual flood pulse between March and October, caused by floods in the Apaporis river [20]. The station is dominated by pristine lowland forests growing in a geographically complex soil that combines acid and clayey soils from different geological ages [20]. According to edaphic, topographic and hydrological differences within the reserve, Defler & Defler [21] described five different landscape units: four on *terra firme* forests and one on floodplain forests. Here, we focused on the three most common: Hilly and Terrace forests (in *terra firme*), and Igapó (in floodplains). Hilly forests are characterized by small hills on clayey soils, Terrace forests are associated to areas that were flooded during the Pleistocene by the Apaporis river, and

Igapó forests are flooded by black water for about eight months each year [20,30].

Data Colection

Tree censuses. Two 1-ha plots were established within each landscape unit. In each plot, all stems ≥10 cm diameter at breast height (DBH), including trees, palms and lianas, were tagged, and measured for DBH. Vouchers were collected from each stem, and identified to species or morpho-species at the ANDES and COAH herbaria in Bogotá, Colombia.

Abiotic variables. In all plots, the topographic profiles were measured within each 10×10 m quadrat using a clinometer (Suunto PM-5, USA). We collected soil samples in all 20×20 m quadrats for a total of 150 samples; each sample consisting in a mixture of topsoil (0 to 10 cm depth). These samples were subsequently analyzed for cation exchange capacity (CEC), clay, sand and silt percentage, and pH in a soil-analysis laboratory in Villavicencio, Colombia. We measured light intensity using a luxometer (Extech 407026, USA) at the center of each 20×20 m quadrat. To calibrate the measures made in the field, we took a reference measure in an open site, and the value for each point measured within the plots was expressed as the percentage of this reference point.

Statistical Analyses

Floristic structure. At each plot, we evaluated species richness using species rarefaction curves, and species diversity using the Fisher's alpha index [31]. Floristic similarity among plots was evaluated calculating the Chao-Jaccard estimator [32] with the package 'vegan' [33] in the R statistical software [34]. This estimator is an abundance-based similarity index that assesses the probability that individuals belong to shared vs. unshared species by accounting for the effect of unseen shared species. In tropical forests, where rare species are frequent and sampling is incomplete, this index is less biased by sample size, and thus more appropriate than other commonly used indices [32]. To characterize floristically each landscape unit, we calculated the Importance Value for each species by accounting by species relative frequency, density and dominance within each plot [35].

To evaluate the importance of environmental variables in determining floristic composition we performed two analyses. Because environmental similarity among plots was correlated with geographical distance ($R_{Mantel} = 0.43$ $P<0.001$), we first evaluated the extent to which environmental similarity accounted for species similarity, while controlling by geographical distance by using Partial Mantel Tests at different spatial scales: within plots, between plots within each landscape unit, and across landscape units. For all three scales, correlations among species similarity, environmental similarity and geographic distance were based on data from 20×20 m quadrats. Thus the spatial extent changed, while the resolution was kept constant. To reduce the multivariate environmental data, we performed a principal component analysis (PCA), and used the scores of the two principal components. The first PCA axis (PC1) was related to CEC and clay percentages in soil samples, and explained 35% of the variance in abiotic variables. The second one (PC2) was related to silt and sand percentages in soil samples, and explained 20% of the variance. Then, we performed a more specific analysis to test how floristic composition was associated to each of the environmental variables measured, by performing a canonical correspondence analysis (CCA). We tested for the significance of this association using an ANOVA like permutation test for CCA from the package 'vegan' [35] in the R statistical software [36]. The CCA analysis was performed at the 20×20 m scale.

Phylogenetic analyses. We constructed a phylogenic tree including all the species occurring in the study plots and those included in the list of the local flora (excluding shrubs and herbaceous plants) [36]. To do so, we used the angiosperm APGIII consensus tree (R20080417) [37,38] from Phylomatic [41] as the backbone super-tree. This tree has family-level resolution, with most species and genera considered as polytomies within genera and families, respectively. Overall, we tagged 3526 individuals, of which 94% were identified to species and the remaining 6% to morpho-species. Since we were certain of the genus for all morpho-species, we included them in the phylogenetic tree. Branch lengths in the tree were adjusted to match clade age estimates reported by Wiksrom et al. [40] using the BLADJ algorithm. We performed several analyses based on this phylogenetic tree. First, we calculated the phylogenetic species richness of each plot using the phylogenetic species richness (PSR) index [41]. The PSR multiplies the number of species in the community by their evolutionary relatedness. This metric is related to the Faith's phylogenetic diversity index, with the difference that the PSR uses more information contained within the phylogeny than does the Faith's index [41]. Second, we evaluated the phylogenetic structure of the local assemblages at two spatial scales (landscape unit and plot level) using the phylogenetic species variability index (PSV) [41]. This metric indicates the degree to which co-occurring species are phylogenetically related to each other by measuring the among-species variance in the value of a hypothetical neutral trait evolving under a Brownian motion model. For a sample of n species,

$$PSV = \frac{ntrC - \sum C}{n(n-1)}$$

where C is the $n \times n$ sample phylogenetic covariance matrix, trC is the sum of diagonal elements of C and $\sum$C is the sum of all elements. As species in the sample become more closely related, the PSV decreases towards zero; and as species become less closely related, the PSV increased towards one; the statistical expectation of PSV is independent of species richness [41]. We also calculated phylogenetic species evenness, PSE, a formulation of the PSV that accounts for species abundance. Because both metrics showed similar trends, we only report the results obtained based on the PSV. The PSV was evaluated at the scale of the landscape unit and at the plot level. In the first case, this metric was calculated as the mean of the two plots within each landscape unit. At the plot level, this metric was calculated as the mean of the 25 20×20 m quadrats of each plot. These values were compared with a frequency distribution based in 1000 iterations generated using two null models: (1) the richness null model shuffles cells within each row so that the number of species within each community is preserved, but the prevalence of species changes across communities; (2) the frequency null model shuffles cells within each column so that the prevalence of each species is preserved but species richness within each community changes [41]. Because our a priori expectations predicted a specific phylogenetic pattern for each landscape unit (see Introduction), the observed means of PSV were compared to the 95% confidence intervals generated by the 1000 iterations based on one-tailed tests. As the presence of phylogenetic signal in species abundance can influence the patterns of phylogenetic structure observed, we tested whether abundant species were randomly distributed across the phylogeny by computing the 'abundance phylogenetic deviation' (APD) (see Table S1) [42]. This metric was not significantly different from

zero, indicating that there was no phylogenetic signal in species abundance.

Finally, to evaluate the pairwise differences in species composition between communities incorporating phylogenetic information, we calculated the phylogenetic community dissimilarity (PCD) index [43]. This metric evaluates how much of the variance among species in the values of a hypothetical trait within a community can be predicted by the known trait values of species from another community [43]. This variance is calculated using the PSV index.

$$PCD = \frac{n_1 PSV_{1|2} + n_2 PSV_{2|1}}{n_1 PSV_1 + n_2 PSV_2}$$

where $PSV_{1|2}$ is calculated for community 2, conditional on information from community 1, and n_i is the number of species in the community i (for a more detailed description see Ives & Helmus [43]). If the PCD is greater than one, then communities tend to be phylogenetically dissimilar; and if the PCD is lower than one, then communities tend to be phylogenetically similar [43]. All the phylogenetic analyses were performed using the package 'picante' [44] in the R statistical software [42].

Results

Diversity and Descriptive Data

Terrace and Hilly forests showed similar stem density and species richness, while Igapó showed lower values in these two attributes (Table 1). Species rarefaction curves did not reach a saturation point; particularly in the *terra firme* plots (Fig. 1). The highest Fisher's index was that of the Hilly forests, followed by Terrace and by Igapó. Patterns of phylogenetic diversity were in agreement with these trends: Igapó forests showed the lower values of PSR, followed by Terrace and by Hilly forests (Table 1).

Each landscape unit was characterized by different dominant species. Based on the Importance Value analysis, Hilly forests were dominated by four species: *Eschweilera coriacea* (Lecythidaceae), *Iryanthera ulei* (Myristicaceae), *Rinorea paniculata* (Violaceae) and *Euterpe precatoria* (Arecaceae). *Micrandra spruceana* (Euphorbiaceae) and *Oenocarpus bataua* (Arecaceae) showed the highest Importance Value in Terrace forests, and *Zygia cataractae* (Fabaceae) was an important species in Igapó forests. Table S2 summarizes the top 10 most important species in each of the six plots.

Abiotic Variables

The partial Mantel partial tests revealed that floristic similarity within and between plots of the same landscape unit was not related to environmental factors. In contrast, at larger scales, floristic similarities among plots were significantly related to environmental factors (Fig. 2). Indeed, the CCA analysis showed that each landscape unit formed an independent floristic unit (axis 1:22.9%, eigenvalue = 0.67, axis 2:14.3%, eigenvalue = 0.42; Fig. 3). Igapó was the one exhibiting the most pronounced divergence in species composition (Fig. 3; Fig. 4). Silt and sand percentages, as well as CEC (Cation Exchange Capacity) and topography, were significantly correlated to the first two CCA axes (ANOVA: $X^2_{Sand} = 0.54$, $P_{Sand} = 0.01$, $F_{Sand} = 2.86$; $X^2_{Silt} = 0.46$, $P_{Silt} = 0.01$, $F_{Silt} = 2.46$; $X^2_{CEC} = 0.22$, $P_{CEC} = 0.01$, $F_{CEC} = 1.18$; $X^2_{Topo} = 0.29$, $P_{Topo} = 0.01$, $F_{Topo} = 1.55$; Fig. 3). The vectors most strongly correlated with species occurrence and relative abundance were the edaphic ones (Fig. 3). In particular, clay and CEC showed a strong correlation with many species in *terra firme* forests. Species composition between Terrace and Hilly forests was differentiated by the silt vector, indicating that Hilly forests grow

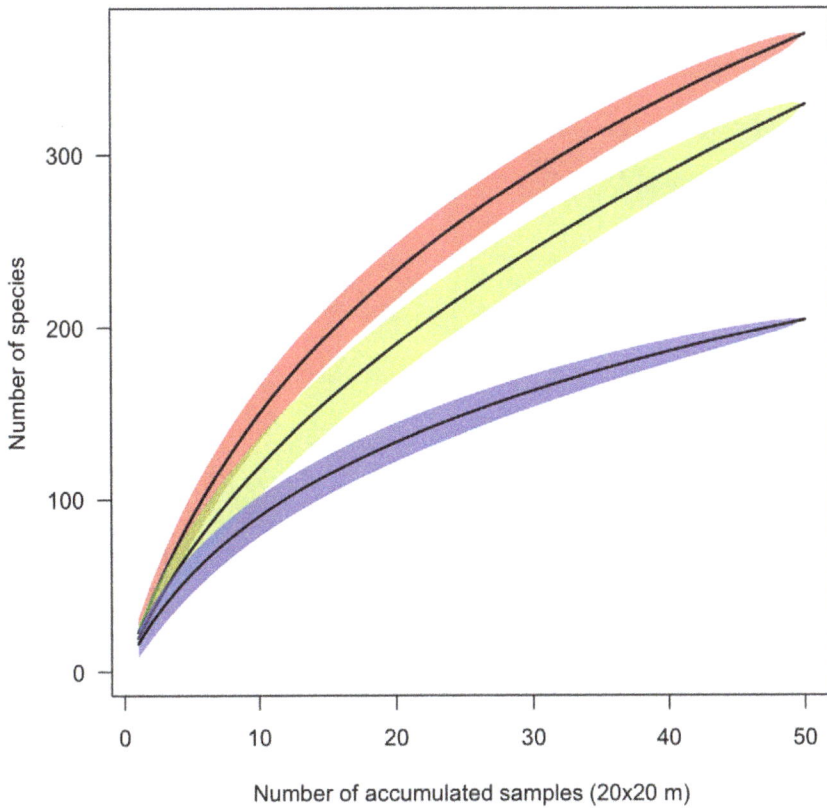

Figure 1. Species rarefaction curves within each landscape unit for trees >10 cm DBH. The red line denotes Hilly forests, the green line denotes Terrace forests and the blue line denotes Igapó forests. The shaded region represents 95% confidence intervals.

over soils richer in silt. Soils of Igapó forests were the most infertile with low contents of silt and CEC. Light and pH were poorly related with species composition.

Phylogenetic Structure

Phylogenetic community dissimilarity was lower between plots from the same landscape unit than between plots of different landscape units. The most dissimilar forests in terms of phylogenetic composition were Hilly and Igapó forests. Terrace forests were more similar to Hilly forests than to Igapó forests (Fig. 4).

As expected, flooded forests showed a phylogenetic clustering whereas *terra firme* forests showed phylogenetic evenness (Table 2).

Table 1. Number of individuals and diversity metrics based on trees ≥10 cm DBH, for each of the six 1-ha plots established at the Mosiro-Itajura Caparú Biological Station (Colombian Amazon).

Plot	Individuals	Species	Fisher's Alpha	Observed PSR
Hilly 1	590	211	124.12	192.45
Hilly 2	641	256	155.27	166.31
Terrace 1	594	220	123.92	131.42
Terrace 2	634	171	78.36	168.60
Igapó 1	553	138	58.05	83.96
Igapó 2	514	116	47.75	103.98

More specifically, in Igapó forests, mean PSV was lower than expected using both null models, but this result was significant only with the frequency null model (PVS_{FN}). In Hilly forests, mean PSV was higher than expected using both null models, but again, this result was significant only with PVS_{FN}. Finally, in Terrace forests, mean PSV was higher than expected using both null models but significantly so only with the richness null model (PVS_{RN}) (Table 2). At the plot level, PSV values were not significantly different from zero in most of the cases, but exhibited similar trends as those found at the landscape unit level, with Igapó showing lower values than Hilly and Terrace forests (Fig. 5; Table S3).

Discussion

Patterns of diversity and species composition showed important variation among landscape units, particularly between flooded and *terra firme* forests. In agreement with previous studies conducted in the Amazonia [22,30,45], *terra firme* exhibited higher levels of diversity than Igapó, and relatively few species were shared between these two types of forests (13% and 20% between Igapó-Hilly and Igapó-Terrace forests, respectively). Such floristic dissimilarity is accompanied by a phylogenetic divergence, suggesting that the sorting due to flood goes beyond the species level. These patterns have been previously reported in SE Asia, were different families were associated with distinct habitats [22]. If so, our results suggest that the low diversity observed in Igapó, and the compositional and phylogenetic differences observed between flooded and non-flooded systems are the outcome of habitat specialization. This pattern, however, is not necessarily

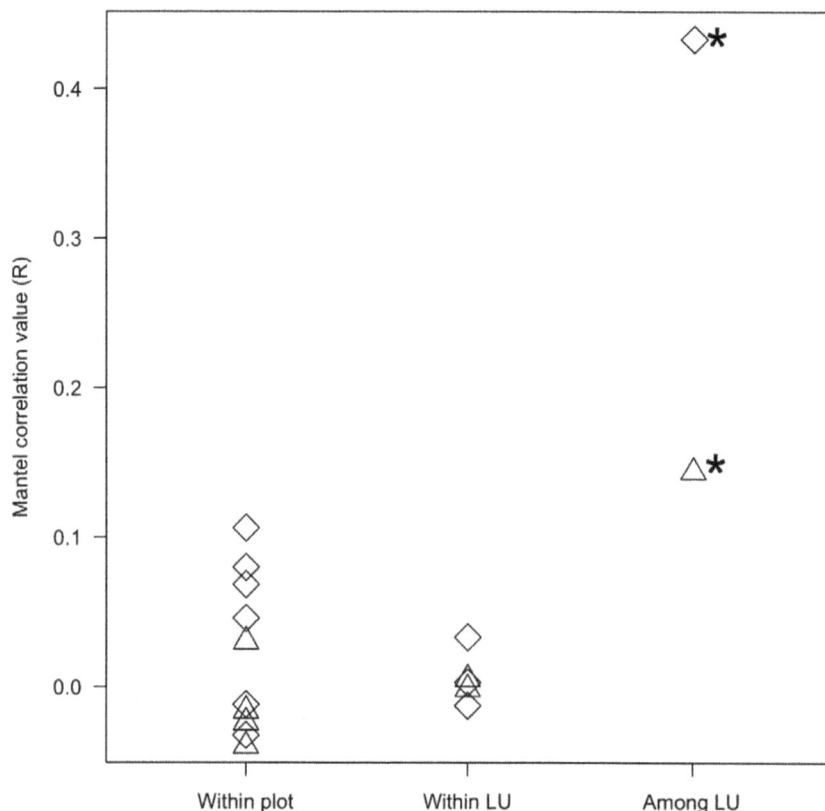

Figure 2. Correlation coefficients of Mantel test (R) relating floristic and environmental similarity, while controlling by distance, at three geographical scales: within plots, within landscape units (LU) and among landscape units. Open represent PC1 values and open represent PC2 values. $P \leq 0.05$ are indicated with asterisks.

driven by differences in competitive abilities among species occurring in distinct habitats. For instance, Fine et al. [46], showed that habitat association patterns in clayey and sandy forests in the Peruvian Amazonia was mediated by differences in antiherbivore defenses among species.

Within *terra firme*, Hilly and Terrace forests shared an important fraction of species (53%), suggesting that a large extent of the floristic divergence observed between these two landscape units relied on species abundance rather that on species incidence. For instance, *O. bataua* and *E. coriacea* occurred in all *terra firme* plots but were more abundant in Terrace and Hilly forests, respectively. These species are known to be generalists, persisting in a wide range of environments across the Amazonia [47], and probably exhibiting a broad range of environmental tolerances. This explains the high similarity between one of the plots in Terrace forests and the Hilly forests plots. However, each landscape unit did exhibit independent floristic units. For instance, *M. spruceana* was particularly dominant in Terrace forests, but was totally absent in Hilly forests. Together, these findings demonstrate that species relative abundance and distribution varies not only between flooded and non-flooded systems but also within *terra firme* forests. Yet, these differences are not reflected by phylogenetic similarity analyses. The PCD index comparing phylogenetic relatedness between Hilly and Terrace forests exhibited values close to one, indicating that, phylogenetically, these stands are not significantly different from communities selected at random from the species pool [43].

Within landscape units, comparing patterns of floristic dissimilarity and phylogenetic turnover brought insightful elements to

understand local community assembly at more local scales. For example, Igapó showed the highest floristic divergence between plots of the same landscape unit, but also showed low phylogenetic turnover. Because these two plots are located at the two opposite shores of the Apaporis river, dispersal limitation may be, in part, the cause of such floristic divergence. Alternatively, as the river stream does not exert the same lateral erosive process at each side, differences in sedimentation, nutrient depletion and deposition might have affected the successional process occurring at each of these locations [48]. Floristic divergence may therefore be the outcome of different successional stages resulting from perturbations that occurred at distinct moments. Finally, it could be the outcome of alternative trajectories that reached different stable states [49]. In any of these scenarios, our findings suggest that Igapó forests are subject to constant disturbance due to flood.

Overall, our findings indicate that each landscape unit harbors relatively different plant communities. The correlation between floristic similarity and geographic distance suggests that differences in species composition among landscape units are the outcome of dispersal limitation. Poor dispersal has been widely reported for tropical trees [50,51] suggesting that spatial processes are important in determining the local abundance of many species [52,53]. Yet, these conclusions need to be taken with caution, as environmental similarity was tightly correlated with geographical distance. Indeed, our findings may also by the result of a spurious effect arising from the geographic location of the landscape units. The distance between the plots established in Igapó, the landscape unit showing the most conspicuous differences in species composition with the other two, is longer than the distance

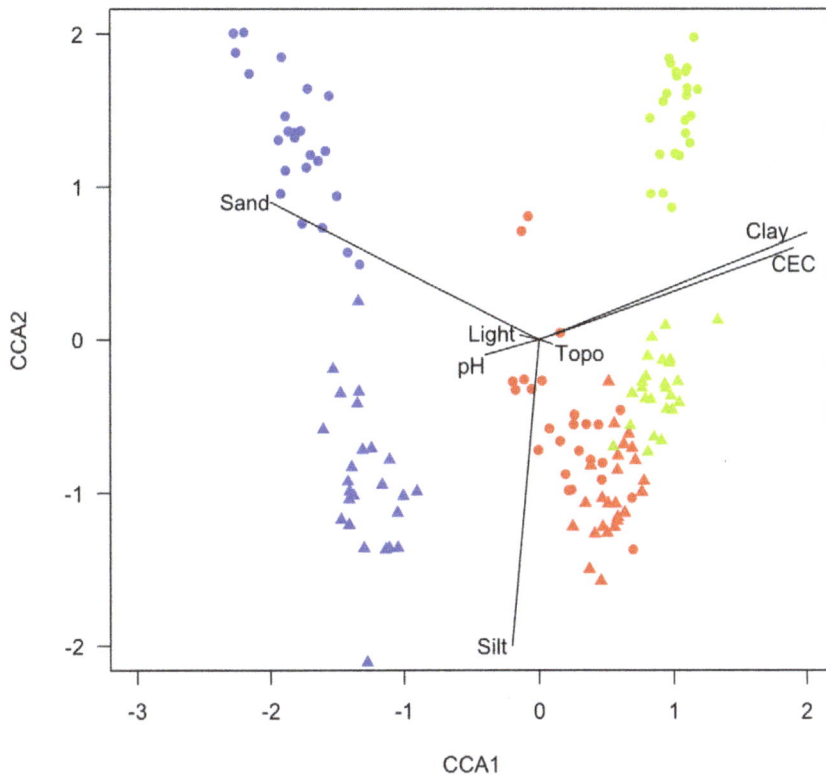

Figure 3. CCA of all tree species occurring in the six 1-ha plots. The arrows correspond to the abiotic variables included in the analysis. Symbols represent species of 20×20 m quadrants and show their association with the abiotic variables. Red triangles correspond to Hilly 1, red circles to Hilly 2; green triangles to Terrace 1, green circles to Terrace 2; blue triangles to Igapó forests plot 1, blue circles to Igapó plot 2.

between the plots of Terrace and Hilly forests. We would need a more extensive sampling in a wider geographical range to address this issue more straightforwardly.

We found that floristic and environmental similarity were significantly correlated only at large scales, indicating that the steeper the gradient in environmental variation, the stronger the influence of environment in species composition [54]. Similar findings have been found in white sand [18] and flooded forests [55] in the Peruvian Amazon, and in Panama [56]. Indeed, many studies have documented habitat association driven by physical factors, in particular by soil variables [2,3,57,58]. Our description of edaphic conditions within each landscape unit showed that the marked differences in floristic composition observed among landscape units were strongly associated with soil characteristics, differentiated by contents of sand, clay and silt. Although these results are globally in agreement with Defler & Defler [21], we found some discrepancies between their study and ours regarding Igapó's edaphic composition. Specifically, we found that Igapó was the sandiest landscape unit, whereas Defler & Defler [21] found very high contents of clay. Because their analyses were based on a low number of replicates within each forest type, they might have overlooked the whole variation in soil composition exhibited within each landscape unit.

Among the other environmental variables studied, only topography seemed to have an effect on species composition in Hilly forests. Similar pH values were found within and across plots, indicating that this factor was irrelevant to discriminate among landscape units. Finally, light availability also appeared to be poorly correlated with species composition. Because *terra firme* forests harbor higher stem density than Igapó forests, one would

have expected shade-tolerant species to be associated with the limiting light conditions in the forest. Yet, such association was not found because light availability did not show strong variation among landscape units. Moreover, light is a limiting factor for plant growth and establishment particularly during early stages [59,60], but at adult stages, it is difficult to detect the footprint of a process that occurred long time ago.

The habitat association patterns observed may be the outcome of different niche-based processes. As predicted, Igapó showed phylogenetic clustering but only under the frequency null model, suggesting that the strong relatedness found among co-occurring species in this landscape unit is driven by nonrandom associations between species among communities [41]. Following, the seminal ideas developed by Webb et al. [16], these results suggest a major role of environmental filtering. Recent findings have highlighted that phylogenetic clustering may also be driven by competition [61]. However, in the light of our results, we believe that community assembly in Igapó is strongly governed by the environmental stress imposed by flooding. At local scales, species did not show any particular trend. This is not surprising since environmental filtering is typically more conspicuous at large spatial scales [62,63].

Conversely, both landscape units in *terra firme* forests showed phylogenetic evenness. Hilly forests showed a significant pattern under the frequency null model, whereas Terrace forests did so under the richness null model. These results suggest that evenness in Hilly forests is driven by nonrandom associations between species among communities, whereas in Terrace forests it is driven by differences in the overall prevalence of species [41]. Together, these findings would suggest that biotic interactions play a major

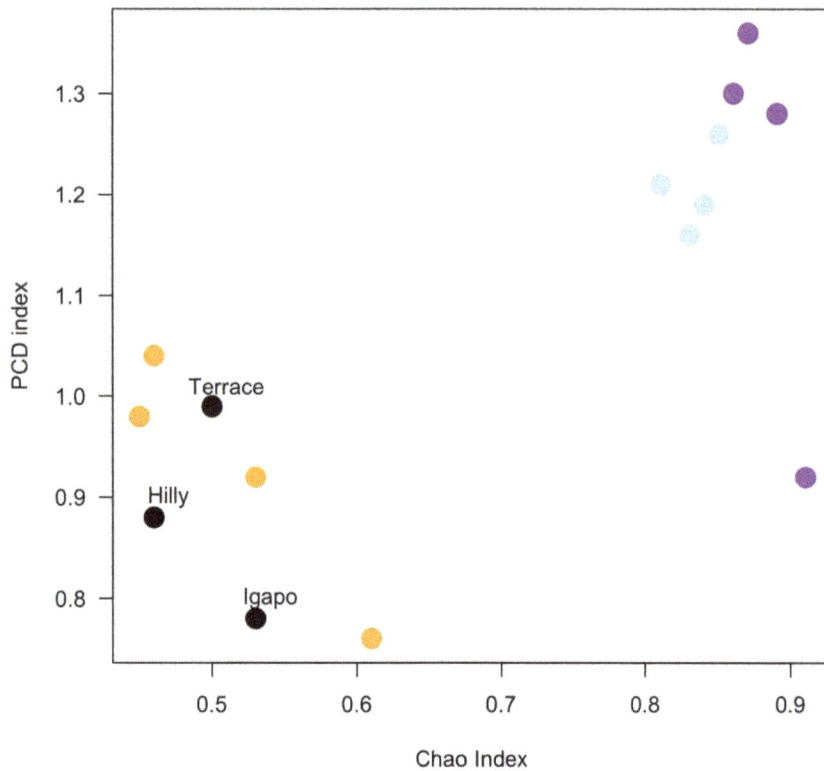

Figure 4. Pearson's correlations between Chao and PCD indices. The orange dots represent coefficients calculated for Hilly and Terrace plots, blue dots represent coefficients calculated for Terrace and Igapó plots, and violet dots represent coefficients calculated for Igapó and Hilly plots. Black dots represent the coefficients for plots from the same landscape unit.

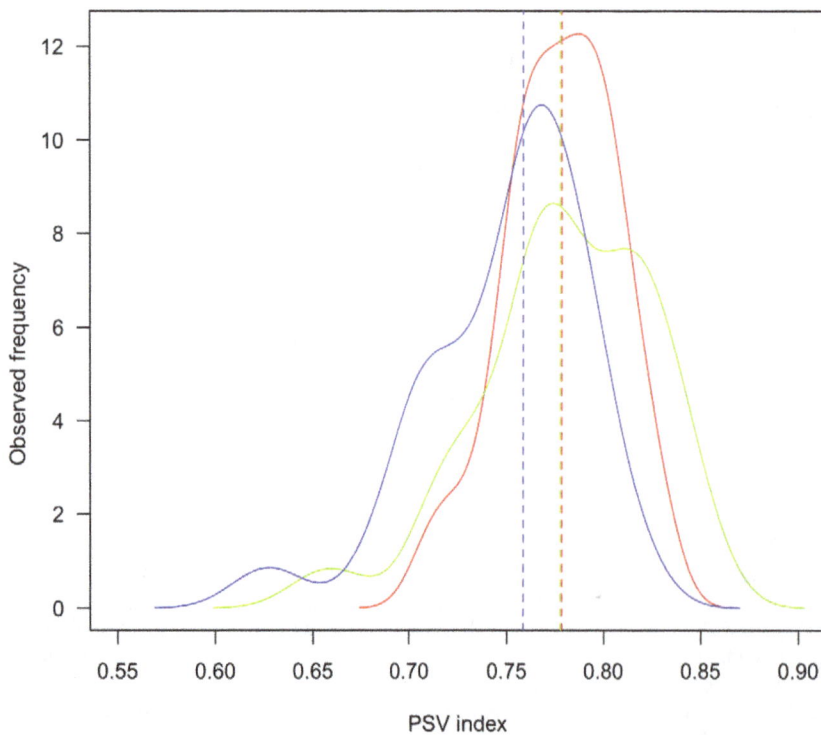

Figure 5. PSV values at 20×20 m scale for each landscape unit. The red line corresponds to Hilly forests, the green line to Terrace forests and the blue line to Igapó forests. The dashed line represents the median of the distribution for each landscape unit.

Table 2. Phylogenetic species diversity (PSV) for each landscape unit.

Plot	Null model	Observed PSV	Confidence intervals of the randomized PSV
Hilly	Richness	0.758	0.749–0.768
Hilly	Frequency	**0.758**	0.756–0.757
Terrace	Richness	**0.774**	0.749–0.773
Terrace	Frequency	0.774	0.773–0.774
Igapó	Richness	0.759	0.746–0.774
Igapó	Frequency	**0.759**	0.761–0.763

Significant values ($P \le 0.05$, one-tailed test) are indicated in bold. Based on our a priori hypotheses, for Hilly and Terrace forests, PSV scores are significant if higher than the 95% quantile of randomized PSV. For Igapó forests, PSV scores are significant if lower than the 5% quantile of randomized PSV.

role in structuring plant communities in *terra firme* forests [16,17]. Yet, at the plot level, most plant assemblages did not show any phylogenetic structure. These results weaken the role of competition in structuring *terra firme* forests, as this ecological process operates at local scales. Because both Hilly and Terrace forests exhibited higher variation in topography (SD = 2.14 and 1.79, respectively) compared to Igapó forests (SD = 0.81), we believe that the observed pattern of phylogenetic evenness may reflect niche differentiation rather than competition. Topography may stand as a proxy of soil resources not measured in this study, such as water availability and drainage, suggesting that forests in *terra firme* might offer a wider diversity of niches, allowing the establishment of species with broad ecological strategies [64]. Also, negative density dependent processes may lead to an even phylogenetic structure, if plant enemies reduce the establishment of individuals closely related to focal species [65]. Yet, this process is more likely to occur at early stages, when negative density dependence is more intense [66].

The recent flourishing of phylogenetic methods has allowed the reappraisal of classical ecological theories. Here we integrate ecological and evolutionary approaches to evaluate the importance of environmental factors in shaping community assembly in a mosaic of landscape units. Edaphic specialization was an important driver of floristic and phylogenetic distinctiveness across the landscape, whereas the role of competition appeared to be weak.

Further understanding of the processes shaping community structure within each landscape unit needs a functional perspective. In particular, root and seed traits may be good indicators of plant ability to establish and prevail in flooded plains [24,25]. Likewise, maximum height could help to understand the role of

local biotic interactions in both flooded and non-flooded forests [67].

Supporting Information

Table S1 APD values for each plot.

Table S2 List of the 10 most important species according to Importance Value.

Table S3 Phylogenetic Species Variability (PSV) within each 1-ha plot.

Acknowledgments

We thank Conservation International Colombia, the SINCHI institute and Universidad de los Andes Herbarium. We are grateful to L. Tanimuka, G. A. Tanimuka, L. Barasano and O. I Yucuna for their help during fieldwork. N. Swenson, J.S. González and J. P. Gómez provided critical assistance with the R programing. Special thanks to A. B. Hurtado for fieldwork assistance and support through this project. Nigel Pitman, C. E. Timothy Paine and two anonymous reviewers, as well as all the members from the PS's laboratory group provided valuable comments on earlier versions of this manuscript.

Author Contributions

Conceived and designed the experiments: MNU NN ÁC PRS. Performed the experiments: MNU NN ÁC PRS. Analyzed the data: MNU NN. Contributed reagents/materials/analysis tools: MNU NN ÁC PRS. Wrote the paper: MNU NN PRS.

References

1. Gentry AH (1988) Changes in plant community diversity and floristic composition on environmental and geographical gradients. Ann MO Bot Gard 75: 1–34.
2. Duque A, Sánchez M, Cavelier J, Duivenvoorden J (2002) Different floristic composition patterns of woody understory and canopy plants in Colombian Amazonian. J Trop Ecol 18: 499–525.
3. Tuomitso H, Roukolainen K, Kalliola R, Linna A, Danjoy W, et al. (1995) Dissecting Amazonian biodiversity. Science 269: 63–66.
4. Condit R, Pitman N, Leigh EG, Chave J, Terborgh J, et al. (2002) Beta-diversity in tropical forest trees. Science 295: 666–669.
5. Paoli GD, Curran LM, Zak DR (2006) Soil nutrients and beta diversity in the Bornean Dipterocarpaceae: evidence for niche partitioning by tropical rain forest trees. J Ecol 94: 157–170.
6. Davidar P, Rajagopal B, Mohandass D, Puyravaud J-P, Condit R, et al. (2007) The effect of climatic gradients, topographic variation and species traits on the beta diversity of rain forest trees. Global Ecol Biogeogr 16: 510–518.
7. Legendre P, Mi X, Ren H, Ma K, Yu M, et al. (2009) Partitioning beta diversity in a subtropical broad-leaved forest of China. Ecology 90: 663–674.

8. Kraft NJB, Valencia R, Ackerly DD (2008) Functional traits reveal niche-based community assembly in an Amazonian forest. Science 322: 580–582.
9. Baraloto C, Hardy OJ, Paine CET, Dexter KG, Graud C, et al. (2012) Using functional traits and phylogenetic trees to examine the assembly of tropical tree communities. J Ecol 100: 690–701.
10. McArthur R, Wilson EO (1967) The theory of island biogeography. Princeton: Princeton Univ Press. 205 p.
11. Ricklefs RE (1987) Community diversity: relative roles of local and regional processes. Science 235: 167–171.
12. Hubbell SP (2001) The unified theory of biodiversity and biogeography. Princeton: Princeton Univ Press. 375 p.
13. Wiens JJ, Donoghue MJ (2004) Historical biogeography, ecology and species richness. Trends Ecol Evol 19: 639–644.
14. Wiens JJ, Pyron RA, Moen DS (2011) Phylogenetic origins of local-scale diversity patterns and the causes of Amazonian megadiversity. Ecol Lett 14: 643–652.
15. Anderson MJ, Crist TO, Chase JM, Vellen M, Inouye BD, et al. (2010) Navigating the multiple meanings of β diversity a roadmap for practicing ecologist. Ecol Lett 14: 19–28.

16. Webb CO (2000) Exploring the phylogenetic structure of ecological communities: an example for Rain Forest trees. Am Nat 156: 145–155.

17. Webb CO, Ackerly DD, McPeek MA, Donoghue MJ (2002) Phylogenies and community ecology. Annu Rev Ecol Syst 33: 475–505.

18. Fine PVA, Kembel SW (2011) Phylogenetic community structure and phylogenetic turnover across space and edaphic gradients in western Amazonian tree communities. Ecography 34: 552–565.

19. Junk WJ (1989) The flood tolerance and tree distribution in central Amazonia. In: Holm-Nielsen LB, Nielsen IC, Balsev H, editors. Tropical forest botanical dynamics: speciation and diversity. London: Academic Press. 47–64.

20. Palacios E, Rodríguez A, Alarcón-Nieto G (2009) Aspectos físicos y biológicos del bajo río Apaporis y la Estación Biológica Mosiro Itajura-Caparú. In: Alarcón-Nieto G, Palacios E, editors. Estación Biológica Mosiro Itajura-Caparú: biodiversidad en el territorio del Yagojé-Apaporis. Bogotá: Conservación Internacional Colombia. 55–97.

21. Defler TR, Defler SB (1996) Diet of a group of *Lagothrix lagotrhicha lagothricha* in southeastern Colombia. Int J Primatol 17: 161–190.

22. Webb CO, Cannon CH, Davies SJ (2008) Ecological organization biogeography and the phylogenetic structure of the tropical forest tree communites. In: Carson WP, Schnitzer SA, editors. Tropical forest community ecology. Oxford: Blackwell Publishing. 79–97.

23. Parolin P, Ferreira LV, Junk WJ (1998) Central Amazonian floodplains: effect of two water types on the wood density of trees. Verh Internat Verein Theor Angew Limnol 26: 1106–1112.

24. Parolin P (2001) Morphological and physiological adjustments to water logging and drought in seedlings of Amazonian floodplain trees. Oecologica 128: 326–335.

25. Parolin P, De Simone O, Haase K, Waldhoff D, Rottenberger S, et al. (2004) Central Amazonian floodplain forests: tree adaptations in a pulsing system. Bot Rev 70: 357–380.

26. Baraloto C, Morneau F, Bonal D, Blanc L, Ferry B (2007) Seasonal water stress tolerance and habitat associations within four Neotropical tree genera. Ecology 88: 478–489.

27. Hutchinson GE (1957) Concluding Remarks. Cold Spring Harbor Symposium. Quant Biol 22: 415–427.

28. Tilman D (1994) Competition and biodiversity in spatially structured habitats. Ecology 75: 2–16.

29. Tilman D (2004) Niche tradeoffs, neutrality, and community structure: a stochastic theory of resource competition, invasion, and community assembly. Proc Nat Acad Sci USA 101: 10854–10861.

30. Cano A, Stevenson PR (2009) Diversidad y composisción florística de tres tipos de bosque en la Estación Biológica Caparú, Vaupés. Revista Colombiana Forestal 12: 63–80.

31. Fisher AA, Cobert AS, Williams CB (1943) The relation between the number of species and the number of individuals in a random sample of an animal population. J Anim Ecol 12: 42–58.

32. Chao A, Chazdon RL, Colwell RK, Shen T (2005) A new statistical approach for assessing similarity of species composition with incidence and abundance data. Ecol Lett 8: 148–159.

33. Oksanen J, Blanchet FG, Kindt R, Legendre P, O'Hara B, et al. (2008) Vegan: Community Ecology Package. P package v. 1.13–8. Available: http://vegan.r-forge.r-project.org/. Accessed 22 August 2012.

34. R Core Team Development (2011) R: a language and environment for statistical computing. R Foundation for Statistical Computing. Coventry, United Kingdom v. 2.14.0. Available: http://www.r-project.org/. Accessed 22 August 2012.

35. Curtis JT, McIntosh RP (1951) An upland forest continuum in the prairie-forest border region of Wisconsin. Ecology 32: 476–496.

36. Clavijo L, Betancur J, Cárdenas D (2009) Las plantas con flores de la Estación Biológica Mosiro Itajura-Caparú, Amazonía Colombiana. In: Alarcón-Nieto G, Palacios E, Editors. Estación Biológica Mosiro Itajura-Caparú: biodiversidad en el territorio de Jagojé-Apaporis. Bogotá: Conservación Internacional Colombia. 55–97.

37. Davies TJ, Barraclough TG, Chase MW, Soltis PS, Soltis DE, et al. (2004) Darwin's abominable mystery: insights from a supertree of the angiosperms. Proc Nat Acad Sci USA 101: 1904–1909.

38. Stevens PF (2008) Angiosperm Phylogeny Website. Version 9. Available: http://www.mobot.org/MOBOT/research/APweb/. Accessed 22 August 2012.

39. Webb CO, Donoghue MJ (2005) Phylomatic: tree assembly for applied phylogenetics. Mol Ecol Notes 5: 181–183.

40. Wikström N, Savolainen V, Chase MW (2001) Evolution of angiosperms: calibrating the family tree. Proc Roy Soc Lond 268: 2211–2220.

41. Helmus MR, Bland TJ, Williams CK, Ives AR (2007) Phylogenetic measures of biodiversity. Am Nat 169: E68–E83.

42. Hardy OJ (2008) Testing the spatial phylogenetic structure of local communities: statistical performances of different null models and statistics on a locally neutral community. J Ecol 96: 914–926.

43. Ives AR, Helmus MR (2010) Phylogenetic metrics of community similarity. Am Nat 176: E128–E142.

44. Kembel S, Ackerly DD, Blomberg S, Cowan P, Helmus MR, et al. (2008) Picante: Phylocom integration, community analyses, null models, traits and evolution in R. Available: http://r-forge.r-project.org/projects/picante/.Ac-Accessed 22 August 2012.

45. Ter Steege H, Sabatier D, Castellanos H, Van Adel T, Duivenvoorden J, et al. (2000) An analysis of the floristic composition and diversity of Amazonian forests including those of the Guiana shield. J Trop Ecol 16: 801–828.

46. Fine PVA, Mesones I, Coley PD (2004) Herbivores promote habitat specialization by trees in Amazonian forests. Science 305: 663–665.

47. Pitman NCA, Terborgh JW, Silman MR, Núñez P, Neill DA, et al. (2001) Dominance and distribution of tree species in upper Amazonian *terra firme* forest. Ecology 8: 2101–2117.

48. Rosales J, Petts G, Salo J (1999) Riparian flooded forest of the Orinoco and Amazon basins: A comparative review. Biodivers Conserv 8: 551–586.

49. Suding KN, Gross KL, Houseman GR (2004) Alternative states and positive feedbacks in restoration ecology. Trend Ecol Evol 19: 46–53.

50. Dalling JW, Hubbell SP (2002) Seed size, growth rate and gap microsite conditions as determinants of recruitment success. J Ecol 90: 557–568.

51. Muller-Landau HC, Wright SJ, Calderón O, Hubbell SP, Foster RB (2002) Assessing recruitment limitation: concepts, methods and examples for tropical forest trees. In: Levey J, Silva WR, Galetti M, editors. Seed dispersal and frugivory: ecology, evolution and conservation. New York: CABI Pub. 35–53.

52. Vormisto J, Svenning J-C, Hall P, Balslev H (2004) Diversity and dominance in palm Arecaceae: communities in *terra firme* forests in the western Amazon basin. J Ecol 92: 577–588.

53. Chust G, Chave J, Condit R, Aguilar S, Lao S, et al. (2006) Determinants and spatial modeling of tree β-diversity in a tropical forest landscape in Panama. J of Veg Sci 17: 83–92.

54. Bazzaz FA (1996) Plants in changing environments: linking physiological, population and community ecology. Cambridge: Cambridge Univ press. 332 p.

55. Phillips OL, Vargas PN, Monteagudo AL, Cruz AP, Zans M-EC, et al. (2003) Habitat association among Amazonian tree species: a landscape-scale approach. J Ecol 91: 757–775.

56. Chust G, Chave J, Condit R, Aguilar S, Lao S, et al. (2006) Determinants and spatial modeling of tree β-diversity in a tropical forest landscape in Panama. J Veg Sci 17: 83–92.

57. Tuomisto H, Ruokolainen K, Poulsen AD, Moral RC, Quintana C, et al. (2002) Distribution and diversity of Pteridophytes and Melastomataceae along edaphic gradients in Yasuní National Park, Ecuadorian Amazonia. Biotropica 34: 516–533.

58. Tuomisto H, Ruokolainen K, Yli-Halla M (2003) Dispersal, environment and floristic variation of western Amazonian forest. Science 299: 241–244.

59. Nicotra AB, Chazdon RL, Iriarte SVB (1999) Spatial heterogeneity of light and woody seedling regeneration in tropical wet forest. Ecology 80: 1908–1926.

60. Montgomery RA (2004) Effects of understory foliage on patterns of light attenuation near the forest floor. Biotropica 36: 33–39.

61. Mayfield MM, Levine JM (2010) Opposing effects of competitive exclusion on the phylogenetic structure of communities. Ecol Lett 13: 1085–1093.

62. Cavender-Bares J, Keen A, Miles B (2006) Phylogenetic structure of floridian plant communities depends on taxonomic and spatial scale. Ecology 87: S109–S122.

63. Willis CG, Halina M, Lehman C, Reich PB, Keen A, et al. (2010) Phylogenetic community structure in Minnesota oak savanna is influenced by spatial extent and environmental variation. Ecography 33: 656–577.

64. Silvertown J (2004) Plant coexistence and the niche. Trends Ecol Evol 19: 605–611.

65. Metz MR, Sousa WP, Valencia R (2010) Widespread density-dependent seedling mortality promotes species coexistence in highly diverse Amazonian rain forest. Ecology 91: 3675–3685.

66. Comita LS, Hubbell S (2009) Local neighborhood and species' shade tolerance influence survival in a diverse seedling bank. Ecology 90: 328–334.

67. Poorter L, Bongers F, Sterck F, Wöll H (2005) Beyond the regeneration phase: differentiation of height-light trajectories among tropical trees species. J Ecol 93: 256–267.

5

Record-Breaking Early Flowering in the Eastern United States

Elizabeth R. Ellwood[1]*, **Stanley A. Temple**[2,3]*, **Richard B. Primack**[1]*, **Nina L. Bradley**[3†], **Charles C. Davis**[4]*

1 Department of Biology, Boston University, Boston, Massachusetts, United States of America, 2 Department of Forest and Wildlife Ecology, University of Wisconsin, Madison, Wisconsin, United States of America, 3 Aldo Leopold Foundation, Baraboo, Wisconsin, United States of America, 4 Department of Organismic and Evolutionary Biology, Harvard University Herbaria, Cambridge, Massachusetts, United States of America

Abstract

Flowering times are well-documented indicators of the ecological effects of climate change and are linked to numerous ecosystem processes and trophic interactions. Dozens of studies have shown that flowering times for many spring-flowering plants have become earlier as a result of recent climate change, but it is uncertain if flowering times will continue to advance as temperatures rise. Here, we used long-term flowering records initiated by Henry David Thoreau in 1852 and Aldo Leopold in 1935 to investigate this question. Our analyses demonstrate that record-breaking spring temperatures in 2010 and 2012 in Massachusetts, USA, and 2012 in Wisconsin, USA, resulted in the earliest flowering times in recorded history for dozens of spring-flowering plants of the eastern United States. These dramatic advances in spring flowering were successfully predicted by historical relationships between flowering and spring temperature spanning up to 161 years of ecological change. These results demonstrate that numerous temperate plant species have yet to show obvious signs of physiological constraints on phenological advancement in the face of climate change.

Editor: Bruno Hérault, Cirad, France

Funding: CCD: National Science Foundation Grant Assembling the Tree of Life (AToL) EF 04-31242; RBP: National Science Foundation Grant DEB-0842749. The funders had no role in study design, data collection and analysis, decision to publish, or preparation of the manuscript.

Competing Interests: The authors have declared that no competing interests exist.

* E-mail: eellwood@bu.edu (ERE); satemple@wisc.edu (SAT); primack@bu.edu (RBP); cdavis@oeb.harvard.edu (CCD)

† Deceased.

Introduction

The sensitivity of flowering times to temperature has proven valuable for investigating the impacts of climate change on plants [1]–[3]. Plant phenology appears to have largely kept pace with warmer temperatures, with numerous species flowering earlier now than in the past. However, recent years have seen record-breaking spring temperatures that are well outside the realm of historical trends [4], [5]. Although flowering dates for many responsive species have greatly advanced with warmer temperatures, at some point plants may no longer flower earlier in response to warming due to photoperiod constraints or unmet winter chilling requirements [6]–[8]. Extreme weather events such as those observed in the eastern United States in 2010 and 2012 provide opportunities to determine if historical phenological responses to rising temperatures are maintained under novel conditions presented by very recent climate change.

Changes in plant phenology have broad implications at the ecosystem level. Flowering and leafing out times signal the start of the growing season, and altered phenology influences associated ecosystem processes such as nutrient cycling and carbon sequestration [9], [10]. Interactions with herbivores, pollinators, and other ecological associates may be compromised and lead to ecological mismatches [11]–[15]. Also, advanced spring phenology, followed by late frost events, can damage flowers and young leaves, which has negative impacts on plant growth and fruit development [16]–[18]. Finally, warmer temperatures can also expose plants to drought, resulting in decreased reproductive success [19].

Two of the best-known American environmental writers initiated extensive phenological observations of flowering times in the eastern United States that encompass 161 years of ecological change. From 1852–1858, Henry David Thoreau, author of *Walden* [20], observed flowering times in Concord, Massachusetts, USA. And from 1935–1945, Aldo Leopold, author of *A Sand County Almanac* [21], recorded flowering times in Dane County, Wisconsin, USA and near the site of his "Shack" in adjacent Sauk County [4]. Several recent re-surveys at these locations [22]–[25], nearly 1500km apart, indicate that many spring-flowering plants now flower much earlier than in the past. This trend appears to be attributable to especially warmer spring (March, April, May) temperatures [25]–[27]. In 2010 and 2012 in Massachusetts [5], and 2012 in Wisconsin [4], spring temperatures were the warmest on record. These long-term datasets thus provide a rare opportunity to investigate if historical relationships between flowering times and spring temperatures apply during these record-breaking years. These observational data are especially timely because recent meta-analyses of flowering phenology [28] have documented that controlled warming experiments greatly under-predict flowering phenology when compared with their responses in natural settings. Thus, historical phenological data, such as those initiated by Thoreau and Leopold, are critical to understanding plant responses to current and future warming, and to test whether increasing temperatures may result in continued earlier flowering.

Results and Discussion

In Concord, Massachusetts, 32 spring flowering native plant species representing a broad phylogenetic diversity were chosen because they were observed in nearly all of the following 29 years: 1852–1858, 1878, 1888–1902, 2004–2006 and 2008–2012 [24] (Fig. 1a; Table 1, and phylogenetic relationships in Figures S1a and S1b). From 1852–1858, when mean spring temperature in the region was 5.5°C, mean first flowering date for these species was 15 May. By 1878–1902 their mean first flowering date had shifted five days earlier to 10 May, when mean spring temperature was 6.3°C. During the past nine years mean first flowering has shifted to 4 May, 11 days earlier than in Thoreau's time and during a period in which mean spring temperature has risen to 8.8°C. Warming in the greater Boston area, which includes Concord, has been attributed to both global warming and the urban heat island [29]. Within the past decade, two years have been record breakers in this region: mean spring temperature in 2010 was the warmest ever recorded at 11.0°C, during which time plants had a mean flowering date of 24 April; and 2012 was the second warmest spring on record at 10.7°C, during which time plants had a mean flowering date of 25 April. In these two years, plants flowered three weeks earlier (i.e., 21 and 20 days in 2010 and 2012, respectively) than when Thoreau observed them in Concord.

Numerous species in Massachusetts have shown remarkable shifts in flowering times in recent years [27], [30]. In 2010, 13 of the 32 species we analyzed had their earliest flowering date on record. In 2012, a different 14 species had their earliest recorded flowering date. Thoreau, for example, observed highbush blueberry (*Vaccinium corymbosum*) flowering in mid-May (11–21 May). In 2012 this species flowered on 1 April, six weeks earlier than observed by Thoreau. Based on our linear regression analysis of these historical phenology and temperature data, plant species flower on average 3.2 days earlier for each 1°C rise in mean temperatures (Figure 2a, $p < 0.001$, $R^2 = 0.75$). Twenty-seven of these 32 species exhibit significantly ($p < 0.05$) earlier flowering times with spring temperatures (Table 1). Our results are robust to phylogenetic relationships: when phylogeny was incorporated into a generalized least squares analysis of phenological response to spring temperature, the results remained highly significant ($P < 0.01$).

In south-central Wisconsin, 23 phylogenetically diverse spring-flowering native plant species have been monitored in each of the following 47 years: 1935–1945 and 1977–2012 (Fig. 1b; Table 1, and phylogenetic relationships in Figures S1a and S1b). During this time, Wisconsin's spring temperatures have warmed dramatically as a result of climate change [31]. During 1935–1945, when mean spring temperature was 7.5°C, the mean flowering date was 7 May. During the most recent 11-year period (2002–2012), when mean spring temperature was 9.3°C, the mean flowering date advanced by 7 days to 1 May. The mean spring temperature in 2012 was 12.2°C, the warmest on record and substantially warmer than the previous high of 11.3°C in 1977. In 2012, mean flowering was 13 April, the earliest date ever recorded, and over 3 weeks earlier (i.e., 24 days) than mean flowering in Leopold's years.

Most species in Wisconsin showed dramatic shifts in their flowering dates during this time. In 2012, 19 of the 23 species equaled or surpassed their previous earliest flowering dates. This response has been especially strong for several species. For example, Leopold recorded the first flower of woodland phlox (*Phlox divaricata*) between 28 April and 27 May; in 2012 it flowered

on 4 April. Likewise, he recorded serviceberry (*Amelanchier arborea*) flowering between 10 April and 9 May; in 2012 it flowered on 25 March. Based on our analyses of these cumulative phenology and temperature data, plants in south-central Wisconsin flower on average 4.1 days earlier for each 1°C rise in mean spring temperature (Figure 2b, $p < 0.001$, $R^2 = 0.88$). All 23 species exhibit significantly ($p < 0.05$) earlier flowering times with warming spring temperatures (Table 1). As in Massachusetts, our results were robust to phylogenetic relationships ($P < 0.05$).

Given the significant relationship between mean spring temperatures and mean first flowering dates, the recent record-breaking warm springs of 2010 and 2012 in Massachusetts and 2012 in Wisconsin provide an opportunity to test whether historical relationships predict mean flowering dates during these exceptionally warm years. Based on regression analyses of pre-2010 data (Massachusetts) and pre-2012 data (Wisconsin), the mean observed first flowering dates for the focal species during 2010 and 2012 fell within the 95% prediction intervals at each location (Figure 2) [32]. These prediction intervals [30] are estimates of the range of dates within which 2010 and 2012 observations of mean first flowering date are expected to fall, within a 95% probability. Results for individual species were also similar (Table 1). For the 32 species in Massachusetts, all but two flowered within the prediction interval for 2010. Marsh marigold [*Caltha palustris*] flowered earlier, and rhodora [*Rhododendron canadense*] flowered later than predicted. In 2012, only early saxifrage [*Saxifraga virginiensis*] flowered earlier than predicted. For Wisconsin, 22 of the 23 species had flowering times in 2012 that were within the 95% prediction intervals. Meadow anemone (*Anemone canadensis*) was the lone outlier, flowering five days earlier than the predicted interval. These results indicate that spring-flowering plants at both locations, whether analyzed as single species or averaged across all species, largely responded to record-breaking warm temperatures as predicted by their historical responses to warming spring temperatures.

These results collectively demonstrate that despite record-breaking warm temperatures in the eastern United States, plants have continued to flower earlier in the face of recent dramatic climate change. While other studies have examined long-term observations with comparable rates of phenological advancement [2], [3], [33], [34], to our knowledge ours is the first to demonstrate the predictive power of such data under unprecedented warm temperatures. In contrast to our results, there is increasing discussion in the literature [6]–[8] that flowering, leaf out, and growth could be delayed for temperate plants that have not experienced lengthened spring photoperiods or extended cool temperatures that satisfy their winter chilling requirements. A delay in phenology caused by insufficient chilling is most likely to be observed first in warm temperate latitudes where winter temperatures are barely adequate for fulfilling chilling requirements for some species [8], [35]. Another scenario is highlighted in a recent study [7] suggesting that individual species thought to be unresponsive to spring temperature were actually responding to both an unsatisfied chilling requirement and warmer spring temperatures resulting in no net change in flowering phenology. Based on our results, there is no indication that the 47 spring flowering plants we studied are delayed in their flowering by insufficient photoperiod or winter chilling requirements. These plants continue to flower earlier apparently in direct response to increasingly warmer mean spring temperatures (R^2 values = 0.75–0.88). Other climatic factors such as late winter temperatures or spring minimum temperatures may exert some effects, but we did not detect them here. This strongly suggests that most of these plants have not yet reached a physiological threshold.

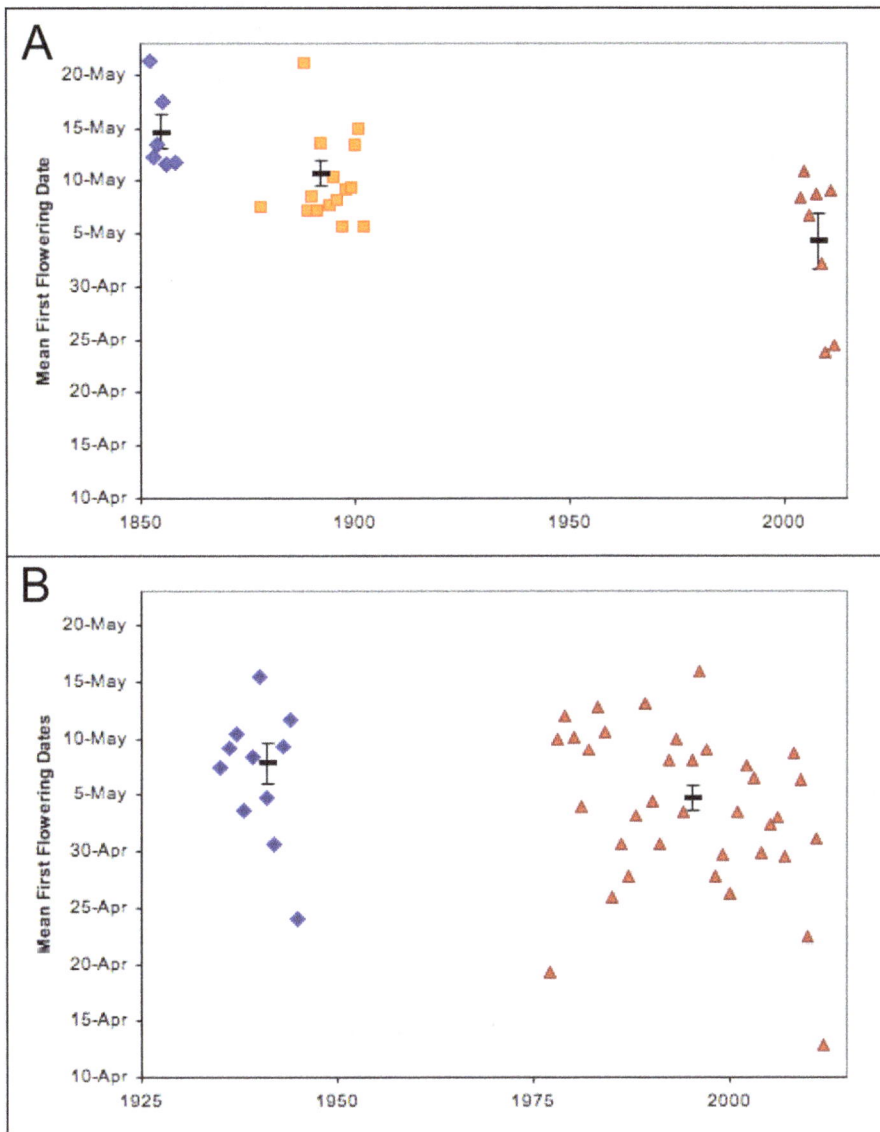

Figure 1. Mean first flowering dates for all species. The annual mean first flowering dates and standard errors of 29 years of data are shown from Massachusetts (a) and 47 years from Wisconsin (b). Blue triangles = Thoreau and Leopold et al.; orange squares = Hosmer; red triangles = Primack et al. and Bradley et al..

By extension, because flowering and leaf-out times are highly correlated for many species [36], [37], we hypothesize that yet earlier flowering times, and potentially leaf out times, will continue to be observed in the face of predicted climate change. In contrast to a number of phenological studies showing nonlinear relationships between phenology and temperature, due largely to unmet chilling and photoperiod requirements, our findings demonstrate the relationship to be linear and to explain most of the variation in flowering. It is possible of course, that these observations are within a fairly linear portion of a relationship that will prove to be nonlinear with future climate change [38] [39]. As temperatures continue to rise in the northeastern United States this linearity of the relationship of flowering time to temperature will be tested. Importantly, on-going ecological monitoring initiated by Thoreau and Leopold will help to clarify the complexities of this system under future change, and to illuminate plant phenological

responses in experimental warming plots and under greenhouse conditions.

Materials and Methods

Phenological and climate data

Observations of first flowering dates of species in Concord, Massachusetts, USA (42°27′37″N, 71°20′58″W) were made by Thoreau during the years 1852–1858, Hosmer for 1878 and 1888–1902, Primack, Miller-Rushing and their associates for 2003–2006, and Primack and his associates for 2008–2012 [24]. Thirty-two spring-flowering native species from a variety of habitats were chosen from a list of over 200 species because of the criterion of being observed in nearly all years. At the Massachusetts site, *Amelanchier arborea* and *A. canadensis* cannot be readily distinguished and flower at the same time; for convenience these combined observations are listed under the name *A. arborea.*

Table 1. List of plant species monitored at each location, along with their phenological responses to temperature (for years prior to 2010 for MA and prior to 2012 for WI) and 95% prediction intervals for 2010 and 2012 for Massachusetts and 2012 for Wisconsin.

Location	Species	n	Temp. Response	95% Prediction Interval	Obs. FFD
Massachusetts	*Amelanchier arborea**	25	$y = -3.24x + 143$	2010: 95–119	2010: 102
	(Serviceberry)		$R^2 = 0.45$***	2012: 96–119	2012: 105
Massachusetts	*Anemone quinquefolia**	25	$y = -1.03x + 123$	2010: 96–127	2010: 102
	(Wood Anemone)		$R^2 = 0.05$*	2012: 97–127	2012: 105
Massachusetts	*Aquilegia canadensis**	25	$y = 0.46x + 112$	2010: 97–138	2010: 105
	(Wild Columbine)		$R^2 = 0.01$	2012: 98–137	2012: 109
Massachusetts	*Aralia nudicaulis*	24	$y = -3.27x + 160$	2010: 113–134	2010: 122
	(Wild Sarsaparilla)		$R^2 = 0.68$***	2012: 114–135	2012: 128
Massachusetts	*Arenaria lateriflora*	23	$y = -4.92x + 178$	2010: 105–142	2010: 127
	(Bluntleaf Sandwort)		$R^2 = 0.43$***	2012: 107–143	2012: 123
Massachusetts	*Caltha palustris**	25	$y = -1.69x + 116$	2010: 75–120	2010: 69
	(Marsh Marigold)		$R^2 = 0.06$**	2012: 76–120	2012: 92
Massachusetts	*Comandra umbellate*	24	$y = -3.26x + 162$	2010: 107–145	2010: 124
	(Bastard Toadflax)		$R^2 = 0.24$***	2012: 108–145	2012: 128
Massachusetts	*Cornus canadensis*	24	$y = -3.27x + 164$	2010: 114–142	2010: 136
	(Dwarf Dogwood)		$R^2 = 0.36$**	2012: 116–143	2012: 138
Massachusetts	*Cypripedium acaule*	25	$y = -3.70x + 165$	2010: 110–138	2010: 124
	(Pink Lady Slipper)		$R^2 = 0.61$***	2012: 112–138	2012: 123
Massachusetts	*Fragaria virginiana**	20	$y = -4.21x + 152$	2010: 85–127	2010: 105
	(Wild Strawberry)		$R^2 = 0.33$***	2012: 87–128	2012: 105
Massachusetts	*Gaylussacia baccata*	24	$y = -5.82x + 174$	2010: 97–123	2010: 115
	(Black Huckleberry)		$R^2 = 0.68$***	2012: 99–125	2012: 109
Massachusetts	*Geranium maculatum**	25	$y = -1.85x + 151$	2010: 117–143	2010: 129
	(Wild Geranium)		$R^2 = 0.16$**	2012: 118–144	2012: 128
Massachusetts	*Houstonia caerulea*	26	$y = -2.70x + 127$	2010: 78–117	2010: 86
	(Bluet)		$R^2 = 0.17$**	2012: 79–118	2012: 92
Massachusetts	*Hypoxis hirsuta*	25	$y = -3.31x + 168$	2010: 117–146	2010: 129
	(Yellow Star-Grass)		$R^2 = 0.34$***	2012: 119–147	2012: 128
Massachusetts	*Krigia virginica*	24	$y = -4.21x + 171$	2010: 99–151	2010: 115
	(Dwarf Dandelion)		$R^2 = 0.22$***	2012: 101–151	2012: 109
Massachusetts	*Potentilla canadensis*	24	$y = 0.21x + 116$	2010: 89–148	2010: 102
	(Dwarf Cinquefoil)		$R^2 = 0.00$	2012: 90–147	2012: 105
Massachusetts	*Prunus pensylvanica*	22	$y = -2.95x + 147$	2010: 100–128	2010: 105
	(Pin Cherry)		$R^2 = 0.29$***	2012: 101–129	2012: 109
Massachusetts	*Prunus serotina**	22	$y = -2.08x + 149$	2010: 91–161	2010: 129
	(Black Cherry)		$R^2 = 0.04$	2012: 92–161	2012: 131
Massachusetts	*Prunus virginiana*	24	$y = -4.06x + 165$	2010: 92–138	2010: 122
	(Chokecherry)		$R^2 = 0.41$***	2012: 94–139	2012: 123
Massachusetts	*Rhododendron canadense*	26	$y = -4.27x + 160$	2010: 104–122	2010: 124
	(Rhodora)		$R^2 = 0.72$***	2012: 106–123	2012: 123
Massachusetts	*Saxifraga virginiensis*	26	$y = 0.81x + 103$	2010: 92–131	2010: 102
	(Early Saxifrage)		$R^2 = 0.02$	2012: 93–130	2012: 92
Massachusetts	*Senecio aureus*	26	$y = -2.36x + 156$	2010: 111–150	2010: 129
	(Golden Ragwort)		$R^2 = 0.13$**	2012: 112–150	2012: 123
Massachusetts	*Silene caroliniana*	26	$y = -3.85x + 169$	2010: 115–137	2010: 129
	(Wild Pink)		$R^2 = 0.58$***	2012: 117–138	2012: 128
Massachusetts	*Smilax rotundifolia*	21	$y = -4.12x + 183$	2010: 109–166	2010: 124
	(Common Greenbriar)		$R^2 = 0.19$***	2012: 111–166	2012: 128

Table 1. Cont.

Location	Species	n	Temp. Response	95% Prediction Interval	Obs. FFD
Massachusetts	*Trientalis borealis*	25	y = −4.43x+165	2010: 103–130	2010: 115
	(Starflower)		$R^2 = 0.53$***	2012: 105–131	2012: 118
Massachusetts	*Trillium cernuum*	25	y = −2.84x+155	2010: 107–142	2010: 122
	(Nodding Trillium)		$R^2 = 0.21$**	2012: 108–142	2012: 131
Massachusetts	*Vaccinium angustifolium*	26	y = −4.41x+152	2010: 88–118	2010: 105
	(Lowbush Blueberry)		$R^2 = 0.63$***	2012: 90–119	2012: 98
Massachusetts	*Vaccinium corymbosum*	26	y = −6.55x+170	2010: 83–113	2010: 97
	(Highbush Blueberry)		$R^2 = 0.66$***	2012: 85–115	2012: 92
Massachusetts	*Viola cucullata*	25	y = −3.28x+140	2010: 86–122	2010: 102
	(Marsh Blue Violet)		$R^2 = 0.27$***	2012: 88–122	2012: 98
Massachusetts	*Viola fimbriatula*	23	y = −2.91x+142	2010: 86–134	2010: 102
	(Arrowleaf Violet)		$R^2 = 0.13$**	2012: 88–135	2012: 105
Massachusetts	*Viola lanceolata*	24	y = −3.17x+150	2010: 100–130	2010: 120
	(Lance-leaved Violet)		$R^2 = 0.33$***	2012: 101–130	2012: 115
Massachusetts	*Viola pedata**	23	y = 2.22x+110	2010: 113–157	2010: 124
	(Birdfoot Violet)		$R^2 = 0.10$	2012: 113–155	2012: 123
Wisconsin	*Amelanchier arborea**	47	y = −4.85x+155	84–108	84
	(Serviceberry)		$R^2 = 0.63$***		
Wisconsin	*Anemone canadensis*	47	y = −4.05x+179	116–144	111
	(Meadow Anemone)		$R^2 = 0.46$***		
Wisconsin	*Anemone patens*	47	y = −3.31x+127	70–103	75
	(Pasque Flower)		$R^2 = 0.30$***		
Wisconsin	*Anemone quinquefolia**	47	y = −4.31x+149	84–109	87
	(Wood Anemone)		$R^2 = 0.55$***		
Wisconsin	*Aquilegia canadensis**	47	y = −3.98x+162	102–125	117
	(Wild Columbine)		$R^2 = 0.55$***		
Wisconsin	*Arabis lyrata*	47	y = −3.84x+140	80–105	80
	(Sand Cress)		$R^2 = 0.49$***		
Wisconsin	*Caltha palustris**	47	y = −2.64x+134	91–112	100
	(Marsh Marigold)		$R^2 = 0.41$***		
Wisconsin	*Dicentra cucullaria*	47	y = −4.46x+140	72–100	76
	(Dutchman's Breeches)		$R^2 = 0.52$***		
Wisconsin	*Dodecatheon meadia*	47	y = −3.73x+163	110–126	122
	(Shooting Star)		$R^2 = 0.70$***		
Wisconsin	*Fragaria virginiana**	47	y = −3.93x+154	90–123	102
	(Wild Strawberry)		$R^2 = 0.38$***		
Wisconsin	*Geranium maculatum**	47	y = −4.52x+165	98–122	111
	(Wild Geranium)		$R^2 = 0.59$***		
Wisconsin	*Hepatica nobilis*	47	y = −4.18x+132	64–98	75
	(Sharp-lobed Hepatica)		$R^2 = 0.40$***		
Wisconsin	*Lithospermum canescens*	47	y = −4.13x+161	96–126	105
	(Hoary Puccoon)		$R^2 = 0.44$***		
Wisconsin	*Oxalis stricta*	47	y = −4.23x+167	104–126	118
	(Wood Sorrel)		$R^2 = 0.62$***		
Wisconsin	*Phlox divaricata*	47	y = −5.38x+167	86–116	94
	(Woodland Phlox)		$R^2 = 0.57$***		
Wisconsin	*Phlox pilosa*	47	y = −3.66x+166	106–136	128
	(Prairie Phlox)		$R^2 = 0.38$***		
Wisconsin	*Prunus serotina**	47	y = −3.74x+167	109–134	126
	(Black Cherry)		$R^2 = 0.49$***		

Table 1. Cont.

Location	Species	n	Temp. Response	95% Prediction Interval	Obs. FFD
Wisconsin	*Rubus allegheniensis* (Common Blackberry)	47	$y = -3.01x+169$ $R^2 = 0.63***$	124–140	129
Wisconsin	*Sanguinaria canadensis* (Bloodroot)	47	$y = -3.55x+129$ $R^2 = 0.40***$	71–100	76
Wisconsin	*Sisyrinchium campestre* (Blue-eyed Grass)	47	$y = -3.83x+165$ $R^2 = 0.58***$	108–129	118
Wisconsin	*Tradescantia ohiensis* (Spiderwort)	47	$y = -3.27x+174$ $R^2 = 0.34***$	119–149	131
Wisconsin	*Trillium grandiflorum* (Large-flowered Trillium)	47	$y = -2.58x+142$ $R^2 = 0.19***$	93–128	99
Wisconsin	*Viola pedata** (Birdfoot Violet)	47	$y = -5.11x+164$ $R^2 = 0.70***$	91–112	101

The samples size is the number of years used for each regression analysis. Species names follow the United States Department of Agriculture Plants Database. Species common to both locations are indicated with an asterisk after the species name. Asterisks that follow R^2 values represent significance at the following levels:
* = p<0.05,
** = p<0.01,
*** = p<0.001.

This dataset includes all species that met these criteria, while non-native species, species with few observations and summer-flowering species were not included in this analysis (Table 1). These data are available on the Primack Lab website (people.bu.edu/primack). Phenological observations were made on both public and private lands; permission was obtained for private land when necessary. No permission was needed for public lands. No destructive tissue sampling was conducted. Temperature data are from Blue Hill Meteorological Observatory in East Milton, Massachusetts and are available through NOAA National Climatic Data Center (http://www.ncdc.noaa.gov/oa/ncdc.html) [40]. Blue Hill Meteorological Observatory is located 33 km southeast of Concord, MA and temperatures between the two nearby locations are highly correlated [27].

Leopold, his family members, and his students collected phenological data from 1935–1945 at locations in Sauk and Dane Counties, Wisconsin, USA, primarily near the Leopold "Shack" (43°33′46″N, 89°39′33″W) and in the University of Wisconsin Arboretum (43°02′48″N, 89°24′58″W). NLB, SAT, and the staff of the Aldo Leopold Foundation collected phenological data from 1977–2012 at locations in Sauk and Dane Counties primarily near the Leopold shack and in Dunlap Hollow (43°12′12″N, 89°45′06″W). Twenty-three spring-flowering native species were chosen from a list of 176, for which observations of first flowering had been made in every year. These data are available by contacting SAT. Permits and approvals were not necessary for the private lands where observations were made in Wisconsin, or for public property of the University of Wisconsin Arboretum. None of the Wisconsin species observed in this study have protective status, and no destructive sampling was conducted. Mean spring temperatures for the south-central Wisconsin climatic region, which includes our study sites, were obtained from the Wisconsin State Climatology Office (http://www.aos.wisc.edu/~sco/clim-history/division/data/temp/WI-08-TEMP.xls).

Statistical analysis and phylogenetic methods

Mean annual temperatures for those months that best predict spring flowering times were used in this analysis (i.e., March, April and May). April and May are the predominant flowering months for these species, and the inclusion of March temperatures strengthened the model. Mean temperatures for this time period provided the strongest model, owing to the fact that plants are accumulating heat and beginning spring growth. While certain studies have shown that the inclusion of winter months improves the relationship between flowering and temperature [41], we did not find that to be the case with this data set. For example, the model of flowering in Concord using only mean monthly April and May temperatures provided a strong model ($R^2 = 0.70$), yet including May temperatures explained an even larger amount of variation ($R^2 = 0.75$). Adding mean February temperature weakened this relationship ($R^2 = 0.71$); using mean monthly temperatures from January through May weakened this relationship further ($R^2 = 0.64$).

We performed all analyses in R 2.15.1 [42]. We calculated linear regressions (mean first flowering date for all species over time as well as mean first flowering date for each species versus mean spring temperature) for all years at both study sites, respectively.

We used mean spring temperature rather than another index of spring (e.g., growing degree days) due to the ease of calculating, displaying, and explaining this variable. Also, this simple measure of spring temperature explained most of the variation in flowering times. To test the linearity of the relationship between temperature and flowering time, we analyzed the residuals of this relationship and found them to be well scattered in a random pattern. This indicates that the relationship is consistent and that flowering is not earlier or later over time other than expected relative to temperature.

We also performed statistical comparisons to account for phylogenetic non-independence. Two highly resolved dated phylogenies were produced for each of the two sites to accomplish this goal (see Text S1 and Figures S1a and S1b). We did not conduct a multiple model regression test, but have previously shown in such an analysis using the Concord data that phenological response and abundance change is most strongly tied to changes in temperature [26]. All phylogenies and data

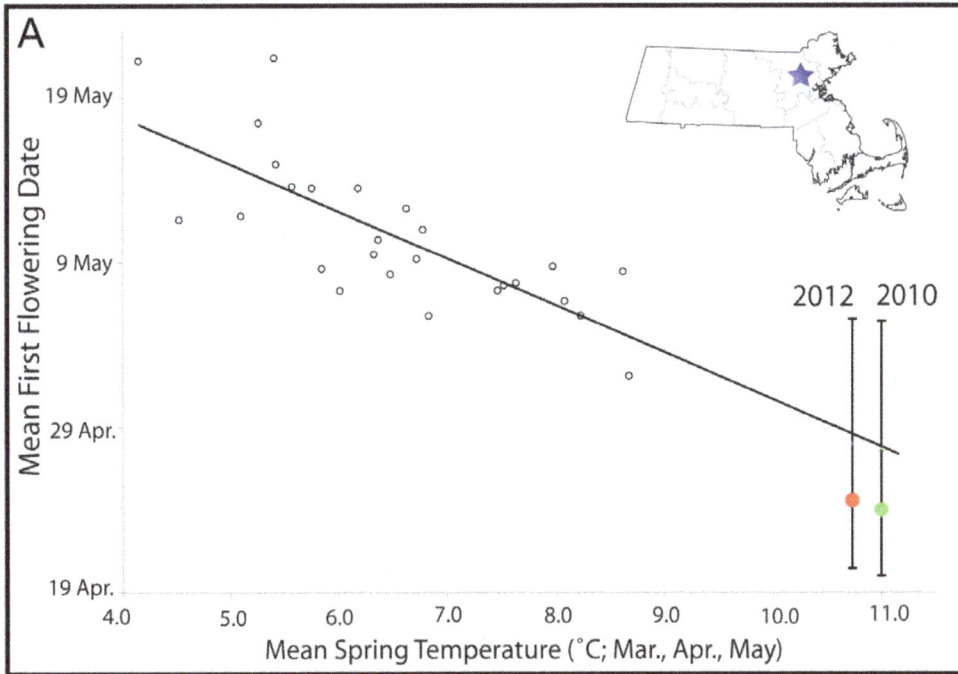

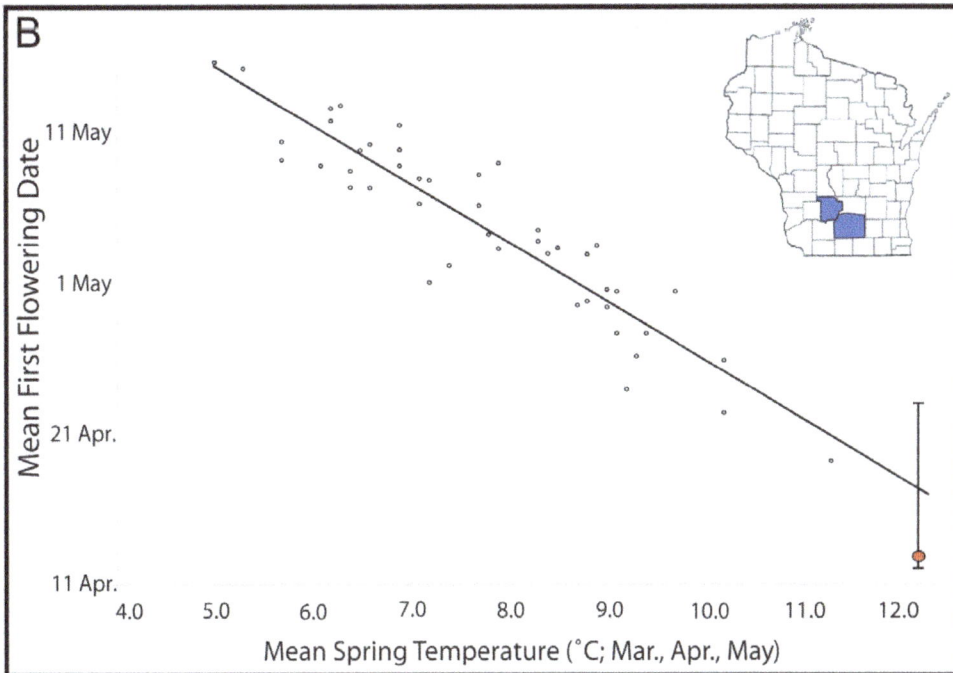

Figure 2. The relationships between mean first flowering dates and mean spring (March, April and May) temperatures. Each dot represents the mean first flowering date of all sampled species for a given year in (a) Massachusetts and (b) Wisconsin. Black regression lines, and 95% prediction intervals, were estimated from pre-2010 data (Massachusetts) and pre-2012 data (Wisconsin). 2012 observed values are shown in solid red, and 2010 (Massachusetts only) in green. The 95% prediction intervals for 2010 and 2012 mean first flowering dates are indicated with vertical lines. Photographs illustrate representative species at each location. Unless specified otherwise, photographs are made available under an Attribution-Share Alike 2.0 License with date and photographer as listed. Massachusetts species: 1) serviceberry (*Amelanchier canadensis*), © 2011 personal collection of R. Primack, 2) marsh marigold (*Caltha palustris*), © 2009 walker_bc, 3) pink lady slipper (*Cypripedium acaule*), © 2012 Graham Hunt, 4) rhodora (*Rhododendron canadense*), © 2012 Andrew Block, 5) nodding trillium (*Trillium cernuum*), © 2008 Ed Post, and 6) highbush blueberry (*Vaccinium corymbosum*), © 2007 Anita363. Wisconsin species: 1) woodland phlox (*Phlox divaricata*), © 2009 Diane DiOhio, 2) shooting star (*Dodecatheon meadia*), © 2006 Frank Mayfield, 3) hoary puccoon (*Lithospermum canescens*), © 2006 cotinis, 4) wild geranium (*Geranium maculatum*), © 2009 aposematic herpetologist, 5) pasque flower (*Anemone patens*) © 2007 Malcom Manners, and 6) sharplobe hepatica (*Hepatica nobilis*) © 2009 Alan J. Hahn.

matrices are available on TreeBase. Traits at both locations did not exhibit phylogenetic conservation as determined by Blomberg's K in the *picante* package version 1.4–2 (K<1.00) [43]. This indicates that the patterns we observed are not caused by groups of related species possessing similar traits. Trait correlations as above were tested using a phylogenetic general linear model as implemented using the pgls function in the *caper* package version 0.5. This model includes a variance-covariance structure based on evolutionary distance to control for phylogenetic non-independence in the data [44].

To determine prediction intervals that excluded recent record-breaking warm years, we recalculated linear regressions using only pre-2010 observations (for Massachusetts) and pre-2012 observations (for Wisconsin). Then, we calculated the 95% prediction intervals for mean first flowering dates for all species and flowering dates for each species for Massachusetts (separately for 2010 and 2012, using only pre-2010 observations) and Wisconsin (for 2012), based on the observed mean spring temperatures for those record-breaking warm years [32]. We then compared the observed mean first flowering dates for all species and flowering dates for each species in 2010 and 2012 (in Massachusetts) and 2012 (in Wisconsin) with those predictions.

Eight species were common to both sites and allow us to compare their responses to temperature (Table 1). An analysis of covariance (ANCOVA) was used to determine if location influenced how first flowering dates varied over time and in response to temperature. We then tested whether the regression lines of the relationship between year and first flowering date were the same between the two locations. This was repeated for the relationship between temperature and first flowering date for these common species. Mean flowering times varied over years in a similar way at both locations (ANCOVA $F_{1, 75} = 2.6$, p = 0.427). However, their responses to temperature differed between locations (ANCOVA $F_{1, 75} = 69.1$, p<0.001). The contrasting responses to temperature may be related to multiple factors,

including local adaptation to temperature and other related climate variables, or sampling issues including changes in species' abundance at each location [45], [46]. Future observational studies and transplant experiments of these species will help us to better understand these differences.

Supporting Information

Figure S1 S1a. Phylogeny of Massachusetts spring-flowering plant species used in the analyses. **S1b.** Phylogeny of Wisconsin spring-flowering plant species used in the analyses.

Text S1 Phylogenetic analysis description and methods.

Acknowledgments

The staff of the Aldo Leopold Foundation, especially Teresa Mayer, helped collect recent phenological data in Wisconsin. Abraham Miller-Rushing, Caroline Polgar and many Boston University students helped to collect phenology data from Concord. We appreciate the suggestion of Don Waller that 2012 might be a unique year for phenology across a wide area. The following individuals provided useful comments on the manuscript: Amanda Gallinat, Wellington Huffaker, Caitlin McDonough MacKenzie, Curt Meine, Abraham Miller-Rushing, Caroline Polgar, Elizabeth Wolkovich, Benjamin Zuckerberg, the Davis lab at Harvard University, and several anonymous reviewers. The A. W. Schorger Fund of the Department of Forest and Wildlife Ecology, University of Wisconsin-Madison, covered part of the publication fees.

Author Contributions

Conceived and designed the experiments: ERE SAT RBP NLB CCD. Performed the experiments: SAT RBP NLB CCD. Analyzed the data: ERE SAT RBP CCD. Contributed reagents/materials/analysis tools: ERE SAT RBP CCD. Wrote the paper: ERE SAT RBP CCD.

References

1. Fitter AH, Fitter RSR (2002) Rapid changes in flowering time in British plants. Science 296(5573): 1689–1691.
2. Parmesan C, Yohe G (2003) A globally coherent fingerprint of climate change impacts across natural systems. Nature 421(6918): 37–42.
3. Amano T, Smithers RJ, Sparks TH, Sutherland WJ (2010) A 250-year index of first flowering dates and its response to temperature changes. Proceedings of the Royal Society B-Biological Sciences 277(1693): 2451–2457.
4. Wisconsin State Climatology Office (2012) Available: http://www.aos.wisc.edu/~sco/clim-history/division/data/temp/WI-08-TEMP.xls. Accessed 2012 Dec 7.
5. NOAA National Climatic Data Center (2012) State of the Climate: National Overview for March 2012, published online April, 2012. Available: http://www.ncdc.noaa.gov/sotc/national/2012/3. Accessed 2012 Dec 7.
6. Cleland EE, Chuine I, Menzel A, Mooney HA, Schwartz MD (2007) Shifting plant phenology in response to global change. Trends Ecol Evol 22(7): 357–365.
7. Cook BI, Wolkovich EM, Parmesan C (2012) Divergent responses to spring and winter warming drive community level flowering trends. Proceedings of the

National Academy of Sciences of the United States of America 109(23): 9000–9005.
8. Schwartz MD, Hanes JM (2010) Continental-scale phenology: warming and chilling. Int J Climatol 30(11): 1595–1598.
9. Piao SL, Ciais P, Friedlingstein P, Peylin P, Reichstein M, et al. (2008) Net carbon dioxide losses of northern ecosystems in response to autumn warming. Nature 451(7174): 49–U43.
10. Menzel A, Sparks TH, Estrella N, Koch E, Aasa A, et al. (2006) European phenological response to climate change matches the warming pattern. Global Change Biology 12(10): 1969–1976.
11. Parmesan C (2006) Ecological and evolutionary responses to recent climate change. Annu Rev Ecol Evol Syst 37: 637–669.
12. Post E, Forchhammer MC (2008) Climate change reduces reproductive success of an Arctic herbivore through trophic mismatch. Philosophical Transactions of the Royal Society B-Biological Sciences 363(1501): 2369–2375.
13. Both C, van Asch M, Bijlsma RG, van den Burg AB, Visser ME (2009) Climate change and unequal phenological changes across four trophic levels: constraints or adaptations? J Anim Ecol 78(1): 73–83.

14. Forrest JRK, Thomson JD (2011) An examination of synchrony between insect emergence and flowering in Rocky Mountain meadows. Ecol Monogr 81(3): 469–491.

15. Durant JM, Hjermann DO, Ottersen G, Stenseth NC (2007) Climate and the match or mismatch between predator requirements and resource availability. Clim Res 33(3): 271–283.

16. Norby RJ, Hartz-Rubin JS, Verbrugge MJ (2003) Phenological responses in maple to experimental atmospheric warming and CO2 enrichment. Global Change Biology 9(12): 1792–1801.

17. Norgaard Nielsen CC, Rasmussen HN (2009) Frost hardening and dehardening in Abies procera and other conifers under differing temperature regimes and warm-spell treatments. Forestry 82(1): 43–59.

18. Inouye DW, McGuire AD (1991) Effects of Snowpack on Timing and Abundance of Flowering in Delphinium nelsonii (Ranunculaceae): Implications for Climate Change. Am J Bot 78(7): 997–1001.

19. Giménez-Benavides L, Escudero A, Iriondo JM (2007) Reproductive limits of a late-flowering high-mountain Mediterranean plant along an elevational climate gradient. New Phytologist 173(2): 367–382.

20. Thoreau HD (1854) Walden; or, Life in the Woods. Boston, MA: Ticknor and Fields.

21. Leopold A (1949) A Sand County Almanac: Oxford University Press.

22. Wright SD, Bradley NL (2008) In:Waller D, Rooney T, editors. The Vanishing Present. Chicago: University of Chicago Press. pp. 42–51.

23. Leopold A, Jones SE (1947) A phenological record for Sauk and Dane Counties, Wisconsin, 1935–1945. Ecol Monogr 17: 81–122.

24. Primack RB, Miller-Rushing AJ (2012) Uncovering, Collecting, and Analyzing Records to Investigate the Ecological Impacts of Climate Change: A Template from Thoreau's Concord. Bioscience 62(2): 170–181.

25. Bradley NL, Leopold AC, Ross J, Huffaker W (1999) Phenological changes reflect climate change in Wisconsin. Proceedings of the National Academy of Sciences 96(17): 9701–9704.

26. Willis CG, Ruhfel B, Primack RB, Miller-Rushing AJ, Davis CC (2008) Phylogenetic patterns of species loss in Thoreau's woods are driven by climate change. Proceedings of the National Academy of Sciences of the United States of America 105(44): 17029–17033.

27. Miller-Rushing AJ, Primack RB (2008) Global warming and flowering times in Thoreau's Concord: A community perspective. Ecology 89(2): 332–341.

28. Wolkovich EM, Cook BI, Allen JM, Crimmins TM, Betancourt JL, et al. (2012) Warming experiments underpredict plant phenological responses to climate change. Nature 485(7399): 494–497.

29. New England Regional Assessment Group (2001) New England Regional Assessment. Durham, New Hampshire, USA : University of New Hampshire, Institute for the Study of Earth, Oceans, and Space.

30. Willis CG, Ruhfel BR, Primack RB, Miller-Rushing AJ, Losos JB, et al. (2010) Favorable Climate Change Response Explains Non-Native Species' Success in Thoreau's Woods. PLoS One 5(1): Article No.: e8878.

31. Wisconsin Initiative on Climate Change Impacts(2011) Wisconsin's Changing Climate: Impacts and Adaptation. Madison, Wisconsin, USA.

32. Aitchison J, Dunsmore IR (1975) Statistical prediction analysis. Cambridge-New York-Melbourne: Cambridge University Press.

33. Beaubien E, Hamann A (2011) Spring flowering response to climate change between 1936 and 2006 in Alberta, Canada. Bioscience 61(7): 514–524.

34. Anderson JT, Inouye DW, McKinney AM, Colautti RI, Mitchell-Olds T (2012) Phenotypic plasticity and adaptive evolution contribute to advancing flowering phenology in response to climate change. Proceedings of the Royal Society B-Biological Sciences 279(1743): 3843–3852.

35. Polgar CA, Primack RB (2011) Leaf-out phenology of temperate woody plants: from trees to ecosystems. New Phytologist 191(4): 926–941.

36. Schwartz MD, Reiter BE (2000) Changes in North American spring. Int J Climatol 20(8): 929–932.

37. Primack RB (1987) Relationships among flowers, fruits, and seeds. Annu Rev Ecol Syst 18: 409–430.

38. Morin X, Roy J, Sonié L, Chuine I (2011) Changes in leaf phenology of three European oak species in response to experimental climate change. New Phytologist 186(4): 900–910.

39. Sparks TH, Jeffree EP, Jeffree CE (2000) An examination of the relationship between flowering times and temperature at the national scale using long-term phenological records from the UK. International Journal of Biometeorology 44(2): 82–87.

40. NOAA National Climatic Data Center (2012) Available: http://www.ncdc. noaa.gov/oa/ncdc.html. Accessed 2012 Dec 7.

41. McEwan RW, Brecha RJ, Geiger DR, John GP (2011) Flowering phenology change and climate warming in southwestern Ohio. Plant Ecology 212(1): 55–61.

42. R Core Development Team (2008) R: A language and environment for statistical computing. R Foundation for Statistical Computing Vienna, Austria.

43. Blomberg SP, Garland T, Ives AR (2003) Testing for phylogenetic signal in comparative data: Behavioral traits are more labile. Evolution 57(4): 717–745.

44. Freckleton RP, Harvey PH, Pagel M (2002) Phylogenetic analysis and comparative data: a test and review of evidence. The American Naturalist 160(6): 712–726.

45. Doi H, Takahashi M, Katano I (2010) Genetic diversity increases regional variation in phenological dates in response to climate change. Global Change Biology 16(1): 373–379.

46. Tryjanowski P, Panek M, Sparks T (2006) Phenological response of plants to temperature varies at the same latitude: case study of dog violet and horse chestnut in England and Poland. Clim Res 32(1): 89–93.

6

New Plant-Parasitic Nematode from the Mostly Mycophagous Genus *Bursaphelenchus* Discovered inside Figs in Japan

Natsumi Kanzaki[1]*, Ryusei Tanaka[2], Robin M. Giblin-Davis[3], Kerrie A. Davies[4]

1 Department of Forest Microbiology, Forestry and Forest Products Research Institute, Tsukuba, Ibaraki, Japan, 2 Division of Parasitology, Faculty of Medicine, University of Miyazaki, Miyazaki, Miyazaki, Japan, 3 Fort Lauderdale Research and Education Center, University of Florida/IFAS, Davie, Florida, United States of America, 4 Centre for Evolutionary Biology and Biodiversity, School of Agriculture, Food and Wine, The University of Adelaide, Waite Campus, Glen Osmond, South Australia, Australia

Abstract

A new nematode species, *Bursaphelenchus sycophilus* n. sp. is described. The species was found in syconia of a fig species, *Ficus variegata* during a field survey of fig-associated nematodes in Japan. Because it has a well-developed stylet and pharyngeal glands, the species is considered an obligate plant parasite, and is easily distinguished from all other fungal-feeding species in the genus based upon these characters. Although *B. sycophilus* n. sp. shares an important typological character, male spicule possessing a strongly recurved condylus, with the "*B. eremus* group" and the "*B. leoni* group" of the genus, it was inferred to be monophyletic with the "*B. fungivorus* group". The uniquely shaped stylet and well-developed pharyngeal glands is reminiscent of the fig-floret parasitic but paraphyletic assemblage of "*Schistonchus*". Thus, these morphological characters appear to be an extreme example of convergent evolution in the nematode family, Aphelenchoididae, inside figs. Other characters shared by the new species and its close relatives, i.e., lack of ventral P1 male genital papilla, female vulval flap, and papilla-shaped P4 genital papillae in males, corroborate the molecular phylogenetic inference. The unique biological character of obligate plant parasitism and highly derived appearance of the ingestive organs of *Bursaphelenchus sycophilus* n. sp. expands our knowledge of the potential morphological, physiological and developmental plasticity of the genus *Bursaphelenchus*.

Editor: John Jones, James Hutton Institute, United Kingdom

Funding: This work was partly supported by Grants-in-Aid for Scientific Research, Nos 22310145, 24658147 and 26292178, Grant-in-Aid for JSPS Fellows, No 259930 from The Ministry of Education, Culture, Sports, Science and Technology, Japan. The funders had no role in study design, data collection and analysis, decision to publish, or preparation of the manuscript.

Competing Interests: The authors have declared that no competing interests exist.

* E-mail: nkanzaki@affrc.go.jp

Introduction

The fig syconium provides a unique and interesting habitat for microbes and microscopic invertebrates. Trees of the genus *Ficus* L. are pollinated by highly specialized fig wasps (Agaonidae). This fascinating relationship has become a model system for studying cospeciation and host switching [1–6]. In this relationship, female wasps carrying pollen enter the young fig through a small hole (ostiole) at the apex of the fig, pollinating it and laying eggs in individual female florets within the fig syconium. After pollination, the syconium develops and the ostiole swells shut during subsequent seed development. Fig wasp larvae feed within infested female florets (seed galls) and develop into winged female and wingless male adults. Males emerge first from their respective seed galls and bore holes into the seed galls housing females for mating access. They then bore exit holes through the syconial wall to allow female wasps carrying the pollen to exit [7,8]. Thus, the fig syconium is often considered a closed environmental niche.

However, regardless of this apparently closed system, many different groups of phoretic and parasitic invertebrates, e.g., nematodes [9–12] and mites [11,13], have been reported from figs and fig wasps. Further, the nematode genus *Parasitodiplogaster*

Poinar, which parasitizes the fig wasps, has been examined as a model system of species radiation and the evolution of pathogenicity [10]. Nevertheless, because of the apparent ubiquity of such associations and the large number of *Ficus* species that occur worldwide (>700 species), the diversity of fig-associated nematodes (and mites) is far from being fully understood, and further intense surveys of diversity are needed.

During a field survey of fig and fig wasp-associated nematodes in Japan, a species of *Bursaphelenchus* was isolated from *F. variegata* Blume. Although two lethal plant pathogens are known [14], members of the genus *Bursaphelenchus* Fuchs is generally regarded as beetle (Coleoptera) or bee (Hymenoptera)-phoretic fungal feeders, and even the plant-parasitic species retain many of the morphological (functional) characters of fungal feeding in their ingestive organs [15]. However, the newly-discovered species appears to be morphologically adapted to being a plant parasite. The nematode is described herein as *B. sycophilus* n. sp., and its molecular phylogenetic status and morphological and biological characters are described and discussed.

Materials and Methods

Nematode isolation

No specific permissions were required for these locations/activities. Field studies did not involve endangered or protected species. The detailed location information is provided in Fig. 1 and supplemental information (Table S1).

A field survey of fig-associated nematodes was conducted during May to June, 2013 at the Ishigaki and Iriomote Islands, Okinawa, Japan. Various stages of fig syconia, i.e., unpollinated (young) and pollinated and developing (mature) ones, were collected from *F. variegata*, *F. septica* Burm. F. and *F. bengtensis* Merrill, and dissected on site using a portable dissecting microscope. The figs were pealed to remove the outer layers containing latex, and cut into small pieces in sterilized distilled water using a sterilized knife. Emerging nematodes were hand-picked with sterilized stainless needles for further analyses. They were heat-killed and fixed in TAF fixative (triethanolamine 2%, formalin 8%) for morphological specimens, or directly fixed in DESS [16] for further morphological and molecular profiling, and all materials were brought back to the laboratory. In addition to the materials collected on site, some syconia were brought back to the laboratory as back-up for additional sampling.

Morphological observation

The TAF-fixed materials were examined under a dissecting microscope and separated into morphotypes. Each morphotype was processed using a glycerin–ethanol series with the modified Seinhorst's method [17], and mounted in glycerin according to the methods of Maeseneer and d'Herde [18]. The mounted specimens were designated as types, and used for morphometrics and morphological observations, micrographs and measurements. The male tail ventral view was observed using the glycerin-processed specimens with the methods provided in Kanzaki [19]. The morphological drawings and measurements (morphometrics) were conducted with the aid of a drawing tube connected to a Nikon Eclipse 80i (Nikon, Tokyo) facilitated with DIC optics. The micrographs were taken and edited with a digital camera system, DS-Ri1 (Nikon, Tokyo) and a computer program, Photoshop Elements v. 3 (Adobe, CA), respectively.

Molecular profiles and phylogeny

For molecular analysis, DESS-fixed materials were washed and rehydrated in the sterilized distilled water, and observed using high magnification light microscopy to determine morphotypes. These observed specimens of *B. sycophilus* n. sp. were then transferred individually to 30 μl of nematode digestion buffer [20,21] and

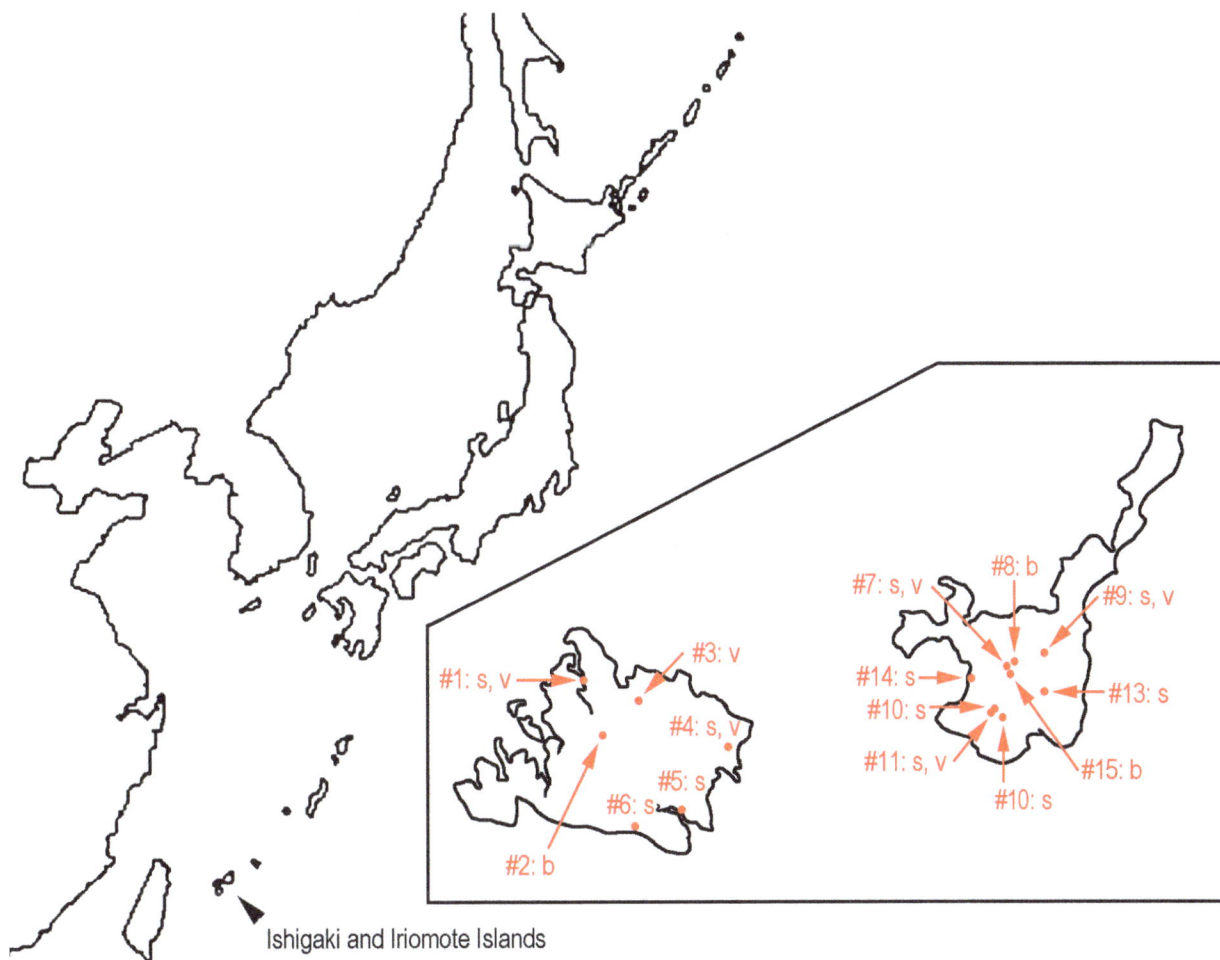

Ishigaki and Iriomote Islands

Figure 1. Outline map of collection localities for the samples examined in this study. For each locality and sampled fig species are suggested by abbreviations (b: *Ficus bengtensis*; s: *F. septica*; v: *F. variegata*).are listed. The GPS of the sites, sampled species and isolated nematode species (morphotype or genotype) \are summarized in Table S1.

Table 1. Morphometric values for *Bursaphelenchus sycophilus* n. sp.

	Male		Female
	Holotype	**Paratypes**	**Paratypes**
n	-	19	20
L	844	840±72 (738–964)	820±79 (666–933)
a	54.7	49.7±5.2 (39.0–59.3)	39.3±4.1 (32.6–45.2)
b	11.8	11.6±5.2 (9.7–13.0)	11.2±1.1 (9.0–13.0)
c	20.0	19.2±1.7 (15.7–22.7)	16.4±1.1 (14.5–18.7)
c'	2.9	2.9±0.3 (2.6–3.5)	4.8±0.7 (3.2–6.0)
T or V	67.2	67.4±5.1 (57.6–78.1)	79.2±1.2 (75.8–81.2)
M	51.9	52.4±2.5 (49.1–60.4)	53.2±1.6 (50–56.7)
Lip diam.	7.5	7.0±0.5 (6.0–7.5)	7.2±0.4 (6.5–8.0)
Lip height	3.5	3.7±0.4 (3.0–4.5)	3.7±0.4 (3.0–4.5)
Stylet conus length	13.9	14.2±1.0 (12.4–15.4)	15.5±0.9 (13.9–17.4)
Total stylet length	26.9	27.2±1.7 (23.9–29.4)	29.2±1.4 (26.7–32.8)
Median bulb length	16.4	17.0±1.5 (14.4–21.4)	17.6±1.5 (15.4–20.4)
Median bulb diam.	10.4	10.6±1.1 (9.5–13.4)	11.1±1.2 (9.5–13.4)
Median bulb length/diam.	1.57	1.61±0.2 (1.26–1.87)	1.59±0.1 (1.38–1.86)
Excretory pore from anterior end	86	87±6.0 (93–97)	85±5.7 (73–96)
Excretory pore from the base of median bulb	16.4	16.3±5.3 (14.4–21.4)	13.8±5.6 (3.0–21.9)
Nerve ring	87	88±3.4 (83–96)	87±3.8 (81–95)
Hemizonid from anterior end	97	100±5.6 (92–114)	97±4.3 (88–104)
Hemizonid from the base of median bulb	27.4	30.1±5.4 (23.9–41.8)	26.7±4.6 (17.9–35.3)
Gonad length (length from cloacal or vulval opening to anterior tip of gonad)	568	564±48 (494–659)	329±46 (198–389)
Cloacal/anal body diam.	14.4	15.0±1.4 (12.9–18.9)	10.6±1.5 (7.5–13.4)
Tail length	42	44±2.2 (40–49)	50±4.3 (43–57)
Spicule length (chord from anterior end of condylus to distal end)	15.6	16.0±0.8 (14.4–17.4)	-
Spicule length (curve from capitulum depression to distal end)	14.7	14.8±0.8 (12.9–15.9)	-
Post-uterine sac length	-	-	47±6.8 (34–60)
Post-uterine sac length per vulva-anus distance in %	-	-	39.3±4.7 (29.9–48.2)

All measurements are in μm in the form, average ± sd (range). The abbreviations for morphometric values are as follows. L: body length; a: body length/maximum body diameter; b: body length/length from anterior end to pharynx-intestine junction (ingestive organ length); c: body length/tail length; c': tail length/cloacal or anal body diameter; T: testis length/body length in %; V: vulval position from anterior end in %; M: conus length to total stylet length in %.

digested at 60°C for 20 min., and the crude DNA solution was used for the PCR template. DNA base sequences of partial ribosomal DNA (ca 1.7-kb near-full-length small subunit [SSU] and 0.7-kb D2/D3 expansion segment of large subunit [D2/D3 LSU]) were determined for *B. sycophilus* n. sp. following the methods of Kanzaki and Futai [22] and Ye et al. [23].

The molecular phylogenetic status of *B. sycophilus* n. sp. was determined based upon SSU and D2/D3 LSU ribosomal RNA gene sequences using Bayesian, Maximum Likelihood (ML) and Maximum parsimony (MP) analyses. The SSU was compared with a wide-range aphelenchids and other infraorder species, and D2/D3 LSU, which is more suitable for lower level phylogenetic comparisons, was compared with those of closely related species. The species (operational taxonomic units: OTUs) compared with *B. sycophilus* n. sp. were determined according to the results of a BLAST homology search (http://blast.ncbi.nlm.nih.gov/Blast.cgi) and the OTUs used in the previous studies on aphelenchid phylogeny [24–27]. The species names and sequence accession numbers used in the present study were summarized in Table S2. Several tylenchid and panagrolaimid nematodes were used as

outgroup species according to the previous studies [24–26]. The compared sequences were aligned using MAFFT [28], and the base substitution model was determined as GTR+I+G using MODELTEST version 3.7 [29] under the AIC model selection criterion. The Akaike-supported model, log likelihood (lnL), Akaike information criterion values, proportion of invariable sites, gamma distribution shape parameters, and substitution rates were used in the analyses. Bayesian analysis was performed using MrBayes 3.2 [30]; four chains were run for 4×10^6 generations. Markov chains were sampled at intervals of 100 generations [31]. Two independent runs were performed, and after confirming the convergence of runs and discarding the first 2×10^6 generations as 'burn in', the remaining topologies were used to generate a 50% majority-rule consensus tree. The PhyML 3.0 online version [32] was employed for the ML analysis. The analysis parameters obtained from the model selection procedure were adopted for the analysis, otherwise the default settings were used. The unweighted MP analysis was performed using PHYLIP 3.69 [33] with default settings. The tree topologies obtained from ML and MP analyses were evaluated with 1000 bootstrap pseudoreplications. The

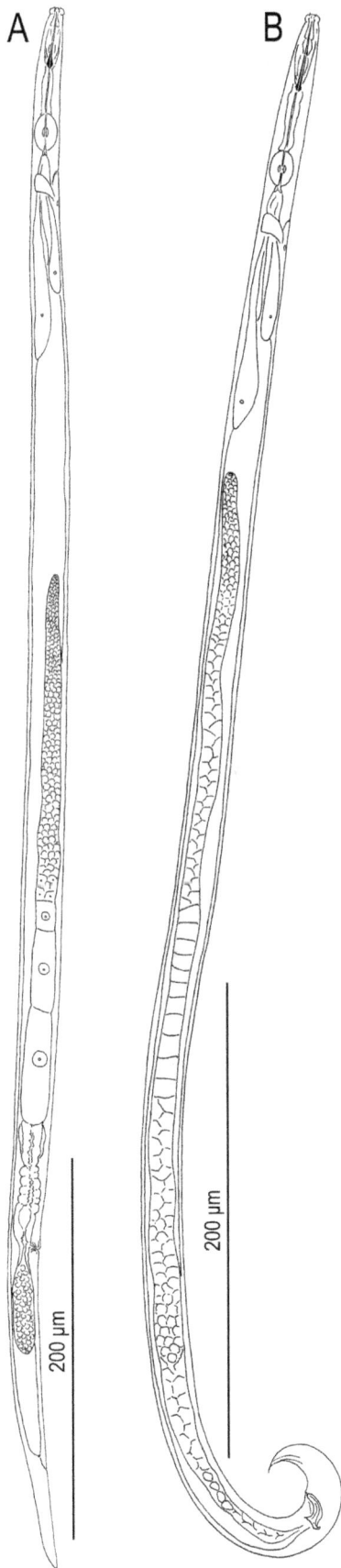

Figure 2. *Bursaphelenchus sycophilus* **n. sp.** A: Adult female; B: Adult male.

results obtained from Bayesian, ML and MP analyses were then compared to evaluate the phylogenetic position of the new species.

Culturing attempt

Attempts were made to culture the nematode using the grey mold *Botrytis cinerea* Pers., a standard feeding resource fungus for mycophagous nematodes, and alfalfa callus (*Medicago sativa* L.), a standard feeding resource for plant-parasitic nematodes. The nematodes were extracted from the additionally-collected samples of *F. variegata* syconia, washed several times with sterilized distilled water, and transferred to fungal lawns on 2.0% malt extract agar (Difco malt extract: 2.0%; Agarose 2.0%) or alfalfa callus donated by Dr. T. Mizukubo (NARO Agricultural Research Center).

The transferred nematodes were kept at 23°C for a month, and were examined under a dissecting microscope every 5–10 days. Culturing attempts were replicated five times for both *B. cinerea* and alfalfa callus.

Nomenclatural acts

The electronic edition of this article conforms to the requirements of the amended International Code of Zoological Nomenclature, and hence the new names contained herein are available under that Code from the electronic edition of this article. This published work and the nomenclatural acts it contains have been registered in ZooBank, the online registration system for the ICZN. The ZooBank LSIDs (Life Science Identifiers) can be resolved and the associated information viewed through any standard web browser by appending the LSID to the prefix "http://zoobank.org/". The LSID for this publication is: urn:lsid:zoobank.org:pub:325625B8-D150-4836-B1A8-FCE3-C8AC311C. The electronic edition of this work was published in a journal with an ISSN, and has been archived and is available from the following digital repositories: PubMed Central, LOCKSS.

Results

Bursaphelenchus sycophilus Kanzaki, Tanaka, Giblin-Davis & Davies n. sp. urn:lsid:zoobank.org:act:6109FB02-E6FA-4570-8959-F918341319A9

Type materials. The holotype male, nine paratype males and 10 paratype females were deposited in the United States Department of Agriculture Nematode Collection (USDANC), Beltsville, Maryland, USA, and 10 paratype males and 10 paratype females were deposited in the Forest Pathology Laboratory Collection, FFPRI, Tsukuba, Japan.

Description

Morphometric values are summarized in Table 1. Morphological illustrations and photographs are shown in Figs. 2–4.

Adults. Intermediate in body size, 738–964 μm and 666–933 μm in length for males and females, respectively. Body cylindrical, slender, weakly ventrally arcuate when killed by heat treatment. Male tail strongly recurved ventrally. Cuticle thin, finely annulated with a lateral field with four incisures. Lip region distinctly offset from body, separated from other body parts by a clear constriction, sub-rectangular or rounded, ca twice as broad as high in lateral view. A cuticular plate present at the anterior end. The edge of plate a little off-set, and appears like two cuticular projections in lateral view. Stylet very well-developed, separated into two parts: a conus occupying approximately 50% of the total stylet length and a shaft with a conspicuous and large basal swelling, but not forming a clear basal knob. Procorpus cylindrical, ca 1.5 metacorpal lengths (= ca 1 stylet length) long,

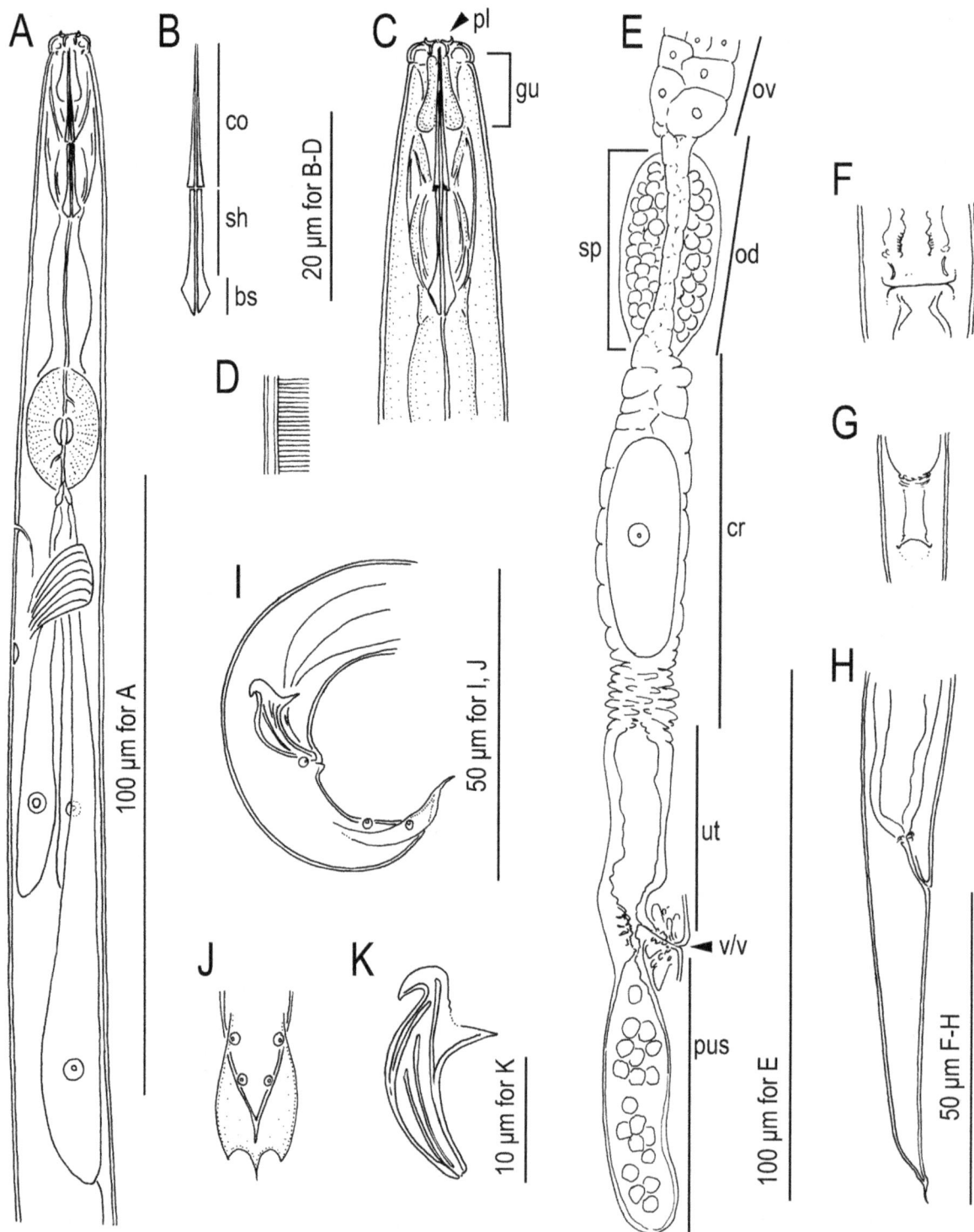

Figure 3. *Bursaphelenchus sycophilus* **n. sp.** A: Anterior region in left lateral view; B: Close-up of stylet (co: conus; sh: shaft; bs: basal swelling); C: Close-up of head region (pl: lip plate; gu: stylet guiding); D: Body surface pattern; E: Female reproductive system in right lateral view (ov: ovary; od: oviduct; sp: spermatheca or *receptaculum seminis*; cr: crustaformeria; ut: uterus; v/v: vagina and vulva; pus: post-uterine sac); F: Female vulval region in ventral view; G: Female anus and rectum in ventral view; H: Female tail in right lateral view; I: Male tail in right lateral view; J: Male tail tip (bursal flap) in ventral view; K: Male spicule in right lateral view.

connected to a well-developed muscular oval-shaped metacorpus (median bulb) occupying ca 90% of the body diameter. Dorsal pharyngeal gland orifice opening into the lumen of the metacorpus midway between the anterior end of the metacorpal valve and the anterior end of the metacorpus. Metacorpal valve conspicuous, ca 1/5 of the metacorpal length, located at the centre or a little posterior to the centre of the metacorpus. Pharyngo-intestinal junction immediately posterior to the posterior end of the

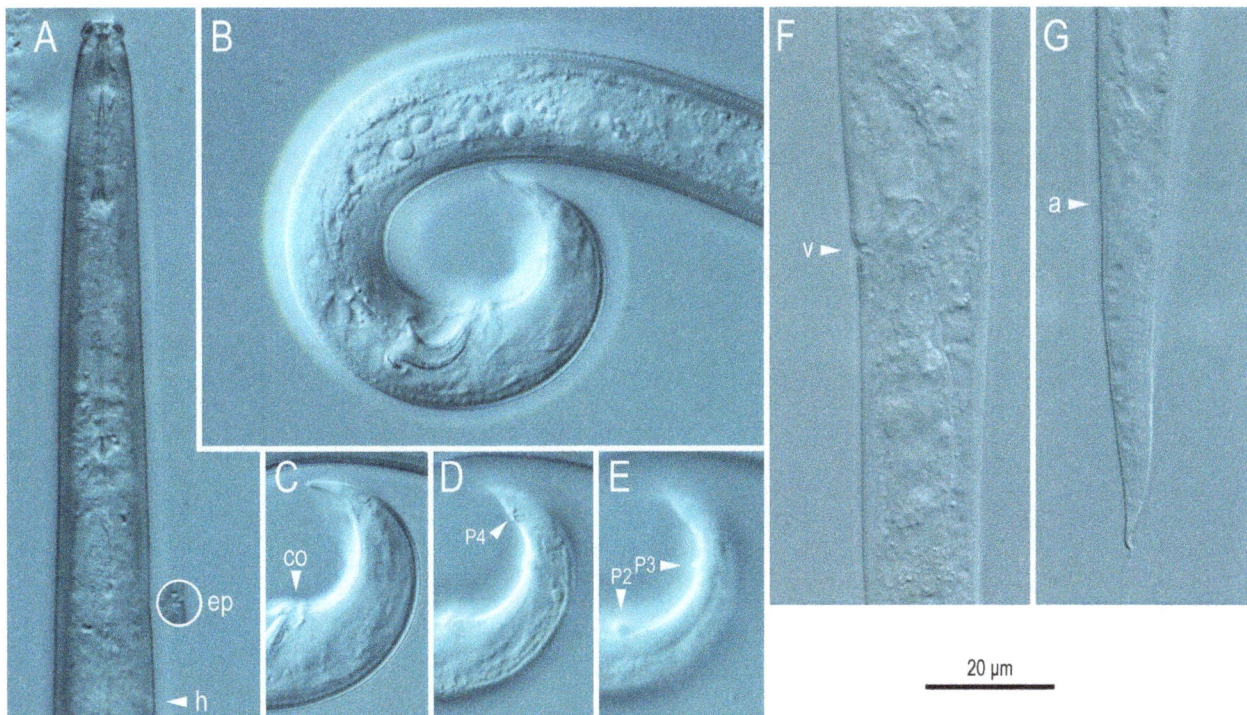

Figure 4. *Bursaphelenchus sycophilus* **n. sp.** A: Anterior region in right lateral view (ep: excretory pore encircled at the corresponding level of the body; h: hemizonid); B: Right lateral view of male tail; C–E: Right lateral view of male tail in different focal plane (co: cloacal opening; P2–P4: genital papillae); F: left lateral view of female vulval region (v: vulva); G: Left lateral view of female tail (a: anus).

metacorpus. Pharyngeal glands well-developed. Dorsal pharyngeal gland ca 5–6 metacorpal lengths long, overlapping the intestine dorsally, ca 50% of the corresponding body diameter at the broadest part. Nerve ring surrounding pharyngeal glands and intestine ca 1 metacorpal length posterior to pharyngo-intestinal junction. Excretory pore visible, located near level of the nerve ring, i.e., varying between the posterior end of the metacorpus to ca 2 metacorpal lengths posterior to the metacorpus. Hemizonid ca 2 metacorpal lengths posterior to the metacorpus.

Male. Gonad single, outstretched in most individuals. Posterior 1/5 of gonad forms *vas deferens*, containing several well-developed sperm. Posterior end of *vas deferens* and intestine fused to form a narrow cloacal tube around the spicule. Sperm amoeboid, spermatocytes arranged in multiple (3–5) rows for anterior 1/5 of testis length, two rows for next 1/5, single row for middle part, with the well-developed sperm packed as 2–5 rows in the posterior part of testis. Tail region strongly arcuate ventrally, terminus claw-like in lateral view. Spicules paired, separate, mitten-shaped, stout, i.e., the length (chord from anterior end of condylus to cucullus) is ca 3 times the length of the widest part of the calomus–lamina complex: condylus short, rounded with anterior tip strongly recurved dorsally. Rostrum triangular, with pointed tip. Capitulum with clear depression immediately anterior to the anterior base of the rostrum. Lamina with two clear lines connecting the dorsal root of the condylus to the blunt tip, smoothly ventrally arcuate. Calomus–lamina complex widest at the posterior end of rostrum and calomus smoothly tapered along with lamina to distal tip. A cuticular limb present between calomus and lamina, extending from the middle of calomus to near the root of condylus. Connection between rostrum and calomus indistinctive, i.e., rostrum and calomus are connected with smooth curvature. Cucullus absent. Six (three pairs) genital papillae present: all are papilla-shaped, i.e., not gland-like (glandular papillae). First pair (P2) subventrally located, adcloacal, *i.e.*, at level of cloacal opening (CO). Second pair (P3) subventral ca 1/2 tail length posterior to CO. Third pair (P4) located ventrally ca 1/2 cloacal body diameter posterior to P3. Bursal flap present, covers the distal part from the level of P3, having oval shape with three projections at the posterior end. The middle projection longer and narrower than the others, appears as hair-like extension in the lateral view.

Female. Reproductive tract composed of ovary, oviduct, spermatheca, crustaformeria, uterus, vagina + vulva and post-uterine sac (branch). Ovary single, anteriorly outstretched, anterior end reflexed once in some individuals. Ovary constructed of flat, plate-like cells. Oocytes present in multiple (2–5) rows in anterior 1/2–4/5 of ovary with a couple of well-developed oocytes in a single row at the posterior end. Oviduct tube-like, constructed of large oval-shaped cells, connecting ovary and crustaformeria, sometimes occupied by well-developed oocytes. Spermatheca (*receptaculum seminis*) constructed of rounded cells, present as a branched overlapping of oviduct, i.e., branching out from anterior end of crustaformeria, slightly irregular oval shape, sometimes filled with well-developed sperm. Crustaformeria not conspicuous, formed of rather large, rounded cells. Uterus short with thick wall, sometimes containing a developing egg and several sperm. A sac-like expansion present on both sides of the uterus, which could be a part of the uterus. Dorsal uterine wall thickened at the uterus/vagina/post-uterine sac junction and a three-celled structure, where each cell has a cuticular (appears like fractal dots in LM observation) pronged structure, present at both right and left sides of the wall, but the structure is rather vague in fixed and mounted materials. Vagina slightly inclined anteriorly. Vulval opening lacking flap apparatus, both anterior and posterior vulval lips slightly expanded, and forming a dome-shaped slit in ventral view.

Post-uterine sac conspicuous, filled with well-developed sperm in many individuals. Rectum present, seemingly functional, ca 1 anal body diameter (ABD) long, intestine–rectum junction constricted by sphincter muscle. Anus a small dome-shaped slit in ventral view; posterior anal lip slightly expanded in lateral view. Tail smoothly tapering to distal part, ca 3–6 ABD long. Tail tip region short, conical, with a hair-like projection.

Diagnosis. Besides the generic characters, *B. sycophilus* n. sp. is characterized by its unusually well-developed stylet, i.e., thick, long and possessing well-developed conus occupying ca 50% of total length and clearly developed basal swelling which forms a diamond-shape in lateral view, well-developed pharyngeal glands, three pairs of papilla-form male genital papillae and male spicule with strongly dorsally recurved condylus. The biological characters of inhabiting a fig syconium of *F. variegata* and being a plant parasite could also be considered diagnostic.

Relationship. Based on the male spicule morphology, i.e., the initial typological character of the genus [34], *B. sycophilus* n. sp. is similar to members of the "*B. eremus* group" (*B. eremus* (Rühm), *B. scolyti* Massey, *B. yongensis* Gu, Braasch, Burgermeister, Brandstetter & Zhang, *B. clavicauda* Kanzaki, Maehara & Masuya, *B. uncispicularis* Zhuo, Li, Li, Yu & Liao) and the "*B. leoni* group" (*B. eidmanni* (Rühm), *B. leoni* Baujard, *B. silvestris* (Lieutier & Laumond) and *B. borealis* Korentchenko) *sensu* Braasch et al. [34] and *B. maxbassiensis* (Massey). The new species and these species share the strongly dorsally arcuate condylus of the male spicule [35–44]. Within these 10 species, *B. maxbassiensis* is most similar to the new species, i.e., these two species share the extremely well-developed stylet [38,41,42]. However, the new species is distinguished from *B. maxbassiensis* by its lip morphology, with cuticular plate *vs* umbrella-like horizontal expansion, pharyngeal glands well-developed and extended vs. relatively short, position of excretory pore, posterior vs. anterior to median bulb, male bursal flap possessing three projections vs. rounded distal end, male spicule, the dorsal curvature in condylus is stronger in *B. sycophilus* n. sp., female post-uterine-branch occupying ca half of vulva-anus distance vs. occupying 2/3 or more of vulva-anus distance, female vulval structure, without any flap apparatus vs. slightly elongated anterior lip forming short flap and female tail, smoothly tapered with short conical distal end vs. smoothly tapered to distal end [38,41,42].

Molecular phylogeny. Because the tree topology and phylogenetic status of *B. sycophilus* n. sp. were consistent among analyses, only Bayesian trees are shown (Figs. 5, 6). In contrast to morphological similarity, *B. sycophilus* n. sp. phylogenetically belongs to clade II of the genus, and is close to *B. willibaldi* Schönfeld, Braasch & Burgermeister, *B. braaschae* Gu & Wang, *B. tadamiensis* Kanzaki, Taki, Masuya & Okabe, *B. kiyoharai* Kanzaki, Maehara, Aikawa, Masuya & Giblin-Davis, *B. thailandae* Braasch & Braasch-Bidasak and *B. parathailandae* Gu, Wang & Chen (Figs. 5, 6). These five species belong to the "*B. fungivorus* group" sensu Braasch et al. [34], and the group is characterized by the male spicule morphology possessing clear dorsal and ventral limbs Braasch et al. [34]. However, the new species is readily distinguished from members of the "*B. fungivorus* group" species based on the well-developed stylet and pharyngeal glands and male spicule morphology possessing a dorsally recurved condylus vs. a small condylus without dorsal curvature [34,45–50].

Type host and locality. The type materials were obtained on June 6, 2013 from a syconium from a *Ficus variegata* tree planted on the Ishigaki Island, Okinawa, Japan (GPS: 24.41′12″N, 124.18′47″E, 59 m a.s.l).

Biological characters. The culturing attempts using *B. cinerea* and alfalfa callus were not successful. *Bursaphelenchus sycophilus* n. sp. showed no feeding behaviour on *B. cinerea* hyphae or on alfalfa callus parenchymal cells, and died out within one month of inoculation. Thus, the feeding preference (host range) of the nematode appears to be narrow, possibly specific to fig syconium tissue. Because a host-specific fig wasp species, *Ceratosolen appendiculatus* (Mayr), was the only insect found with the syconium from which type materials were obtained, the wasp is hypothesized to be the carrier (or host) insect of the nematode.

Etymology. The species was named after its characteristic habitat, the fig syconia.

Discussion

The morphological characters and molecular phylogenetic status of *B. sycophilus* n. sp. are contradictory to each other, i.e., the new species shares a characteristic recurved condylus with members of the "*B. eremus* group", "*B. leoni* group" and *B. maxbassiensis*, but is molecular phylogenetically close to members of the "*B. fungivorus* group" (Figs. 5, 6). Therefore, the recurved condylus found in the "*B. eremus* group", the "*B. leoni* group" and the new species is considered as an analogous or convergent morphological character, i.e., the condylus morphology of the new species is a species-specific apomorphy.

Comparing the spicule and male tail morphology among members of the "*B. eremus* group", the "*B. fungivorus* group" and *B. sycophilus* n. sp., three characters appear shared between the "*B. fungivorus* group" and *B. sycophilus* n. sp., i.e., the lack of a ventral single papilla anterior to cloacal opening (P1), lack of a cucullus on the spicule and possession of a ventral limb of the spicule (Fig. 7). The secondary loss of the P1 papilla is considered as an apomorphic character of some members of the "*B. fungivorus* group" [48,49], and the presence of the ventral limb of the spicules is considered a shared character with members of the "*B. fungivorus* group" (Fig. 7). In addition to male tail morphology, *B. sycophilus* n. sp. and its close relatives are distinguished from the *B. eremus* group by the lack of a flap apparatus on the female vulva vs. possessing a short vulval flap (referred to as a side flap [51]) and the number of lateral lines being four vs. three [34,52]. Thus, the presence of a ventral limb in the male spicule, papilla-shaped P4 papillae of males, lack of a ventral P1 papilla, and the lack of a female vulval flap are the synapomorphic characters shared by *B. sycophilus* n. sp. and its close relatives, i.e., the characters are congruent with the inferred molecular phylogenetic relationships.

The lip and stylet morphology of the new species is very unusual within the genus, and is similar to *B. maxbassiensis*, which was originally described as *Omemeea maxbassiensis* Massey because of its unusual lip and stylet morphology [41]. The large basal knobs are often found in the plant-parasitic tylenchids [53], and the extremely well-developed basal swellings of the stylet of the new species may be the result of convergent evolution of traits adaptive for plant parasitism in a fig sycone. Because the molecular phylogenetic status of *B. maxbassiensis* has not been established, we cannot determine whether their similarity is due to sharing a recent common ancestor or not. *Bursaphelenchus maxbassiensis* was isolated from the galleries of a bark beetle species, *Hylesinus californicus* (Swaine), infesting green ash, *Fraxinus pennsylvanica* Marsh. from North Dakota [41]. *Hylesinus californicus* sometimes bores into living trees [41], i.e., providing the nematode species with opportunities to encounter live plant tissue. Thus *B. maxbassiensis* may also have the ability to feed on live tree tissue. Re-isolation of *B. maxbassiensis* and detailed molecular and morphological comparison are necessary to examine the origin of plant parasitism of *B. sycophilus* n. sp.

The inferred molecular phylogenetic relationships and morphological characters are not congruent to each other in the

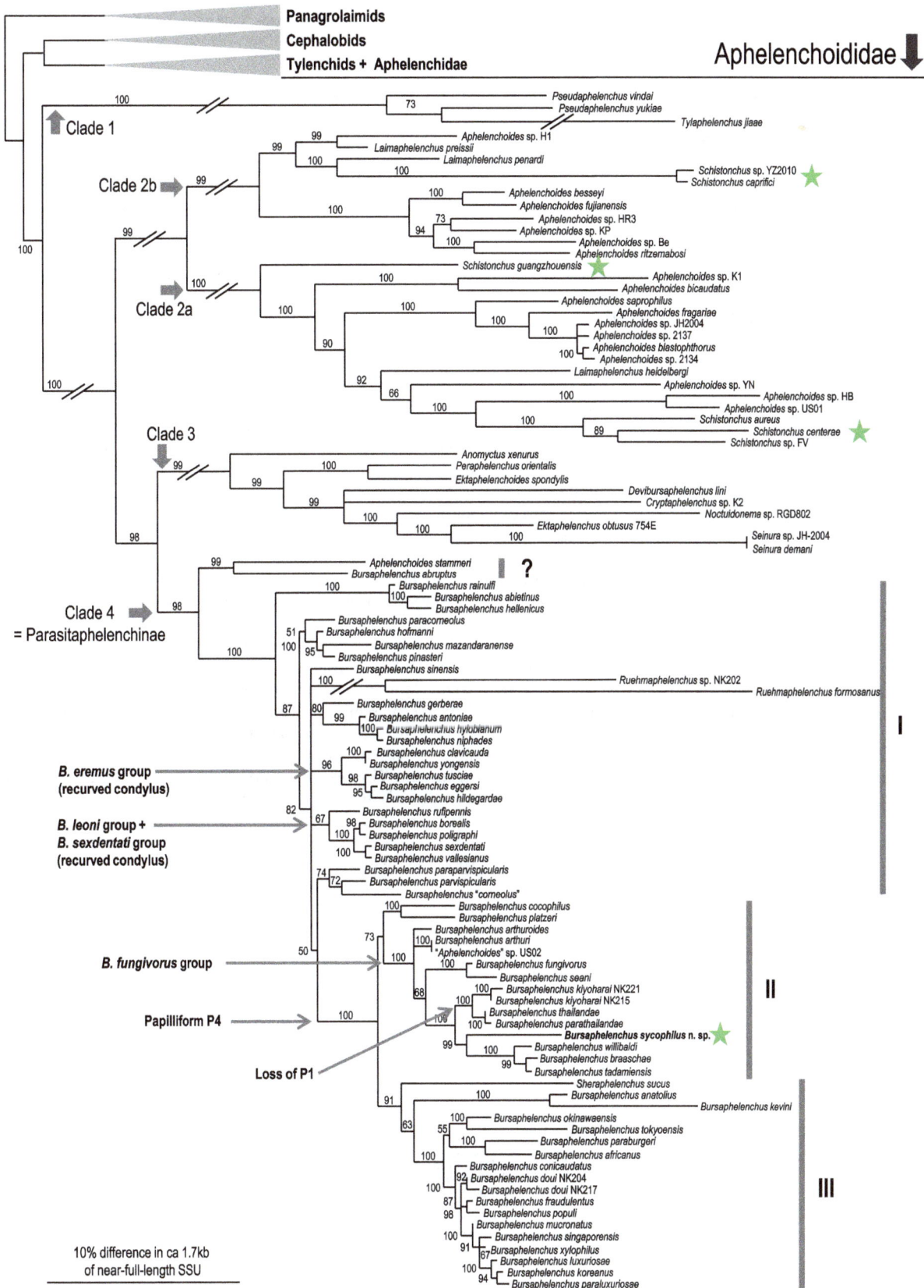

Figure 5. Molecular phylogenetic relationship among aphelenchid nematodes. The 10001st Bayesian tree inferred from near-full-length SSU under GTR+I+G model (lnL = 34563.4258; freqA = 0.2468; freqC = 0.1869; freqG = 0.2517; freqT = 0.3145; R(a) = 1.203; R(b) = 2.9772; R(c) = 1.1044; R(d) = 0.8365; R(e) = 4.1891; R(f) = 1; Pinva = 0.1959; Shape = 0.582). Posterior probability values exceeding 50% are given on appropriate clades. The

blanch lengths for subfamily Parasitaphelenchinae which includes new species were expanded to show the topology clearly. The phylogenetic groups within the family Aphelenchoididae and within the subfamily Parasitaphelenchinae following Kanzaki et al. (2013) were indicated with thick arrows and bars, respectively. Parsimonious explanation on male genital papillae characters were indicated with thin arrows. The biological character, fig-association was indicated by stars.

subfamily Parasitaphelenchinae, which contains *Bursaphelenchus*, *Parasitaphelenchus* Fuchs, *Sheraphelenchus* Nickle and *Ruehmaphelenchus* Goodey [15]. Because of the morphological and biological diversity of *Bursaphelenchus*, the other three genera are included in the genus as subclades [24,25,54]. Therefore, a taxonomic revision is needed where all genera should be lumped together using their only common character, arrangement of the male genital papillae [15], or alternatively they should be separated into many small genera. If the new species and *B. maxbassiensis* shared a common ancestor, then the resurrection of the genus *Omemeea* Massey may be justified, if the genus (subfamily) is split into many small genera.

In the present study, culturing attempts using *B. cinerea* and alfalfa callus as food were not successful. The feeding resource preferences of several different nematodes, including *Bursaphelenchus* spp., have been previously studied, and in many cases, several different species of plant callus and ascomycete fungi, including *B. cinerea* were considered to be suitable food for aphelenchid nematodes [38,55,56]. Thus, although more replications using multiple species of fungi may be necessary, *B. sycophylus* n. sp. does not appear to feed on fungus, and the species is hypothesized to be an obligate plant parasite.

The obligate plant parasite, *B. cocophilus* can be cultured using fresh palm tissue [14,57–59], although more detailed culturing attempts, e.g., using several different plant callus in including alfalfa, has not been conducted. A similar methodology, e.g., fig callus and/or fresh tissue of *Ficus variegata* may be available for the culture of *B. sycophilus* n. sp.

Several obligate and facultative plant parasites (pathogens) are known in the genus *Bursaphelenchus*, and they are phylogenetically distant from each other (Fig. 7), e.g., *B. xylophilus* (Steiner & Buhrer), the pathogen of pine wilt disease [60], *B. cocophilus* Cobb, the pathogen of red ring disease [57–59] and *B. sexdentati* Rühm, which has moderate to strong pathogenicity to pine trees [61]. This pattern of multiple lineages of plant parasitism may suggest the physiological plasticity of *Bursaphelenchus* nematodes in their feeding abilities, e.g., digestive enzymes. However, the morphology of the ingestive/digestive organs in these plant-parasites, including the obligate plant-parasite, *B. cocophilus* [62], are basically identical to that of other mycophagous *Bursaphelenchus* species [63–67]. The highly derived morphology of *B. sycophilus* n. sp. may

represent the potential for morphological and developmental plasticity in the genus *Bursaphelenchus*.

Although *B. sycophilus* n. sp. clearly belongs to the genus *Bursaphelenchus*, its biological and morphological characters are similar to those of *Schistonchus* spp. The genus *Schistonchus* Cobb is known as to parasitize fig syconia, and are phoretically/parasitically associated with fig wasps [68]. Morphologically, the stylet of *Schistonchus* spp. is very similar to that of *B. sycophilus*, i.e., it has a long conus and distinct basal swellings. Interestingly, the genus is clearly paraphyletic [69–71] (Fig. 5). The similar morphology and life cycle of these fig-associated nematodes appears to have emerged from fungal feeding aphelenchoidid nematodes at least four times, inclusive of *B. sycophilus* n. sp. (Fig. 5).

Some biological characters of *B. sycophilus* n. sp. have not been clarified so far, e.g., insect interaction (parasitic or phoretic, and which developmental stage of nematode is carried by insect), detailed host range and distribution range. However, because the genotypes of *B. sycophilus* n. sp. isolated from several different locations on the Ishigaki and Iriomote Islands were identical, it is considered to be commonly distributed on these two islands. The distribution of *F. variegata* is widespread, from Ishigaki Island, Japan through South Eastern Asia to Northern Australia. However, regardless of multiple surveys of *F. variegata* in Northern Australia, *B. sycophilus* n. sp. has not been isolated from the area [69]. Therefore, the species is rare or absent from the region. To determine the distribution range of *B. sycophilus*, more surveys in South Eastern Asia and Northern Australia are needed.

In the present study, three species of figs were examined for their nematode association. The new species was isolated only from *F. variegata*, and was not found from the other two species, *F. septica* and *F. bengtensis*. Thus, *F. variegata* is considered as the specific fig host of *B. sycophilus* n. sp.

Center et al. [72] examined parasitized syconium tissue and suggested that each *Schistonchus* species has potential tissue specificity, which could lead to partitioning of the microhabitat inside the syconium. In the present study, detailed histological analysis was not conducted. Because *B. sycophilus* n. sp. often shares the same syconium with a currently undescribed *Schistonchus* species (Table S1), similar niche partitioning may be present contingent upon competition and other evolutionary pressure.

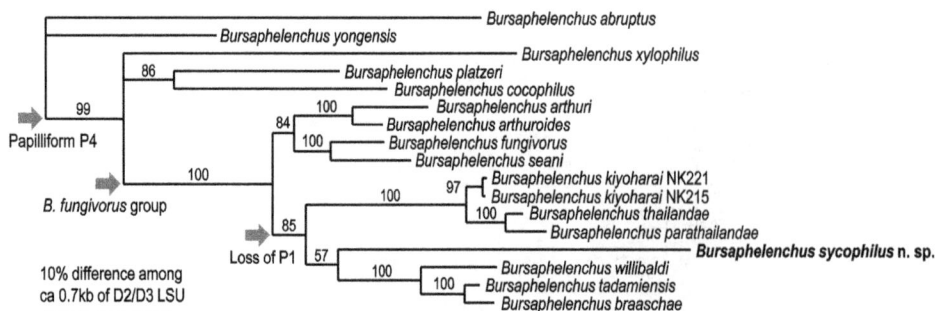

Figure 6. Molecular phylogenetic relationship among *Bursaphelenchus* nematodes belonging to '*B. fungivorus* group'. The 10001st Bayesian tree inferred from D2/D3 LSU under GTR+I+G model (lnL = 4855.5322; freqA = 0.1828; freqC = 0.1894; freqG = 0.3464; freqT = 0.2813; R(a) = 0.7168; R(b) = 2.1523; R(c) = 0.925; R(d) = 0.2398; R(e) = 4.6466; R(f) = 1; Pinvar = 0.322; Shape =). Posterior probability values exceeding 50% are given on appropriate clades.

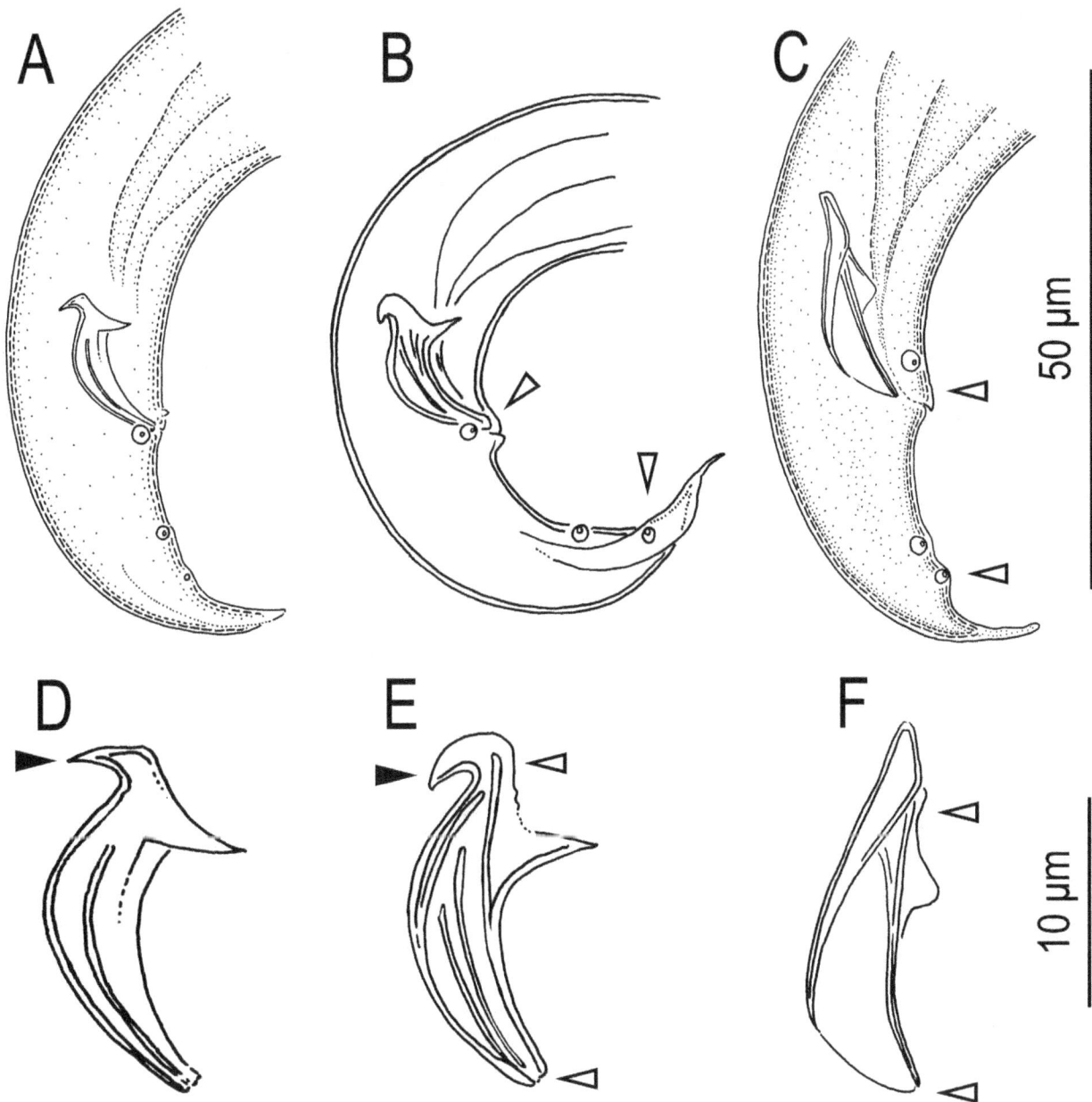

Figure 7. Comparison of male tail character among *Bursaphelenchus sycophilus* **n. sp. '***B. eremus* **group' and '***B. fungivorus* **group'.** A: Male tail of *B. clavicauda* (*B. fungivorus* group) in right lateral view; B: Male tail of *B. sycophilus* n. sp. in right lateral view; C: Male tail of *B. tadamiensis* ("*B. fungivorus* group") in right lateral view; D: Spicule of *B. clavicauda* in right lateral view; E: Spicule of *B. sycophilus* n. sp. in right lateral view; F: Spicule of *B. tadamiensis* in right lateral view. Convergent character (condylus) is suggested in black arrowhead and homologous characters (lack of P1 papilla, papilla-formed P4 papillae, presence of ventral limb and lack of cucullus) are suggested with white arrowhead.

Detailed analyses are necessary to clarify the biological interaction between *B. sycophilus* n. sp. and fig tissues, and *B. sycophilus* n. sp. and other fig-associated nematodes.

Because fig syconia represent an exclusive habitat relative to nematode immigration, i.e., only fig wasps and parasitic wasps enter the young figs, the history of multiple aphelenchoidid lineage introductions into different *Ficus* lineages is unclear. The simplest explanation might be through fig wasp host switching and lineage sorting for recent mixing. However, the convergence in feeding morphology between *B. sycophilus* n. sp. and the two or three different paraphyletic *Schistonchus* lineages suggest other possible scenarios, such as rogue introductions of fungal feeding nematodes

by other fig-associating insects or other unknown aspects concerning fig biology. For example, the new species is close to the species associated with ambrosia beetles (*B. kiyoharai*) [48] and stag beetles (*B. tadamiensis*) [49], and some ambrosia beetles invade the petiole of living trees in tropical region [73,74]. Given the relatively short branch length (genetic distance) between *B. sycophilus* n. sp. and its related *Bursaphelenchus* spp., the new species seems to have adapted to this closed environment relatively rapidly. More detailed analyses, e.g., comparative genomic analyses using the new species and its close relatives, may yield interesting information concerning the genes involved in switching feeding habitats and the development of plant parasitism.

Supporting Information

Table S1 GPS coordinates of sampling sites and nematode species (morphotypes and/or genotypes) isolated from the materials.

Table S2 Accession numbers and species names of nematodes used in the phylogenetic analysis.

Acknowledgments

We thank Dr. Takayuki Mizukubo, National Agriculture and Food Research Organization, Agricultural Research Center, Tsukuba Japan for providing alfalfa callus materials, and Ms. Ami Akasaka and Ms. Noriko Shimoda, FFPRI for preparing materials and assistance in molecular sequencing.

Author Contributions

Conceived and designed the experiments: NK. Performed the experiments: NK RT RGD KAD. Analyzed the data: NK. Contributed reagents/materials/analysis tools: NK. Wrote the paper: NK RGD KAD.

References

1. Hembry DH, Kawakita A, Gurr NE, Schmaedick MA, Baldwin BG, et al. (2013) Non-congruent colonizations and diversification in a coevolving pollination mutualism on oceanic islands. Proc R Soc B 280: 20130361.

2. Cruaud A, Rønsted N, Chanterasuwan B, Chou LS, Clement WL, et al. (2012) An extreme case of plant-insect co-diversification: figs and fig-pollinating wasps. Syst Biol 61: 1029–1047.

3. Machado CA, Robbins N, Gilbert TP, Herre EA (2005) Critical review of host specificity and its coevolutionary implications in the fig/fig-wasp mutualism. Proc Nat Acad Sci USA 102: 6558–6565.

4. Molbo D, Machado CA, Sevenster JG, Keller L, Herre EA (2003) Cryptic species of fig-pollinating wasps: implications for the evolution of the fig-wasp mutualism, sex allocation, and precision of adaptation. Proc Natl Acad Sci USA 100: 5867–5872.

5. Murray EA, Carmichael AE, Heraty JM (2013) Ancient host shifts followed by host conservatism in a group of ant parasitoids. Proc R Soc B 280: 20130495.

6. Rønsted NAH, Weiblen GD, Cook JM, Salamin N, Machado CA, et al. (2005) 60 million years of co-divergence in the fig-wasp symbiosis. Proc R Soc B 272: 2593–2599.

7. Wiebes JT (1979) Co-evolution of figs and their insect pollinators. Annu Rev Entomol 10: 1–12.

8. Weiblen GD (2002) How to be a fig wasp. Annu Rev Entomol 47: 299–330

9. Giblin-Davis RM, Center BJ, Nadel H, Frank JH, Ramirez W(1995) Nematodes associated with fig wasps, Pegoscapus spp. (Agaonidae), and syconia of native Floridian figs (Ficus spp.). J Nematol 27: 1–14.

10. Herre EA (1993) Population structure and the evolution of virulence in nematode parasites of fig wasps. Science 259: 1442–1445.

11. Jauharlina J, Lindquist EE, Quinnell RJ, Robertson HG, Compton SG (2012) Fig wasps as vectors of mites and nematodes. Afr Entomol 20: 101–110.

12. Poinar GO (1979) Parasitodiplogaster sycophilon gen. n. sp. n. (Diplogastridae, Nematoda), a parasite of Elisabethiella stuckenbergi Grandi (Agaonidae, Hymenoptera) in Rhodesia. Proc Koninklijke Nederlandse Akad Wetenschappen C 82: 375–381.

13. Walter DE (2000) First record of a fig mite from the Australian Region: Paratarsonemella giblindavisi sp. n. (Acari: Tarsonemidae). Austral J Entomol 39: 229–232.

14. Giblin-Davis RM (1993) Interactions of nematodes with insects. In: Khan MW ed. Nematode interactions. London: Chapman and Hall. pp 302–344.

15. Kanzaki N, Giblin-Davis RM (2012) Aphelenchoidea. In: Manzanilla-López RH, Marbán-Mendoza N eds. Practical plant nematology. Jalisco: Colegio de Postgraduados and Mundi-Prensa, Biblioteca Básica de Agricultura. pp 161–208.

16. Yoder M, De Ley IT, King IW, Mundo-Ocampo M, Mann J, et al. (2006) DESS: A versatile solution for preserving morphology and extractable DNA of nematodes. Nematology 8: 367–376.

17. Minagawa N, Mizukubo T (1994) A simplified procedure of transferring nematodes to glycerol for permanent mounts. Jpn J Nematol 24: 75.

18. Hooper DJ (1986) Handling, fixing, staining and mounting nematodes. In: Southey, JF ed. Laboratory methods for work with plant and soil nematodes. London: Her Majesty's Stationery Office. pp 59–80.

19. Kanzaki N (2013) Simple methods for morphological observation of nematodes. Nematol Res 43: 9–13.

20. Kikuchi T, Aikawa T, Oeda Y, Karim N, Kanzaki N (2009) A rapid and precise diagnostic method for detecting the pinewood nematode Bursaphelenchus xylophilus by loop-mediated isothermal amplification (LAMP). Phytopathology 99: 1365–1369.

21. Tanaka R, Kikuchi T, Aikawa T, Kanzaki N (2012) Simple and quick methods for nematode DNA preparation. Appl Entomol Zool 47: 291–294.

22. Kanzaki N, Futai K (2002) A PCR primer set for determination of phylogenetic relationships of Bursaphelenchus species within xylophilus group. Nematology 4: 35–41.

23. Ye W, Giblin-Davis RM, Braasch H, Morris K, Thomas WK (2007) Phylogenetic relationships among Bursaphelenchus species (Nematoda: Parasitaphelenchidae) inferred from nuclear ribosomal and mitochondrial DNA sequence data. Mol Phylogenet Evol 43: 1185–1197.

24. Kanzaki N, Tanaka R (2013) Sheraphelenchus sucus n. sp. (Tylenchina: Aphelenchoididae) isolated from sap flow of Quercus serrata in Japan. Nematology 15: 975–990.

25. Kanzaki N, Taki H, Masuya H, Okabe K, Chen C-Y (2013) Description of Ruehmaphelenchus formosanus n. sp. (Tylenchina: Aphelenchoididae) isolated from Euwallacea fornicates from Taiwan. Nematology 15: 895–906.

26. Kanzaki N, Tanaka R, Ikeda H, Taki H, Sugiura S, et al. (2013) Phylogenetic status of and insect parasitism in the subfamily Entaphelenchinae Nickle with description of Peraphelenchus orientalis n. sp. (Tylenchomorpha: Aphelenchoididae). J Parasitol 99: 639–649.

27. Zhao Z, Ye W, Giblin-Davis RM, Li D, Thomas WK, et al. (2008) Morphological and molecular analysis of six aphelenchoidoids from Australian conifers and their relationship to Bursaphelenchus (Fuchs, 1937). Nematology 10: 663–678.

28. Katoh K, Misawa K, Kuma K, Miyata T (2002) MAFFT: a novel method for rapid multiple sequence alignment based on fast Fourier transform. Nucleic Acids Res 30: 3059–3066.

29. Posada D, Crandall KA (1998) Modeltest: testing the model of DNA substitution. Bioinformatics 14: 817–818.

30. Huelsenbeck JP, Ronquist F (2001) MR BAYES: Bayesian inference of phylogenetic trees. Bioinformatics 17: 1754–755.

31. Larget B, Simon DL (1999) Markov chain Monte Carlo algorithms for the Bayesian analysis of phylogenetic trees. Mol Biol Evol 16: 750–759.

32. Guindon S, Dufayard JF, Lefort V, Anisimova M, Hordijk W, et al. (2010) New algorithms and methods to estimate maximum-likelihood phylogenies: assessing the performance of PhyML 3.0. Syst Biol 59: 307–321.

33. Felsenstein J (2005) PHYLIP (Phylogeny Inference Package) version 3.6. Distributed by the author. Department of Genome Sciences, University of Washington, Seattle. (http://evolution.genetics.washington.edu/phylip.html)

34. Braasch H, Burgermeister W, Gu J (2009) Revised intrageneric grouping of Bursaphelenchus Fuchs, 1937 (Nematoda: Aphelenchoididae). J Nematode Morphol Syst 12: 65–81.

35. Baujard P (1980) Trois nouvelle espèces de Bursaphelenchus (Nematoda: Tylenchida) et remarques sur le genre. Rev Nématol 3: 167–177.

36. Gu J, Braasch H, Burgermeister W, Brandstetter M, Zhang J (2006) Description of Bursaphelenchus yongensis sp. n. (Nematoda: Parasitaphelenchidae) isolated from Pinus massoniana in China. Russ J Nematol 14: 91–99.

37. Kanzaki N, Giblin-Davis RM, Center BJ (2009) Redescription of four species of North American Bursaphelenchus Fuchs, 1937 (Nematoda: Parasitaphelenchinae) from Massey's type material. Nematology 11: 129–150.

38. Kanzaki N, Maehara N, Masuya H (2007) Bursaphelenchus clavicauda n. sp. (Nematoda: Parasitaphelenchidae) isolated from Cryphalus sp. emerged from a dead Castanopsis cuspidata (Thunb.) Schottky var. sieboldii (Makino) Nakai in Ishigaki Island, Okinawa, Japan. Nematology 9: 759–769.

39. Korenchenko EA (1980) New species of nematodes from the family Aphelenchoididae, parasites of stem pests of the Dahurian Larch. Zoologichesky Zhurnal 59: 1768–1780.

40. Lieutier F, Laumond C (1978) Nématodes parasites et associés à Ips sexdentatus et Ips typographus (Coleoptera, Scolytidae) en région parisienne. Nematologica 24: 184–200.

41. Massey CL (1971) Omemeea maxbassiensis n. gen., n. sp. (Nematoda: Aphelenchoididae) from galleries of the bark beetle Lepersinus californicus Sw. (Coleoptera: Scolytidae) in North Dakota. J Nematol 3: 189–291.

42. Massey CL (1974) Biology and taxonomy of nematode parasites and associates of bark beetle in the United States. Washington DC: US Government Printing Office. 233 p.

43. Rühm W (1956) Die Nematoden der Ipiden. Parasitologische Schriftenreihe 6: 1–437.

44. Zhuo K, Li X, Li D, Yu S, Liao J (2007) Bursaphelenchus uncispicularis n. sp. (Nematoda: Parasitaphelenchidae) from Pinus yunnanensis in China. Nematology 9: 237–242.

45. Braasch H, Braasch-Bidasak R (2002) First record of the genus Bursaphelenchus Fuchs, 1937 in Thailand and description of B. thailandae sp. n. (Nematoda: Parasitaphelenchidae). Nematology 4: 853–863.

46. Gu J, Wang J (2010) Description of Bursaphelenchus braaschae sp. n. (Nematoda: Aphelenchoididae) found in dunnage from Thailand. Russ J Nematol 18: 59–68.

47. Gu J, Wang J, Chen X (2012) *Bursaphelenchus parathailandae* sp. n. (Nematoda: Parasitaphelenchidae) in packaging wood from Taiwan. Russ J Nematol 20: 53–60.

48. Kanzaki N, Maehara N, Aikawa T, Masuya H, Giblin-Davis RM (2011) Description of *Bursaphelenchus kiyoharai* n. sp. (Tylenchina: Aphelenchoididae) with remarks on the taxonomic framework of the Parasitaphelenchinae Rühm, 1956 and Aphelenchoidinae Fuchs, 1937. Nematology 13: 787–804.

49. Kanzaki N, Taki H, Masuya H, Okabe K (2012) *Bursaphelenchus tadamiensis* n. sp. (Nematoda: Aphelenchoididae), isolated from a stag beetle, *Dorcus striatipennis* (Coleoptera: Lucanidae), from Japan. Nematology 14: 223–233.

50. Schönfeld U, Braasch H, Burgermeister H (2006) *Bursaphelenchus* spp. (Nematoda: Parasitaphelenchidae) in wood chips from sawmills in Brandenburg and description of *Bursaphelenchus willibaldi* sp. n. Russ J Nematol 14: 119–126.

51. Giblin-Davis RM, Kanzaki N, Ye W, Center BJ, Thomas WK (2006) Morphology and systematics of *Bursaphelenchus gerberae* n. sp. (Nematoda: Parasitaphelenchidae), a rare associate of the palm weevil, *Rhynchophorus palmarum* in Trinidad. Zootaxa 1189: 39–53.

52. Kanzaki N (2008) Taxonomy and systematics of the nematode genus *Bursaphelenchus* (Nematoda: Parasitaphelenchidae). In: Zhao BG, Futai K, Sutherland J, Takeuchi Y, eds. Pine wilt disease. Tokyo: Springer Japan. pp 44–66.

53. Siddiqi MR (2000) Tylenchida parasites of plants and insects. Wallingford: CABI Publishing. 833p.

54. Kanzaki N (2013) Phylogenetic and taxonomic relationship among the genera belonging to subfamily Parasitaphelenchinae (Aphelenchoididae). Nematol Res 43: 43.

55. Giblin-Davis RM, Kaya HK (1984) Host, temperature and media additive effects on the growth of *Bursaphelenchus seani*. Rev Nématol 7: 13–17.

56. Kanzaki N, Futai K (2002) Life history of *Bursaphelenchus conicaudatus* (Nematoda: Aphelenchoididae) in relation to the yellow-spotted longicorn betle, *Psachothea hilaris* (Coleoptera: Cerambycidae). Nematology 3: 473–479.

57. Giblin-Davis RM, Faleiro JR, Jacas JA, Peña JE, Vidyasagar PSPV (2013) Coleoptera: biology and management of the red palm weevil, *Rhynchophorus ferrugineus*. In: Peña JE, ed. Potential invasive pests of agricultural crop species. Wallingford: CABI Publishing. pp. 1–34.

58. Giblin-Davis RM, Gerber K, Griffith R (1989a) *In vivo* and *in vitro* culture of the red ring nematode, *Rhadinaphelenchus cocophilus*. Nematropica 19:135–142.

59. Griffith R (1987) Red ring disease of coconut palm. Plant Disease 71: 193–196.

60. Futai K (2013) Pine wood nematode, *Bursaphelenchus xylophilus*. Ann Rev Phytopathol 51: 61–83.

61. Skarmoutsos G, Michalopoulos-Skarmoutsos H (2000) Pathogenicity of *Bursaphelenchus sexdentati*, *Bursaphelenchus leoni* and *Bursaphelenchus hellenicus* on European pine seedlings. For Pathol 30: 149–156.

62. Baujard P (1989) Remarques sur les genres des sous-familles Bursaphelenchinae Paramonov, 1964 et Rhadinaphelenchinae Paramonov, 1964 (Nematoda: Aphelenchoididae). Rev Nématol 12: 323–324.

63. Braasch H (2001) *Bursaphelenchus* species in conifers in Europe: distribution and morphological relationships. Bull EPPO 31: 127–142.

64. Giblin-Davis RM, Mundo-Ocampo M, Baldwin JG, Gerber K, Griffith R (1989b) Observations on the morphology of the red ring nematode, *Rhadinaphelenchus cocophilus*. Rev Nématol 12: 285–292.

65. Lange C, Burgermeister W, Metge K, Braasch H (2007) Phylogenetic analysis of isolates of the *Bursaphelenchus sexdentati* group using ribosomal intergenic transcribed spacer DNA sequences. J Nematode Morphol Syst 9: 95–109.

66. Mamiya Y, Kiyohara T (1972) Description of *Bursaphelenchus lignicolus* n. sp. (Nematoda: Aphelenchoididae) from pine wood and histopathology of nematode-infested trees. Nematologica 18: 120–124.

67. Ryss A, Vieira P, Mota MM, Kulinich O (2005) A synopsis of the genus *Bursaphelenchus* Fuchs, 1937 (Aphelenchida: Parasitaphelenchidae) with keys to species. Nematology 7: 393–458.

68. Vovlas N, Inserra RN, Greco N (1992) *Schistonchus caprifici* parasitizing caprifig (*Ficus carica sylvestris*) florets and the relationship with its fig wasp (*Blastophaga psenes*) vector. Nematologica 38: 215–226.

69. Davies KA, Bartholomaeus F, Ye W, Kanzaki N, Giblin-Davis RM (2010) *Schistonchus* (Aphelenchoididae) from Ficus (Moraceae) in Australia, with description of *S. aculeata* sp. n. Nematology 12: 935–958.

70. Zeng Y, Ye W, Huang J, Li C, Giblin-Davis RM (2013) Description of *Schistonchus altissimus* n. sp. (Nematoda: Aphelenchoididae), an associate of *Ficus altissima* in China. Zootaxa 3717: 598–600.

71. Zeng Y, Ye W, Li C, Wang X, Du Z, et al. (2013) Description of *Schistonchus superbus* n. sp. (Nematoda: Aphelenchoididae), an associate of *Ficus superba* in China. Nematology 15: 771–781.

72. Center BJ, Giblin-Davis RM, Herre EA, Chung-Schickler GC (1999) Histological comparisons of parasitism by *Schistonchus* spp. (Nemata: Aphelenchoididae) in neotropical *Ficus* spp. J Nematol 31: 393–406.

73. Andersen HF, Jordal BH, Kambestad M, Kirkendall LR (2011) Improbable but true: the invasive inbreeding ambrosia beetle *Xylosandrus morigerus* has generalist genotypes. Ecol Evol 2: 247–257.

74. Jordal BH, Kirkendall LR (1998) Ecological relationships of a guild of a tropical beetles breeding in *Cecropia* petioles in Costa Rica. J. Tropic Ecol 14: 153–176.

7

Congruence and Diversity of Butterfly-Host Plant Associations at Higher Taxonomic Levels

José R. Ferrer-Paris[1,2,3], **Ada Sánchez-Mercado**[1,2,3]*, **Ángel L. Viloria**[4], **John Donaldson**[1,2]

1 Kirstenbosch Research Centre, South African National Biodiversity Institute, Cape Town, Western Cape, Republic of South Africa, 2 Botany Department, University of Cape Town, Cape Town, Western Cape, Republic of South Africa, 3 Centro de Estudios Botánicos y Agroforestales, Instituto Venezolano de Investigaciones Científicas, Maracaibo, Estado Zulia, Venezuela, 4 Centro de Ecología, Instituto Venezolano de Investigaciones Científicas, Caracas, Distrito Capital, Venezuela

Abstract

We aggregated data on butterfly-host plant associations from existing sources in order to address the following questions: (1) is there a general correlation between host diversity and butterfly species richness?, (2) has the evolution of host plant use followed consistent patterns across butterfly lineages?, (3) what is the common ancestral host plant for all butterfly lineages? The compilation included 44,148 records from 5,152 butterfly species (28.6% of worldwide species of Papilionoidea) and 1,193 genera (66.3%). The overwhelming majority of butterflies use angiosperms as host plants. Fabales is used by most species (1,007 spp.) from all seven butterfly families and most subfamilies, Poales is the second most frequently used order, but is mostly restricted to two species-rich subfamilies: Hesperiinae (56.5% of all Hesperiidae), and Satyrinae (42.6% of all Nymphalidae). We found a significant and strong correlation between host plant diversity and butterfly species richness. A global test for congruence (Parafit test) was sensitive to uncertainty in the butterfly cladogram, and suggests a mixed system with congruent associations between Papilionidae and magnoliids, Hesperiidae and monocots, and the remaining subfamilies with the eudicots (fabids and malvids), but also numerous random associations. The congruent associations are also recovered as the most probable ancestral states in each node using maximum likelihood methods. The shift from basal groups to eudicots appears to be more likely than the other way around, with the only exception being a Satyrine-clade within the Nymphalidae that feed on monocots. Our analysis contributes to the visualization of the complex pattern of interactions at superfamily level and provides a context to discuss the timing of changes in host plant utilization that might have promoted diversification in some butterfly lineages.

Editor: Hans Henrik Bruun, University Copenhagen, Denmark

Funding: This work was supported by Instituto Venezolano de Investigaciones Científicas (IVIC), and by a postdoctoral fellowship "Threatened species program" from South African National Biodiversity Institute (SANBI) and University of Cape Town (UCT) to ASM and JRFP. The funders had no role in study design, data collection and analysis, decision to publish, or preparation of the manuscript.

Competing Interests: The authors have declared that no competing interests exist.

* E-mail: asanchez@ivic.gob.ve

Introduction

Plant feeding insects make up a large part of the earths total biodiversity so that explaining mechanisms behind the diversification of these groups could promote the understanding of global biodiversity [1]. A seminal paper about coevolution between butterflies and host plants by Ehrlich and Raven [2] triggered intensive discussions about the role of biotic interactions in the evolutionary processes that led to radiation in species numbers.

There are two key predictions in Ehrlich and Raven's coevolution scenario. The first is that related butterflies tend to feed on related host plants as a consequence of a stepwise coevolutionary process in which plants evolve defenses against herbivores and these herbivores, in turn, evolve new capacities to cope with the defenses. Insects that manage to colonize plants with novel defenses would enter a new adaptive zone and could in turn diversify onto the relatives of this plant, because they will be chemically similar. The second prediction is that there should be a general correlation between host diversity and herbivore species richness as a consequence of the adaptive radiation and enhanced diversification experienced by insect lineages due to the adaptation to diverse, chemically distinct plant clades [3].

Later on it was recognized that other evolutionary scenarios could also explain the patterns observed. Herbivores and plants can radiate in separate bursts following the evolution of novel defenses and counter-defenses (escape-radiate scenario), or follow a sequence of independent host diversification followed by colonization and radiation of herbivores (sequential evolution). Both scenarios might result in some degree of congruence between the cladograms of insects and their host plants, but strict congruence appears to be rare among insect herbivores [3,4]. This is probably because plant diversification preceded herbivore radiation and insect plant recognition mechanisms might focus on phytochemical cues that are not necessarily related to host plant taxonomy [5,6].

More recently, a broad-scale phylogenetic analysis of butterflies [7] found that host shifts were more common between closely related plants and that there is a higher tendency to recolonize ancestral hosts. These results led them to propose the oscillation hypothesis as an alternative mechanism to explain the patterns in host plant associations [8]. They argue that dynamic oscillations in host range, instead of a steady process of specialization and cospeciation, is the principal driver of the high diversity of plant feeding insects. However, the assumptions and predictions of the oscillation hypothesis have been tested in only one butterfly family [7,9].

Besides the mechanism for diversification, the direction of evolution of host plant associations is profoundly dependent on the ancestral character [5]. Ehrlich and Raven [2] proposed a unique ancestral host plant for true butterflies (Papilionoidea, but excluding Hesperiidae and Hedylidae) and it was most likely a primitive angiosperm in the lineage of the Aristolochiaceae. Later revision of host plant associations from different regions suggested a common ancestral plant clade near the Malvaceae that would explain the range of host plants used by butterflies in the families Hedylidae, Hesperiidae and Nymphalidae, but not the associations of Pieridae and Papilionidae [10]. More recently, Janz and Nylin [7] proposed that the ancestral host plant of Papilionoidea appeared to be within a highly derived clade in the plant subclass Rosidae, including the family Fabaceae.

Tests to determine whether hypotheses about the evolution of insect-host plant associations and ancestral host plant are generally applicable, or even if they apply to the butterfly lineages from which support has previously been found, has been limited because of the scarcity of extensive datasets and comprehensive phylogenies [11]. The first general and global account of butterfly host plant associations outlined by Ehrlich and Raven [2] was purely qualitative. Some authors have provided quantitative or semi-quantitative analyses focused on describing taxonomic or regional patterns in host plant use for particular butterfly families or regions [12–14]. Semi-quantitative data in the form of binary association indices have been used in several phylogenetic analyses, sometimes removing uncommon observations [15–18]. Recent efforts to compile several data sources [19–21] and provide access to these compilations in on-line databases and other web-based resources, have improved the availability of the data [e.g. HOST, Caterpillar, and FUNET databases]. However, there have been few published quantitative analyses based on these sources [9,22,23], probably because this kind of dataset needs to be carefully revised and validated to avoid negative effects of biased or incomplete information [9,14,23].

In this paper we provide an updated quantitative summary of host plant associations for all butterfly families, based on updated and validated data from different sources. We focus on higher taxonomic levels (butterfly subfamilies and Angiosperm orders) in order to evaluate whether macro-evolutionary patterns of host plant associations can be detected in a large-scale analysis encompassing the phylogenetic relationships of all butterfly families [24]. Specifically, we want to evaluate: (1) is there a general correlation between host diversity and butterfly species richness? (2) whether evolution of host plant use has followed consistent patterns across butterfly lineages, and (3) what is the common ancestral host plant for each butterfly lineage?

Methods

Butterfly Phylogeny, Taxonomy and Host Plant Associations

Traditionally the clade "Rhopalocera" was considered as a monophyletic group within the Lepidoptera, comprising three distinct superfamilies: Papilionoidea (five families of "true butter-flies"), Hesperioidea ("skippers", one family) and Hedyloidea ("butterfly moths", one family) [25]. Recent combined morpho-logical and molecular analysis suggests that the "true butterflies" are paraphyletic and the superfamily Papilionoidea has been redefined to include all seven families [26,27]. For simplicity we will refer to all seven families collectively as "butterflies".

We compiled a tentative global checklist of butterfly species from different sources, including authoritative checklists that have been published or made available in electronic format by several authors (e.g. GloBIS/GART, http://www.globis.insects-online. de/species; The Lepidoptera Taxome Project, http://www.ucl.ac. uk/taxome/; Nymphalidae.net, http://www.nymphalidae.net/ home.htm; Afrotopical butterflies, http://www.atbutterflies.com/ index.htm) and published catalogues [28,29]. For several taxo-nomic groups not yet included in such lists, we used information from the best available sources (Encyclopedia of life, EOL, http:// www.eol.org; Lepidoptera Phylogeny, LepTree, http://www. leptree.net/; Tree of Life, http://tolweb.org/tree/; Lepidoptera and some other life forms at FUNET, ftp://www.nic.funet.fi/ index/Tree_of_life/intro.html) and carefully checked to remove duplicates or inconsistent nomenclature. All species were assigned to one of five regions according to distributional information obtained from the previous sources and the Global Biodiversity Information Facility (http://www.gbif.org/). These broad regions reflect a very crude approximation to the major biogeographical division of butterflies [30–32] and were used here only as a reference of geographical zones where butterfly research can be summarized consistently: Oriental (OR), Nearctic (NC), Neotrop-ical (NT), Afrotropical (AT) and Palearctic (PA). Species with their main distribution in one region and only marginally represented in another region were assigned to the main region. When it was not possible to determine a main region, or when the species was present in more than two regions, we classified it as "widespread" (W).

We used four types of sources to compile a list of butterfly-host plant associations. The first source was the *Lepidoptera Host Plant* database (http://www.nhm.ac.uk/hosts) that made a systematic compilation of information from literature references worldwide. The second source was *FUNET*, which also provides several summarized, well-documented, literature-based records at world-wide scale. The third source was a series of study-site databases that have been compiled from field rearing records of caterpillars and their host plants. These include the *Caterpillar Data Base* (http://caterpillars.unr.edu/) and the project *Inventory of the macrocaterpillar fauna and its food plants and parasitoids of Area de Conservación Guanacaste* (http://janzen.sas.upenn.edu) that together comprise information from Costa Rica, Ecuador, Brazil, and the United States. Finally, we digitalized host plant records from published sources for selected species and regions that were underrepresented in other sources [10,31,33–37].

The initial compilation comprised all records listed in the referenced sources, including angiosperm and non-angiosperm plants, detritus and animal food sources. We validated and updated plant names at species, genus or family level by using the taxonomic and nomenclatural information tools provided on the *Phylomatic* home page (http://www.phylodiversity.net/ phylomatic/), The *Plant List* (http://www.theplantlist.org/), and additional information on the Angiosperm Phylogeny Website (http://www.mobot.org/MOBOT/Research/APweb/welcome. html). Taxonomic validation for butterfly names was based on the previously compiled checklist of butterfly species. This compilation includes records with different levels of taxonomic resolution for both the host plant (order, family, genus, species), and the butterfly (genera, species), but in this analysis we focus on higher-level relationships and thus summarize the information at the level of plant orders and butterfly subfamilies.

Phylogenies

We used the updated phylogeny of angiosperm plant orders (APGIII) provided by The Angiosperm Phylogeny Group [38]. In this APGIII, the Aristolochiaceae of Ehrlich and Raven [2] is located in the order Piperales within the magnoliid clade, the Malvales of Ackery [10] and the rosid clade of Janz and Nylin [7] correspond loosely to the malvid and fabid clades within the rosids.

For butterflies, we combined information from higher level classification of families [25,26] and lower level classification of subfamilies (from LepTree and TOL) to build three tentative cladograms that reflect the current views derived from traditional classifications (mostly based on adult and early stage morphology) [12,25], and recent phylogenetic analyses based on a combination of morphological and molecular data [26,39–42].

The recent proposal to combine all seven families in a single superfamily [27] is based on the work of Heikkilä et al. [26], which proposes Papilionidae as a basal group to a clade formed by Hesperiidae (skippers) and Hedylidae (butterfly moths), and the four remaining families. Riodinidae and Lycaenidae have been confirmed as close but distinct sister groups, but the position of Pieridae is ambiguous, suggesting two alternative hypotheses: that Pieridae is the sister group to Lycaenidae+Riodinidae ("alternative 1" cladogram in Fig. 1A); or that Pieridae is the sister group to Nymphalidae+Lycaenidae+Riodinidae ("alternative 2" cladogram in Fig. 1B). For the sake of comparison, the traditional view of three separate superfamilies, with Papilionidae and Pieridae families as basal clades within the Papilionoidea [25], is represented as a "traditional" cladogram (Fig. 1C).

In the lower level classification we followed current views in most groups, except in some tribes with distinct host plant associations. Thus we retained the traditional Morphinae (Morphini and Brassolini tribes) as a sister clade of Satyrinae, and the subfamily status for Danainae, Ithominae and Tellervinae; we also retained the Pyrrhopyginae (Oxynetrini, Passovini, Pyrrhopygini and Zoniini tribes) as a sister group to Pyrginae, and Megathyminae as a distinct subfamily.

For all cladograms we computed branch lengths using the method of Grafen [43]. We provide a dataset (Dataset S1) with the summaries of host plant associations per butterfly genus and subfamily and the final phylogenies of the plant orders and butterfly subfamilies used in the current analysis.

Analysis

Representativeness and biases. We evaluated representativeness and biases of the compiled information by measuring three aspects: (1) proportion of butterfly species with host plant information across regions and butterfly families; (2) number of erroneous or discarded records including typing errors, non-resolved taxonomy, or records with general terms such as "grasses" or "palms", or ambiguous references to orders (or other higher level classification terms) that might have changed in circumscription; and (3) number of plant families recorded, and the plant families, genera and species more frequently used.

Association matrices. For the analysis we built association matrices between plant orders (rows) and butterfly subfamilies (columns) and a single measure of association strength in each cell [44]. We use upper case bold letters to denote the association matrix and lower case italic letters to refer to the index of association strength.

For most analyses we consider two association matrices, either matrix **A** based on a binary association index a_{ij}, which simply measures absence (0) or presence (1) of association, or matrix **C** based on a quantitative measure of association strength c_{ij} representing the number of butterfly species from subfamily j feeding on host plant order i.

To compare the relative importance of host plant orders for each butterfly subfamily, we calculated a matrix of proportions **Z**,

based on the index $z_{ij} = c_{ij}/S_j$, where S_j is the number of butterfly species in subfamily j that have at least one host plant record in the compilation. It is important to note that since many species were polyphagous, and can use host plants from more than one order, the sum of z_{ij} values for a particular subfamily does not necessarily add up to one. We consider that an order i was important for a subfamily j if $z_{ij} > 0.1$, and the term "most important resource" was used for the order with the highest value of z_{ij} for a particular subfamily j. Cases where an order was used by most species in a butterfly subfamily ($z_{ij} > 0.9$) were further recognized and are referred to as a "primary resource" even if many species in that subfamily might use additional orders as well.

For some analyses we used matrix **X**, based on a binary index x_{ij} that represents only the "important" associations between host plant orders and butterfly subfamilies, and is equal to 1 if $z_{ij} > 0.1$ and 0 otherwise.

Host plant diversity and species richness. We estimated host plant diversity by three different methods. First we estimated the total number of host plant species ($h =$ sum of columns in association matrix **A**) used by all the members of each butterfly subfamily. Second, we fitted a Fisher's log-series to the columns of the association matrix **C** and estimated the value of the parameter α [14]. These measures do not take the phylogenies of plant orders into account. Third, we calculated a Faith's index of Phylogenetic Diversity (PD) based on the binary association matrix **A** and the branch lengths of the phylogeny for plant orders [45]. We compared the calculated value of PD with the expected PD value of a sample of plant orders of equivalent size drawn at random from the plant phylogeny [46].

We calculated Pearson's product moment correlation between each measure of host plant diversity with the logarithm of species richness for each butterfly subfamily (R_j as defined above), using phylogenetically independent contrasts calculated from the butterfly cladograms and scaled with their expected variance [47].

Congruence in phylogenies. We used the ParaFit test to measure the congruency between host plant and butterfly phylogenies [48]. Congruence refers to the degree to which the herbivores and their hosts occupy corresponding positions in the phylogenetic trees. The test is based on a binary association matrix and contrasts the observed pattern against the null hypothesis of independent evolution (ParaFitGlobal).

We used a jackknife method to test the significance of individual links against the null hypothesis of random association (ParaFitLink2). We applied the test to the unweighted and weighted binary interaction matrices (**A** and **X**).

Ancestral character estimation. We grouped butterfly subfamilies according to the main patterns in host plant use and we estimated the ancestral character state using a maximum likelihood method [49]. We assigned each butterfly subfamily to the resource used by most species: non-angiosperms, magnoliids, monocots, basal eudicots, and core-eudicots (fabids, malvids, and asterids), and animal (entomophagous). We consider that non-angiosperm hosts and animal resources are derived states [2; but see 50], with transition rates in one direction from angiosperm to the derived states, but the transition rates among angiosperms might be variable [7]. We considered three models to tests this hypothesis: the null model with constant transition rates among angiosperm groups (*one single rate*); a full model with different transition rates within basal groups (magnoliids, monocots and basal eudicots), from basal groups to core-eudicots, and from core-eudicots to basal groups (*three rates*); and a simplified model where the transition rates from core-eudicots to the basal groups and within basal groups are constant, but the transition rates from

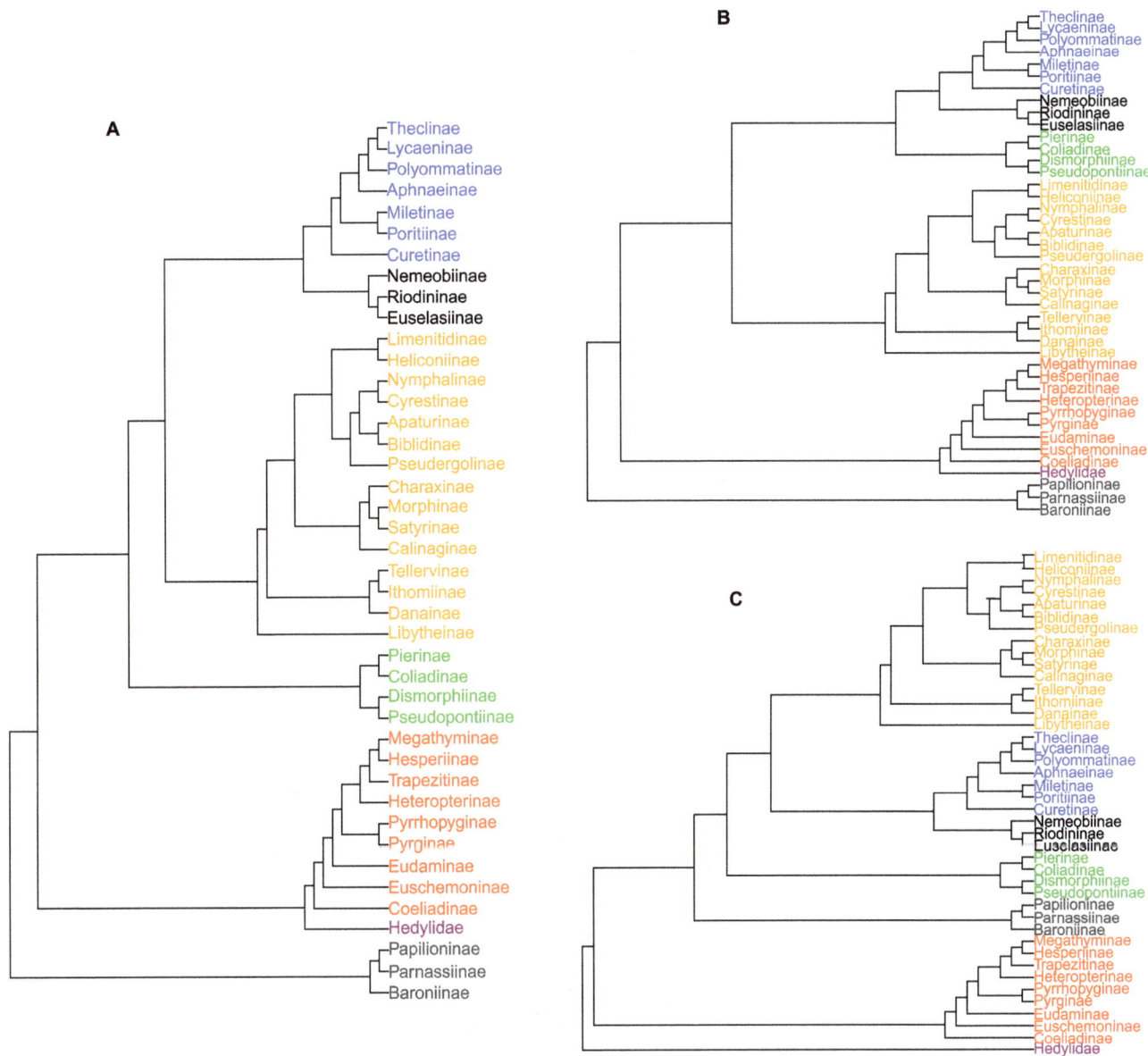

Figure 1. Three alternative phylogenetic relationships among butterfly families and subfamilies. Based on Heikkilä et al. [27] and Kristensen et al. [26]. **A**) Alternative 1 cladogram, **B**) alternative 2 cladogram, **C**) traditional cladogram.

basal groups to core-eudicots are different (*two rates*). We used Akaike Information Criterion (AIC) to compare models [51].

All the statistical analyses were performed with the free statistical software R [http://cran.r-project.org/, version 2.5.14], and Phylocom [52], and R-packages *picante*, *ape* and *vegan* [52–54].

Results

Representativeness and Biases of the Database

The global checklist compiled for this work includes 17,854 species from 1,804 genera (Table 1). Except for the Hedylidae, all butterfly families were represented worldwide, but with regional differences in species richness (Fig. 2). The Nymphalidae was the largest of all butterfly families with 5,921 species worldwide (5,339 with distribution information), but better represented in NT (40.3% of the species) and AT (23.4%). Most subfamilies were present in NT, but Satyrinae, Ithominae and Biblidinae were the

most important. In contrast, only eight subfamilies were represented in AT, with Limenitidinae, Satyrinae, Heliconiinae and Charaxinae being the most important. The subfamilies with the most restricted distribution within Nymphalidae were Tellervinae, with one species in OR, and Calinaginae with eight species between OR and PA.

Lycaenidae was the second largest butterfly family, with 5,076 species (4,109 with distribution information), most of them present in AT (33.7%), and OR (26.1%) regions. All subfamilies were present in AT except Curetinae, and most species were in the Poritinae, Theclinae and Polyommatinae subfamilies, while in OR and NT Theclinae were clearly dominant.

Hesperiidae was a medium-sized family (3,968 species, 3,562 with distribution information) with a large proportion in NT (61.7%). Within NT, Hesperiidae and Pyrginae were richer in species, but Pyrrhophyginae, Heteropteriinae and Eudaminae were also well represented. In all other regions the Hesperiinae

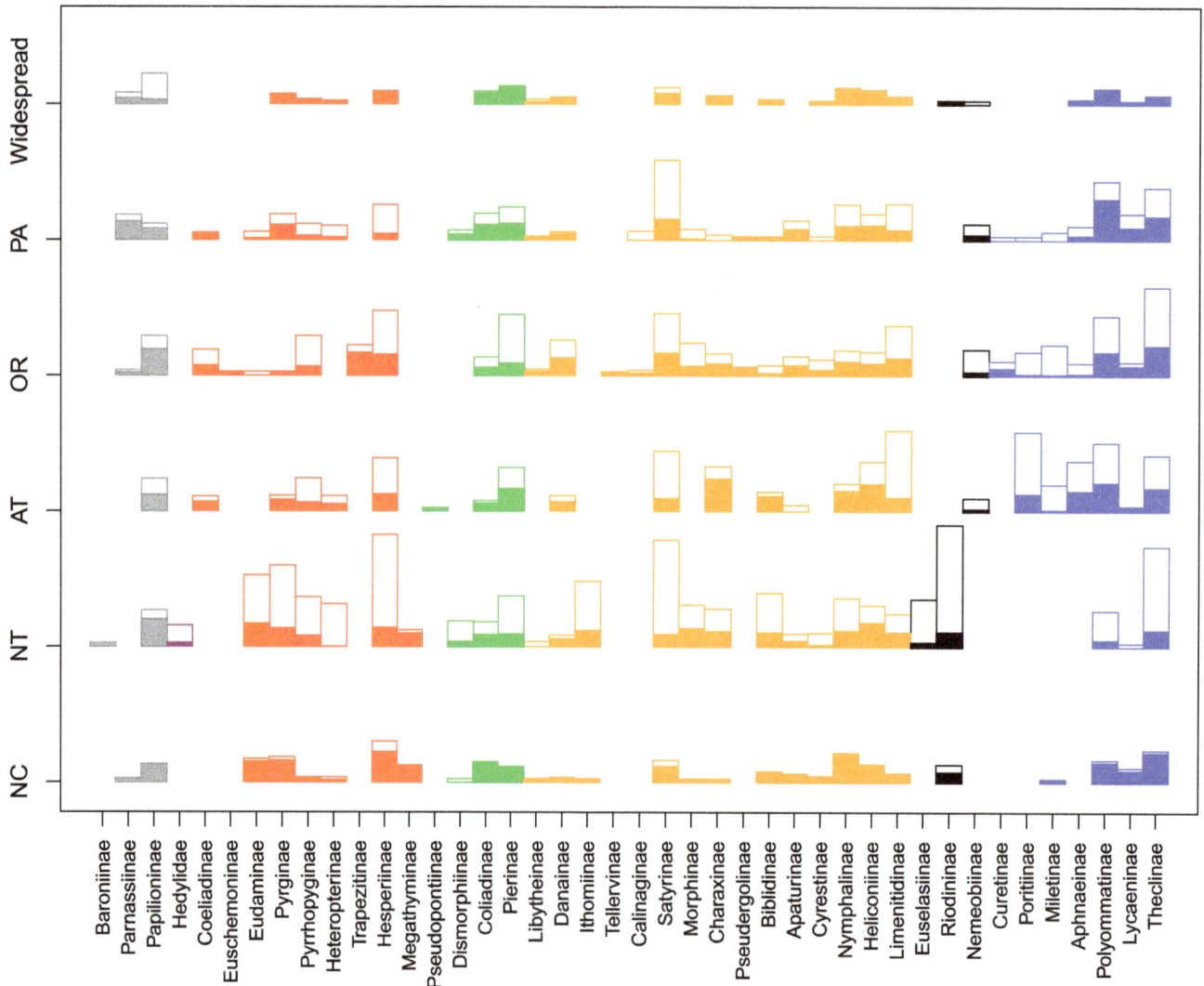

Figure 2. Geographical and taxonomical representativeness of host plant association data. Block height is proportional to the square root of the number of butterfly species among regions and subfamilies. Solid blocks represent the number of species with host plant records. Open blocks represent the number of species without host plant records. Grey: Papilionidae. Dark red: Heylidae. Red: Hesperiidae. Green: Pieridae. Orange: Nymphalidae. Blue: Lycaenidae. Black: Riodinidae.

was the most important subfamily, while the Trapezitinae, Euchemoninae and Coeliadinae were mainly distributed in, or restricted to, the OR region.

Riodiniidae (1,391 species, 1,381 with distribution information) was mostly restricted to a single region, with up to 92.2% of the species in NT, and only 107 species in the other regions, including 51 in OR region.

The majority of Pieridae (1,000 species, 984 with distribution information) were distributed in OR (30.2%) and NT (28.8%), with most species in the subfamily Pierinae. Papilionidae (462 species, 444 with distribution information) were also mainly distributed in OR (25.2%) and NT (21.8%), but they also had an important number of widespread species (17%), with Papilioninae being the most important subfamiliy. Hedylidae was barely represented by 36 species restricted to the NT region (Fig. 2).

The Neotropical region had a high number of species with host plant records (1,500), but they represent only 40.9% of the fauna of the region. On the other hand, NC had the highest proportion of species with host plant records (92%). Among butterfly families,

Papilionidae was the best represented with 59% of the species with information, while there were records for only 14% of Riodinidae (Fig. 2).

The present compilation included 51,425 records, of which 44,593 have valid information on butterfly-host plant associations (valid butterfly names at species level and valid host plant names at family, genus or species level), and a further 226 records refer to non-plant resources (detritivore or insectivore). The remaining records (6,606) are incomplete, dubious or generic records. Among the valid records, 58% had complete taxonomic information of plants (at species level), while an additional 35% had information at genus level.

The valid records included 5,146 butterfly species from 1,193 genera, that corresponds to 29% of the butterfly species and 66% of the genera estimated to occur worldwide, according to this compilation (Table 1). In general, all subfamilies were well represented (above 60% of the genera reported worldwide), except Satyrinae, Heteropterinae and Pyrginae (54–55%) and the

Table 1. Taxonomic representation of butterflies in the compilation.

Family	Subfamily	Number of genera			Number of species		
		World wide*	Compilation	Proportion	World wide*	Compilation	Proportion
Hedylidae		1	1	1.000	36	6	0.167
Hesperiidae	Coeliadinae	8	6	0.750	89	33	0.371
	Eudaminae	50	43	0.860	430	159	0.370
	Euschemoninae	1	1	1.000	1	1	1.000
	Hesperiinae	314	188	0.599	2,020	462	0.229
	Heteropterinae	11	6	0.545	182	15	0.082
	Megathyminae	5	5	1.000	39	36	0.923
	Pyrginae	86	62	0.721	642	209	0.326
	Pyrrhopyginae	67	37	0.552	490	100	0.204
	Trapezitinae	18	14	0.778	75	52	0.693
Papilionidae	Baroniinae	1	1	1.000	1	1	1.000
	Papilioninae	20	20	1.000	400	237	0.593
	Parnassiinae	8	7	0.875	61	43	0.705
Pieridae	Coliadinae	18	15	0.833	180	112	0.622
	Dismorphiinae	7	5	0.714	58	14	0.241
	Pierinae	59	46	0.780	761	258	0.339
	Pseudopontiinae	1	1	1.000	1	1	1.000
Lycaenidae	Aphnaeinae	17	13	0.765	286	92	0.322
	Curetinae	1	1	1.000	18	8	0.444
	Lycaeninae	6	4	0.667	110	60	0.545
	Miletinae	13	12	0.923	188	40	0.213
	Polyommatinae	121	93	0.769	1,477	523	0.354
	Poritiinae	56	35	0.625	721	109	0.151
	Theclinae	216	137	0.634	2,276	607	0.267
Riodinidae	Euselasiinae	5	2	0.400	171	16	0.094
	Nemeobiinae	13	6	0.462	82	15	0.183
	Riodininae	122	51	0.418	1,138	155	0.136
Nymphalidae	Apaturinae	19	16	0.842	87	43	0.494
	Biblidinae	39	27	0.692	275	95	0.345
	Calinaginae	1	1	1.000	10	1	0.100
	Charaxinae	20	17	0.850	342	180	0.526
	Cyrestinae	3	3	1.000	46	13	0.283
	Danainae	12	9	0.750	167	76	0.455
	Heliconiinae	43	37	0.860	562	275	0.489
	Ithomiinae	43	29	0.674	339	81	0.239
	Libytheinae	2	2	1.000	10	5	0.500
	Limenitidinae	48	37	0.771	1,023	232	0.227
	Morphinae	36	25	0.694	245	84	0.343
	Nymphalinae	55	47	0.855	509	254	0.499
	Pseudergolinae	4	4	1.000	7	7	1.000
	Satyrinae	233	126	0.541	2,292	441	0.192
	Tellervinae	1	1	1.000	7	1	0.143
Totals		1,804	1,193	0.661	17,854	5,152	0.289

Riodinidae (40–46%, Table 1). Plant records include 6,008 host plant species, 2,289 genera and 212 plant families.

Butterfly species have been reported feeding on 204 angiosperm plant families that represent the most species rich plant families in the world (comprising about 94% of the species and 92% of the genera reported worldwide; [38]). However only 20% of these plant genera were actually recorded. In general, Fabaceae (by 1,007 butterfly species), and Poaceae (by 811 species) were the plant families most frequently used. At generic level, *Acacia* (by 155 spp.), *Poa* (by 125 spp.), *Citrus* (by 102 spp.), and *Quercus* (by 100 spp.) were the most frequently used host plant genera. At species level, the most frequently reported host plants were mostly widespread or cultivated plants such as *Oryza sativa* (by 56 spp.), *Saccharum officinarum* (by 52 spp.), *Poa annua* (by 44 spp.), *Cocos nucifera* (by 44 spp.), and *Medicago sativa* (by 42 spp.). Only 276 species have recorded associations with non-angiosperm plants, or non-plant resources.

Phylogenetic Pattern in Host Plant Association

There was a notable disparity in host plant associations among butterfly subfamilies, even those that belong to the same family. Six butterfly families used magnoliids to some extent, but these plants only seem to be an important resource for three subfamilies: Papilioninae (on Piperales, Magnoliales and Laurales), Parnasiinae (Piperales), and Charaxinae (Laurales). The only species of Euschemoninae, as well as one of the five species of Lybiteinae, feed on Laurales (Fig. 3).

Six families used monocots, especially Poales, which is used by 891 butterfly species and is the second most used plant order overall. Poales was the primary resource for Satyrinae and Heteropterinae, the most important resource for Hesperiinae and Trapezitinae, and of some importance for Morphinae and few species in Lybiteinae and Nemeobiinae. The order Asparagales

was the primary resource for Megathyminae, and was an important resource for Trapezitinae. Arecales was the most important host plant order for Morphinae, but was also of some importance for Hesperiinae. Zingiberales was important for Morphinae and Hesperiinae whereas Dioscoreales was important for Pyrrhopyginae. Records on basal eudicots were sparingly distributed, but Sabiaceae was important for Coeliadiinae and Pseudergoliinae, and Ranunculales was the most important order for Parnasiinae.

All seven families, and 36 of 41 subfamilies feed on rosids (fabids+malvids), including more than 90% of the records for Apaturinae, Baroninae, Biblidinae, Calinagynae, Curetinae, Dismorphiinae and Hedylidae. There were, however, two important gaps: the groups feeding on monocots, and the danaine clade (Danainae, Ithomiinae and Tellervinae) of Nymphalidae that fed on lamids (see below). Three of the four most frequently used orders were in the fabid clade: Fabales (by 1,009 spp.), Malpighiales (by 693 spp.) and Rosales (by 522 spp.). Fabales was the primary resource for Baroninae, Curetinae and Dismorphinae, and was the main resource for Coliadinae, Eudaminae, Polyomatinae, Charaxinae, Riodiniinae, and Theclinae. Plants of the Malpighiales were the main resource for Heliconiinae, Biblidinae, Coeliadinae, and Limenitidinae. Rosales was the primary resource for Calinaginae, Lybiteinae and Cyrestinae, and was the main resource for Apaturinae and Pseudergolinae.

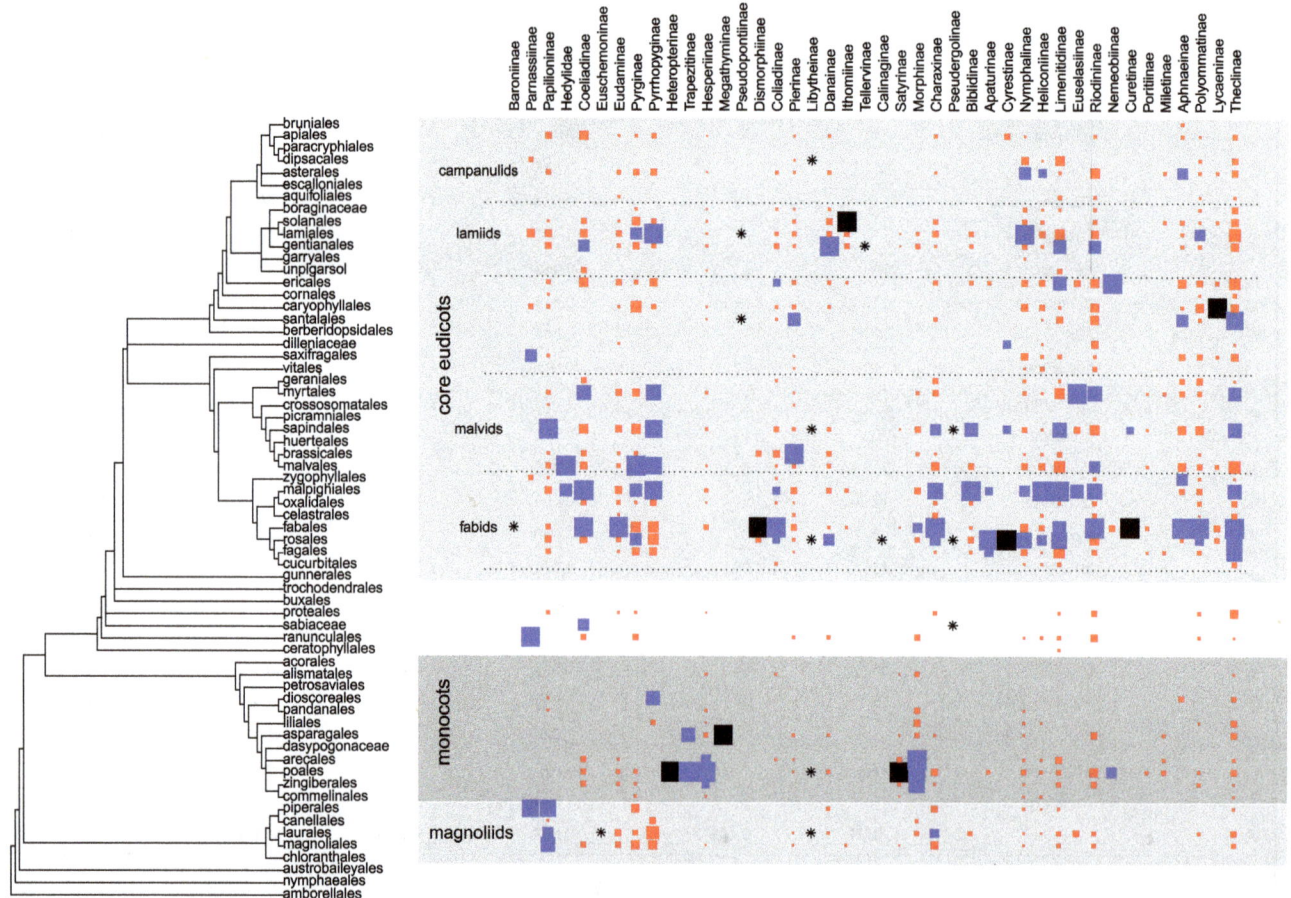

Figure 3. Graphical representation of the butterfly host plant association matrix. The squares represent the proportion of butterfly species in each subfamily that feed on a plant order (z_{ij}). Only important resources are shown, colors denote values between $0.1 < zij \le 0.5$ (red), $0.5 < zij \le 0.9$ (blue), and $zij > 0.9$ (black). The stars (*) denoted subfamilies with 15 or less species.

Within the malvids, the orders Sapindales (420 spp.), Malvales (281 spp.), and Brassicales (204 spp.) were amongst the ten most used plant groups, but only a few butterfly subfamilies use them as the most important resource: Euselasiinae on Myrtales, Pierinae on Brassicales, Papilioninae on Sapindales, and Pyrginae and the family Hedylidae on Malvales.

Within basal asterids, the Santalales, Caryophyllales and Ericales were used by ca. 200 species each. Santalales was used by the only species of Pseudopontinae and was also important for the Pierinae and the Theclinae. Caryophyllales was the primary resource for Lycaeninae, while Ericales was the main resource for Nemeobiinae, and was also important for Limenitidinae and Coeliadinae.

Within Lamiids, Gentianales was used by 204 butterfly species, and Lamiales was used by 421 species. Gentianales was used by the only species of Tellervinae, was the main resource for Danaiinae, and was also important for Limenitidinae, Coeliadinae and Riodiniinae. Lamiales was used by the only species of Pseudopontinae and was the main resource for Nymphalinae and Pyrrhopyginae, but also important for Polyommatinae and Pyrginae. Solanales was the primary resource for Ithomiinae.

Many butterfly subfamilies have single records on Capanulids, but only the Asterales was important for Nymphalinae, Aphnaeinae, and Heliconiinae, and the Dipsacales was used by one species of Lybitheinae.

Relation between Host Plant Diversity and Butterfly Species Richness

All measures of host plant diversity were higher for intermediate to high values of butterfly species richness. Typically a subfamily with 500 or more species would use >25 host plant orders, but since many of these are either used by few species or are closely related, the values of α and PD are between six and nine (Table 2). Only Satyrinae, and to some extent Hesperiinae, showed lower host plant diversity with high species richness. However, for all subfamilies the observed values of PD were either similar or significant lower (p<0.05) than the value of PD expected from a random sample with a similar value of h (Table 2).

In general there was a significant (p<0.001) and strong positive correlation between host plant diversity measures and the logarithm of butterfly species richness. Correlations, based on number of taxa (h), were lower than those based on phylogenetic information (PD) or the association matrix $\mathbf{C}$ (α). Similarly, using phylogenetic independent contrasts resulted in higher correlation, and these results were similar for alternative phylogenies (Table 3).

Congruence Analysis

The global test for congruence for matrix $\mathbf{A}$ was not significant ($p = 0.157$), but 17% of the 570 links were apparently significant (p<0.05), as might be expected for systems with a mixed structure containing a partial coevolutionary structure with additional random shifts in hosts use. However, in this situation the tests of individual links have inflated type I error, and an adjusted significance level should be used to identify truly significant links [50]. With $p < 0.03$ the number of significant links reduces to only three, suggesting that these relationships are almost completely spurious.

Fitting the model to the matrix of important links, $\mathbf{X}$ (more than 10% of the species in each subfamily, 113 links), resulted in a significant global test ($p = 0.004$). In this situation, the nominal significance level for the link-tests are valid [48], ($p < 0.05$), and 56.6% of the associations were found to be significant according to the parameter ParaFit2.

Congruent links were found between the Papilionidae-magnoliids, Hesperiidae-monocots (including Pyrrhopyginae-Dioscoreales), Pieridae with asterids, and Nymphalidae, Riodinidae and Lycaenidae with rosids and some asterids (Fig. 4). Interestingly, Baroninae, Hedylidae, and the basal Hesperiidae, and the danaine clade of Nymphalidae do not show significant congruent links.

Results with a traditional phylogeny were very similar (global test p = 0.132, 4% of significant links for matrix $\mathbf{A}$ and p = 0.002, 47.8% of significant links for matrix $\mathbf{X}$), but with the alternative 2 phylogeny, both matrices were significantly congruent (global test p = 0.042 with 39.8% of significant links for matrix $\mathbf{A}$ and p = 0.003 with 42.5% of significant links for matrix $\mathbf{X}$).

Ancestral Character Estimation (ACE)

The simplified model was slightly favored by the AIC-criterion ($\mathrm{AIC}_{simple} = 140.6$ vs. $\mathrm{AIC}_{full} = 142.6$ and $\mathrm{AIC}_{null} = 148.9$). In the selected model, the transition rate towards core-eudicots was the highest, with very low rates towards the basal groups (Table 4). The models for the other butterfly phylogenies were very similar in AIC support and rate estimates (Table 4) and resulted in similar estimates of ancestral character. We therefore only present the results for the first alternative.

There was no conclusive evidence for a common ancestral state with the alternative 1 phylogeny (scaled likelihood around 0.25 for all four groups), but there seem to be at least three different lineages: 1) the most likely ancestral state for Papilionidae was equally likely to be the magnoliids or the basal eudicots (0.451); 2) Hesperiidae-Hedylidae were more likely to be originally associated with monocot- (0.445) or magnoliid-feeding (0.269), with a later shift to core-eudicots; 3) The ancestral character remained unresolved in the Nymphalidae, but with a slightly higher likelihood (0.295) of core-eudicots compared to the basal groups; 4) for all other groups the ancestral character estate was most likely within core-eudicots: 0.751 for Pieridae, 0.493 for Lycaenidae and 0.403 for Riodinidae (Fig. 5).

Discussion

The present analysis provides a first step for a comprehensive and quantitative review of butterfly diversity and their associations with host plants at the level of plant orders and butterfly subfamilies. The pioneering work by Ehrlich and Raven [2], and the broad-scale phylogenetic analysis of Janz and Nylin [7] considered around 400–450 taxa (including a mixture of species and genera), while the present compilation includes almost three times as many butterfly genera, representative of all bioregions and all currently recognized subfamilies.

A key result from this effort was that, despite the frequently mentioned incompleteness of host plant information for tropical species, we were able to compile records for an important proportion of species in the three tropical regions analyzed (NT, OR and AT). Although NT was the region with the most incomplete dataset, it was also the region with the highest absolute numbers of species with host plant information (Table 1. Fig. 2). Gaps in knowledge are more striking precisely in species-rich taxa and regions, where rare species make up a large proportion of the species pool [55]. In these cases, the lack of field observations might lead to underestimates of host plant use, but even so the data are likely to be representative of larger patterns. For example, Satyrinae is one of the most speciose subfamilies among Nymphalidae, with 2,292 species known worldwide [25,56], but despite its high diversity it has only been recorded on eleven plant orders (Fig. 3). The 414 species of Satyrinae compiled in this study represent one of the largest absolute values for any subfamily,

Table 2. Host shift transition rates (+/− S.E.) among plant orders and non-plant resources for the three possible butterfly phylogenies.

	Animal resources	Non angiosperm	magnoliids	monocots	basal eudicots	core eudicots
Alternative 1						
Animal resources				fixed at 0		
Non angiosperm						
magnoliids						
monocots		0.132+/−0.066		0.619+/−0.244		7.346+/−2.845
basal eudicots						
core eudicots						
Alternative 2						
Animal resources				fixed at 0		
Non angiosperm						
magnoliids						
monocots		0.137+/−0.068		0.617+/−0.244		7.301+/−2.868
basal eudicots						
core eudicots						
Alternative 3						
Animal resources				fixed at 0		
Non angiosperm						
magnoliids						
monocots		0.122+/−0.061		0.612+/−0.249		7.265+/−2.825
basal eudicots						
core eudicots						

which provides a good representation of the taxonomic diversity of this group (49% of the known genera), even though they result in a low proportion of the subfamily total (18%; Table 1). Fieldwork in tropical areas like the ACG in Costa Rica confirms the predictions of previous authors that most rare Satyrinae would turn out to feed on grasses [2,14].

Clearly the completeness of the present database was only possible thanks to the availability of digital resources, which represent an important opportunity for the analysis of biotic associations [57]. Host plant-associations and distribution records, tools for validation of taxonomic and nomenclatural information, and detailed phylogenies for both taxonomic groups, were all available in different sources thanks to the contribution of several individuals and research groups. However, validating large amounts of isolated data and keeping this information up to date represent major challenges for online services [58]. The heterogeneity in the quality of data compiled required careful revision and checking in order to combine them into a useful quantitative dataset. Nevertheless, the results are useful for evaluating the role of host plant diversity in butterfly diversification and for addressing questions regarding the macroevolutionary patterns in host plant association.

Correlation between Host Diversity and Butterfly Species Richness

If herbivore species richness has been promoted by the diversification of the plants they interact with, there should be a general correlation between host plant diversity and butterfly species richness [17]. Indeed, a significant and strong correlation between host plant diversity and butterfly species richness was found, and this was even higher when phylogenetic relationships among butterflies was considered (Table 3). Characteristic examples of this correlation are evident in the Theclinae, Nymphalinae and Pierinae (Table 2). Hesperiinae and Satyrinae are important outliers in this general trend: both had extraordinary species richness (represent 56.5% of all hesperiids, and 42.6% of all nymphalids respectively), combined with very low host plant diversity that was mainly restricted to monocots. The importance of Hesperiinae and Satyrinae has been clearly understated in most discussions on butterfly diversification and host plant diversity (in fact, Janz et al. [17] reduced Satyrinae to a single clade in their analysis), and deserves more attention in the future. Even considering these two important outliers, the correlation between butterfly species richness and host plant diversity seems to be more robust than initially believed [17].

Host plant diversity can be both a cause and a consequence of butterfly species diversification [8], and this association should be analyzed in a phylogenetic and historical context in order to quantify the relative contribution of biotic interactions [59], climate change [41] and biogeographical history [50]. We will attempt to evaluate two macroevolutionary questions with the compiled information: whether evolution of host plant use has followed consistent patterns across butterfly lineages, and if there is a common ancestral host for all butterfly lineages.

Macroevolutionary Patterns in Host Plant Association

Our results suggest that, under the current view of butterfly phylogeny, there are significant congruencies with the phylogenies of plant orders. We were able to identify three main groups of congruent links: (1) Papilionidae with magnoliids, (2) Hesperidae

Table 3. Correlation between measures of host plant diversity with butterfly species richness.

Family	Subfamily	Number of species	H	α Mean	α SE	PDobs	PDrand Mean	PDrand SD	p (PDobs ≠ PDrand)
Papilionidae	Baroniinae	1	1	0	–	1	–	–	–
	Parnassiinae	61	7	2.277	1.036	3.889	3.781	0.712	0.565
	Papilioninae	400	26	6.306	1.418	6.19	8.676	1.025	0.008
Hedylidae	Hedylidae	36	4	0.935	0.863	1.397	2.556	0.604	0.032
Hesperiidae	Coeliadinae	89	21	9.966	2.848	5.968	7.55	0.97	0.041
	Euschemoninae	1	1	0	–	1	–	–	–
	Eudaminae	430	26	7.325	1.731	6.54	8.595	1.054	0.025
	Pyrginae	642	26	6.708	1.562	6.365	8.645	1.024	0.011
	Pyrrhopyginae	490	20	6.426	1.726	4.952	7.327	0.967	0.007
	Heteropterinae	182	1	0.241	0.276	1	–	–	–
	Trapezitinae	75	2	0.409	0.324	1.079	1.347	0.569	0.194
	Hesperiinae	2,020	25	5.438	1.228	6.111	8.409	1.036	0.013
	Megathyminae	39	1	0.191	0.212	1	–	–	–
Pieridae	Pseudopontiinae	1	2	0	–	1.254	1.392	0.565	0.395
	Dismorphiinae	58	3	1.090	0.775	1.286	2.038	0.571	0.084
	Coliadinae	180	20	5.494	1.486	5.254	7.310	0.996	0.018
	Pierinae	761	29	7.62	1.642	6.571	9.207	1.04	0.004
Nymphalidae	Libytheinae	10	5	4.632	3.325	3.635	3.026	0.651	0.824
	Danainae	167	19	6.192	1.711	5.46	7.126	0.985	0.042
	Ithomiinae	339	7	1.774	0.774	2.889	3.803	0.676	0.085
	Tellervinae	7	1	0	–	1	–	–	–
	Calinaginae	10	1	0	–	1	–	–	–
	Satyrinae	2,292	12	2.031	0.679	4.111	5.357	0.866	0.068
	Morphinae	245	18	5.477	1.536	5.286	6.823	0.977	0.048
	Charaxinae	342	25	6.312	1.456	5.698	8.437	1.041	0.005
	Pseudergolinae	7	3	1.989	1.651	1.889	2.042	0.576	0.332
	Biblidinae	275	11	3.023	1.065	3.254	5.100	0.839	0.010
	Apaturinae	87	5	1.383	0.724	2.571	3.023	0.631	0.201
	Cyrestinae	46	4	1.594	1.001	2.381	2.572	0.595	0.300
	Nymphalinae	509	33	8.02	1.601	6.651	10.065	1.046	0.001
	Heliconiinae	562	29	7.088	1.51	7.143	9.222	1.078	0.032
	Limenitidinae	1,023	31	8.21	1.713	7.73	9.672	1.107	0.043
Riodinidae	Euselasiinae	171	5	2.212	1.273	2.873	3.055	0.627	0.39
	Riodininae	1,138	30	8.66	1.863	7.873	9.471	1.026	0.06
	Nemeobiinae	82	3	1.128	0.807	2.444	2.028	0.562	0.75
Lycaenidae	Curetinae	18	2	0.797	0.708	1.238	1.334	0.562	0.44
	Poritiinae	721	6	–	–	–	–	–	–
	Miletinae	188	7	–	–	–	–	–	–
	Aphnaeinae	286	19	5.897	1.615	4.460	7.129	0.998	0.005
	Polyommatinae	1,477	32	6.732	1.338	7.413	9.815	1.085	0.014
	Lycaeninae	110	8	2.328	0.971	2.762	4.164	0.747	0.022
	Theclinae	2,276	39	7.974	1.43	8.413	11.259	0.99	0.004

h = simple richness of host plant orders. α = Fishers's alpha. PD = Faith's index of Phylogenetic Diversity based on plant phylogeny, with values observed (obs) and expected under random sampling of the phylogeny (rand).

with monocots, and (3) Pieridae, Lycaenidae, Riodinidae and Nymphalidae with the eudicots, particularly fabids and malvids, and few asterids (Fig. 4). These were also recovered as the most probable ancestral states (Fig. 5). As other authors have previously pointed out, a strict congruence does not necessarily mean that a continual association has occurred between two clades [3,5]. This

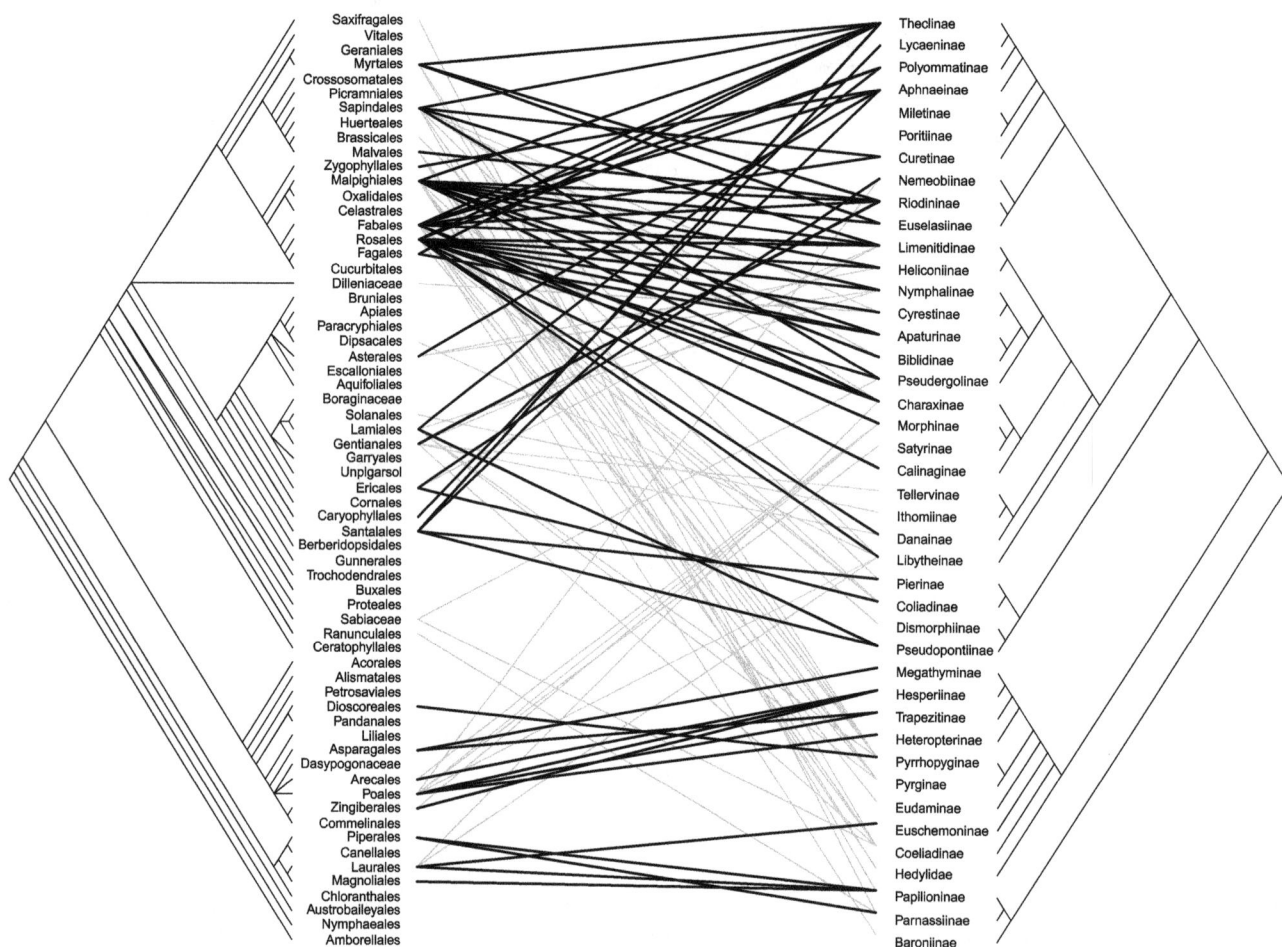

Figure 4. Congruence among plant (right) and butterfly (left) phylogenies. Lines between the phylogenies indicate associations based on the interaction matrix of important links (**X**), black lines represent congruent links (p<0.05) according to the ParaFitLink2 test. Based on the alternative 1 cladogram.

Table 4. Pearson's product moment correlation between logarithm of butterfly richness and three measures of host plant diversity using raw data and phylogenetic independent contrasts.

| | Normal correlation | Phylogenetic contrast | | |
		Alternative 1	Alternative 2	Traditional
df	38	37	37	37
h	0.782	0.754	0.802	0.800
α	0.695	0.959	0.958	0.920
PD	0.792	0.979	0.979	0.980

df = degrees of freedom for the correlation test. h = simple richness of host plant orders. α = Fishers's alpha. PD = Faith's index of Phylogenetic Diversity based on plant phylogeny. All correlations were significant (p<0.05).

at least requires that the two clades be of similar age [3]. The relative timing of adaptive radiations in host plants and butterfly is controversial. Although the major angiosperm radiation occurred ~140 to 100 million years ago (Mya), and fossil data suggest that angiosperm feeding Lepidoptera were already present ~97 Mya, butterflies probably radiated long after their host plants (~75 Mya) [26,60,61]. This hypothesis of recent butterfly origin necessarily implies a very limited role, if any, for stepwise coevolution in butterfly diversification [62,63]. However, others posit a much older age of butterflies (~100 Mya), with speciation influenced by angiosperm evolution and the breakup of the supercontinent Gondwana [50,64,65].

Beside the incongruences in timing of diversification between host plants and butterflies, the high frequency of apparently random host plant shifts – represented by a large number of marginal associations (<10% of the species in each subfamily), and >40% of non-significant links in the Parafit analysis – also points to a more complex scenario of ancestral relationships and makes the interpretation of congruence patterns more difficult. Nylin and

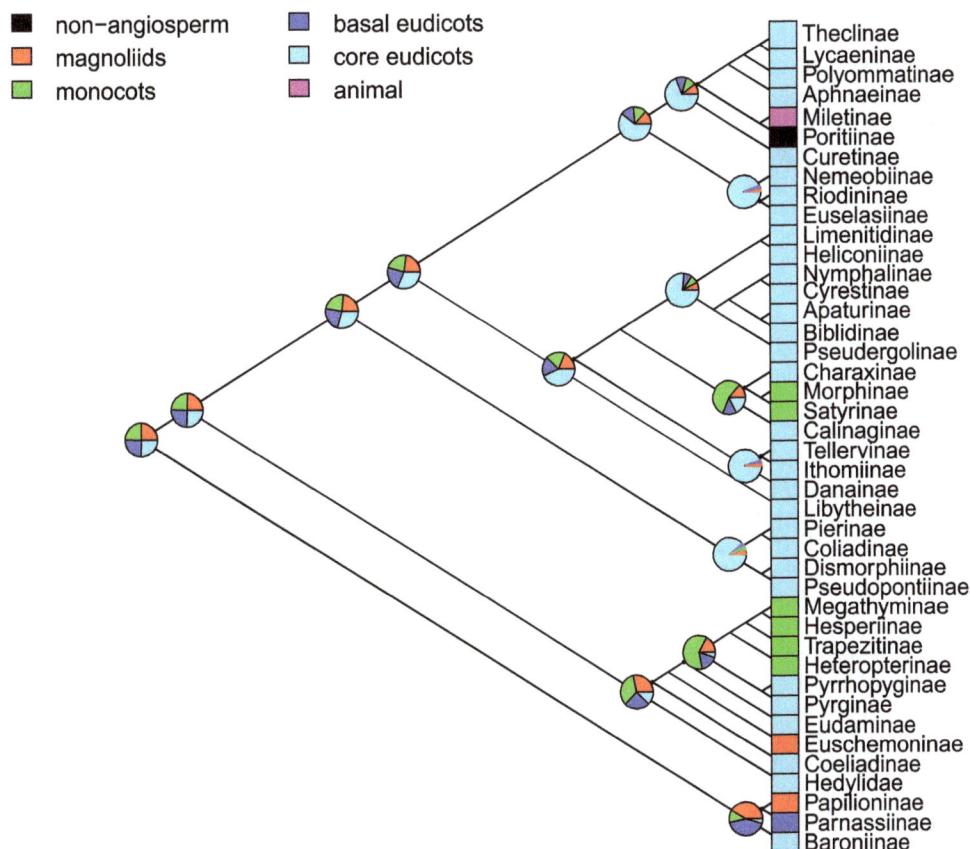

Figure 5. Likelihood of ancestral host plant in the butterfly phylogeny. Blocks on the right represent the observed states for each subfamily, piecharts represents the scaled likelihood of each potential ancestral character at selected nodes in the phylogeny. Based on the alternative 1 cladogram.

Wahlberg [66] suggested that some shifts are more probable, either because of an easier return to the ancestral state, or because a group of hosts is more favorable. Our results from ACE models showed a large difference in the transition rates from the other angiosperms toward the eudicots, with only one major shift from eudicots to monocots (Table 4). This result agrees with those reported by Janz and Nylin [7] and provides support for the oscillation hypothesis as an alternative explanation for butterfly diversification.

Alternative topologies had large effects on estimates of congruence, but not on the estimation of ancestral characters. Analyses based on modern butterfly phylogeny (alternative 2 cladogram), suggest more significant congruencies, with 39–42% of significant links. Clearly a deeper knowledge of butterfly family-level relationships is necessary to resolve these discrepancies and highlights the importance of developing comprehensive phylogenetic studies combining molecular and morphological data [26,39,67].

Our approach to reconstruct ancestral states is based on the most commonly used resource for each subfamily. This may not be the original host if, for example, a clade of butterflies has colonized and radiated on an apomorphic resource. In fact, the basal groups within the Papilionidae and the Hesperiidae-Hedylidae clades show different associations from the most diverse clades (Figs. 4 and 5) and this can lead to different interpretations (see below). Future analysis should combine this dataset with genus- and species-level butterfly phylogenies to shed more light on this issue.

The Larger Picture

Our study contributes to the visualization of the complex pattern of interactions at family level and provides a context to discuss the potential mechanisms that might explain the macroevolutionary pattern of host plant association observed at lower levels. Detailed studies at family or subfamily levels highlight the role of host plant association in the diversification of specific groups, and reveal the importance of the timing of host shifts and changes in paleoclimate and paleohabitat.

The most likely ancestral host of Papilionidae is in the Aristolochiaceae (order Piperales within the magnoliids, Fig. 5) [68], although the basal position of the Baroniinae has been used as an argument to suggest fabid-feeding as the original state for this family [2,7,10]. This family shows a prominent latitudinal gradient in species richness and host plant specialization [69], but a detailed phylogenetically integrated approach has shown that diversification of tropical species was more related to climate than to host plant association, whereas both factors seem to affect diversification in temperate clades [68].

The biggest discrepancy between our analysis and previous results is about the ancestral host of the Hedylidae/Hesperiidae clade. The relationship between Hedylidae and Hesperiidae has only been pointed out in a recent analysis of the redefined Papilionoideae [26], but the associations of Hedylidae and basal Hesperiidae were already used as an argument in favor of malvales as an ancestral host plant for all butterflies [10]. However, we found that feeding on monocots is a more likely ancestral state (Fig. 5). The host plant relationships of Hesperiidae were included as characters in a phylogenetic analysis of the group by Warren et al. [67] and the resulting phylogeny implied a single major switch from dicot to monocot feeding among the Hesperiidae (presumably by the ancestor of Heteropterinae, Trapezitinae and Hesperiinae). The host switch was accompanied by considerable diversification, especially in the New World Moncini and Hesperiini. Under this scenario, there have been just a few secondary gains of monocot feeding among dicot-feeding lineages, and only a few reversals back to dicot feeding among monocot-feeding lineages [67]. However the authors only distinguished between monocot and dicot (eudicot+magnolids) feeding and did not include complete and quantitative data on host plant associations to test this assumption explicitly. Our observations suggest that host range in Hesperiidae is very diverse, including 44 orders across the whole plant phylogeny (Fig. 3), and thus the estimation of the ancestral state is more difficult (Fig. 5). A more detailed assessment of the associations within this clade is needed, especially to account for the scattered records of basal Hesperiidae in the magnoliids and monocots, including the only species of Euschemoniinae on Laurales and several records of Pyrrhopyginae on magnoliids and dioscoreales (Fig. 3).

In the remaining components of the butterfly phylogeny, the core-eudicots dominate as host plants and most likely represent the ancestral host for each group, with only one major shift toward monocots and a few particular shifts to other hosts (Figs. 4 and 5) [13,59,66,68]. A series of host-shifts within the Pieridae appears to be linked to extraordinary radiation of the subfamily Pierinae [40] and involve an initial diversification on Brassicaceae, followed by a second and probably larger diversification on parasitic plants in the order Santalales (basal asterids), and later colonization of the hosts of these parasitic plants. The host plant associations of many Pierinae remain unknown, but it seems that the larger genera *Delias*, *Catasticta* and *Mylothris* are mostly restricted to Santalales [42,65,70]. However, diversification in these large genera is probably only partially related to host plant use [71] and much more due to geographical isolation in tropical mountains during periods of climatic change [40].

The Nymphalidae include several families with both low and high diversity of species and restricted or generalized host plant associations [12,17,26,72]. The subfamily Nymphalinae shows an elevated diversity in host plant use, which could be caused by ancestral polyphagy [73], and it has been proposed that the evolutionary trend is actually towards increased generalization rather than specialization [17]. In contrast, the diversification of Satyrinae seems to have followed a shift to feeding on monocots and may be linked to the radiation and expansion of Poales as a dominant plant form after climatic changes created suitable new habitats for colonization by grasses [50]. Current estimates of the tentative time frames of these events confirm this is a plausible sequence (origin of Poales, radiation of Poales, origin and diversification of Satyrinae), and could explain the diversification of some of the most complex Satyrinae groups (tribes, subtribes and genus-groups) [41].

Finally, within the Lycaenidae the extreme diversification in the Theclinae has been previously linked to their strong associations with ants, which might also be partly responsible for frequent host shifts [1,74]. This in turn could explain the higher host plant diversity for Theclinae that was found in this study and previous studies [14,74–76], and may also explain the species diversity in other subfamilies in the Lycaenidae and Riodinidae [77]. Recently, Megens et al. [78] suggested that the timing of a basal radiation in *Arhopala* (the most speciose genus of Theclinae, with 9% of the species in Southeast Asia) coincided with major climate changes commencing during the middle Miocene. These climatic changes could have produced massive floristic changes in the rainforest of the Southeast Asian tropics, dominated by trees of the family Dipterocarpaceae. Preadapted *Arhopala* species may have been able to fully exploit the newly formed dipterocarp rain forest emerging some 10–15 Mya, resulting in massive speciation in this genus of butterflies.

Conclusion

The data compiled here represent host records for nearly one third of all butterfly species (~29%) and 58% of these records had complete taxonomic information on host plants (at species level). Despite limitations in the dataset, it is an important step towards assembling and analysing standardized information about host plant association for this important group of insects. As such, it can be used to evaluate macroecological hypotheses such as tests of latitudinal gradients in species richness and patterns of host specialization (monophagy vs polyphagy). Here we give the first quantitative account of host plant associations for all seven butterfly families at a global scale and describe macroevolutionary patterns in host plant associations.

We found a positive correlation between host plant diversity and butterfly diversification and a congruent association between the phylogenies of plants and butterflies. However, we also detected a high number of random associations that could be interpreted as host shifts that might have helped to promote the diversification of certain butterfly lineages [8]. The congruent associations are also within the most likely ancestral hosts of each butterfly clade and tend to show a large agreement with previous analyses [13,59,66,68]. The one exception is Hesperiidae where the ancestral host seems to be within the monocots and not the dicots [18]. These results should be combined with studies of selected clades to assess the relative importance of changes in host plant associations through evolutionary time.

Supporting Information

Dataset S1 Compressed R-data file with objects used in the analysis. The file contains the association matrices (Aij, Cij Zij and Xij), the butterflies phylogenies (Alternative1.tree, Alternative2.tree and Traditional.tree), plant phylogeny (APGorders.tree), and summary table (Summary.table).

Dataset S2 File in comma separated value format used to build Figure 2. The file contains the number of butterfly species and number of butterfly species with host plant records among regions and subfamilies.

Text S1 Text file with example of R-code. The file contains commented R-code to use with the Dataset S1.

Author Contributions

Conceived and designed the experiments: JRFP ASM. Performed the experiments: JRFP ASM. Analyzed the data: JRFP. Wrote the paper: JRFP ASM ALV JD.

References

1. Pierce NE, Braby MF, Alan Heath A, Lohman DJ, Mathew J, et al. (2002) The ecology and evolution of ant association in the Lycaenidae (Lepidoptera). Annu Rev Entomol 47: 733–771.

2. Ehrlich PR, Raven PH (1964) Butterflies and plants: A study in coevolution. Evolution 18: 586–608.

3. Janz N (2011) Ehrlich and Raven revisited: Mechanisms underlying codiversification of plants and enemies. Annu Rev Ecol Evol Syst 42: 71–89.

4. Farrell BD, Mitter C (1998) The timing of insect/plant diversification: might Tetraopes (Coleoptera: Cerambycidae) and Asclepias (Asclepiadaceae) have coevolved? Biol J Linn Soc 63: 553–577.

5. Futuyma DJ, Agrawal AA (2009) Macroevolution and the biological diversity of plants and herbivores. PNAS 106: 18054–18061.

6. Miller JS (1992) Host-plant associations among prominent moths. Bioscience 42: 50–57.

7. Janz N, Nylin S (1998) Butterflies and plants: A phylogenetic study. Evolution 52: 486–502.

8. Janz N, Nylin S (2008) The oscillation hypothesis of host plant-range and speciation. In: Tilmon KJ, editor. Specialization, speciation, and radiation: The evolutionary biology of herbivorous insects. Berkeley, California, USA: University of California Press. 203–215.

9. Slove J, Janz N (2011) The relationship between diet breadth and geographic range size in the butterfly subfamily Nymphalinae – A study of global scale. PLoS One 6: e16057.

10. Ackery PR (1991) Hostplant utilization by African and Australian butterflies. Biol J Linn Soc 44: 335–351.

11. Lewinsohn TM, Novotny V, Basset Y (2005) Insect on plant: Diversity of herbivore assemblages revisited. Annu Rev Ecol Syst 36: 597–620.

12. Miller JS (1987) Host-plant relationships in the Papilionidae (Lepidoptera): Parallel cladogenesis or colonization? Cladistics 3: 105–120.

13. Ackery PR (1988) Host plants and classification: A review of nymphalid butterflies. Biol J Linn Soc 33: 95–203.

14. Fiedler K (1998) Diet breadth and host plant diversity of tropical- vs. temperate-zone herbivores: South-East Asian and West Palaearctic butterflies as a case study. Ecol Entomol 23: 285–297.

15. Wahlberg N (2001) The phylogenetics and biochemistry of host-plant specialization in Melitaeine butterflies (Lepidoptera: Nymphalidae). Evolution 55: 522–537.

16. Braby MF (2006) Evolution of larval food plant associations in Delias Hübner butterflies (Lepidoptera: Pieridae). Entomol Sci 9: 383–398.

17. Janz N, Nylin S, Wahlberg N (2006) Diversity begets diversity: Host expansions and the diversification of plant feeding insects. BMC Evol Biol 6: DOI 10.1186/1471-2148-6-4.

18. Warren AD, Ogawa JR, Brower AVZ (2008) Phylogenetic relationships of subfamilies and circumscription of tribes in the family Hesperiidae (Lepidoptera: Hesperiodea). Cladistics 24: 642–676.

19. Robinson GS, Ackery PR, Kitching IJ, Beccaloni GW, Hernández LM (2001) Hostplants of the moth and butterfly caterpillars of the Oriental Region. 744 p.

20. Robinson GS, Ackery PR, Kitching IJ, Beccaloni GW, Hernández LM (2002) Hostplants of the moth and butterfly caterpillars of America north of Mexico: Memoirs of the American Entomological Institute. 824 p.

21. Beccaloni GW, Viloria AL, Hall SR, Robinson GS (2008) Catálogo de las plantas huésped de las mariposas neotropicales; Milenio mm-MT, editor. Zaragoza, España: Sociedad Entomológica Aragonesa-CYTED, IVIC-RiBES, Natural History Museum, London 536 p.

22. Symons FB, Beccaloni GW (1999) Phylogenetic indices for measuring the diet breadths of phytophagous insects. Oecologia 119: 427–434.

23. Beccaloni GW, Symons FB (2000) Variation of butterfly diet breadth in relation to host-plant predictability: Results from two faunas. Oikos 90: 50–66.

24. Menken SBJ, Boomsma JJ, van Nieukerken EJ (2009) Large-scale evolutionary patterns of host plant associations in the Lepidoptera. Evolution 64: 1098–1119.

25. Kristensen NP, Scoble MJ, Karsholt O (2007) Lepidoptera phylogeny and systematics: The state of inventorying moth and butterfly diversity. Zootaxa 1668: 699–747.

26. Heikkilä M, Kaila L, Mutanen M, Peña C, Wahlberg N (2011) Cretaceous origin and repeated tertiary diversification of the redefined butterflies. Proc Roy Soc Lond B doi:10.1098/rspb.2011.1430.

27. Van Nieukerken EJ, Kaila L, Kitching IJ, Kristensen NP, Lees DC, et al. (2011) Order Lepidoptera Linnaeus, 1758. In: Zhang, Z.-Q. (Ed.) Animal biodiversity: An outline of higher-level classification and survey of taxonomic richness. Zootaxa 3148: 212–221.

28. Scoble MJ (1990) A catalogue of the Hedylidae (Lepidoptera: Hedyloidea), with descriptions of two new species. Entomol Scand 21: 113–119.

29. Lamas G (2004) Atlas of eotropical Lepidoptera. Checklist: Part 4A, Hesperioidea - Papilionoidea; Heppner JB, editor. Florida, USA: Scientific Publishers. 439 p.

30. Robbins CS, Bystrak D, Geissler PH (1997) The Breeding Bird Survey: Its first fifteen years, 1965-1979. Washington, DC: United States Department of the Interior Fish and Wildlife Service.

31. Larsen TB (2005) Butterflies of West Africa. Stenstrup, Denmark: Apollo Books. 270 p.

32. Lamas G (2008) La sistemática sobre mariposas (Lepidoptera: Hesperoidea y Papilionoidea) en el mundo: Estado actual y perspectivas futuras. In: Llorente Bousquets J, Lanteri A, editors. Contribuciones taxonómicas en ódenes de insectos hiperdiversos. III Reunión anual de la Red Iberoamericana de Biogeografía y Entomología Sistemática, La Plata, Argentina. La Plata, Argentina: Las Prensas de Ciencias, UNAM. México D. F. 57–70.

33. Braby MF (2005) Afrotropical mistletoe butterflies: Larval food plant relationships of Mylothris Hübner (Lepidoptera: Pieridae). J Nat Hist 39: 499–513.

34. Braby MF, Nishida K (2007) The immature stages, larval food plants and biology of Neotropical mistletoe butterflies. I. The Hesperocharis group (Pieridae: Anthocharidini). J Lepid Soc 61: 181–195.

35. Kroon DM (1999) Lepidoptera of Southern Africa. Host-plants and other associations. A Catalogue. Sasolburg, South Africa: Lepidopterists' Society of Africa. 160 p.

36. Woodhall S (2005) Field guide to butterflies of South Africa. Cape Town, South Africa: Struik Publishers. 464 p.

37. Viloria AL, Pyrcz TW, Orellana A (2010) A survey of the Neotropical montane butterflies of the subtribe Pronophilina (Lepidoptera, Nymphalidae) in the Venezuelan Cordillera de la Costa. Zootaxa 2622: 1–41.

38. The Angiosperm Phylogeny Group (2009) An update of the Angiosperm Phylogeny Group classification for the orders and families of flowering plants: APG III. Bot J Linn Soc 161: 105–121.

39. Wahlberg N, Braby MF, Brower AVZ, de Jong R, Lee MM, et al. (2005) Synergistic effects of combining morphological and molecular data in resolving the phylogeny of butterflies and skippers. Proc Roy Soc Lond B 272: 1577–1586.

40. Braby MF, Trueman JWH (2006) Evolution of larval host plant associations and adaptive radiation in pierid butterflies. J Evol Biol 19: 1677–1690.

41. Peña C, Wahlberg N (2008) Prehistorical climate change increased diversification of a group of butterflies. Biol Lett 4: 274–278.

42. Braby MF, Nishida K (2010) The immature stages, larval food plants and biology of Neotropical mistletoe butterflies (Lepidoptera: Pieridae). II. The Catasticta group (Pierini: Aporiina). J Nat Hist 44: 1831–1928.

43. Grafen A (1989) The phylogenetic regression. Philos Trans R Soc Lond Ser B 326: 119–157.

44. Ives AR, Godfray HCJ (2006) Phylogenetic analysis of trophic associations. Am Nat 168: E1–E14.

45. Faith DP (1992) Conservation evaluation and phylogenetic diversity. Biol Conserv 61: 1–10.

46. Proches S, Wilson JRU, Cowling RM (2006) How much evolutionary history in a 10×10m plot? Proc Roy Soc Lond B 273: 1143–1148.

47. Felsenstein J (1985) Phylogenies and the comparative method. Am Nat 125: 1–15.

48. Legendre P, Desdevises Y, Bazin E (2002) A statistical test for host–parasite coevolution. Syst Biol 51: 217–234.

49. Pagel MD (1994) The adaptationis wager. In: Eggleton P, Vane-Wright R, editors. Phylogenetics and ecology. London, UK: Academic Press. 29–51.

50. Viloria AL (2003) Historical biogeography and the origins of the satyrine butterflies of the tropical Andes (Lepidoptera: Rhopalocera). In: Llorente J, Morrone JJ, editors. Una perspectiva latinoamericana de la biogeografía. México: Universidad Autónoma de México. 247–261.

51. Anderson DR (2008) Model based inference in the life sciences. New York, USA: Springer.

52. Webb CO, Ackerly DD, Kembel SW (2008) Phylocom: Software for the analysis of phylogenetic community structure and trait evolution. Bioinformatics 24: 2098–2100.

53. Kembel SW, Cowan PD, Helmus MR, Cornwell WK, Morlon H, et al. (2010) Picante: R tools for integrating phylogenies and ecology. Bioinformatics 26: 1463–1464.

54. Oksanen J, Blanchet FG, Kindt R, Legendre P, O'Hara RB, et al. (2010). vegan: Community Ecology Package. v. 1.17–4.

55. Collen B, Ram M, Zamin T, McRae L (2008) The tropical biodiversity data gap: Addressing disparity in global monitoring. Trop Cons Sci 1: 75–88.

56. Lamas G (2008) Contribuciones taxonómicas en órdenes de insectos hiperdiversos. In: Bousquets JL, Lanteri A, editors. La sistemática sobre mariposas (Lepidoptera: Hesperioidea y Papilionoidea) en el mundo: Estado actual y perspectivas futuras. Mexico, DF: Las Prensas de Ciencias, UNAM.

57. Mulder C (2011) World wide food webs: Power to feed ecologists. AMBIO 40: 335–337.

58. Wilson EO (2003) The encyclopedia of life. Trends Ecol Evol 18: 77–80.

59. Megens H-J, De Jong R, Fiedler K (2005) Phylogenetic patterns in larval host plant and ant association of Indo-Australian Arhopalini butterflies (Lycaenidae: Theclinae). Biol J Linn Soc 84: 225–241.

60. Labandeira CC, Dilcher DL, Davis DR, Wagner DL (1994) Ninety-seven million years of angiosperm-insect association: Paleobiological insights into the meaning of coevolution. Proc Nat Acad Sci USA 91: 12278–12282.

61. Magallón SA, Sanderson MJ (2005) Angiosperm divergence times: The effect of genes, codon positions, and time constraints. Evolution 59: 1653–1657.

62. de Jong R (2003) Are there butterflies with Gondwanan ancestry in the Australian region? Invertebr Syst 17: 143–156.

63. Vane-Wright D (2004) Butterflies at that awkward age. Nature 428: 477–479.

64. Miller JY, Miller LD (2001) New perspectives on the biogeography of west Indian butterflies: A vicariance model. In: Woods CA, Sergile FE, editors. Biogeography of the the West Indies: Patterns and Perspectives. Boca Raton, FL, USA: CRC Press. 127–150.

65. Braby MF, Trueman JWH, Eastwood R (2005) When and where did troidine: Gondwana in the Late Cretaceous. Invertebr Syst 19: 113–143.

66. Nylin S, Wahlberg N (2008) Does plasticity drive speciation? Host-plant shifts and diversification in nymphaline butterflies (Lepidoptera: Nymphalidae) during the Tertiary. Biol J Linn Soc 94: 115–130.

67. Warren AD, Ogawa JR, Brower AVZ (2009) Revised classification of the family Hesperiidae (Lepidoptera: Hesperioidea) based on combined molecular and morphological data. Syst Entomol 34: 467–523.

68. Condamine FL, Sperling FA, Wahlberg N, Rasplus JY, Kergoat GJ (2012) What causes latitudinal gradients in species diversity? Evolutionary processes and ecological constraints on swallowtail biodiversity. Ecol Lett 15: 267–277.

69. Scriber JM (2002) Latitudinal and local geographic mosaics in host plant preferences as shaped by thermal units and voltinism in Papilio spp. (Lepidoptera). Eur J Entomol 99: 225–239.

70. Braby MF, Pierce NE (2007) Systematics, biogeography and diversification of the Indo-Australian genus *Delias* Hübner (Lepidoptera: Pieridae): Phylogenetic evidence supports an 'out-of-Australia' origin. Syst Entomol 32: 2–25.

71. Wheat CW, Vogel H, Wittstock U, Braby MF, Underwood D, et al. (2007) The genetic basis of a plant-insect coevolutionary key innovation. Proc Nat Acad Sci USA 104: 427–431.

72. Nylin S, Nygren GH, Soderlind L, Stefanescu C (2009) Geographical variation in host plant utilization in the comma butterfly: The roles of time constraints and plant phenology. Evol Ecol 23: 807–825.

73. Janz N, Nyblom K, Nylin S (2001) Evolutionary dynamic of host-plant specialization: A case study of the tribe Nymphalini. Evolution 55: 783–796.

74. Fiedler K (1994) Lycaenid butterflies and plant: Is myrmecophyly associated with amplified hostplant diversity? Ecol Entomol 19: 79–82.

75. Fiedler K (1995) Lycaenid butterflies and plants: Is myrmecophily associated with particular hostplant preferences? Ethology, Ecol & Evol 7: 107–132.

76. Fiedler K (1996) Host-plant relationships of lycaenid butterflies: Large-scale patterns, interactions with plant chemistry, and mutualism with ants. Entomol Exp Appl 80: 259–267.

77. Eastwood R, Pierce NE, Kitching RL, Hughes JM (2006) Do ants enhance diversification in lycaenid butterflies? Phylogeographic evidence from a model myrmecophile, *Jalmenus evagoras*. Evolution 60: 315–327.

78. Megens H-J, van Moorsel CHM, Piel WH, Pierce NE, de Jong R (2004) Tempo of speciation in a butterfly genus from the Southeast Asian tropics, inferred from mitochondrial and nuclear DNA sequence data. Mol Phylogen Evol 31: 1181–1196.

8

Multiple Inter-Kingdom Horizontal Gene Transfers in the Evolution of the Phosphoenolpyruvate Carboxylase Gene Family

Yingmei Peng[1,4,9], Jing Cai[2,3,9], Wen Wang[1]*, Bing Su[1]*

1 State Key Laboratory of Genetic Resources and Evolution, Kunming Institute of Zoology, Chinese Academy of Sciences, Kunming, PR China, **2** Shenzhen Key Laboratory for Orchid Conservation and Utilization, National Orchid Conservation Center of China and Orchid Conservation and Research Center of Shenzhen, Shenzhen, China, **3** Center for Biotechnology and BioMedicine, Graduate School at Shenzhen, Tsinghua University, Shenzhen, China, **4** University of Chinese Academy of Sciences, Beijing, PR China

Abstract

Pepcase is a gene encoding phosphoenolpyruvate carboxylase that exists in bacteria, archaea and plants, playing an important role in plant metabolism and development. Most plants have two or more pepcase genes belonging to two gene sub-families, while only one gene exists in other organisms. Previous research categorized one plant pepcase gene as plant-type pepcase (PTPC) while the other as bacteria-type pepcase (BTPC) because of its similarity with the pepcase gene found in bacteria. Phylogenetic reconstruction showed that PTPC is the ancestral lineage of plant pepcase, and that all bacteria, protistpepcase and BTPC in plants are derived from a lineage of pepcase closely related with PTPC in algae. However, their phylogeny contradicts the species tree and traditional chronology of organism evolution. Because the diversification of bacteria occurred much earlier than the origin of plants, presumably all bacterialpepcase derived from the ancestral PTPC of algal plants after divergingfrom the ancestor of vascular plant PTPC. To solve this contradiction, we reconstructed the phylogeny of pepcase gene family. Our result showed that both PTPC and BTPC are derived from an ancestral lineage of gamma-proteobacteriapepcases, possibly via an ancient inter-kingdom horizontal gene transfer (HGT) from bacteria to the eukaryotic common ancestor of plants, protists and cellular slime mold. Our phylogenetic analysis also found 48other pepcase genes originated from inter-kingdom HGTs. These results imply that inter-kingdom HGTs played important roles in the evolution of the pepcase gene family and furthermore that HGTsare a more frequent evolutionary event than previouslythought.

Editor: Ross Frederick Waller, University of Melbourne, Australia

Funding: These authors have no support or funding to report.

Competing Interests: The authors have declared that no competing interests exist.

* E-mail: sub@mail.kiz.ac.cn (BS); wwang@mail.kiz.ac.cn (WW)

⑨ These authors contributed equally to this work.

Introduction

Following wide acceptance of Darwin's theory of evolution, the tree of life became a well accepted representation of the evolutionary relationships among organisms. Recent findings of the horizontal gene transfer (HGT) in the genomes of many species [1,2,3,4,5,6] strongly challenge this certainty. HGT, though, is still thought as rare event and genes that originated from HGT account for a tiny proportion in each genome, while vertical descent of genes remains the major mechanism of evolution.Moreover, all HGT genes are treated as noise when species phylogeny is constructed. Here, for the first time, we found 48 members from well supported inter-kingdom HGT in a single gene family coding phosphoenolpyruvate carboxylase. This case demonstratesthe means by which the evolution of a single gene family can form a complex web via horizontal gene transfer, and likewise suggests that the previously ignored contribution of HGT to the evolution pattern would strongly enhance our understanding of the evolution as a tree of life to more rich and diversified web of life that revealsthe unexpected complexity of evolution.

Phosphoenolpyruvate carboxylase (PEPC) is an important enzyme that catalyzes the carboxylation reaction of phosphoenol-pyruvate into oxalacetate, which is then used by the citric cycle. This reaction is also used by C4 and crassulacean acid metabolic pathway and is an important step to store and concentrate carbon dioxide for photosynthesis. In 2003, Sanchez and Cejudo found a PEPC gene in Arabidopsis and rice with close homologs with PEPCs in bacteria [7]. Since then, the plant PEPC gene family has been categorized in to plant-type (PTPC) and bacteria-type (BTPC) subfamilies. Despite this organization, the actual evolution of the whole gene family has not been discussed in any detail. Only O'Leary et al.'s [8] recent review included a constructed phylogeny of PEPC gene family including members from Archaea, Bacteria, protists and plants. In this tree, the BTPC were clustered with bacteria PEPCs forming a clade as a sister group of protist PEPC. This phylogeny showed that the ancestor of all bacteria PEPCs, protists PEPCs and BTPCs originated from a duplication event in the lineage of PTPC to algae after its divergence with vascular plant PTPCs. This gene phylogeny has many inconsistencies with the accepted species tree constructed by multiple gene

Table 1. Sequences used in the phylogenetic reconstruction.

Taxon	GenBank or Uniprot ID
Acidimicrobium ferrooxidans	256007505
Acidobacterium capsulatum	225874618
Algoriphagus sp.	311746515
Arabidopsis thaliana g1	15232442
Arabidopsis thaliana g2	30697740
Arabidopsis thaliana g3	240254631
Arabidopsis thaliana g4	15219272
Arabidopsis thaliana g5	222423984
Archaeoglobus fulgidus	11499081
Aureococcus anophagefferens	323453325
Babesia bovis	156084500
Capsaspora owczarzaki	320168251
Chlamydomonas reinhardtii	51701320
Chlorobaculum parvum	193085694
Chloroflexus sp.	222450523
Cryptosporidium hominis	67594757
Cryptosporidium muris	209881885
Cryptosporidium parvum	66357588
Deinococcus deserti	226355772
Dictyoglomus thermophilum	206740030
Dictyostelium discoideum	66806573
Dictyostelium fasciculatum	328865638
Dictyostelium purpureum	330798819
Ectocarpus siliculosus	299117425
Emiliania huxleyi	223670909
Escherichia coli	15804552
Gemmatimonas aurantiaca	226229154
Haemophilus influenzae	16273525
Halobacterium sp.	15791074
Lentisphaera araneosa	149200328
Leptospira biflexa	167780286
Methanosarcina acetivorans	229017561
Methanothermobacter thermautotrophicus	15678963
Mycoplasma penetrans	26554388
Myxococcus xanthus	108759396
Nitrosomonas europaea	30248603
Oryza sativa g1	222622510
Oryza sativa g10	115476100
Oryza sativa g11	15022444
Oryza sativa g2	51091643
Oryza sativa g3	222617602
Oryza sativa g4	115440043
Oryza sativa g5	115434082
Oryza sativa g6	115435200
Oryza sativa g7	50251800
Oryza sativa g8	9828445
Oryza sativa g9	222619275
Phaeodactylum tricornutum g1	219120583
Phaeodactylum tricornutum g2	327343197

Table 1. Cont.

Taxon	GenBank or Uniprot ID
Physcomitrella patens g1	168044057
Physcomitrella patens g2	168010333
Physcomitrella patens g3	168027443
Physcomitrella patens g4	168042979
Physcomitrella patens g5	168016115
Physcomitrella patens g6	168061648
Picrophilus torridus	48478036
Pirellula staleyi	283779027
Plasmodium berghei	68071185
Plasmodium chabaudi	70950271
Plasmodium falciparum	124808830
Plasmodium knowlesi	221060224
Plasmodium vivax	156102026
Plasmodium yoelii	83282693
Polysphondylium pallidum	281207688
Pseudomonas aeruginosa	347303632
Pyrobaculum aerophilum	18314050
Pyrococcus furiosus	18978347
Rhodospirillum centenum	209965727
Selaginella moellendorffii g1	302800171
Selaginella moellendorffii g2	302783266
Selaginella moellendorffii g3	302817036
Selaginella moellendorffii g4	302795803
Streptobacillus moniliformis	269123480
Streptococcus thermophilus	89143166
Sulfolobus solfataricus	15899028
Synechococcus sp.	87284805
Thalassiosira pseudonana g1	224000774
Thalassiosira pseudonana g2	223998678
Verrucomicrobium spinosum	171911854
Vibrio cholerae	227082762
Volvox carteri g1	302835908
Volvox carteri g2	302830816
Halobacterium salinarum	CAPPA HALSA (Q9HN43)
Archaeoglobus fulgidus	CAPPA ARCFU (O28786)
Archaeoglobus veneficus	F2KS60 ARCVE (F2KS60)
Caldivirga maquilingensis	CAPPA CALMQ (A8MBK0)
Candidatus Caldiarchaeum	E6N9G7 9ARCH (E6N9G7)
Candidatus Kuenenia	Q1PXR4 9BACT (Q1PXR4)
Candidatus Methylomirabilis	D5MHI6 9BACT (D5MHI6)
Clostridium cellulovorans	D9SUK0 CLOC7 (D9SUK0)
Clostridium perfringens g1	B1RBJ1 CLOPE (B1RBJ1)
Clostridium perfringens g2	B1BWT1 CLOPE (B1BWT1)
Clostridium perfringens g3	CAPPA CLOPE (Q8XLE8)
Clostridium perfringens g4	CAPPA CLOPS (Q0STS8)
Clostridium perfringens g5	B1RT70 CLOPE (B1RT70)
Clostridium perfringens g6	B1RJT6 CLOPE (B1RJT6)
Clostridium perfringens g7	CAPPA CLOP1 (Q0TRE4)
Clostridium perfringens g8	B1BFT5 CLOPE (B1BFT5)

Table 1. Cont.

Taxon	GenBank or Uniprot ID
Clostridium perfringens g9	B1V5L0 CLOPE (B1V5L0)
Desulfonatronospira thiodismutans	D6SP11 9DELT (D6SP11)
Desulforudis audaxviator	B1I2W1 DESAP (B1I2W1)
Dictyoglomus thermophilum	B5YCF7 DICT6 (B5YCF7)
Ferroglobus placidus	D3S0D1 FERPA (D3S0D1)
Halobacterium salinarum	CAPPA HALS3 (B0R7F9)
Ignicoccus hospitalis	A8A9C2 IGNH4 (A8A9C2)
Ignisphaera aggregans	E0SSB1 IGNAA (E0SSB1)
Lactobacillus brevis	C2D3X1 LACBR (C2D3X1)
Lactobacillus buchneri	C0WSM6 LACBU (C0WSM6)
Lactobacillus hilgardii	C0XL21 LACHI (C0XL21)
Leptospirillum ferrodiazotrophum.	C6HVN3 9BACT (C6HVN3)
Leptospirillum rubarum.	A3EQI3 9BACT (A3EQI3)
Leptospirillum sp.	B6AN75 9BACT (B6AN75)
Leuconostoc citreum	B1N089 LEUCK (B1N089)
Leuconostoc gasicomitatum	D8ME72 LEUGT (D8ME72)
Leuconostoc kimchii	D5T4D7 LEUKI (D5T4D7)
Leuconostoc mesenteroides	C2KKA6 LEUMC (C2KKA6)
Leuconostoc mesenteroides	CAPPA LEUMM (Q03VI7)
Metallosphaera sedula	CAPPA METS5 (A4YES9)
Methanohalobium evestigatum	D7E7Q5 METEZ (D7E7Q5)
Methanoplanus petrolearius	E1RII9 METP4 (E1RII9)
Methanopyrus kandleri	CAPPA METKA (Q8TYV1)
Methanosarcina acetivorans	CAPPA METAC (Q8TMG9)
Methanosarcina barkeri	CAPPA METBF (Q469A3)
Methanosarcina mazei	CAPPA METMA (Q8PS70)
Methanospirillum hungatei	CAPPA METHJ (Q2FLH1)
Methanothermobacter marburgensis	D9PXG9 METTM (D9PXG9)
Methanothermobacter thermautotrophicus	CAPPA METTH (O27026)
Methanothermus fervidus	E3GXT0 METFV (E3GXT0)
Oenococcus oeni g1	A0NKU8 OENOE (A0NKU8)
Oenococcus oeni g2	D3LBW5 OENOE (D3LBW5)
Oenococcus oeni g3	CAPPA OENOB (Q04D35)
Picrophilus torridus	CAPPA PICTO (Q6L0F3)
Pyrobaculum aerophilum	CAPPA PYRAE (Q8ZT64)
Pyrobaculum arsenaticum	CAPPA PYRAR (A4WJM7)
Pyrobaculum calidifontis	CAPPA PYRCJ (A3MVZ5)
Pyrobaculum islandicum	CAPPA PYRIL (A1RR50)
Pyrococcus abyssi	CAPPA PYRAB (Q9V2Q9)
Pyrococcus furiosus	CAPPA PYRFU (Q8TZL5)
Pyrococcus horikoshii	CAPPA PYRHO (O57764)
Sulfolobus acidocaldarius	CAPPA SULAC (Q4JCJ1)
Sulfolobus islandicus g1	CAPPA SULIA (C3N0D7)
Sulfolobus islandicus g2	CAPPA SULIY (C3N8C3)
Sulfolobus islandicus g3	CAPPA SULIL (C3MJE5)
Sulfolobus islandicus g4	F0NMR2 SULIH (F0NMR2)
Sulfolobus islandicus g5	CAPPA SULIN (C3NJA0)
Sulfolobus islandicus g6	CAPPA SULIM (C3MTS7)
Sulfolobus islandicus g7	D2PDY7 SULID (D2PDY7)

Table 1. Cont.

Taxon	GenBank or Uniprot ID
Sulfolobus islandicus g8	CAPPA SULIK (C4KJI5)
Sulfolobus islandicus g9	F0NG17 SULIR (F0NG17)
Sulfolobus solfataricus g1	CAPPA SULSO (Q97WG4)
Sulfolobus solfataricus g2	D0KUQ4 SULS9 (D0KUQ4)
Sulfolobus tokodaii	CAPPA SULTO (Q96YS2)
Thermococcus barophilus	F0LK16 THEBM (F0LK16)
Thermococcus sibiricus	C6A2T7 THESM (C6A2T7)
Thermofilum pendens	CAPPA THEPD (A1RZN3)
Thermoproteus neutrophilus	B1YBY2 THENV (B1YBY2)
Thermoproteus uzoniensis g1	F2L305 THEU7 (F2L305)
Thermoproteus uzoniensis g2	F2L5Y2 9CREN (F2L5Y2)
Vulcanisaeta distributa	E1QNA4 VULDI (E1QNA4)
Acidobacterium capsulatum	C1F4Y2 ACIC5 (C1F4Y2)
Cellulomonas flavigena	D5UGP1 CELFN (D5UGP1)
Chitinophaga pinensis	C7PRS5 CHIPD (C7PRS5)
Dokdonia donghaensis	A2TNK9 9FLAO (A2TNK9)
Erythrobacter sp. g1	A5P918 9SPHN (A5P918)
Erythrobacter sp. g2	A3WAI8 9SPHN (A3WAI8)
Flavobacteria bacterium	A3J3B3 9FLAO (A3J3B3)
Flavobacteriales bacterium	A8UJQ6 9FLAO (A8UJQ6)
Flavobacterium johnsoniae	A5FG47 FLAJ1 (A5FG47)
Geobacter sp.	B9M086 GEOSF (B9M086)
Gramella forsetii	A0M1G5 GRAFK (A0M1G5)
Haladaptatus paucihalophilus	E7QR15 9EURY (E7QR15)
Halalkalicoccus jeotgali	D8JA44 HALJB (D8JA44)
Haloarcula marismortui	Q5V4H5 HALMA (Q5V4H5)
Haloferax volcanii	D4GUG0 HALVD (D4GUG0)
Halogeometricum borinquense	E4NPR5 HALBP (E4NPR5)
Halomicrobium mukohataei	C7NYU1 HALMD (C7NYU1)
Haloquadratum walsbyi	Q18FG1 HALWD (Q18FG1)
Halorhabdus utahensis	C7NNW9 HALUD (C7NNW9)
Halorubrum lacusprofundi	B9LS13 HALLT (B9LS13)
Haloterrigena turkmenica g1	D2RVU2 HALTV (D2RVU2)
Haloterrigena turkmenica g2	D2S2A1 HALTV (D2S2A1)
Haloterrigena turkmenica g3	D2S1E1 HALTV (D2S1E1)
Kordia algicida	A9E081 9FLAO (A9E081)
Kribbella flavida	D2PKN1 KRIFD (D2PKN1)
Leeuwenhoekiella blandensis	A3XNY5 LEEBM (A3XNY5)
Microbacterium sp.	B1NEZ1 9MICO (B1NEZ1)
Natrialba magadii	D3SY20 NATMM (D3SY20)
Physcomitrella patens	A9SLH0 PHYPA (A9SLH0)
Polaribacter irgensii	A4BW74 9FLAO (A4BW74)
Polaribacter sp.	A2TXN6 9FLAO (A2TXN6)
Populus trichocarpa	B9PBR9 POPTR (B9PBR9)
Ricinus communis	B9T8D2 RICCO (B9T8D2)
Riemerella anatipestifer g1	E4T920 RIEAD (E4T920)
Riemerella anatipestifer g2	F0TPC5 RIEAR (F0TPC5)
Riemerella anatipestifer g3	E6JHS7 RIEAN (E6JHS7)
Tetrahymena thermophila	Q23YQ3 TETTH (Q23YQ3)

Table 1. Cont.

Taxon	GenBank or Uniprot ID
uncultured haloarchaeon g1	A5YSL4 9EURY (A5YSL4)
uncultured haloarchaeon g2	A7U0W6 9EURY (A7U0W6)
Zunongwangia profunda	D5BFE2 ZUNPS (D5BFE2)

analysis and can only be explained by multiple gene transfer from the common ancestor of all BTPC, protists PEPC and bacteria PEPC to the ancestor of protists and bacteria. There is one remaining problem: the diversification of bacteria is a very ancient event,predating the divergence between algae and vascular plants. In theory, the duplicated copy of the ancestral PTPC which postdates the divergence of vascular and algal plant PTPC can by no means be transferred to the ancestors of the bacteria. Reconciliation between the gene tree and species tree is then almost impossible. This phylogeny must be reconsidered with caution.

We searched the GenBank and UniProt to explore the entire range of existent PEPC genes in all organisms sequenced in the database. We identified possible inter-kingdom HGT candidates in PEPC family, and constructed the gene family phylogeny with genes from representative taxa and those identified inter-kingdom HGT candidates in order to clarify the evolution of this gene family and validate the suspected inter-kingdom HGT events.

Results and Discussion

We searched the GenBank by BLASTP and tBLASTn using PEPCs as a query and found that PEPC is a widely spread gene in archaea, prokaryotes and eukaryotes. In eukaryotes, PEPC exists mostly in plants, protists and slime mold. Only two hits were found in animals:The first was found in the genome of the black-legged tick, *Ixodesscapularis*. The 164-amino-acid fragment on the C-terminus of a 193-amino-acid protein(gene ID: 8031581) has 100% identity with pepcase from an alpha-proteobacterium, *Rhodobacterales bacterium* HTCC2255. Because this peptide is very short and possibly non-functional, it may be the relic of a recent unsuccessful horizontal gene transfer. The second was found in the genome of platypus, *Ornithorhynchusanatinus*. This is a peptide of 374 amino-acid (gene ID: 345310721) coded on a short contig of 1,614 base pair in the genome assembly. This gene has its closest homolog (e value, 3e-98) in a parasite, *Babesiabovis*. This may be a result of gene transfer from the parasite to the host, but we cannot exclude the possibility of parasitic genome pollution during genomic DNA preparation of the sequencing project.

We confirmed our suspicion of parasite contamination after reviewing the gene family information in Pfam database, in which we found two PEPC gene families, PEPcase (PF00311) and PEPcase_2 (PF14010). PEPcase is distributed in bacteria and eukaryotes including plants, protists and slime mold, while PEPcase_2 is mainly distributed in Archaea. However, there are also members within the two gene families whose taxonomy positions are incongruent with the main distribution, potentially due to an inter-kingdom HGT. From the maximum likelihood phylogenetic tree based on the curated seed alignment of PF00311 (Figure S1), we saw that plant PEPC is clustered with a group of PEPCs from gamma-proteobacteria,forming a sister group to other bacteria PEPCs. This phylogeny supported the idea that plant PEPCs is a lineage derived from ancestral bacteria PEPCs by

means of an ancestral inter-kingdom HGT, contrary to the previous understandings that bacteria PEPCs originated from plant PEPCs. However, the plant PEPCs in the seed alignment all belong to the so-called BTPC group and many important eukaryotic taxa that are not plant, such as the protist and cellular slime mold, were not included in the seed alignment. To identify the origin of PTPC and PEPCs in the non-plant eukaryotic taxa, we carried out further phylogeny reconstruction of PEPCs from representative taxa in bacteria, archaea, plant and non-plant eukaryotes.

To explore the possible existence of inter-kingdom HGT in PEPC, we screened the full curation of PF00311 and PF14010 in the Pfam database to find inter-kingdom HGT candidates and included those candidates in the sequences for the following phylogenetic reconstruction. We searched the Pfam "full" tree to find the PEPC sequences from different kingdoms with the branches surrounding it. As no PEPC is found in fungi and only two are found in animals, we focused on divisions of the plants, bacteria and archaea. In total, we found 29 sequences from non-archaea organisms in the full tree of PF14010, 49 sequences from non-plant organisms and 30 sequences from non-bacteria organisms in the plant and bacteria divisions of the PF00311 full tree, respectively. Because the phylogeny of PF00311 contain 2976 sequences and many alignments of short fragments are represented on the tree and many internal branches have low bootstrap support value, we removed dubious candidates from short fragment of peptide (less than 300 amino acids), and used the remaining 21 sequences from non-plant organisms and 19 sequences from non-bacteria organisms to carry out further phylogenetic analysis.

Having collected the inter-kingdom HGT candidates from plant and bacteria, we carried out phylogeny reconstruction in combination with the sequences of the non-plant eukaryotic taxa, BTPC and PTPC from several plants and representative bacteria PEPCs curated in the seed alignment (Table 1). In total, we used 122 PEPCs for gene phylogeny reconstruction. For the inter-kingdom HGT in archaeaphylogenetic reconstruction, we used the sequences of all 77 members of PEPcase_2 and four bacteria PEPCs as outgroups. We first aligned the sequences and then adopted a program MUMSA to assess the quality in order to find the best alignment by calculating the multiple overlap score (MOS) that indicates the overall inter-consistency with other alignments (see Materials and Methods). The alignment with the highest MOS was selected as the best alignment, and those alignments were then used to carry out phylogeny reconstruction.

We constructed the phylogenetic tree using three methods: maximum likelihood, neighbor joining and maximum parsimony. The protein substitution model used in maximum likelihood was selected by calculating the likelihood score under all 20 available models implemented in RAxML, and then we selected the model with the highest score. To avoid artificial resultscaused by improper construction methods, we combined the three trees to build a consensus tree that only contained branches supported by all the three methods. By inspecting this final consensus tree manually, we confirmed that there are 19 non-bacteria sequences clustered within the bacteria branches, a single non-plant sequence clustered within the plant branches (Figure 1) and 29 non-archaea sequences clustered within the archaea branches (Figure 2). To avoid artificial results due to uncertainty of alignment, we also repeated the phylogenetic analysis with the second best alignments and found no contradictory evidence (data not shown). To further exclude the possibility of artifacts due to alignment, we used GUIDANCE [9] to carry out alignment and bootstrap assessment of the alignment confidence and used only the high confidence

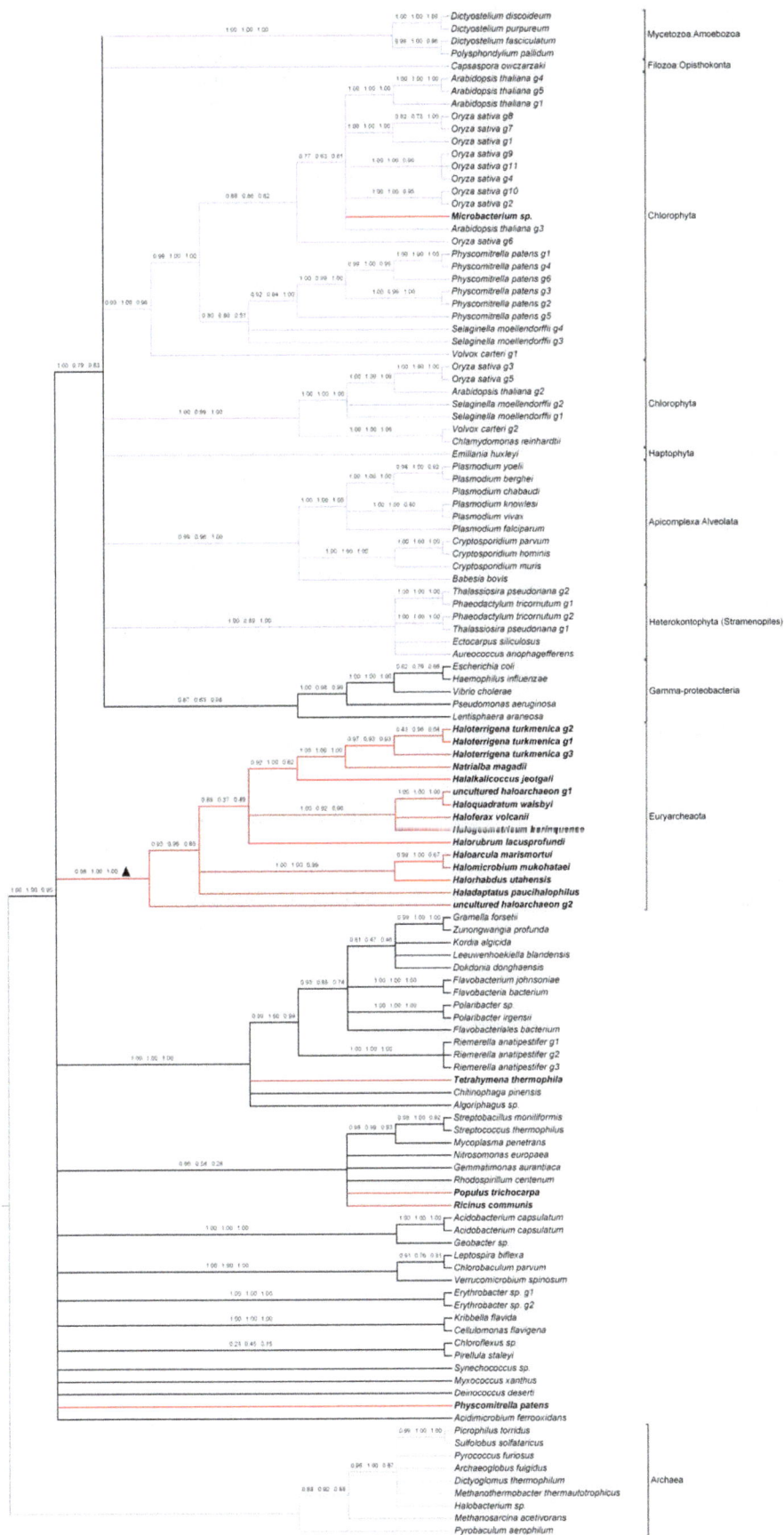

Figure 1. Phylogeny of bacteria and eukaryotic PEPcase and inter-kingdom HGT candidates. Phylogeny of inter-kingdom HGT candidates and PEPcase sequences from representative taxa in bacteria and eukaryotes were reconstructed. Nine archaea sequences were included as outgroups. HGT candidates confirmed in this phylogeny are in bold letters with red branches. The branches of outgrouparcheae are in grey and all eukaryotic branches are in blue. The bootstrap values of 100 replicates in the three different methods were labeled on each branch in order of maximum likelihood, neighbor joining and maximum parsimony. Ancient HGT events are marked with a triangle on the branch.

columns (with bootstrap scores greater than 0.93) in the alignment to reconstruct the phylogeny. The results also showed no contradictory evidence with our major conclusion (See Figure S2 and S3). In Figure S2, the monophyly of all eukaryotic genes is supported by ML, NJ, MP with bootstrap value of 0.43, 0.64 and 0.17, respectively. However, the relation between eukaryotic groups (plant, protist, slime mold) is not consistent among three methods and most of the nodes are of low confidence. And for the pepcase_2 tree in Figure S3, the topology of NJ tree and MP tree are mostly consistent and those consensus nodes also receive high bootstrap support in NJ tree. The ML tree differs with the other two trees in the branch order of the basal branches. In the ML tree, thegroup of HGTs in Clostridium split first with the other archeae groups, while in NJ and MP trees a group of *Crenarchaeota* containing *Ignicoccushospitalis* diverges first with the other archeae groups. And also the NJ tree received the highest bootstrap support of those consensus nodes for pepcase_2. Compared with the computational cost of the ML and MP method, NJ seems to be the most efficient method among them.

And we also checked the genomic location of those candidates to exclude the possibility of sequence pollution for those un-clustered HGT genes. The result showed that most genes are from long genomic scaffolds except for the HGT genes in poplar and *Microbaterium* sp. which are from short fragments of 1,312 bp and 2,913 bp (Table S1). However, because the HGT gene in *Microbacterium* sp. clustered together with genes from seed plants and the possibility of genomic contamination of microbial genome library from multi-cellular organism is very low.We believe that the HGT in *Microbatierium* sp. is probably not the result of contamination. Further experiment is needed to exclude the possibility of genomic contamination for the HGT candidates in *Populustrichocarpa*. Collectively, in the evolution of phosphoenol-pyruvate carboxylase gene family, we found 48 sequences originated from inter-kingdom HGTs. We also found that there three separate ancient HGT events,one from bacteria to archaea and the other two from archaea to bacteria,that respectively contributed to 15, 10 and 14 genes (Figure 1 and 2).

As for the origin of BTPC and PTPC, our phylogeny supported the idea that each type of PEPCs form a monophyletic group and both originated from ancestral bacteria lineage. That said, there is still uncertainty as to the precise relationship between these two groups and other eukaryotic PEPCs, due to inconsistency between different methods and low bootstrap support. This is consistent with the reality that the deep phylogeny of eukaryotes is still surrounded by controversy. Hopefully, further research on the basal phylogeny of eukaryotes will shed light on some of the controversy and further help explain the evolution of BTPC and PTPC. And our results also provide some information concerning the large scale phylogeny of the three life domains: Eukaryote, Eubacteria and Archeae. The well accepted phylogeny based on small-subunit (SSU) rDNA showed that Eukaryote and Archeae form a sister group with Eubacteria as the outgroup. However, many operational genes in Eukaryote are found to be more similar with homologs in Eubacteria while most eukaryotic informational genes is closer to their homologs in Archeae. And many hypothesis of symbiotic origin of Eukaryote are formed based on this finding. PEPC in Eukaryote is another gene originated via the horizontal

gene transfer from bacteria symbiont (probably the ancestor of chloroplast) to the nucleus of the ancestral eukaryotic host [10,11].

On a broader level, HGT was thought to be a relatively rare event in evolution. As more and more genome sequences become available, we continue to find many genes in the genome originated from HGT [12,13,14]. To date, however, there are no well-supported cases of multiple HGT events occurring in one gene family. One potential reason is that HGT was thought of as rare event, unlikely to hit a single gene family more than once. Consequently, little systematic research looking for HGT events in one gene family has been done. Our research provides the first case of multiple inter-kingdom HGTs in a single gene family and furthermore suggests that HGTsare much more frequent and important than previously expected. There is also research showing that HGT is more frequent between closely related organisms [15]. Here we opted to only look into the inter-kingdom HGT because HGTs between different kingdomsaremore readily identified when the intra-kingdom phylogeny of many species based on well recognized orthologs is not available. However, the frequency of all HGTs should be much higher than that of inter-kingdom HGT which we found in this study.

Successful HGTs involve two processes: the physical transfer of the genetic material into the recipient genome of another species, and the fixation of the gene in the population of the species by selection forces. Our findings are consistent with the fact that HGTs were found to be biased toward operational genes as opposed to informational gene because the operational gene can function and bring out fitness advantages with less interaction with other genes [11,16]. PEPC is an operational gene that can function in many metabolic and developmental pathways but does not need many partner genes. We can only speculate that this may be the reason there are so many HGT events surrounding the evolution of this gene.

Materials and Methods

We downloaded the protein sequences, alignment and phylogenetic trees of PEPcase (PF00311) and PEPcase_2 (PF14010) from the Pfam database [17]. Phylogenetic tree viewing and editing was done in the tree editor Archaeopteryx (0.960 beta A48) [18]. We cut the kingdom specific sub-trees for both bacteria and plant from Pfam full tree of PF00311. For archaea, we use the full Pfam tree of PF14010. Base on those kingdom specific tree, we use home-made scripts to find out the inter-kingdom HGT candidate, which is wrapped in the branches belong to a different kingdom in the Pfam tree. First, the taxonomy codes of all leaves were extracted from the sub-trees of bacteria, plant and archaea and searched in the UniProt taxonomy database [19]. We then inspected the taxonomy search results to find the taxa whose lineages do not contain the bacteria, plant or archaea. Finally, we extracted the full protein sequences and aligned fragments of those taxa from Pfam database; aligned fragments shorter than 300 amino acids were excluded from candidate list.

To validate the phylogenetic relationship between those HGT candidates and other members of PEPcase gene family and get a panorama of the gene family evolution in plant and bacteria, we collected the HGT candidates' full sequences and PEPcase sequences from representative taxa, totally 122 protein sequences

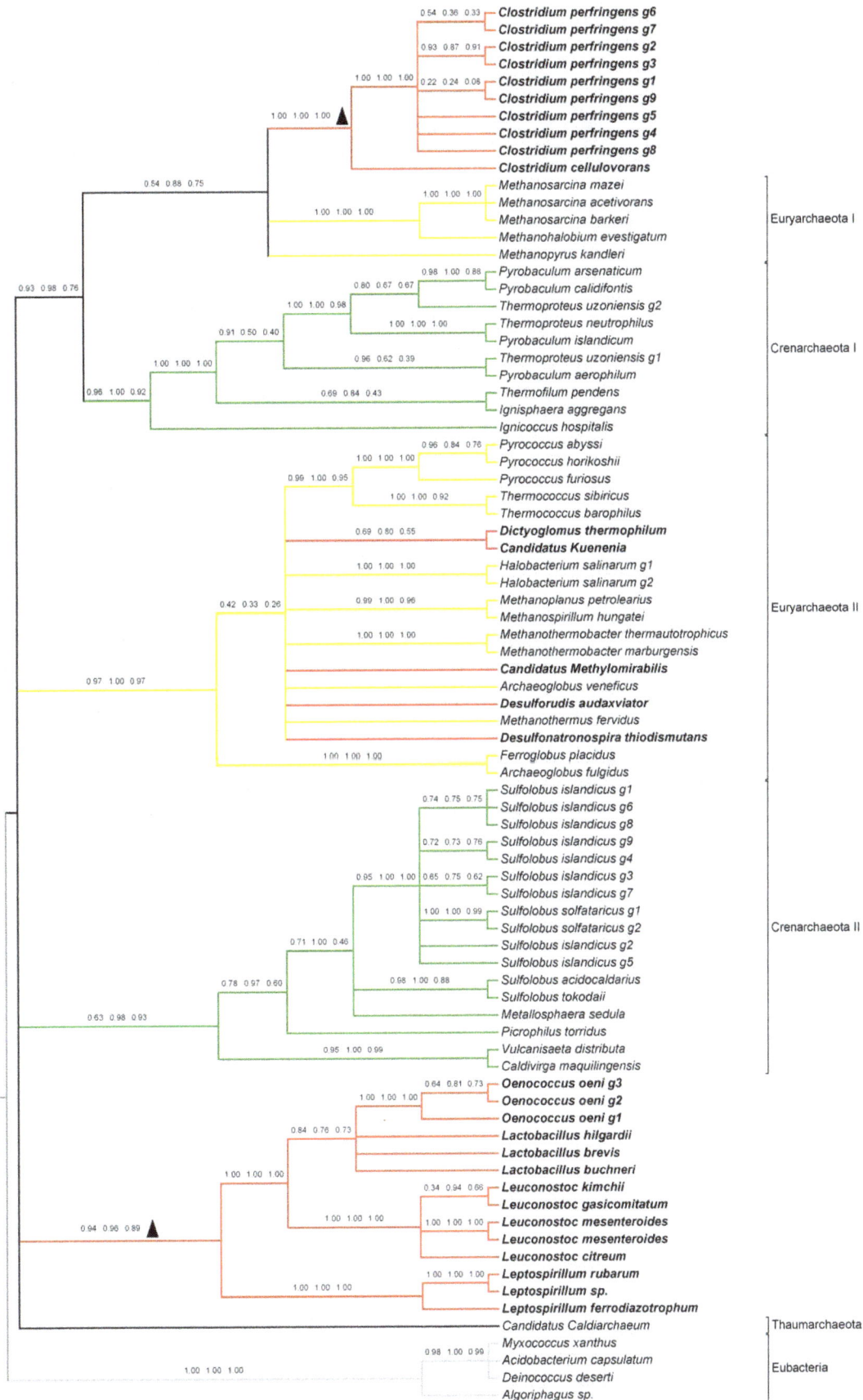

Figure 2. Phylogeny of archaea PEPcase and inter-kingdom HGT candidates. Phylogeny of PEPcase sequences from PF14010 were reconstructed. Four bacteria sequences were included as outgroups and their branches are in grey. HGT candidates confirmed in this phylogeny are in bold letters with red branches. The bootstrap values of 100 replicates are labeled in the same manner as Figure 1. Ancient HGT events were marked with triangles. Euryarchaeota branches were drawn in yellow while Crenarchaeota branches were in green.

to reconstruct the phylogeny of the gene family. For archaea, we used the full sequences of all PF14010 members. We applied four programs (T-Coffee, MAFFT, MUSCLE and ClustalW) to align the sequences and then assessed the quality of the alignments with Mumsa (online server at http://msa.sbc.su.se/cgi-bin/msa.cgi) [20,21,22,23,24]. All alignment programs were run using the default parameters, except T-Coffee where we used the "expresso" option.

The sequences in all alignments were sorted into the same order with MEGA5 [25] and then submitted to the Mumsa server to get the quality scores. Mumsa program calculates the MOS score of each alignment (See [24] for the detail of the algorithim). Briefly, the aligned residues shared by many alignments are more reliable, and the alignment with the largest number of such residues is supposed to be the closest to the true alignment [24]. We then selected the alignment with best quality to carry out phylogeny reconstruction with maximum likelihood, neighbor-joining and maximum parsimony methods. For maximum likelihood tree, we first use RAxML and a wrapperPERL script proteinmodelselection.pl to find the substitution model with highest likelihood score for the protein alignment, and then we used this substitution model with GAMMA model of rate heterogeneity and carried out rapid bootstrap test of 100 replicates [26]. The neighbor-joining tree was inferred using MEGA5 with distances calculated with Possion correction and bootstrap test of 100 replicates. The maximum parsimony tree was also inferred using MEGA5 with the Close-Neighbor-Interchange algorithm and bootstrap test of 100 replicates. We combined the consensus trees of three methods using TreeGraph2 and deleted the different methods' contradictory nodes [27]. Finally, inter-kingdom HGT genes were identified by manual inspection of the combined phylogenetic tree.

To further test our conclusion against alignment artifacts, we used the GUIDANCE webserver [9] to carry out alignment and assessment of the alignment accuracy. The analysis was carried out with default parameters, using MAFFT as the aligner and GUIDANCE as the algorithms for evaluating confidence scores, which measures the robustness of the alignment to guide-tree uncertainty. Then the high confidence columns of the alignments were extracted from the result with threshold of score 0.93. Then the filtered alignments were further used to reconstruct the phylogeny with three different methods (same as the above).

Supporting Information

Figure S1   Maximum likelihood tree of PF00311 seed alignment. Phylogenetic tree of PF00311 seed alignment were

downloaded from Pfam database and then midpoint-rooted and visualized with the tree viewer, Archaeoptertx 0.960 beta A48. All sequences were labeled in the Pfam style (UniProt protein ID+UniProt taxonomy ID+coordinates of beginning and ending of alignment). Bootstrap support values are labeled by the nodes. Plant PEPCs are marked with a curly bracket.

Figure S2   Phylogeny of bacteria and eukaryotic PEPcase and inter-kingdom HGT candidatesbaced on filtered aligment with GUIDANCE. Phylogeny of inter-kingdom HGT candidates and PEPcase sequences from representative taxa in bacteria and eukaryotes were reconstructed based on the filtered alignment result of GUIDANCE using three methods: a. Maximum Likelihood; b. Neighbor-Joining; c. Maximum Parsimony. Nine archaea sequences were included as outgroups. HGT candidates are in bold letters with red branches. The branches of outgrouparcheae are in grey and all eukaryotic branches are in blue. The bootstrap values of 100 replicate are labeled on the branches. The branch line widths were set with the support value.

Figure S3   Phylogeny of archaeaPEPcase and inter-kingdom HGT candidatesbaced on filtered aligment with GUIDANCE. Phylogeny of PEPcase sequences from PF14010 were reconstructed based on the filtered alignment result of GUIDANCE using three methods: a. Maximum Likelihood; b. Neighbor-Joining; c. Maximum Parsimony. Four bacteria sequences were included as outgroups and their branches are in grey. HGT candidates are in bold letters with red branches. The bootstrap values of 100 replicate are labeled on the branches. The branch line widths were set with the support value. Euryarchaeota branches were drawn in yellow while Crenarchaeota branches were in green.

Table S1   Genomic information on singular HGT candidates.

Author Contributions

Conceived and designed the experiments: YP JC BS. Performed the experiments: YP JC. Analyzed the data: YP JC. Contributed reagents/materials/analysis tools: JC. Wrote the paper: YP JC WW BS.

References

1. Garcia-Vallve S, Romeu A, Palau J (2000) Horizontal gene transfer in bacterial and archaeal complete genomes. Genome Res 10: 1719–1725.
2. Huang J, Mullapudi N, Lancto CA, Scott M, Abrahamsen MS, et al. (2004) Phylogenomic evidence supports past endosymbiosis, intracellular and horizontal gene transfer in Cryptosporidium parvum. Genome Biol 5: R88.
3. Khaldi N, Collemare J, Lebrun MH, Wolfe KH (2008) Evidence for horizontal transfer of a secondary metabolite gene cluster between fungi. Genome Biol 9: R18.
4. Moustafa A, Beszteri B, Maier UG, Bowler C, Valentin K, et al. (2009) Genomic footprints of a cryptic plastid endosymbiosis in diatoms. Science 324: 1724–1726.
5. Rumpho ME, Worful JM, Lee J, Kannan K, Tyler MS, et al. (2008) Horizontal gene transfer of the algal nuclear gene psbO to the photosynthetic sea slug Elysia chlorotica. Proc Natl Acad Sci U S A 105: 17867–17871.
6. Nikoh N, McCutcheon JP, Kudo T, Miyagishima SY, Moran NA, et al. (2010) Bacterial genes in the aphid genome: absence of functional gene transfer from Buchnera to its host. PLoS Genet 6: e1000827.
7. Sanchez R, Cejudo FJ (2003) Identification and expression analysis of a gene encoding a bacterial-type phosphoenolpyruvate carboxylase from Arabidopsis and rice. Plant Physiol 132: 949–957.
8. O'Leary B, Park J, Plaxton WC (2011) The remarkable diversity of plant PEPC (phosphoenolpyruvate carboxylase): recent insights into the physiological functions and post-translational controls of non-photosynthetic PEPCs. Biochem J 436: 15–34.
9. Penn O, Privman E, Ashkenazy H, Landan G, Graur D, et al. (2010) GUIDANCE: a web server for assessing alignment confidence scores. Nucleic Acids Res 38: W23–28.

10. Henze K, Badr A, Wettern M, Cerff R, Martin W (1995) A nuclear gene of eubacterial origin in Euglena gracilis reflects cryptic endosymbioses during protist evolution. Proc Natl Acad Sci U S A 92: 9122–9126.

11. Jain R, Rivera MC, Lake JA (1999) Horizontal gene transfer among genomes: the complexity hypothesis. Proc Natl Acad Sci U S A 96: 3801–3806.

12. Fitzpatrick DA, Logue ME, Butler G (2008) Evidence of recent interkingdom horizontal gene transfer between bacteria and Candida parapsilosis. BMC Evol Biol 8: 181.

13. Gladyshev EA, Meselson M, Arkhipova IR (2008) Massive horizontal gene transfer in bdelloid rotifers. Science 320: 1210–1213.

14. Faguy DM, Doolittle WF (1999) Lessons from the Aeropyrum pernix genome. Curr Biol 9: R883–886.

15. Wagner A, de la Chaux N (2008) Distant horizontal gene transfer is rare for multiple families of prokaryotic insertion sequences. Mol Genet Genomics 280: 397–408.

16. Lercher MJ, Pal C (2008) Integration of horizontally transferred genes into regulatory interaction networks takes many million years. Mol Biol Evol 25: 559–567.

17. Punta M, Coggill PC, Eberhardt RY, Mistry J, Tate J, et al. (2012) The Pfam protein families database. Nucleic Acids Res 40: D290–301.

18. Han MV, Zmasek CM (2009) phyloXML: XML for evolutionary biology and comparative genomics. BMC Bioinformatics 10: 356.

19. Magrane M, Consortium U (2011) UniProt Knowledgebase: a hub of integrated protein data. Database (Oxford) 2011: bar009.

20. Di Tommaso P, Moretti S, Xenarios I, Orobitg M, Montanyola A, et al. (2011) T-Coffee: a web server for the multiple sequence alignment of protein and RNA sequences using structural information and homology extension. Nucleic Acids Res 39: W13–17.

21. Edgar RC (2004) MUSCLE: multiple sequence alignment with high accuracy and high throughput. Nucleic Acids Res 32: 1792–1797.

22. Katoh K, Toh H (2010) Parallelization of the MAFFT multiple sequence alignment program. Bioinformatics 26: 1899–1900.

23. Larkin MA, Blackshields G, Brown NP, Chenna R, McGettigan PA, et al. (2007) Clustal W and Clustal X version 2.0. Bioinformatics 23: 2947–2948.

24. Lassmann T, Sonnhammer EL (2005) Automatic assessment of alignment quality. Nucleic Acids Res 33: 7120–7128.

25. Tamura K, Peterson D, Peterson N, Stecher G, Nei M, et al. (2011) MEGA5: molecular evolutionary genetics analysis using maximum likelihood, evolutionary distance, and maximum parsimony methods. Mol Biol Evol 28: 2731–2739.

26. Stamatakis A, Ludwig T, Meier H (2005) RAxML-III: a fast program for maximum likelihood-based inference of large phylogenetic trees. Bioinformatics 21: 456–463.

27. Stover BC, Muller KF (2010) TreeGraph 2: combining and visualizing evidence from different phylogenetic analyses. BMC Bioinformatics 11: 7.

9

Variation in Stem Anatomical Characteristics of Campanuloideae Species in Relation to Evolutionary History and Ecological Preferences

Fritz Hans Schweingruber[1], Pavel Říha[2,3], Jiří Doležal[2,3]*

1 Swiss Federal Research Institute WSL, Birmensdorf, Switzerland, 2 Section of Plant Ecology, Institute of Botany, Academy of Sciences of the Czech Republic, Třeboň, Czech Republic, 3 Department of Botany, Faculty of Science, University of South Bohemia, České Budějovice, Czech Republic

Abstract

Background: The detailed knowledge of plant anatomical characters and their variation among closely related taxa is key to understanding their evolution and function. We examined anatomical variation in 46 herbaceous taxa from the subfamily Campanuloideae (Campanulaceae) to link this information with their phylogeny, ecology and comparative material of 56 woody tropical taxa from the subfamily Lobelioideae. The species studied covered major environmental gradients from Mediterranean to Arctic zones, allowing us to test hypotheses on the evolution of anatomical structure in relation to plant competitive ability and ecological preferences.

Methodology/Principal Findings: To understand the evolution of anatomical diversity, we reconstructed the phylogeny of studied species from nucleotide sequences and examined the distribution of anatomical characters on the resulting phylogenetic tree. Redundancy analysis, with phylogenetic corrections, was used to separate the evolutionary inertia from the adaptation to the environment. A large anatomical diversity exists within the Campanuloideae. Traits connected with the quality of fibres were the most congruent with phylogeny, and the *Rapunculus* 2 ("phyteumoid") clade was especially distinguished by a number of characters (absence of fibres, pervasive parenchyma, type of rays) from two other clades (*Campanula* s. str. and *Rapunculus* 1) characterized by the dominance of fibres and the absence of parenchyma. Septate fibres are an exclusive trait in the Lobelioideae, separating it clearly from the Campanuloideae where annual rings, pervasive parenchyma and crystals in the phellem are characteristic features.

Conclusions/Significance: Despite clear phylogenetic inertia in the anatomical features studied, the ecological attributes and plant height had a significant effect on anatomical divergence. From all three evolutionary clades, the taller species converged towards similar anatomical structure, characterized by a smaller number of early wood vessels of large diameter, thinner cell-walls and alternate intervessel pits, while the opposite trend was found in small Arctic and alpine taxa. This supports the existing generalization that narrower vessels allow plants to grow in colder places where they can avoid freezing-induced embolism, while taller plants have wider vessels to minimize hydraulic resistance with their greater path lengths.

Editor: Keping Ma, Institute of Botany, Chinese Academy of Sciences, China

Funding: The study was funded by research grants GAČR 13-13368S, GAČR P505/11/1617 of the Grant Agency of the Czech Republic, institutional long-term research plan AV0Z60050516, and the grant OP Education for Competitiveness CZ.1.07/2.3.00/30.0048. The funders had no role in study design, data collection and analysis, decision to publish, or preparation of the manuscript.

Competing Interests: The authors have declared that no competing interests exist.

* E-mail: jiriddolezal@gmail.com

Introduction

The detailed knowledge of plant anatomical characters and their variation among closely related taxa is key to understanding their evolution and function [1]. Variation in anatomical structure is a result of several forces such as: the adaptation of species to the prevailing conditions in their habitats [2], phenotypic plasticity as an ability of individuals with an identical genotype to develop differently - based on specific conditions during their ontogeny [3,4], and evolutionary constraints (phylogenetic inertia) in which taxa that share part of their evolutionary history possess similar 'blue-prints' [5–7]. One of the critical features of comparative studies on plant trait variations in relation to ecological adaptations is therefore the extent of phylogenetic relatedness among taxa [8], which makes them partly dependent in any statistical inference. In other words, part of the explanatory power uncovered by relating the anatomical traits to ecological preferences might be alternatively explained by phylogenetic inertia affecting both the similarity of anatomical traits among closely related taxa and the similarity of ecological niches that such taxa occupy [9].

Understanding the evolution of plant structures requires separation of evolutionary inertia from a true adaptation to the environment. This is commonly done by comparing analyses made with and without phylogenetic corrections [10]. This approach is based on discounting all of the variation that could

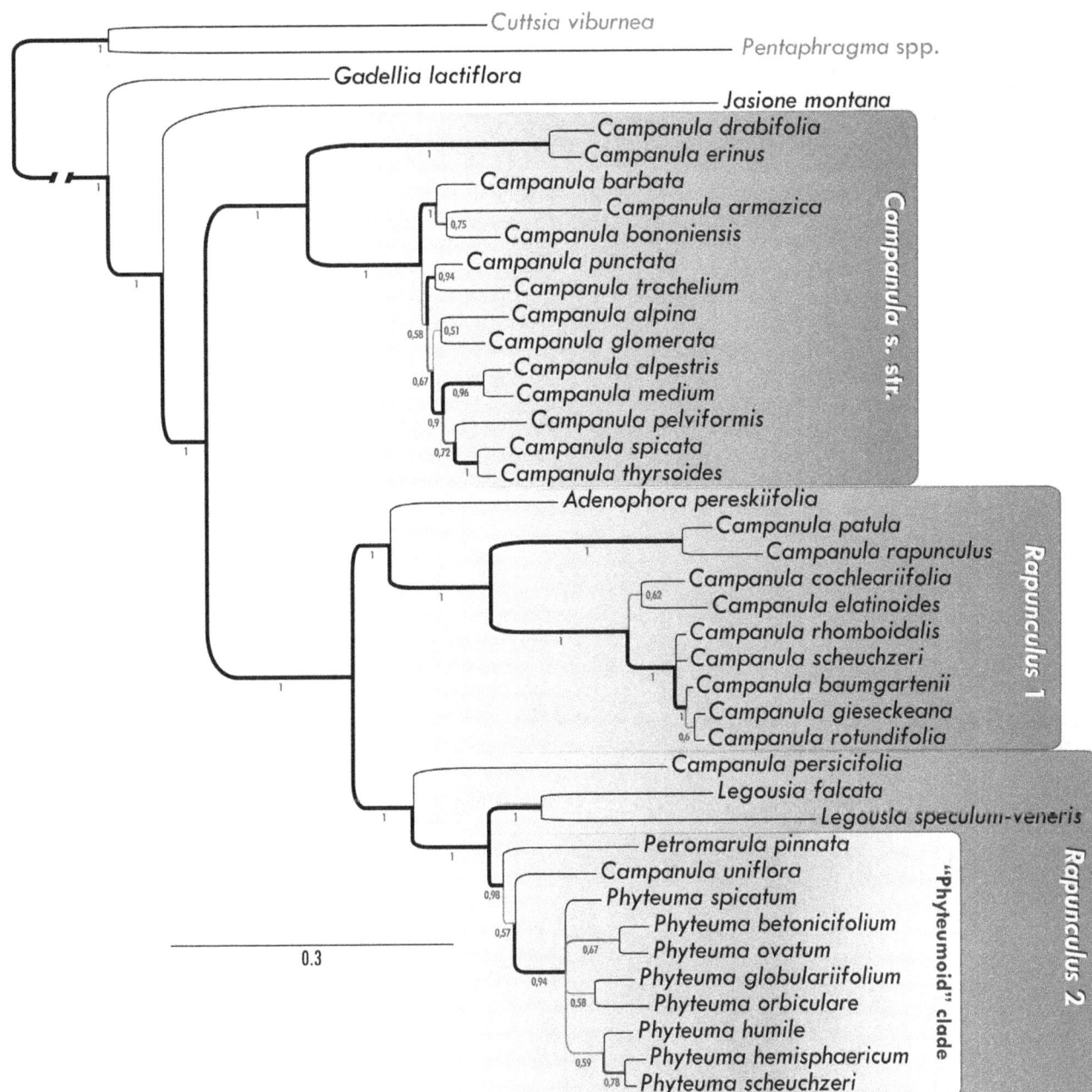

Figure 1. Majority rule consensus of trees sampled by the Bayesian analysis of a concatenated partitioned dataset consisting of ITS, *trnL-trnT* **spacer,** *matK, rbcL* **and** *petB-petD* **spacer sequences.** Numbers below the branches indicate Bayesian posterior probability (BPP) values. For better orientation, major clades mentioned in the text are delimited by grey boxes.

possibly be explained by phylogenetic relatedness of the studied species [9], before studying the effect of other potential predictors such as the species' ecological preferences. The phylogenetic relatedness is presented by a cladogram, the quality of which can affect inferences about adaptations. For instance, phylogenetic trees derived from DNA sequences give more accurate information than traditional taxonomy based on morphological data. The phylogenetic tree is then turned into a distance matrix, representing the distance of any pair of taxa measured along the branches of the tree; this value represents the distance to the nearest common ancestor of the two taxa being compared. The distance matrix is then used as a set of descriptors which can be used as covariates in

analyses that need to account for the effect of phylogenetic relatedness first, before focusing on the effect of ecological variables [7,11].

In this paper, we aim to provide new insights into anatomical stem variation in closely related plant taxa from the subfamily Campanuloideae (Campanulaceae family, belonging to the order of Asterales) and to link this information with their phylogeny and ecology. The Campanulaceae family includes 84 genera with 2400 species [12]. In Europe there are 14 genera representing 209 endemic species [13], with the genera *Campanula* (142 species) and *Phyteuma* (24 species) being most common. Eddie et al. [14] characterized the morphological and phylogenetic features of this

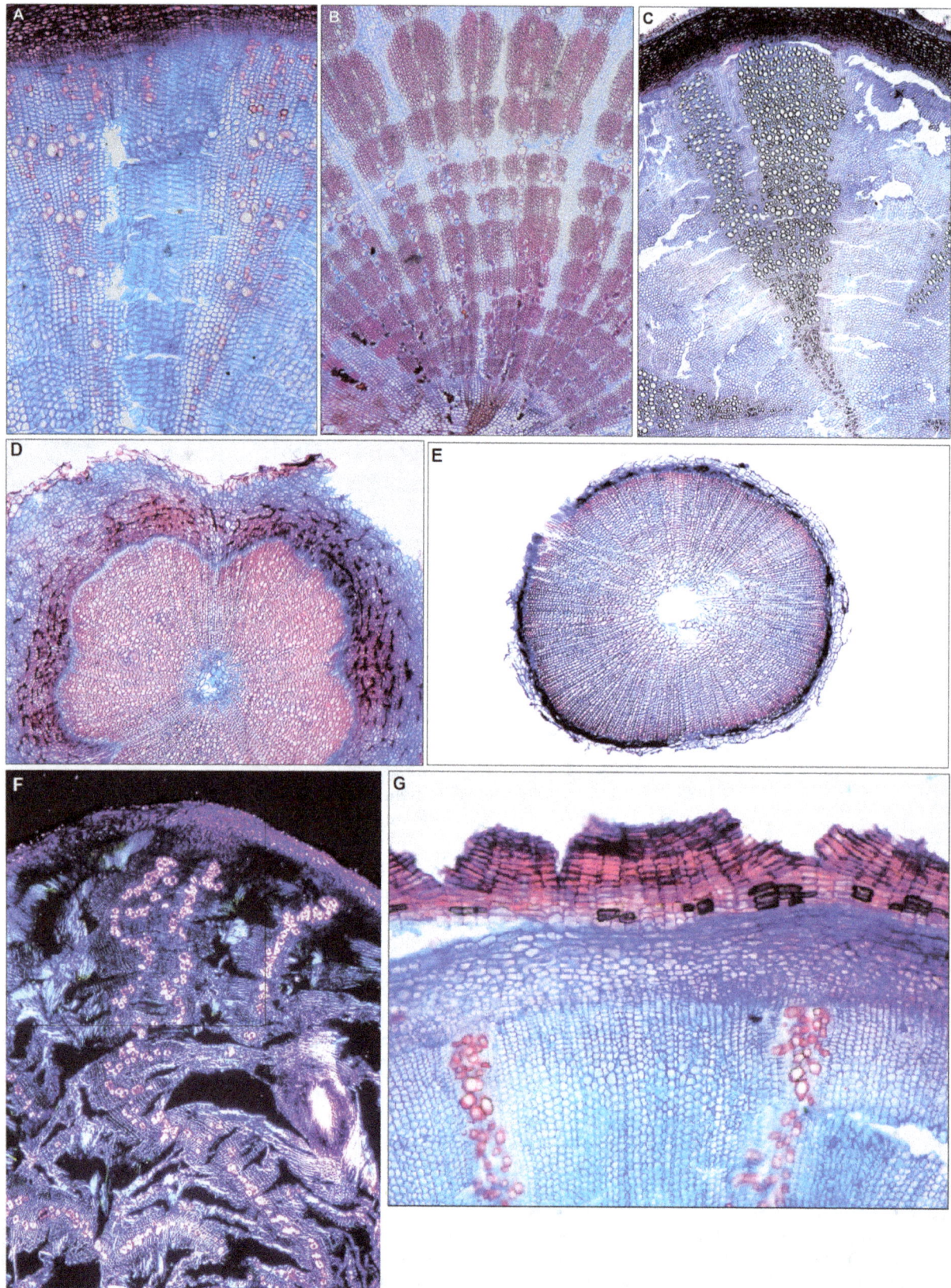

Figure 2. Annual ring boundaries and ray-like structures in herbaceous Camapanulaceae. (A) Semi-ring porous xylem without fibres in root collar of *Campanula rotundifolia*, 20 cm tall perennial herb, dry meadow, montane zone of the Swiss Alps. Expanding ray-like, vessel free zones separate radial strips of vessels which are surrounded by pervasive parenchyma. (B) Diffuse porous xylem with marginal, not lignified parenchymatous zones in root collar of *Gadellia lactiflora*, 80 cm tall perennial herb, meadow, subalpine zone of the Georgian Caucasus. Unlignified ray-like structures separate the radial vessel/fibre/parenchyma strips. (C) Diffuse porous xylem without fibres and tangential cracks (ring shakes) in the

vessel-free, ray-like radial zone in root collar of *Campanula rotundifolia*, 10 cm tall perennial herb, meadow, montane zone of the Swiss Alps. Ring boundaries do not exist within the fibre/parenchyma zones. Primary vascular bundles keep their form over many years. Annual species with mandatory fibre formation, very small vessels, absent axial parenchyma and rays and different bark structures. (D) A large zone consisting of lignified fibres and vessels surrounds the pith. Dark spots in the large phloem represent collapsed sieve tubes in root collar of a *Campanula erinus*, 5 cm tall herb in a dry meadow of the Mediterranean zone, Crete, Greece 40x. Cells in the active cambium are not lignified (blue). (E) As (D) but with a small and not distinctly structured phloem. The cambium is no longer active. Root collar of a *Legousia falcata*, 20 cm tall herb, meadow of the hill zone in Switzerland. 40x. Perennial species with dense phellem belts. (F) Squeezed xylem due to high tensile strength of the phellem. The originally radial vessel/parenchyma strips are bent. Root collar of *Phyteuma orbiculare*, 20 cm tall herb, meadow, subalpine zone of the western Alps of France. Polarized light. 40x. (G) The phellem consists of rectangular cork cells, which are produced by an active phellogen. Root collar of *Campanula elatinoides*, 20 cm tall herb, Botanical garden Bern, hill zone of Switzerland.

family, with representatives of the Campanuloideae subfamily mostly concentrated in the Northern Hemisphere and widely distributed from subtropical Mediterranean to temperate and alpine-Arctic regions. Target species of this study included common taxa from all these habitats, allowing us to test several hypotheses on the evolution of plant structure and function. In general, variations in plant construction should lead to differences in plant physiological function. These differences in morphological structure and physiological function should allow differential tolerance to changes in environmental settings. For instance, in colder places smaller vessels have repeatedly evolved to enable plants to cope with freezing-induced embolism and cavitation [15]. Xylem cavitation diminishes a plant's capacity to transport water from the soil to the leaves. This reduction in xylem hydraulic conductivity can impair the carbon fixation rate by inducing stomatal closure to prevent further cavitation and desiccation of leaf tissues. In less hostile environments, taller plants should have larger vessels which will, in part, minimize hydraulic resistance by their greater path lengths [2,16]. The evolutionary and ecological implications of anatomical character variation in different environments have mostly been studied in conifers and deciduous broadleaved trees [17–19], with herbaceous plants remaining somewhat neglected.

Very few studies exist on the anatomy of Campanulaceae stems: Metcalfe and Chalk [20] studied two European herbaceous species (*Campanula pyramidalis*, *Asyreunema limonifolium*) of the subfamily Campanuloideae; Carlquist [1] described 56 species of Lobelioideae (summarized by Lammers [12]). Shulkina et al. [21] related anatomical structures of 15 Russian Campanuloideae species to life forms and evolution and Schweingruber et al. [22] made an anatomical survey of 36 stems from Campanuloideae species.

The present study has four goals: a) Describing the stem-anatomical structures of herbaceous plants from seasonal climates within the subfamily Campanuloideae, b) Constructing a phylogenetic tree for target species from nucleotide sequences obtained from genBank, c) Relating stem anatomical features to phylogeny and ecology, and d) Comparing these specimens with the anatomy of shrubs and trees of the tropical subfamily Lobelioideae described by Carlquist [1].

Materials and Methods

Target Species

The 46 species analyzed in this study were recently collected from their native habitats, mainly in Western Europe. Forty common and relatively widespread species were collected in the Alps and southern Europe, three in Georgia, two in Greenland and one in China along altitudinal gradients ranging from 20 to 3000 m a.s.l. The species mainly represent the European plants (Figure 1) from the Mediterranean zone and within an alpine altitudinal transect north and south of the Alps. Plant size varied between 5 cm and 100 cm. The detailed information on the species studied can be found in Table S1, available online. We

classified each species into one of the eight habitat categories representing species growth optima: Arctic zone, alpine meadows, alpine rocks (e.g. screes, rocks, outcrops), low-elevation meadows, low-elevation rock habitats, deciduous mixed forests and the Mediterranean zone. Anatomical codes and photographs of most of the species can be found in an online database (http://www.wsl. ch/dendro/xylemdb/index.php). Plant identification was based on the following references: Europe [13,23,24], China [25], Greenland [26], and Georgia [27]. No specific permits were required for the described field collections, the locations were not privately-owned or protected in any way and the field studies did not involve endangered or protected species.

Anatomical Sections

Transverse, tangential and radial sections were cut from a total of 122 individuals (see Figure 2 for examples). Since anatomical differences exist between roots, bulbs, root collars and annual flower stalks (Figure 3a–d), comparisons of anatomical sections were exclusively based on sections within the transition between the hypocotyl and the primary root (root collar). In this zone all annual rings of perennial plants do exist and the reaction to mechanical stress seems to be reduced to a minimum. All samples were stored in 40% ethanol before being sectioned with a sliding microtome. Sections were simultaneously stained with Safranin and Astrablue, dehydrated with ethanol and xylene, and mounted in Canada balsam [28]. The anatomical descriptions of the xylem are based on the IAWA List of microscopic features for hardwood identification [29] and specific xylem and phloem features of herbs based on Schweingruber et al. [30].

Phylogenetic Analysis

All the sequences used in this study were obtained from genBank (www.ncbi.nlm.nih.gov). Our goal was to get maximum overlap with the pre-existing morphological dataset and yet not to clutter the alignment with excessive unknown positions. A combination of five loci satisfied this condition: internal transcribed spacer (ITS), trnT-trnL intergenic spacer, matK+trnK region, the gene for rubisco large subunit (rbcL) and petB-petD intergenic spacer. Hence the full-length ITS locus was not always available and sometimes a concatenate of ITS1+ ITS2 was used; all the accession numbers can be found in Table S2, available online. The L-INS-i algorithm implemented in the online version of MAFFT 6 [31] was employed to align the sequence datasets. Partial alignments were concatenated, manually adjusted in BioEdit [32] and subdued to the *automated1* algorithm in trimAll software [33] to exclude highly divergent and gap-rich regions.

Prior to the phylogenetic analysis, the best-fit model was selected by Kakusan4 [34], where the baseml software [35] served as the computational core and both non-partitioned and partitioned models were evaluated. According to the Bayesian information criterion [36], we finally used the GTR model with rate variation across sites simulated by discrete gamma distribution ($\Gamma 8$), autocorrelated by the AdGamma rates prior and unlinked for

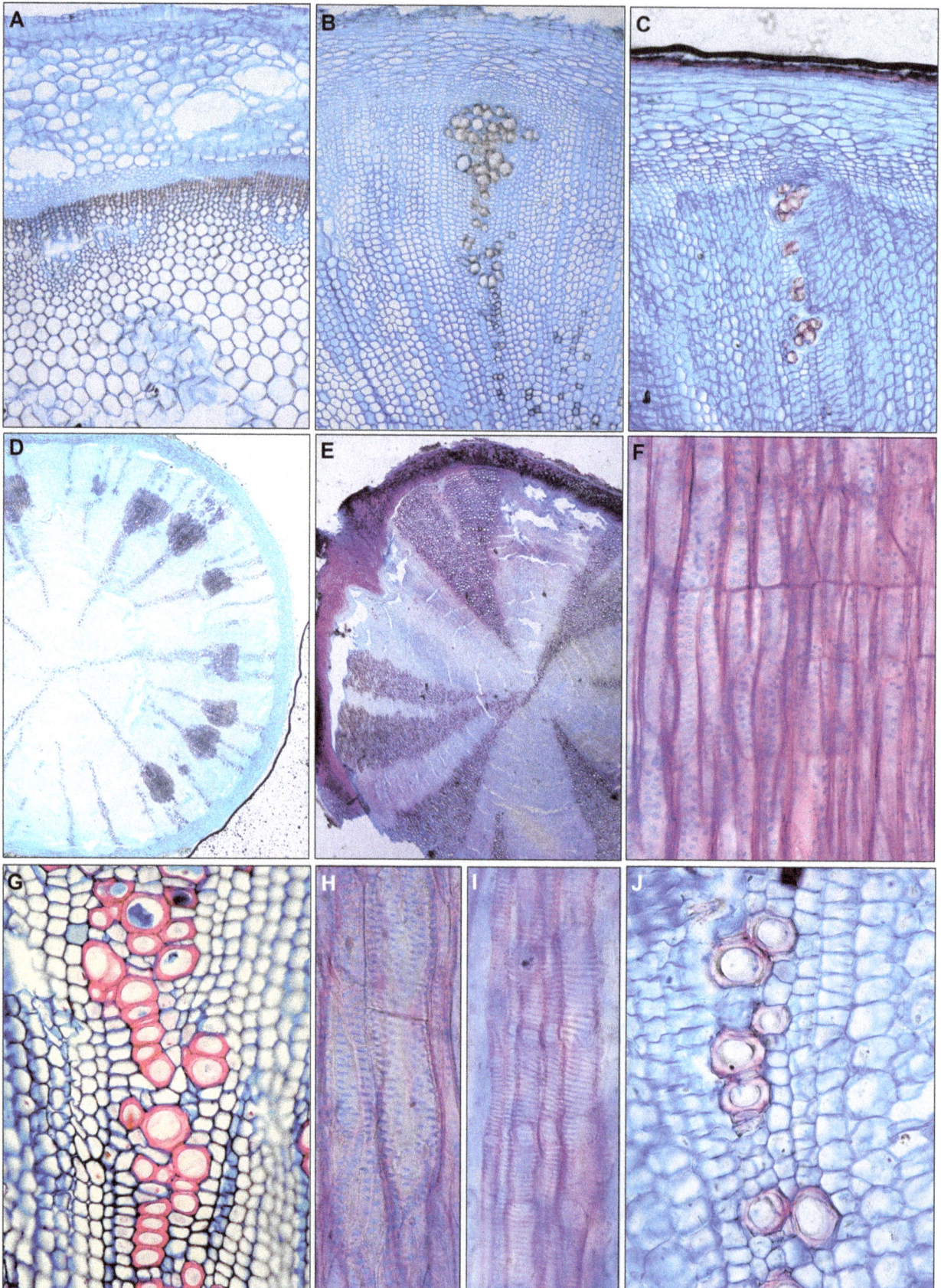

Figure 3. *Phyteuma ovatum*, **40 cm tall perennial herb, meadow of the subalpine zone in the Austrian Alps.** (A) Pith, xylem and bark of a flower stalk. The xylem consists mainly of lignified fibres. 100x. (B) Pith, xylem and bark of the upper part of a polar root (root collar). The xylem consists mainly of unlignified parenchyma cells and radial arranged vessel groups. Fibres are absent. 100x. (C) Pith, xylem and bark of a bulb. The

xylem consists mainly of unlignified parenchyma cells and small radially arranged vessel groups. Fibres are absent. Representative of a species with optional fibre formation. (D) Groups of fibres occuring in the third ring of root collar of *Campanula rotundifolia*, 20 cm tall, perennial herb, cultivated under optimal conditions, garden, Zürich, Switzerland. (E) Fibre-free stem in root collar of *Campanula rotundifolia*, 20 cm tall perennial herb, dry meadow, montane zone of the Swiss Alps. The radial wedges consist of parenchyma and vessels. (F) Scalariform perforations with 2 and 3 bars in root collar of a *Campanula drabifolia*, 5 cm tall perennial herb in a dry meadow of the Mediterranean zone, Crete, Greece. (G) Thick-walled vessels arranged in irregular groups in root collar of *Campanula glomerata*, 20 cm tall perennial herb, dry meadow, hill zone of Switzerland. (H) Round inter-vessel pits in a vessel within a fibre-zone. Root collar of *Campanula medium*, 40 cm tall perennial herb, meadow, Mediterranean zone of France. (I) Scalariform intervessel pits in vessels within a zone of pervasive parenchyma. Root collar of *Campanula medium*, 40 cm tall perennial herb, meadow, Mediterranean zone of France. 400x. (J) Thick-walled vessels within a thin-walled parenchyma zone of root collar of *Campanula medium*, 40 cm tall perennial herb, meadow, Mediterranean zone of France.

particular gene partitions. To reflect the increased probability of transitions over transversions in non-coding loci, we set the substitution rates prior (revMatPr) for the ITS, trnT-L and petB-D partition to the Dirichlet function with values 1 and 3.

The phylogenetic analysis in itself was represented by the Bayesian inference (BI), conducted in MrBayes version 3.1.2 [37]. This comprised two independent runs with four Metropolis-coupled MCMC chains of 1×10^7 generations sampled after every 1000th generation. In every run, one Markov chain was cold and three were incrementally heated by a parameter of 0.3. To eliminate trees sampled before reaching apparent stationarity, the first 25% of entries were discarded as burn-in and the rest was used to compute the majority-rule consensus (Figure 1).

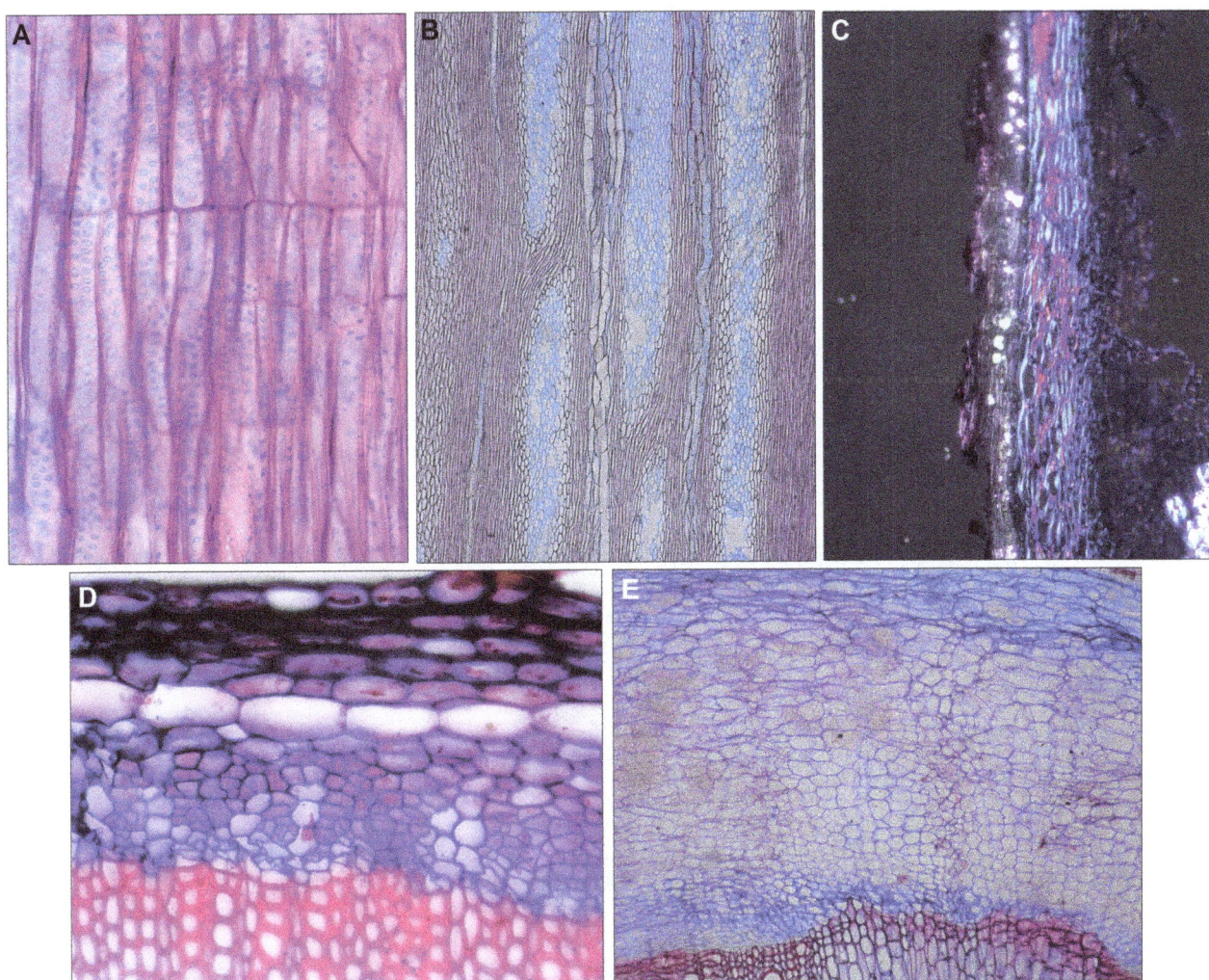

Figure 4. (A) Radial section of a uniseriate ray with upright cells. Root collar of a *Campanula drabifolia*, 5 cm tall perennial herb in a dry meadow of the Mediterranean zone, Crete, Greece. (B) Tangential section of very large rays, partially confluent with the axial tissue. Root collar of *Gadellia lactiflora*, 80 cm tall perennial herb, meadow, subalpine zone of the Georgian Caucasus. (C) Crystal sand in the phellem. Root collar of *Campanula giesekiana*, 5 cm tall perennial herb, on rock, arctic zone, Greenland. Polarized light. (D) Phloem with small groups of sieve tubes and companion cells. Root collar of *Campanula rapunculoides*, 40 cm tall perennial herb, meadow, hill zone of Switzerland. (E) Phloem with small groups of sieve tubes and companion cells between large parenchyma zones. Root collar of *Petromaerula pinnata*, 80 cm tall perennial herb, meadow, Mediterranean zone, Crete, Greece.

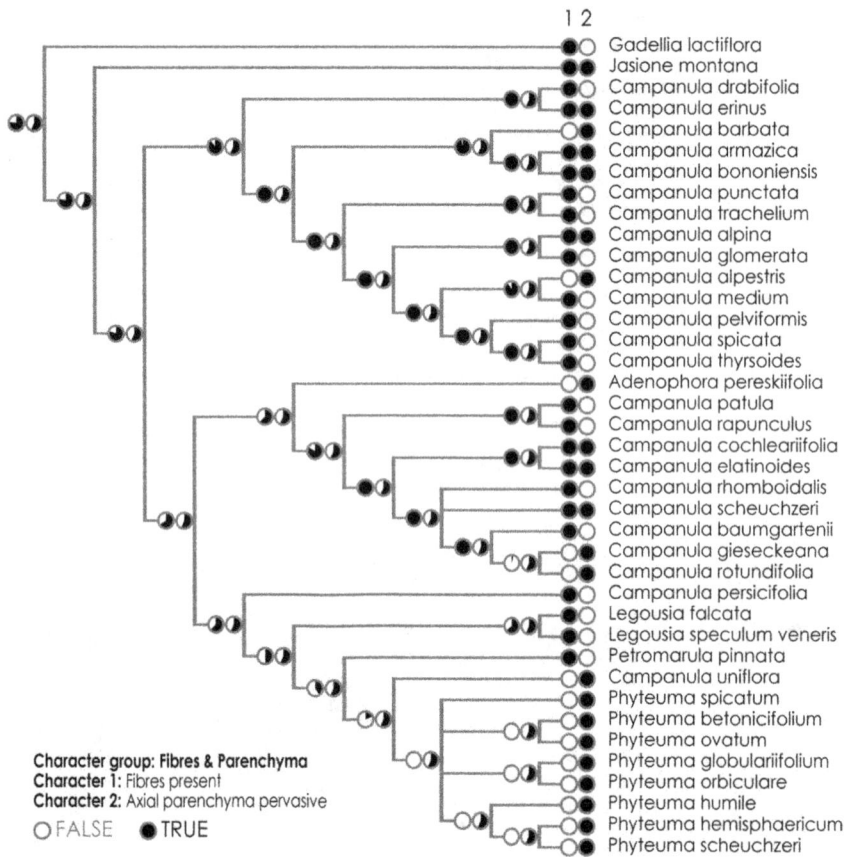

Character group: Fibres & Parenchyma
Character 1: Fibres present
Character 2: Axial parenchyma pervasive
○ FALSE   ● TRUE

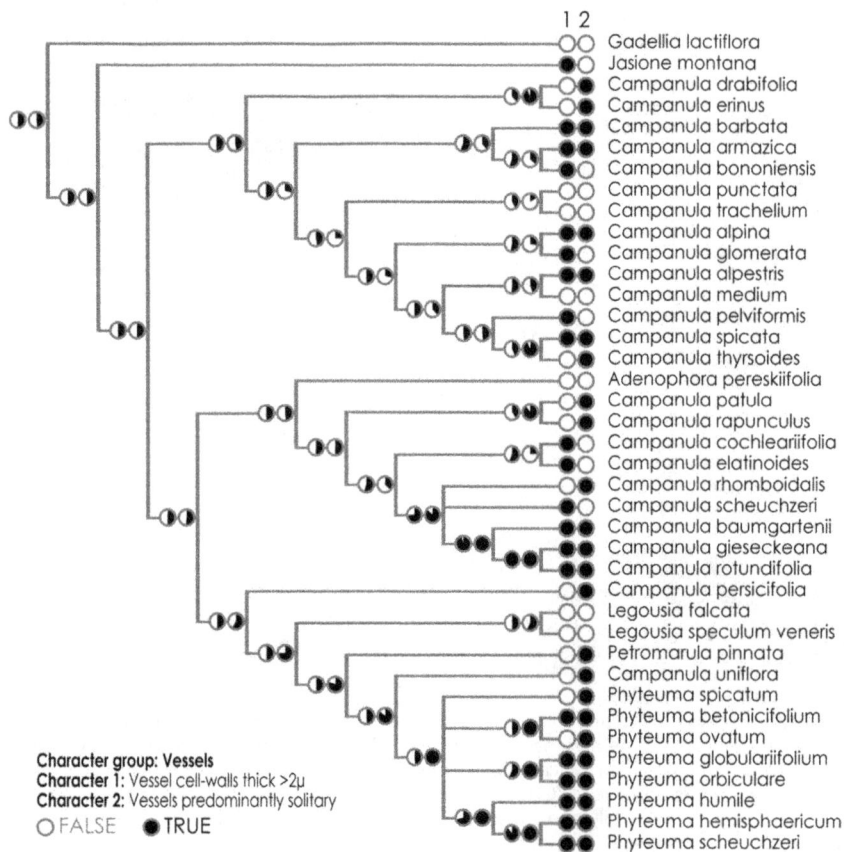

Character group: Vessels
Character 1: Vessel cell-walls thick >2µ
Character 2: Vessels predominantly solitary
○ FALSE   ● TRUE

Figure 5. Maximum likelihood reconstruction of three best fitting characters: fibres present/absent (RI = 0.69), axial parenchyma pervasive (RI = 0.44), vessel cell-walls thick >2 μ (RI = 0.47), vessels predominantly solitary (RI = 0.31). The calculation was done in accordance with the asymmetrical 2-parameter Markov k-state model as implemented in Mesquite 2.6 (Maddison and Maddison, 2009).

The tree inferred by MrBayes served as groundwork for further evaluation of the anatomical trait dataset. All the anatomical traits were coded as binary data (49 traits in total) and confronted with the tree in Mesquite version 2.6 [38]. Various criteria of character fitting were observed and will be discussed later, from the retention index RI [39], to the tracing of character history. In the latter case, two unequal rates of transition between character states, depending on the direction of transition, were assumed (asymmetrical 2-parameter Markov k-state model) [40], and the branch lengths were taken into account (Figure 1).

Relationships between Anatomy, Ecological Preferences and Phylogeny

Relationships between the anatomical structure of the studied species and their ecological preferences (habitat type, elevation) and competitive ability (expressed by plant height at the adult stage, which is an important trait in asymmetric competition for light) were evaluated using linear models, redundancy analysis RDA; see Legendre and Legendre [41], fitted for both (i) individual anatomical traits (single response variable model) and, (ii) all anatomical traits analyzed together as response variables in a global multivariate test. Type I errors were estimated using non-parametric Monte Carlo permutation tests, based on the F statistic, with 1999 random permutations. Since part of the explanatory power uncovered by relating the anatomical traits to ecological preferences might be alternatively explained by phylogenetic inertia affecting both the similarity of anatomical traits among closely related taxa and the similarity of ecological niches that such taxa occupy, we decided to use so-called phylogenetic corrections. The method of Diniz-Filho et al. [42], as modified by Desdevises et al. [11], was used. Here, the variation explained by the phylogenetic relatedness of species was removed from the model, using species coordinates on selected axes of a principal coordinate analysis (PCoA) calculated from a patristic distance matrix corresponding to the MrBayes phylogenetic tree described above.

Selected principal coordinates, which were used as covariates during tests including phylogenetic correction, were also used as predictors for individual anatomical traits to estimate the amount of variation in the trait values explained by species phylogeny [11]. Particular attention was paid to the relationships between the anatomical traits (selected based on the highest fit in the global RDA analysis) and plant height and elevation. All statistical methods were applied using Canoco 5 [43], including the estimation of linear models (performed using RDA with a single response variable). The family-wise error rate was accounted for by Bonferroni correction of significance values.

Results

Anatomical Structure

All of the analysed annual and perennial Campanulaceae had secondary radial growth in the root collar zone. Growth rings of varying distinctness occurred in most perennial species (Figure 2), but this varied within species and between individuals. For example, rings could be very distinct at the periphery of the stem but absent at the centre. Ring boundaries were expressed by semi-ring porosity (Figure 2a), marginal parenchyma (Figure 2b) or ring shake in the large rays (Figure 2c). The ages of plants with distinct rings varied. Four of the species were annual and therefore had only one ring, 15 species had 2 to 4 rings, ten species had 5–10 rings and ten species had 10 to 21 rings. Annual rings tended to be very narrow: 0.15 to 0.5 mm in 25 species, 0.55 to 0.9 mm in 7 species, and 1 to 2 mm in 7 species.

Two principal stem constructions occurred within the studied taxa (Figures 2 and 3). One group had no fibres (Figure 2 a–c) or, if present, fibre zones were sporadic. This group included all *Phyteuma* and many *Campanula* species. Species that lacked fibre in their stem centres may have had peripheral fibre zones. In the 12 species where this occurred, the central stem was classified as juvenile, while the peripheral section was classified as adult. This division is slightly problematic as fibre-formation within a species seems to relate to the precise position in the root collar zone (Figure 3 d–e).

The second group consisted of stems composed mainly of fibres and vessels (Figure 2 d–e), such as *Legousia sp.* and tall *Campanula spp.* Vessel diameters of all the measured species were small. Early wood vessels of plants under 20 cm in height had a diameter of 15–25 μm and those of larger plants 25–50 μm. The length of early wood vessels varied between 80 and 600 μm. Minimum vessel lengths averaged between 80 and 100 μm in small arctic and alpine plants (e.g., *Campanula giesekiana*, *Phyteuma globulariifolia*). Maximum vessel length averaged between 200 to 300 μm in the 5 cm tall *Campanula erinus*. Species with heights from 10 to 80 cm (N = 16) had vessel lengths between 100 and 200 μm. In species with large amounts of parenchyma (Figure 2 a–c) (*Phyteuma spp.*, *Jasione montana*, alpine species such as *Campanula alpina*, *C. alpestris*, *C. barbata*), vessels tended to be solitary or grouped into short radial or irregular clusters (Figure 3a and 3g), whereas long radial multiples occurred in species with a large proportion of fibres (Figure 2a and 2f–g).

Perforation plates of all of the species were simple, with the exception being *Campanula drabifolia* and occasionally *Campanula erinus*. Among the many simple perforations a few plates had 1 to 3 horizontal bars (Figure 3f). Two distinct types of inter-vessel pits occurred. Pits were round or slightly laterally enlarged in vessels of fibre-rich zones (as in most *Campanula spp.* from lower elevation meadows and dry mediterranean sites such as *Campanula glomerata*, *C. persicifolia*, *C. pelviformis*, *C. drabifolia*, but also *Legousia falcata*), and scalariform in areas with pervasive parenchyma (*Phyteuma spp.*, *Jasione montana, and Campanula* spp. from alpine zone). Therefore, species with large, fibre-rich zones contained exclusively round pits and those without fibres contained only scalariform inter-vessel pits. Since some species contained fibre-rich and fibre-less zones, both pit types existed in the same individuals (Figure 3h–i). In addition, small thick-walled vessels were combined with the presence of pervasive parenchyma in the absence of fibres (Figure 2a–c and 3g).

Fibres were thin or thin-to-thick walled, and had small (<3 μm) oblique, slit-like pits without small pit borders (libriform fibres). Peripheral fibre belts occurred in the genera *Campanula*, *Petromarula* and *Jasione* but not in the genus *Phyteuma*.

Axial parenchyma was mainly pervasive. In fibre-less zones the unlignified parenchyma cells were thin-walled and surrounded vessels (Figure 3i–i); parenchyma was mostly absent in fibre-rich zones (Figure 2d–e) and the only species containing vasicentric paratracheal parenchyma was *Gadellia lactiflora* (Figure 2b). Typical rays were rare in the present material. Uniseriate rays occurred in

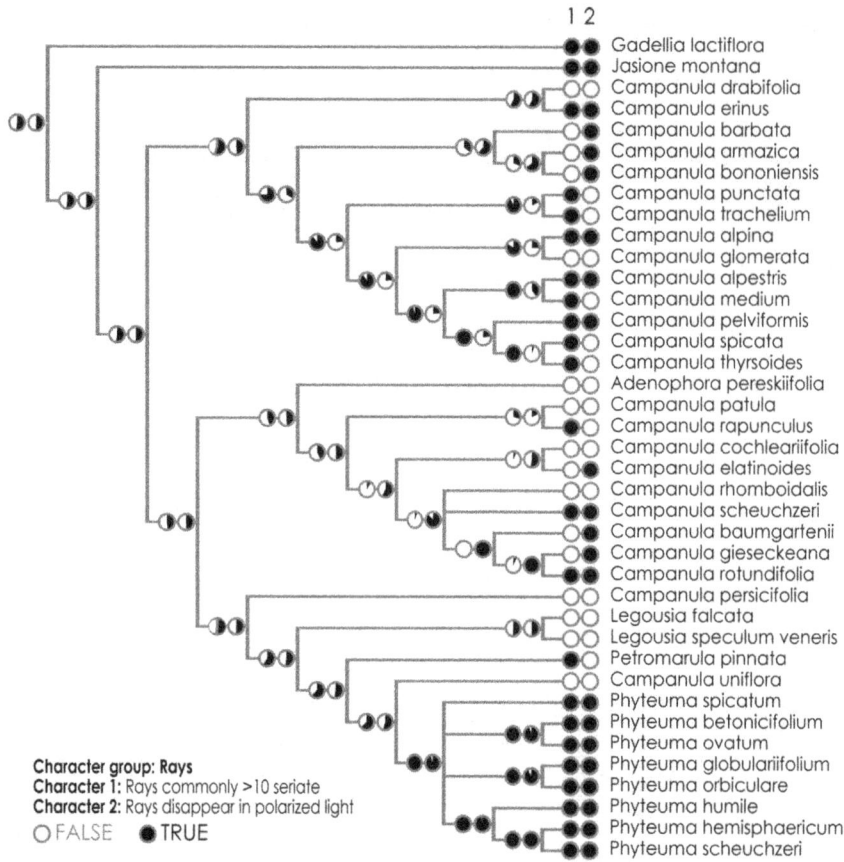

Character group: Rays
Character 1: Rays commonly >10 seriate
Character 2: Rays disappear in polarized light
○ FALSE ● TRUE

Character group: Other
Character 1: Growth rings distinct and recognizable
Character 2: Crystals present in phellem
○ FALSE ● TRUE

Figure 6. Maximum likelihood reconstruction of three best fitting characters: rays commonly >10 seriate (RI = 0.47), rays disappear in polarized light (RI = 0.44), growth rings distinct and recognizable (RI = 0.33), crystals present in phellem (RI = 0.38).

some individuals of species with fully lignifed stem centres (*Campanula erinus*) or large fibre belts (e.g., *C. drabifolia*, Figure 4b). Upright cells of uni-seriate rays appeared extremely small in tangential sections and multi-seriate rays occurred in *Gadellia lactiflora* (Figure 2b and 4a). Large parenchymatic zones between vascular bundles appeared as rays in cross-sections (Figure 2 and 3b,d), but they cannot be defined as rays in tangential sections. Rays within vessel-fibre strips were entirely absent.

The basic phloem-structure of all species consisted of parenchyma cells. The phloem included locally smaller and larger round or radial groups of sieve tubes and companion cells. They remained turgid for the lifespan of the plant (Figure 4d–e) or collapsed during the first growing season (Figure 2d). The phellem of many species formed a dense belt consisting of rectangular cork cells (Figure 2f), and its tensile force often exceeded the turgor of parenchymatous cells in the xylem. Therefore, the radially-organized xylem structure was deformed (Figure 2g), making ring counting and ring-width measurements problematic. Laticifers were absent or were anatomically not differentiated from other cells in the phloem and the cortex of the Campanuloideae. Finally, crystal sand occurred in all genera and most species, but only in the phellem (Figure 4c).

Evolutionary Trends in Anatomical Traits

We included 39 Campanuloidae species in the phylogenetic evaluation - those for which anatomical traits were known together

with at least one sequence of interest. In congruence with recent research [44–46], the genus *Campanula* was found to be deeply polyphyletic, forming two major clades accompanied by several somewhat isolated lineages (*C. uniflora*, *C. persicifolia* and *Campanula/ Gadellia lactiflora*). In accordance with Wendling et al. [46], the clade involving *C. erinus* and *C. bononensis* (supported by the 1.00 Bayesian posterior probability) is flagged as *Campanula* s. str., the second large cluster (containing e.g. *C. rapunculus* and *C. scheuchzeri*; BPP = 1.00) can be identified with *Rapunculus* 1 clade and it groups firmly with *Adenophora pereskiifolia* (BPP = 1.00). *Campanula uniflora*, along with *Petromarula pinnata*, has strong adherence to the genus *Phyteuma* ("phyteumoid clade", BPP = 0.98); together with *Legousia* spp. and possibly also *C. persicifolia*, this group (BPP = 1.00) matches clade *Rapunculus* 2 of Wendling et al. [46].

The dataset consisting of anatomical traits contained a very low phylogenetic signal (Retention Index is 0.343 for displayed tree) due to their highly plastic nature, both developmental and ecological. Yet, some evolutionary trends can be seen within the main lineages with the "phyteumoid" clade especially distinguished by quite a large number of characters (absence of fibres, pervasive parenchchyma, and type of rays). The reconstructions of ancestral states for several chosen characters are shown in Figures 5 and 6. Traits connected with the quality of fibres are among the most congruent with phylogeny and the retention index for several of them is above 0.5 (RI = 0.69 in the case of presence/absence of fibres); they firmly support the close relationship between *C. uniflora* and *Phyteuma spp*. Presence/absence of fibres is moreover

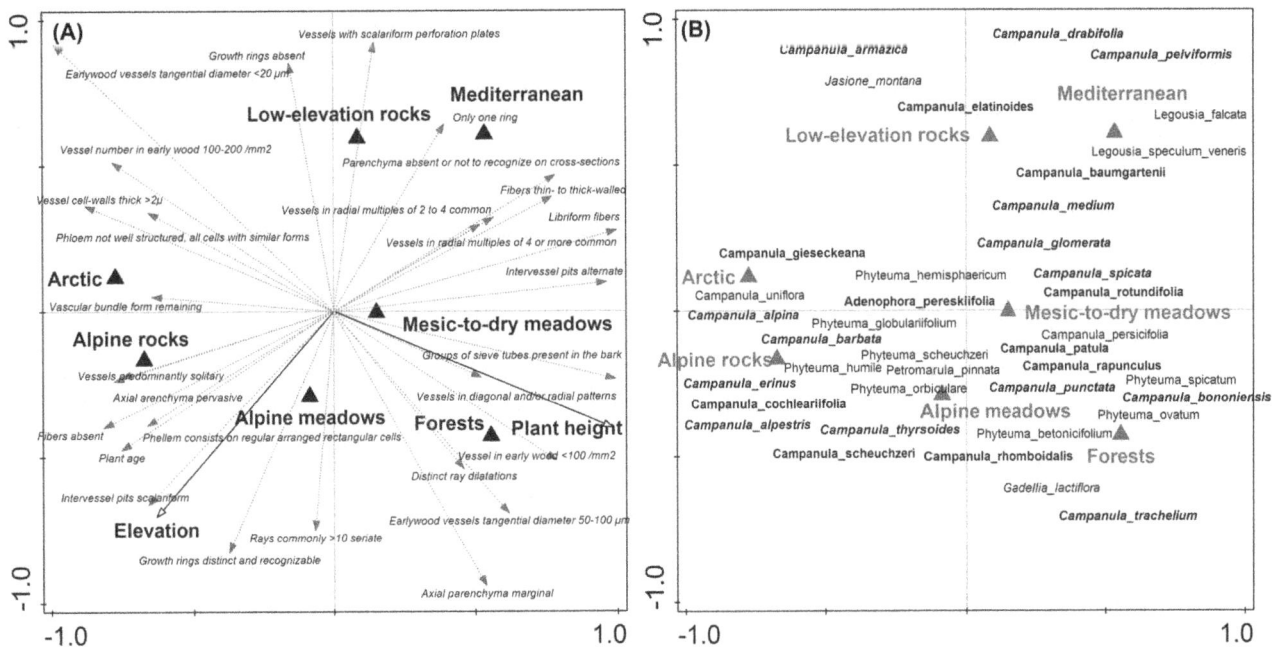

Figure 7. Ordination diagrams with the first two partial RDA axes (after accounting for phylogenetic relations among species using the methods of Desdevises *et al.* (2003)) showing how anatomical traits of plant species are related to their habitats, elevation and adult plant height. Arrows in the left diagram (A) point in the direction of increasing expected values of corresponding anatomical traits, symbols in the right diagram (B) represent individual plant species, and the triangles show the type of habitat (ecological niches) that individual taxa occupy. Relative values of individual species for a trait can be deduced by a perpendicular projection of the species symbols onto the trait arrow; the Pearson correlation coefficient is approximated by the cosine of the angle between the arrows of two traits being compared [40]. Different styles were used to separate the groups of species belonging to individual clades (bold italic for Campanula s. str., bold for Rapunculus 1, and normal for Rapunculus 2 clade).

Table 1. Results of partial tests of relationships between selected individual anatomical traits of plants (in rows) and their adult plant height and elevation.

Anatomical traits	Evol (%)	Plant height				Elevation			
		Res	AH	PC	Ecol (%)	Res	AH	PC	Ecol (%)
Plant age	24.7	−	0.03	ns	4.6	+	0.004	0.011	16.3
Vessel diameter <20 μm	21.9	−	0.000	0.001	39.9		ns	ns	0.1
Vessel diameter 20–50 μm	63.1		ns	ns	0.6		ns	ns	0.9
Vessel diameter 50–100 μm	26.7	+	0.02	0.045	8.1		ns	ns	0.06
Intervessel pits sclariform	39.2		ns	ns	0	+	0.001	0.003	24.4
Thick-walled vessels >2 μm	23.4	−	0.004	0.047	10.7		ns	ns	5.9
Fibres absent	29.1	−	0.007	0.012	17.3	+	0.002	0.03	13.5
Parenchyma pervasive	31.9	−	0.064	0.062	9.9	+	0.002	0.005	21.7
Vascular bundles remaining	56.1	−	0.013	0.024	15.4		ns	ns	0.1
Libriform fibres	11.2	+	0.007	0.002	20.9	−	0.009	0.002	18.5
Fibres absent in the stem centre	64.1		ns	ns	1.45		0.054	ns	4.8
Parenchyma absent	47.2	+	0.046	ns	6.46	−	0.01	0.006	18.5
Bark with groups of sieve tubes	25.2	+	ns	0.007	20.9		ns	ns	2.4
Phellem cells rectangular	35.8	−	0.053	ns	1.62	+	0.091	0.006	20.6
Crystals in phellem	33.6		ns	ns	0.12		ns	ns	4.6

AH headed columns refer to ahistorical comparisons (i.e without phylogenetic corrections), while PC headed columns refer to models with a correction for phylogenetic relatedness. Individual cells in AH and PC columns show Type I error probability estimates (adjusted by Bonferroni correction within test families) or ns when adjusted with P≥0.10. The percentage of trait variation explained by phylogenetic relatedness (Evol) between the species, and ecological preferences (Ecol) in the phylogeny-corrected analysis, are shown. Response (Res) columns show positive (+) or negative (−) relationships between the anatomic trait value and the predictor.

the only character for which the ancestral state for the whole Campanuloidae subfamily can be unequivocally reconstructed under the decision threshold of 1.0 (Figure 5); the Proportional Likelihoods are 0.77:0.23 in favour of their presence and all the cases of fibre loss should be therefore considered as secondary. Other well fitting characters can be found also amongst traits describing the morphology of vessels and rays (Figure 6), although the variability within major clades is quite high.

Relationships between Plant Anatomical Traits, their Ecological Preferences, Adult Plant Height and Phylogeny for the Displayed Tree

The ecological attributes (habitat type, elevation) and adult plant height had a significant effect on the divergence of anatomical traits, and altogether explained 34.9% of the total variation in anatomical traits between species. This relationship remained significant after accounting for phylogenetic relatedness among taxa in the phylogeny-corrected analysis ($P < 0.01$). The results of comparisons without phylogenetic corrections (hereafter ahistorical comparison) and phylogeny-corrected analysis showed similar patterns. Therefore, only results from the latter analysis are presented in the RDA ordination diagram (Figure 7). The main anatomical differences along the first (horizontal) ordination axis are associated with decreasing adult plant height from the arctic to temperate zones, i.e., the first axis separated taller plants such as *Campanula trachelium*, *C. persicifolia*, *C. rapunculus*, *Phyteuma spicatum* and *P. ovatum*, typical of temperate meadows and forests from smaller taxa such as *Campanula gieseckeana*, *C. cochleariifolia*, *C. uniflora*, *C. alpine* and *Phyteuma humile* which occupy a rocky alpine and arctic environment. The taller plants included distant taxa from all three evolutionary clades (*Campanula* s. str., *Rapunculus* 1 and *Rapunculus* 2) that converged towards a similar anatomical structure characterized by large vessel tangential diameters of 50–

100 μm, smaller numbers of vessels in early wood (<100/mm²), alternate intervessel pits and groups of sieve tubes in phloem. Smaller Arctic and alpine taxa, however, had more vessels in early wood (100–200/mm²), with smaller tangential diameter <20 μm and thicker vessel cell-walls >2 μm, pervasive axial parenchyma and poorly structured phloem with all cells with similar forms. The changes along the second axis seemed to be associated with habitat preferences along elevational gradients from dry and warm Mediterranean and temperate, open rocky sites to cool and moist alpine meadows and screes. The second axis separated species such as *Campanula pelviformis*, *C. baumgartenii*, *Jasione montana*, *C. elatinoides* and *Legousia falcata*, typical for the Mediterranean and lower elevation temperate habitats, from alpine taxa such as *Campanula scheuchzeri*, *C. alpestris*, *Phyteuma orbiculare* and *P. humile*. The anatomical features for these alpine meadow taxa included distinct growth rings, the absence of fibres, greater longevity (higher number of annual rings), scalariform intervessel pits, and phellem consisting of regularly arranged rectangular cells. Compared to the alpine species, those taxa typical for the dry and warm Mediterranean sites were shorter-lived, with thin- to thick-walled fibres present at the periphery of the stem but absent in the centre. Vessels were arranged in diagonal and/or radial patterns and were missing parenchyma.

Table 1 gives an overview of partial RDA analyses with a single response variable and single predictor, together with the fraction of variation in the measured anatomical traits that can be accounted for by evolutionary history, elevation and plant height. Stepwise selection of principal coordinates, representing components of phylogenetic relatedness that significantly explain differences in anatomical properties, indicated that five axes should be considered and these explained 27.2% of the total variation in all anatomical traits. When only plant age was analyzed as a single response variable, the evolutionary history accounted for 24.7% of the variation, the plant height 4.6% of the variation and elevations 16.3%. The significant negative relationship between plant age and height was lost after phylogenetic correction, while the positive relationship with elevation was retained. A significant negative relationship was found for the presence of fibres and plant height, with a positive relationship found between the presence of fibres and elevation. This effect manifested itself both in phylogeny-corrected and ahistorical (uncorrected) tests ($P<0.05$). The largest amount of variation explained by evolutionary history concerned vessel diameter and fibres.

Discussion

To our knowledge, this is one of the first studies where the anatomical features of so many species from wide-ranging habitats have been related to morphological and ecological traits and phylogenetic history data derived from DNA sequences. We acknowledge that the selection of species in a subfamily as large as Campanuloideae (the *Camplanula* genus itself includes over 500 species) is difficult. In this paper, the availability of material has been the basis for inclusion which means that some parts of the subfamily are overrepresented and some underrepresented. At the same time, however, the studied species do represent major clades of the subfamily. Both significant variability in anatomical traits and major habitats, from low-elevation Mediterranean to high-elevation alpine and Arctic areas, are well represented. We believe that carefully interpreted inference about species' anatomical adaptation to the environment from a smaller, but nonetheless representative, set of common species covering a wide range of habitats and observed variability in traits can significantly contribute to the ecological understanding of the evolution of plant structures. This is particularly true when such inference is based on a modern phylogenetic tree constructed for target species from nucleotide sequences. We found no major conflict between previous phylogenetic works on Campanulaceae [14,47] and our phylogeny. It is quite interesting to note that after ten years, using other methods and different data sources (loci), we achieved consonance with Eddie et al. [14]. We added several new species to the picture using a combination of five loci, also including those unlinked with chloroplast genes, which are countable as independent data sources.

Phylogenetic and ecological assessments of anatomical features within the Campanulaceae family have so far been limited by the fact that most species in the literature are tropical shrubs and trees belonging to the subfamily Lobelioideae [1]. However, in this study all the species described are annual or perennial herbs from seasonal temperate regions. Until now, most herbaceous species of Lobelioideae have not been characterized, thus comparisons of the present material is only possible with the well-documented material of Carlquist [1]. As a systematic study of herbaceous species of Lobelioideae is lacking, phylogenetic and ecologic interpretations for the entire family are still under debate.

Based on the findings in this study a number of features seem to be primarily related to phylogeny. The family is characterized by vessels with simple perforation plates. However, some herbs and shrubs of the Campanuloideae and Lobelioideae have subdivided types with one to three bars [1,21]. Scalariform perforation plates do not separate any systematic groups. The occurrence of libriform fibres is not a reliable feature for separating subfamilies, as they form the basic tissue in all shrub and tree Lobelioideae, which also occurs within the Campanuloideae. The fibre-rich and parenchyma-lean tissue of the species *Legousia speculum-veneris*, *Petromarula pinnata* and some *Campanula spp.* separate them from all

Phyteuma spp. ("Phyteumoid" clade) within the Campanuloideae, although septate fibres [1] occur only in the subfamily Lobelioideae. Distribution patterns of parenchyma cells are specific for subfamilies as well as for species. Parenchyma is absent in *Legousia*, in the fibre-rich zones of all Campanuloideae species and in the herb *Lobelia syphilitica*. Pervasive parenchyma occurs only in herbaceous Campanuloideae species. The presence of vasicentric parenchyma, which is specific for all shrub- and tree like Lobeliaceae [1], cannot be confirmed with the material prepared in this study.

The presence of rays and their structure appear to be of taxonomic value. Large, distinct rays separate Lobelioideae shrubs and trees, and the large herb *Gadellia lactiflora* from all other species at the very least. What appear as large rays in the xylem of herbaceous Campanuloideae actually represent vessel-free parenchymatous zones between vascular bundles and not true rays, and rays are not present within vascular bundles either. A characteristic for the Campanuloideae is the presence of crystal sand in the phellem. Limited stem-anatomical material, and the lack of consensus among various researchers on a classification below the Campanuloideae and Lobelioideae [12] prohibit further taxonomic conclusions.

A number of features that evolved repeatedly during evolution in 'unrelated' taxa from the Campanuloideae subfamily seem to be primarily related to habit adaptation and ecology. The taller, low-elevation plants from all tree evolutionary clades (*Campanula* s. str., *Rapunculus 1* and *Rapunculus 2*) tended to converge towards similar anatomical structure characterized by large vessel tangential diameter, smaller number of vessels in early wood and alternate intervessel pits. Small-stature alpine and Arctic taxa developed, on the other hand, distinct growth rings, an absence of fibres, greater longevity (higher number of annual rings), and scalariform intervessel pits. Formation of annual rings is a response to intra-annual weather and climate changes. In the present material, rings occured mostly in plants from highly seasonal climates compared to other Campanulaceae species from tropical zones. Explaining annual rings of small plants based on their growth habit and environmental factors is problematic because several stem components, and not only the xylem, optimize stems for water transport, stability and nutrition storage. Despite these restrictions, some relational observations can be made. Plant size determines ring width, where small plants have smaller rings than large plants and annual rings are generally smaller in plants found at high altitudes than those growing at low altitudes. However, since plant size is related to altitude, which is more than simply a temperature gradient effect, the dominant influencing factor remains unexplained. Interestingly, age trends occur in perennial herbs, as average ring width normally decreases with age, which is commonly observed in shrub and tree species.

Conclusions

The large anatomical variability of the subfamily Campanuloideae found within the studied material was partitioned into phylogenetic relatedness and ecological adaptation. The main evolutionary trend concerns the separation of the "Phyteumoid" clade as distinguished by quite a large number of characters (absence of fibres, pervasive parenchyma). Despite clear phylogenetic inertia in the anatomical features studied, several important links between these traits and adult plant height and habitat preferences were found after removing the effects of phylogenetic relatedness. The observed variability in early wood vessel parameters (diameter, cell-wall thickness, number) seems to be best predicted by adult plant height (as a proxy of species competitiveness) as related to the gradient of productivity. This is represented in our study by species occurrence in habitats from mesic forests and meadows to alpine and Arctic rocky sites. Variability in plant age, parenchyma and fibre type is related to species optima along altitudinal gradients from low-elevation Mediterranean to high-elevation alpine meadows. The large anatomical variability of the family Campanulaceae found within the available material can mainly be explained through taxonomic differences between the subfamilies Lobelioideae and Campanuloideae. One of the questions that remains, however, is how many features are life form specific? Hence, it would be interesting to repeat such a comparative study with the focus on similar life forms from both subfamilies, not limiting the choice just to trees and shrubs [1]. We hope that our novel, comparative approach, which attempts to understand the evolution of plant anatomical structures through separation of evolutionary inertia from a true adaptation to the environment, will prompt new research in this area.

Supporting Information

Table S1 Summary table of plant species studied.

Table S2 The accession numbers of nucleotide sequences (internal transcribed spacer (ITS), trnT-trnL intergenic spacer, matK+trnK region, the gene for rubisco large subunit (rbcL) and petB-petD intergenic spacer) obtained from GenBank (www.ncbi.nlm.nih.gov/nuccore/).

Acknowledgments

We are grateful to Jodi Axelson and Dr. Brian G. McMillan for linguistic improvements.

Author Contributions

Conceived and designed the experiments: FS JD PR. Performed the experiments: FS JD PR. Analyzed the data: JD PR. Contributed reagents/materials/analysis tools: FS. Wrote the paper: JD FS PR.

References

1. Carlquist S (1969) Wood anatomy of Lobelioideae (*Campanulaceae*). Biotropica 1: 47–72.
2. Niklas K J (1985) The evolution of tracheid diameter in early vascular plants and its implications on the hydraulic conductance of the primary xylem strand. Evolution 39: 1110–1122.
3. Sultan SE (2000) Phenotypic plasticity for plant development, function and life history. Trends Plant Sci 5: 537–542.
4. Pigliucci M (2005) Evolution of phenotypic plasticity: where are we going now? Trends Ecol Evol 20: 481–486.
5. Peat HJ, Fitter AH (1994) Comparative analyses of ecological characteristics of British angiosperms. Biol Rev 69: 95–115.
6. Ackerly DD (2000) Taxon sampling, correlated evolution, and independent contrasts. Evolution 54: 1480–1492.
7. Šmilauerová M, Šmilauer P (2007) What youngsters say about adults: seedling roots reflect clonal traits of adult plants. J Ecol 95: 406–413.
8. Dubuisson JY, Hennequin S, Bary S, Ebihara A, Boucheron-Dubuisson E (2011) Anatomical diversity and regressive evolution in trichomanoid filmy ferns (*Hymenophyllaceae*): a phylogenetic approach. C R Biol 334 (12): 880–95.
9. Harvey PH, Reader AF, Nee S (1995) Why ecologists need to be phylogenetically challenged. J Ecol 83: 535–536.
10. Klimešová J, Doležal J, Sammul M (2011) Evolutionary and organismic constraints on the relationship between spacer length and environmental conditions in clonal plants. Oikos 120: 1110–1120.
11. Desdevises Y, Legendre P, Azouzi L, Morand S (2003) Quantifying phylogenetically structured environmental variation. Evolution 57: 2647–2652.
12. Lammers TG (2007) Campanulaceae. In: Kubitzki K. eds. The families and genera of vascular plants: Flowering Plants: Eudicots, Asterales. Vol. VIII. Heidelberg: Springer Verlag. 26–59.
13. Tutin TG, Heywood VH, Burges NA, Moore DM, Valentine DH et al. (1964–1980) *Flora Europaea*, Vol. IV. Cambridge, London, New York, Melbourne: Cambridge Univ Press.
14. Eddie WMM, Shulkina T, Gaskin J, Haberle RC, Jansen RK (2003) Phylogeny of Campanuloideae inferred from its sequences of nuclear ribosomal DNA. Ann Mo Bot Gard 90: 554–575.
15. Foster AS, Gifford EM (1974) Comparative Morphology of Vascular Plants (2nd ed.). San Francisco: WH Freeman.
16. Sperry J S (2003) Evolution of water transport and xylem structure. Int J Plant Sci 164: 115–127.
17. Pittermann J, Sperry J (2003) Tracheid diameter is the key trait determining the extent of freezing-induced embolism in conifers. Tree Physiol 23: 907–914.
18. Willson CJ, Jackson RB (2006) Xylem cavitation caused by drought and freezing stress in four co-occurring *Juniperus* species. Physiol Plantarum 127: 374–382.
19. Sevanto S, Holbrook NM, Ball MC (2012) Freeze/thaw-induced embolism: probability of critical bubble formation depends on speed of ice formation. Front Plant Sci 3: 107.
20. Metcalfe CR, Chalk L (1957) Anatomy of the Dicotyledons. Oxford: Clarendron Press.
21. Shulkina TV, Zikov SE (1980) The anatomical structure of the stem in the family Campanuloideae in relation to evolution and life forms. Bot Zh SSSR 65: 627–639 (in Russian, english summary).
22. Schweingruber FH, Börner A, Schulze ED (2012) Atlas of stem anatomy in herbs shrubs and trees. Vol. 2. Heidelberg, Dorecht, London, New York: Springer.
23. Lauber K, Wagner G, Gygax A (2012) Flora Helvetica. Bern: Paul Haupt.
24. Jahn R, Schönfelder P (1995) Exkursionsflora für Kreta. Stuttgart:Verlag Eugen Ulmer.
25. Da-chang Zhao (2007) Botanical atlas of Changbai mountain. Shenyang: Shenyang Press House.
26. Böcher TW, Holmen K, Jakobsen K (1968) The Flora of Greenland. Copenhagen: P. Haase and Son publishers.
27. Gagnidze R (2005) Vascular Plants of Georgia: A Nomenclatural Checklist.
28. Schweingruber FH (2012) Microtome sectionioning of small plant stems. IAWA J 33: 457–460.
29. Wheeler EA, Baas P, Gasson PE (1989) IAWA list of microscopic features for hardwood identification. IAWA Bull n.s. 10: 219–332.
30. Schweingruber FH, Börner A, Schulze ED (2011) Atlas of stem anatomy in herbs shrubs and trees. Vol.1. Heidelberg, Dorecht, London, New York: Springer.
31. Katoh K, Toh H (2008) Recent developments in the MAFFT multiple sequence alignment program. Brief Bioinform 9: 286–298.
32. Hall T (1999) BioEdit: a user-friendly biological sequence alignment editor and analysis program for Windows 95/98/NT. Nucl Acid S 41: 95–98.
33. Capella-Gutierrez S, Silla-Martínez JM, Gabaldón T (2009) trimAl: a tool for automated alignment trimming in large-scale phylogenetic analyses. Bioinformatics 25: 1972–1973.
34. Tanabe A (2011) Kakusan4 and Aminosan: two programs for comparing nonpartitioned, proportional and separate models for combined molecular phylogenetic analyses of multilocus sequence data. Mol Ecol Resour 11: 914–21.
35. Adachi J, Hasegawa M (1996) MOLPHY Version 2.3: Programs for molecular phylogenetics based on maximum likelihood. Computer science monographs 28: 1–150.
36. Schwarz G (1978) Estimating the dimension of a model. Annals of Statistics 6: 461–464.
37. Ronquist F, Huelsenbeck JP (2003) MRBAYES 3: Bayesian phylogenetic inference under mixed models. Bioinformatics 19: 1572–1574.
38. Maddison WP, Maddison DR (2009) Mesquite: a modular system for evolutionary analysis, version 2.6. Available: http://mesquiteproject.org.
39. Kitching IJ, Forey PL, Humphries CJ, Williams DM (1998) Cladistics: the theory and practice of parsimony analysis. 2nd ed. Oxford: Oxford University Press.
40. Lewis PO (2001) A likelihood approach to estimating phylogeny from discrete morphological character data. Syst Biol 50: 913–925.
41. Legendre P, Legendre L (1998) Numerical Ecology, 2nd edn. Amsterdam: Elsevier Science.
42. Diniz-Filho JAF, de Sant'Ana CER, Bini LM (1998) An eigenvector method for estimating phylogenetic inertia. Evolution 52: 1247–1262.
43. ter Braak CJF, Šmilauer P (2012) Canoco reference manual and users's guide: sofware for ordination (version 5.0). Ithaca, NY, USA: Microcomputer Power.
44. Haberle RC, Dang A, Lee T, Penaflor C, Cortes-Burns H et al. (2009) Taxonomic and biogeographic implications of a phylogenetic analysis of the *Campanulaceae* based on three chloroplast genes. Taxon 58: 715–734.
45. Roquet CL, Sanmartín I, Garcia-Jacas N, Saéz L, Susanna A, et al. (2009) Reconstructing the history of *Campanulaceae* with a Bayesian approach to molecular dating and dispersal-vicariance analyses. Mol Phylogenet Evol 52: 575–587.

46. Wendling BM, Galbreath KE, DeChaine EG (2011) Resolving the evolutionary history of *Campanula* (*Campanulaceae*) in Western North America. PLoS ONE 6: e23559.

47. Mansion G, Parolly G, Crowl AA, Mavrodiev E, Cellinese N, et al. (2012) How to handle speciose clades? Mass taxon-sampling as a strategy towards illuminating the natural history of Campanula (Campanuloideae). PLoS ONE 7(11): e50076.

10

Evolutionary Association of Stomatal Traits with Leaf Vein Density in *Paphiopedilum*, Orchidaceae

Shi-Bao Zhang[1,2✪], Zhi-Jie Guan[1,3✪], Mei Sun[2], Juan-Juan Zhang[1], Kun-Fang Cao[2], Hong Hu[1]*

1 Key Laboratory of Economic Plants and Biotechnology, Kunming Institute of Botany, Chinese Academy of Sciences, Kunming, Yunnan, China, **2** Key Laboratory of Tropical Plant Ecology, Xishuangbanna Tropical Botanical Garden, Chinese Academy of Sciences, Kunming, Yunnan, China, **3** State Key Laboratory of Plant Physiology and Biochemistry and College of Agronomy and Biotechnology, China Agricultural University, Beijing, China

Abstract

Background: Both leaf attributes and stomatal traits are linked to water economy in land plants. However, it is unclear whether these two components are associated evolutionarily.

Methodology/Principal Findings: In characterizing the possible effect of phylogeny on leaf attributes and stomatal traits, we hypothesized that a correlated evolution exists between the two. Using a phylogenetic comparative method, we analyzed 14 leaf attributes and stomatal traits for 17 species in *Paphiopedilum*. Stomatal length (SL), stomatal area (SA), upper cuticular thickness (UCT), and total cuticular thickness (TCT) showed strong phylogenetic conservatism whereas stomatal density (SD) and stomatal index (SI) were significantly convergent. Leaf vein density was correlated with SL and SD whether or not phylogeny was considered. The lower epidermal thickness (LET) was correlated positively with SL, SA, and stomatal width but negatively with SD when phylogeny was not considered. When this phylogenetic influence was factored in, only the significant correlation between SL and LET remained.

Conclusion/Significance: Our results support the hypothesis for correlated evolution between stomatal traits and vein density in *Paphiopedilum*. However, they do not provide evidence for an evolutionary association between stomata and leaf thickness. These findings lend insight into the evolution of traits related to water economy for orchids under natural selection.

Editor: Giovanni G. Vendramin, CNR, Italy

Funding: This work is financially supported by the National Natural Science Foundation of China (31170315, 30770226). The funders had no role in study design, data collection and analysis, decision to publish, or preparation of the manuscript.

Competing Interests: The authors have declared that no competing interests exist.

* E-mail: huhong@mail.kib.ac.cn

✪ These authors contributed equally to this work.

Introduction

Plants often exhibit considerable variations in their functional traits that affect the capture and utilization of resources and enable them to adapt to changing environments [1,2]. The development of leaf cuticles and stomata might be linked to the success of terrestrial plants because they resolve two conflicting physiological requirements: increasing CO_2 uptake *vs.* reducing water loss [3,4]. Much of the evolutionary history of land plants involves leaf activities for obtaining water and preventing transpirational water losses, thereby improving their photosynthetic carbon gain and survival in dry habitats [5]. Both environment and evolutionary history are important to shape the hydraulic properties that determine how plants respond to water shortages [6]. Evolutionary pressures that drive such conservation strategies favor the coupling of the cuticle with the development of stomata [7]. Consequently, one might expect a correlated evolution between leaf attributes and stomatal traits [8]. However, little work has been done on such coordination within an evolutionary context even though one could gain valuable insights into ecological and evolutionary principles [8,9].

Water is transpired from the leaf surface through either the outer epidermal cell walls or the stomata. Although cuticles can reduce water loss from the leaf to the atmosphere, they also slow the CO_2 diffusion in the reverse direction [10]. Therefore, stomata can effectively regulate gas exchange where water vapor leaves the plant and CO_2 enters. The potential transpirational demand is primarily determined by both stomatal aperture and density [11]. Over time, stomata have changed markedly in their size and numbers since first appearing on the leaf surface approximately 411 million years ago [12]. Stomatal density (SD) is negatively correlated with atmospheric CO_2 concentration, while size is positively correlated [3,13,14]. Although the level of atmospheric CO_2 is a main selective agent, SD is also related to water availability, light intensity, and temperature [13,15,16,17]. Water deficits lead to more densely packed but smaller stomata [17,18]. The efficiency with which CO_2 is taken up and water loss restricted appears to be partially a function of stomatal size [19,20]. Small stomata enable the leaf to attain high and rapid diffusive conductance under favourable conditions, and they afford greater water-use efficiency (WUE) in dry habitats because they can react more quickly to environmental stimuli [14]. By contrast, large stomata are slower to close. Although they are less able to prevent hydraulic dysfunction in dry habitats, this lag in response may be advantageous in cool, moist, or shaded environments [19,20].

Leaf venation provides mechanical support and carboxylate transport, and aids in replacing the water transpired during photosynthesis [21,22]. Vein density (VD) is correlated with SD, maximum hydraulic conductance, maximum photosynthetic rate, and WUE [11,22,23]. Vein patterns are highly diverse across species, and have a significant phylogenetic signal [5,24,25]. Historically, the evolution of VD resulted in high photosynthetic capacity during early angiosperm diversification, and promoted species diversity among angiosperms [5]. This feature can also serve as an environmental proxy [24]. For example, Dunbar-Co et al. have found that Hawaiian *Plantago* taxa in drier regions have higher VD values [9]. Loss of hydraulic conductance is accompanied by stomatal closure under water deficits [26]. The density of major veins plays a role in determining leaf drought tolerance [27].

Leaf structural traits determine how plants adapt to changes in water availability [15,28]. For example, gametophyte morphology can influence water-holding capacity in ferns [29]. A leaf with a high mass per unit area is better able to store water and maintain more stable hydraulic functioning during droughty periods [30]. Consequently, leaf thickness tends to increase with site aridity [18,28,31]. The potential transpirational demand by plants is primarily determined by stomata. However, when water is severely limited and the stomata reach their minimum aperture, water loss from a leaf is mainly determined by epidermal conductance [32]. The cuticle is a hydrophobic and flexible membrane composed of cutin and associated solvent-soluble lipids. One of its functions is to protect against water loss from the leaf interior [33]. Cuticular property is often correlated with transpirational demand [33,34]. Although a thick cuticle can help prevent water loss when moisture is limited [28,35], thickness alone is not a good predictor of a species' drought tolerance because it is not always correlated with cuticular water permeability [4,36].

Leaf structure can also reflect the plant response to environmental stresses, such as a low supply of soil nutrients. Evolutionary pressures usually favour investment toward chemical and structural defences in stressed plants [31]. This drought response is often similar to that for nutrient limitations, i.e., the production of small leaves with thick cuticles [31,37]. In fact, the thickened cuticles of sclerophylls can serve as a sink for excess photosynthate because those membranes do not require phosphorus or nitrogen to form cutin, suberin, and waxes [38]. Consequently, the sclerophyll protects against leaf herbivory and abiotic physical damage [37].

The well-known genus *Paphiopedilum* within Orchidaceae comprises 66 species, with plants usually occurring in limestone or mountainous forests of tropical and subtropical zones from Asia to the Pacific islands [39]. These species vary in their growing environments, developmental habit, and leaf morphology. The low capacity for water storage in the shallow soil layer of karst areas limits water supplies. Plants in this genus manifest three contrasting growth habits: terrestrial, facultative epiphytic or obligatory epiphytic. For epiphyte species, the amount of available moisture is a factor in determining the best sites for growth. Although periodic water deficit is a main environmental stressor that limits plant growth and survival within that genus [2], some species can adapt to relatively dry, calcareous regions [40]. Drought tolerance by *Paphiopedilum* is linked to leaf anatomy [2], which is evergreen and fleshy, with distinct epidermal cuticles, but no guard cell chloroplasts [2,40]. This lack of guard cell

chloroplasts slows the induction of photosynthesis, and is considered an ecophysiological adaptation to water shortage [41,42]. Therefore, the wide range of morphological and ecological variations among *Paphiopedilum* species provides a valuable research system for understanding morphological evolution related to water-use traits [2,42].

Plants adapt to challenging conditions through simultaneous configurations of multiple traits [9]. Their leaf vein network, stomatal design, leaf structure and cuticle are ordinately linked to water transport, regulation, storage and conservation, respectively. Here, we investigated the stomatal traits and leaf attributes of 17 species in *Paphiopedilum* when all plants were tested in the same growing environment. Our objectives were to assess the effect of phylogeny on leaf structure and stomatal traits, and to examine any correlated evolution between them. Because the responsiveness to environmental changes is generally more similar among closely related species than among those more distantly related, we expected that stomatal traits would manifest a correlated evolution with leaf attributes.

Phylogenetic signals of SL, SA, UCT, and TCT were >1.0, demonstrating that these traits were phylogenetically conserved (Table 3). However, the K values for SD and SI were <0.5, indicating that these *Paphiopedilum* relatives resembled each other less than expected, under the Brownian model, along the phylogenetic tree. These results were confirmed by our phylogenetic distribution (Fig. 1).

Materials and Methods

Ethics Statement

None of these experimental materials was collected from national parks or other protected areas. No tested species are under first- or second-class state protection, and they are not listed in the Inventory of Rare and Endangered Plants of China (http://zrbhq.forestry.gov.cn/portal/zrbh/s/3053/content-457748.html), or the Key Protected Inventory of Wild Plants of China (http://zrbhq.forestry.gov.cn/uploadfile/zrbh/2010-10/file/2010-10-14-bb296addeaa047798d6b6c476aaa1da9.doc). These plants were used for only scientific research as permitted by the Wildlife Protection and Administration Office under the Forestry Department of Yunnan Province.

Plant Materials

Sample plants representing 17 species of *Paphiopedilum* were collected from their natural habitats and grown in a greenhouse at Kunming Institute of Botany, CAS (elev. 1990 m, E102°41′, N25°01′). Applying similar culturing practices largely helped to minimize any plastic differences among species in functional traits that might have resulted from environmental heterogeneity. Thus, any variations would likely reflect the role of a genetic component. Conditions included 30 to 40% of full sunlight controlled by shade nets and an ambient temperature of 20 to 25°C. Before the sample plants were analyzed, these plants were watered as needed, and were then cultivated for two to three years to ensure that their adaptation to a new environment was complete.

Leaf Attributes

Six mature, undamaged leaves were evaluated from individual plants of each species. Leaf area (LA) was measured with a Li-Cor 3000A area meter (Li-Cor Inc., Lincoln, NE, USA). Each leaf was then divided along the midrib. One half was re-measured with the area meter, then oven-dried at 70°C for 48 h to obtain its dry weight. Specific leaf weight was expressed as leaf dry mass per unit area (LMA). The other half was cleaned for 1 h in a 5% NaOH

Table 1. Leaf carbon stable isotope ratios ($\delta^{13}C$) and stomatal traits of 17 *Paphiopedilum* species.

Species	Growth habit	$\delta^{13}C$	SL	SW	SA	SD	SI
malipoense	facultative	−27.24±0.05	73.43±0.88	63.74±0.57	3681.7±64.9	17.41±1.17	11.67±0.59
emersonii	facultative	−23.93±0.07	56.39±0.42	53.93±0.66	2391.7±39.0	34.06±1.47	12.45±0.53
micranthum	facultative	−27.53±0.02	56.23±0.47	46.76±0.50	2066.9±30.7	27.57±1.31	11.48±0.52
armeniacum	facultative	−26.52±0.09	63.64±0.56	53.65±0.52	2686.6±42.6	29.14±1.99	13.47±0.72
bellatulum	facultative	−26.89±0.01	48.73±0.59	47.30±0.52	1815.8±37.0	40.87±2.16	16.81±0.55
concolor	facultative	−26.60±0.08	50.14±0.61	45.58±0.59	1800.4±40.2	37.47±1.55	16.64±0.57
hirsutissimum	facultative	−23.32±0.16	54.27±0.64	45.22±0.44	1927.2±28.9	38.23±1.61	13.65±0.56
tigrinum	terrestrial	−24.00±0.07	58.06±0.54	49.24±0.85	2249.5±49.8	37.47±1.33	16.26±0.50
henryanum	facultative	−24.32±0.12	56.16±0.58	50.72±0.63	2243.3±43.5	55.26±2.03	18.97±0.56
charlesworthii	epiphytic	−26.06±0.03	49.79±0.43	43.74±0.52	1711.2±25.9	55.25±2.11	16.56±0.56
villosum	epiphytic	−25.32±0.03	57.70±0.57	49.95±0.52	2268.3±38.8	48.82±2.27	18.92±0.74
gratrixianum	facultative	−24.02±0.07	54.61±1.05	50.97±0.66	2204.3±65.7	66.23±2.46	20.53±0.72
insigne	terrestrial	−23.42±0.04	56.56±1.36	46.19±0.60	2054.8±60.4	34.82±1.65	15.32±0.61
dianthum	epiphytic	−25.12±0.08	62.10±1.43	63.63±0.67	3113.3±87.2	38.23±1.61	19.57±0.61
wardii	terrestrial	−24.64±0.02	71.11±0.84	54.12±0.57	3017.7±39.7	21.19±1.30	13.02±0.72
appletonianum	terrestrial	−24.42±0.10	77.95±0.55	57.10±1.26	3497.7±81.6	18.54±1.03	15.15±0.79
purpuratum	terrestrial	−23.52±0.03	68.86±0.72	59.68±0.38	3225.3±35.4	17.03±0.96	10.91±0.57

SL, stomatal length (µm); SW, stomatal width (µm); SA, stomatal area (µm); SD, stomatal density (number mm^{-2}); SI, stomatal index (%).

aqueous solution. Three sections of leaf lamina were excised from the top, middle, and bottom portions, stained with 1% safranin, and mounted in glycerol to obtain the vein density (VD). Samples were photographed at 10× magnification with an Olympus U-CMAD3 light microscope (Olympus Inc., Tokyo, Japan). Vein lengths were determined from digital images via the IMAGEJ program (http://rsb.info.nih.gov/ij/). Values for VD were recorded as vein length per unit area (mm mm^{-2}). Leaf stable carbon isotope ratio ($\delta^{13}C$) was analyzed using an IsoPrime100 isotope ratio mass spectrometer (Isoprime Ltd., Cheadle Hulme, UK).

Table 2. Leaf structural traits of 17 *Paphiopedilum* species. LMA, leaf mass per unit area (g m^{-2}).

Species	LMA	UET	UCT	LET	LCT	MT	LT	LA	VD
malipoense	116.2±5.1	258.8±12.0	23.88±0.69	84.27±4.42	15.22±0.47	466.0±22.6	847.2±35.7	44.82±5.35	0.919±0.049
emersonii	181.9±10.3	295.0±7.3	22.50±0.61	77.12±1.64	15.56±0.81	734.2±22.2	1144.4±21.7	40.84±3.92	0.880±0.040
micranthum	165.8±3.7	162.8±5.1	25.87±1.14	70.14±1.64	12.32±0.68	927.4±21.2	1198.5±20.9	20.13±2.02	1.186±0.057
armeniacum	139.3±3.1	285.7±8.1	24.09±0.65	81.54±2.87	15.15±0.50	561.1±16.1	967.6±15.0	21.41±1.65	1.183±0.042
bellatulum	130.2±6.2	553.9±11.7	24.78±1.16	59.18±1.60	13.74±0.75	560.7±13.0	1212.3±21.9	18.42±1.91	1.063±0.145
concolor	116.8±7.4	455.9±8.3	23.04±2.02	66.88±1.73	12.38±0.78	601.0±26.1	1159.2±25.8	18.47±1.30	1.207±0.099
hirsutissimum	157.9±7.2	304.3±6.3	13.42±0.51	67.49±2.28	9.86±0.60	492.0±27.1	887.1±29.3	39.03±1.83	1.328±0.037
tigrinum	107.3±6.8	206.6±18.2	21.30±0.95	51.99±1.61	12.38±0.65	348.2±7.3	640.5±12.7	45.31±4.17	1.225±0.045
henryanum	139.7±12.1	271.2±17.4	23.18±0.95	61.28±2.06	14.74±0.41	586.6±39.8	957.0±58.6	30.69±1.34	1.213±0.065
charlesworthii	125.5±3.8	374.9±32.9	18.59±0.67	43.81±1.40	13.11±0.39	428.8±15.4	879.2±39.5	16.43±2.68	1.496±0.046
villosum	121.5±12.5	149.4±6.1	12.82±0.68	69.38±2.02	10.68±0.38	393.6±10.1	635.9±17.1	64.48±7.20	1.195±0.068
gratrixianum	115.1±4.9	104.7±3.3	14.47±0.62	55.46±1.50	10.13±0.69	513.6±5.1	698.4±5.8	41.35±6.34	1.191±0.071
insigne	134.1±6.0	198.5±7.7	12.46±0.50	59.00±2.02	10.68±0.53	558.7±16.1	839.2±16.8	44.04±2.53	1.020±0.058
dianthum	237.4±15.1	606.2±46.3	24.69±1.88	66.46±1.50	12.41±0.87	824.1±40.7	1533.8±71.4	74.91±5.89	0.971±0.038
wardii	100.8±1.9	246.9±9.2	16.11±0.62	71.25±3.63	11.77±0.57	405.4±12.8	751.4±19.6	26.89±1.34	0.796±0.047
appletonianum	138.4±15.8	241.2±6.0	14.28±0.56	91.54±2.80	11.86±0.56	522.4±29.6	881.3±34.2	33.76±3.38	0.651±0.031
purpuratum	97.0±3.8	318.5±16.4	17.57±0.55	57.08±1.12	12.37±0.64	512.8±27.9	918.2±39.1	26.82±2.28	0.628±0.033

UET, upper epidermal thickness (µm); UCT, upper cuticle thickness (µm); LET, lower epidermal thickness (µm); LCT, lower cuticle thickness (µm), MT, mesophyll thickness (µm); LT, leaf thickness (µm); LA, leaf area (cm^{-2}); VD, vein density (mm mm^{-2}).

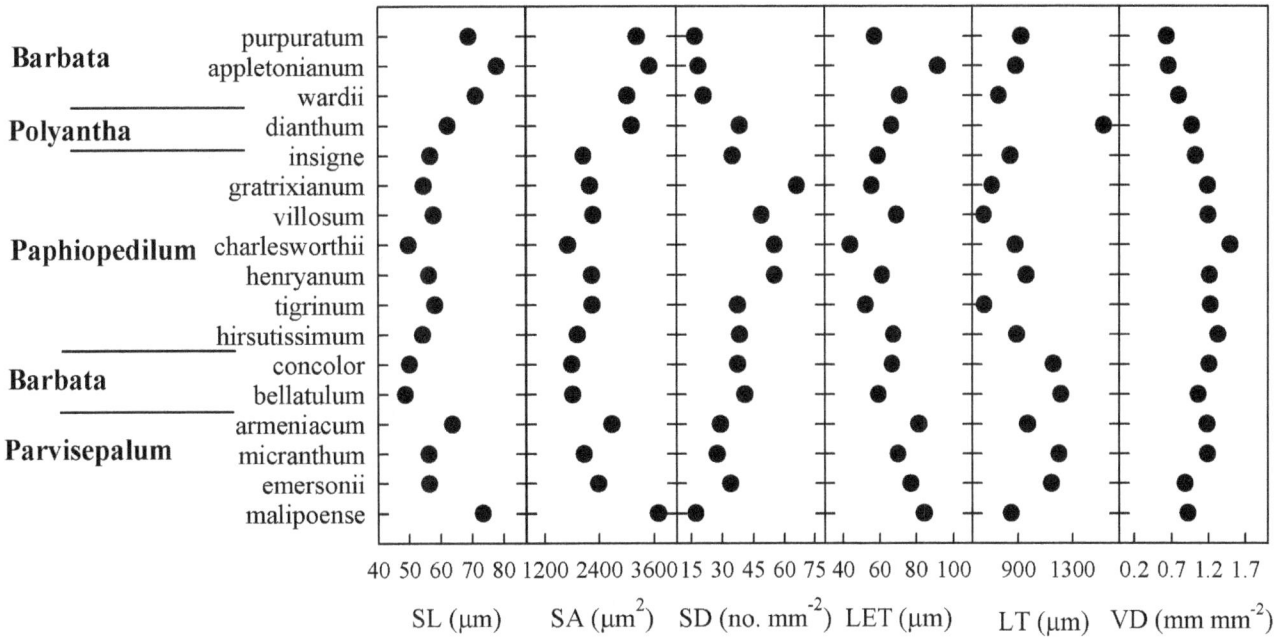

Figure 1. Values for leaf traits and stomatal straits in *Paphiopedilum* species. SL, stomatal length; SA, stomatal area; SD, stomatal density; LET, lower epidermal thickness; LT, leaf thickness; and VD, vein density. Names of subgenera are at left, and are based upon nuclear rDNA ITS trees from Cox *et al.* [46].

Histological Observations

From samples of all 17 species, the middle portions of mature leaves were fixed in FAA (formalin, glacial acetic acid, ethanol, and distilled water; 10:5:50:35, v:v:v:v) for at least 24 h. They were then dehydrated in an ethanol series and embedded in paraffin for sectioning. Transverse sections, made on a Leica RM2126RT rotary microtome (Leica Inc., Bensheim, Germany), were mounted on glass slides. These tissues were examined and photographed under an Olympus U-CMAD3 light microscope. Thicknesses of the upper cuticle (UCT, μm), upper epidermis (UET, μm), palisade tissue (PTT, μm), spongy tissue (STT, μm), lower epidermis (LET, μm) and lower cuticle (LCT, μm) were measured at the midpoint of each transverse section with Adobe Photoshop 8.0 (Adobe Systems Inc., California, USA). For each species, six leaves were taken from different plants.

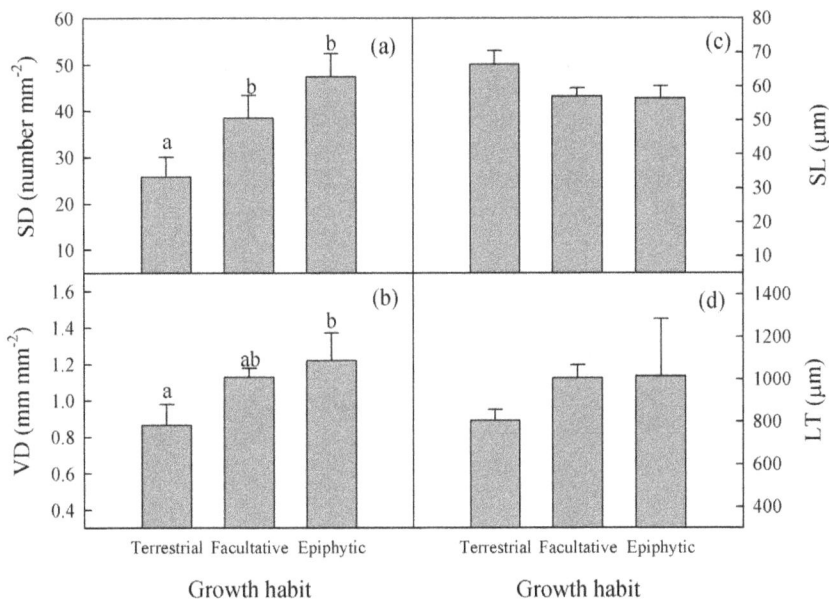

Figure 2. Differences in stomatal traits and leaf thickness of *Paphiopedilum* due to growth habit. SD, stomatal density; VD, vein density; SL, stomatal length; and LT, leaf thickness. Different letters above bars for each component indicate statistically different mean values (*p*≤0.05), as determined by LSD multiple comparison tests.

Table 3. Phylogenetic signal (K) of leaf attributes and stomatal traits in 17 *Paphiopedilum* species.

	K	p
SL	1.215	0.001
SW	0.761	0.010
SA	1.078	0.001
SD	0.199	0.796
SI	0.283	0.573
UET	0.836	0.017
UCT	1.100	0.004
PTT	0.626	0.082
STT	0.547	0.118
LET	0.582	0.066
LCT	0.771	0.018
MT	0.532	0.161
LT	0.772	0.019
TCT	1.207	0.004
LMA	0.723	0.037
LA	0.569	0.068
VD	0.729	0.016

K <1 indicate that relatives resemble each other less than expected under Brownian motion evolution along the phylogenetic tree; while K >1 show that close relatives are more similar than expected. SL, stomatal length; SW, stomatal width; SA, stomatal area; SD, stomatal density; SI, stomatal index; UET, upper epidermal thickness; UCT, upper cuticular thickness; PTT, palisade tissue thickness; STT, spongy tissue thickness; LET, lower epidermal thickness; LCT, lower cuticular thickness, MT, mesophyll thickness; LT, leaf thickness; TCT, total cuticular thickness; LMA, leaf mass per unit area; LA, leaf area; and VD, vein density.

Stomatal Observations

The adaxial and abaxial epidermises were peeled from the middle portions of fresh, mature leaves, and images were made under an Olympus U-CMAD3 light microscope. For each species, six leaves from different plants were used for stomatal observations. Their stomata were tallied in 30 randomly selected fields. Stomatal density (SD) was calculated as the number per unit leaf area. Stomatal size was represented as the guard cell length, possibly indicating the maximum potential opening of the pore [43]. Stomatal length (SL, μm) and stomatal width (SW, μm) were measured from 30 stomata selected randomly. Stomatal area (SA) was calculated as $1/4 \times \pi \times$ SL $\times$ SW [44]. Stomatal index (SI) was estimated as the ratio of stomatal numbers per given area divided by the total number of stomata and other epidermal cells within the same area.

Data Analysis

A phylogenetic signal (K) can be used to express the conservatism of traits. Cases where K<1 indicate convergent traits, K = 1 implies that closely related species have trait values that completely agree with a Brownian model, and K>1 represents traits more conserved than presumed from a Brownian expectation [45]. Our phylogenetic tree of *Paphiopedilum*, based on nuclear rDNA ITS sequences, was obtained from a previous report by Cox *et al.* [46]. The K value for each trait was calculated using 'picante', based on the R package 2.14 [47].

A principal component analysis (PCA) was performed with the 'prcomp' function of the R package 'vegan' to characterize the

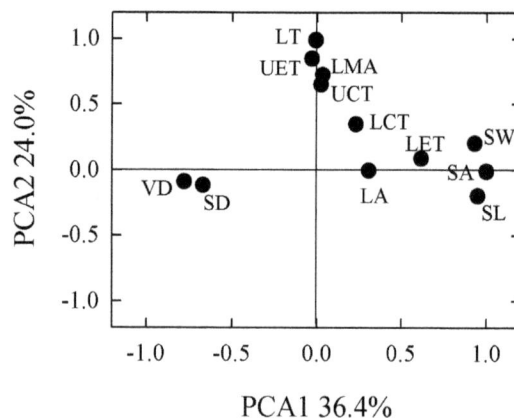

Figure 3. Factor-loading for stomatal and leaf traits along 2 axes of principal component analysis (PCA). SL, stomatal length; SW, stomatal width; SA, stomatal area; SD, stomatal density; UET, upper epidermal thickness; UCT, upper cuticular thickness; LET, lower epidermal thickness; LCT, lower cuticular thickness; LMA, leaf mass per unit area; LA, leaf area; VD, vein density; and MT, mesophyll thickness.

associations among leaf attributes and stomatal traits. Relationships among variables were analyzed using both Pearson regressions in R package 2.14 and phylogenetically independent contrasts (PICs). Possible evolutionary associations were assessed via PIC analysis, utilizing molecular phylogenetic trees [46]. This PIC analysis was evaluated with the "analysis of traits" (AOT) module in Phylocom, a program that calculates the internal node values for continuous traits [48,49].

Results

None of the species tested within *Paphiopedilum* had pubescent leaves, and all were hypostomatic. Although leaf and stomatal traits varied considerably across species (Tables 1, 2, Fig. 1), the magnitudes of variation were generally smaller for the stomata. Among species, fluctuations in SL, SW, SI, LCT and TCT were less than 2.0-fold, while those in VD, LMA, LA, SA, SD, UET, UCT, PTT, STT, LET and MT differed by 2.1- to 5.7-fold. For stomatal traits, the magnitude of variation was largest for SD (3.9-fold) and smallest for SW (1.4-fold). For leaf attributes, UET exhibited the largest variation (5.7-fold) across species while LCT showed the smallest range. The stable carbon isotope ratio (δ^{13}C) ranged from −27.24‰ to 23.32‰ (Table 1). Values for SD and VD differed significantly among growth habits, whereas the other traits showed no significant differences. Both SD and VD tended to increase from terrestrial to facultative and epiphytic orchids (Fig. 2).

All stomatal traits (SL, SW, SA and SD), plus VD, LET, and LA, loaded mainly on the first PCA axis, explaining 36.4% of the total variation (Fig. 3). By contrast, SD and VD loaded in the opposite direction on that axis. Leaf attributes, including LT, LMA, UET, MT, UCT, and LCT, loaded on the second axis, explaining 24.0% of the total.

Vein density was correlated with SL, SW, SA, SD, and LET; after phylogeny was considered, VD was still correlated with SL and SD (Fig. 4). Values for LET were correlated positively with SL, SW and SA, but negatively with SD (Fig. 5). After eliminating any phylogenetic effects via PICs, those correlations of LET with SW, SA, and SD became insignificant. Stomatal index was not correlated with any leaf structural straits.

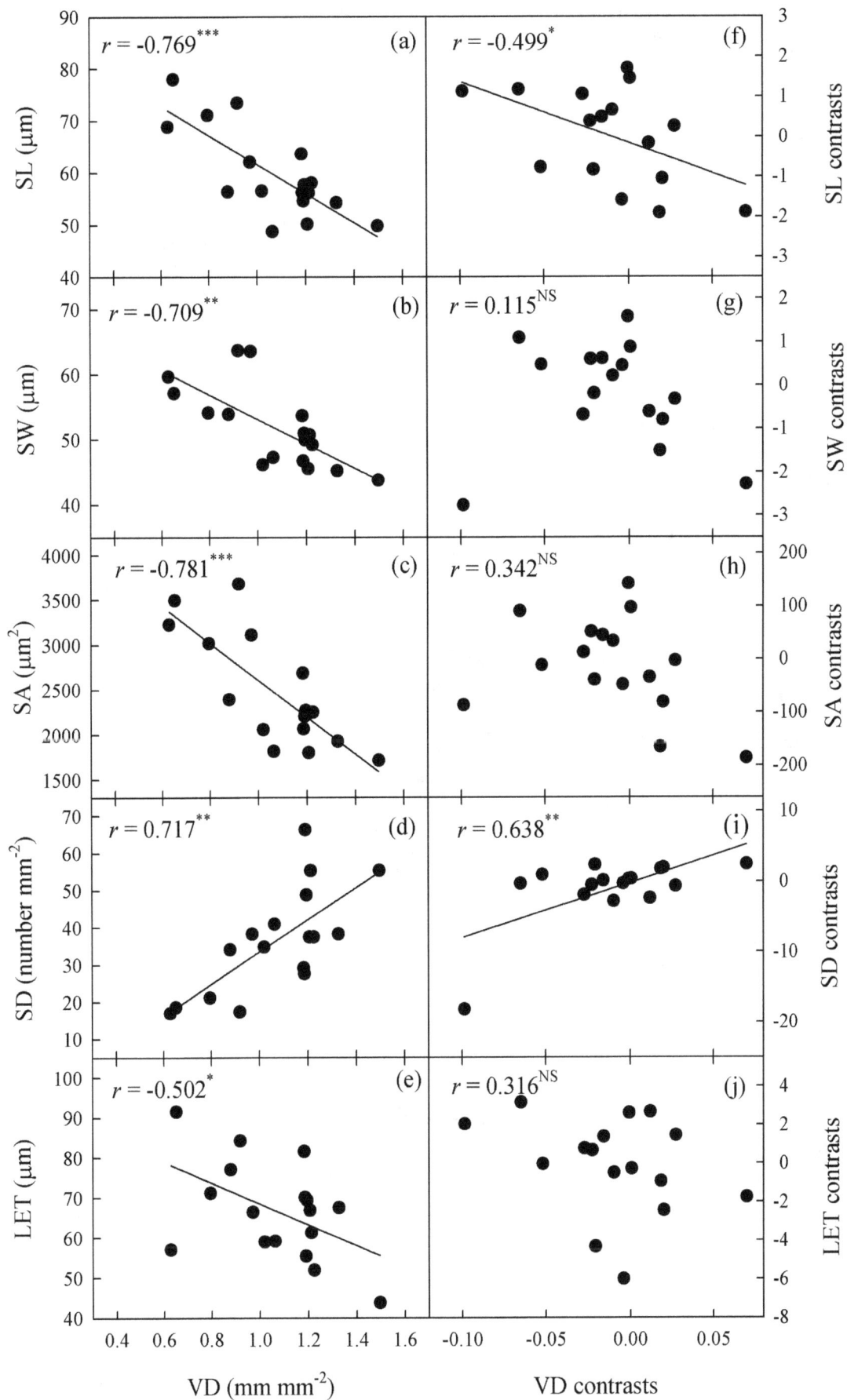

Figure 4. Correlations vein density with stomatal traits or lower epidermal thickness. Plate (a) to (e), Pearson's regressions; and plate (f) to (j), phylogenetically independent contrast correlations. VD, leaf vein density; SL, stomatal length; SW, stomatal width; SA, stomatal area; SD, stomatal density; and LET, lower epidermal thickness.

The UET was not correlated with SD when phylogeny was not considered, but a significant correlation was found between them after phylogenetic correction (Fig. 6). Conversely, stomatal density was positively correlated with stomatal length when a Pearson regression was used, but that correlation became insignificant after correction (Fig. 7). Neither leaf size nor thickness was correlated with SD or VD under any circumstances.

Discussion

The evolutionary coordination of stomatal density with leaf thickness has been assessed in numerous species [8]. Here, we took a phylogenetically comparative approach to examine the correlated evolution between stomatal traits and leaf attributes from closely related species of *Paphiopedilum* grown under controlled conditions. Vein density had an evolutionary association with stomatal density and size, but traits for stomata and leaf thickness showed independent evolution.

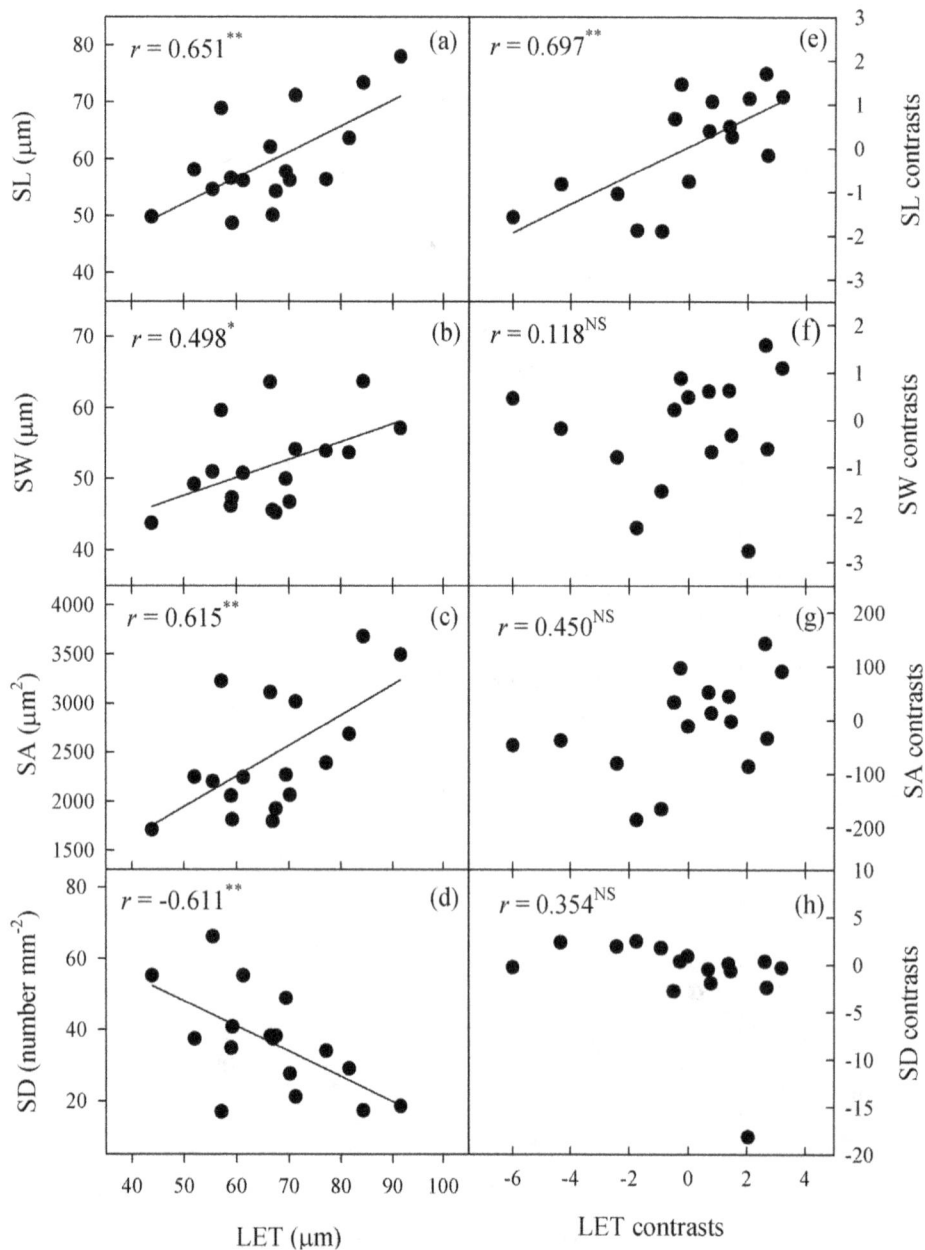

Figure 5. Correlations of lower epidermal thickness (LET) with stomatal traits. Plate (a) to (d), Pearson's regressions; and plate (e) to (h), phylogenetically independent contrast correlations. SL, stomatal length; SW, stomatal width; SA, stomatal area; and SD, stomatal density.

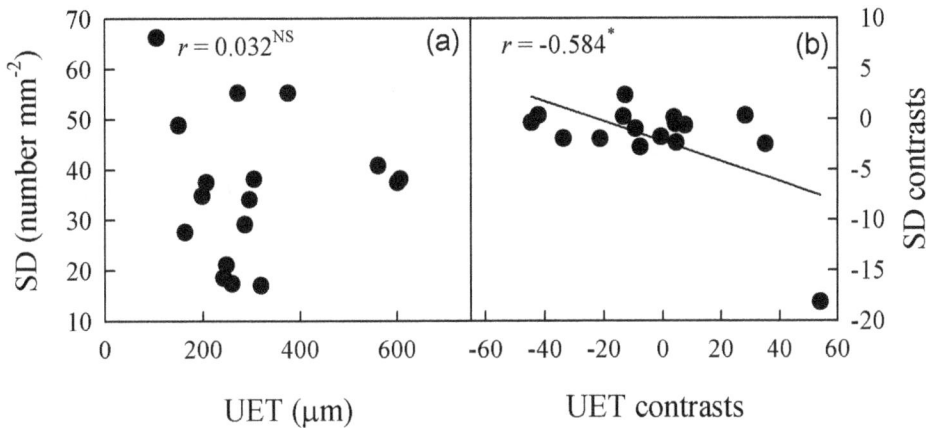

Figure 6. Correlation of upper epidermal thickness (UET) with stomatal density (SD). (a) Pearson's regression, and (b) phylogenetically independent contrast correlation.

Leaf Attributes and Stomatal Traits in *Paphiopedilum*

Leaves were fleshy and had cuticles on both sides. These characters are common among xeromorphic plants. Growth habit had no obvious influence on LMA, LT or cuticle thickness (Fig. 2). Samples from all species were hypostomatic, and their stomata were sunken into the leaf epidermis. This adaptive feature shields exerophytic plants from the effects of desiccating winds, and can help prevent excessive transpiration losses [50]. Compared with data reported from other angiosperms, *Paphiopedilum* members had relatively lower VD and SD, but larger stomata [9,51]. In fact, previous study has suggested that the species in Orchidaceae have, relatively, the lowest SD values in the entire plant kingdom [40]. We noted that epiphytic *Paphiopedilum* had higher VD and SD than the terrestrial species (Fig. 2). Dunbar-Co *et al.* have also found that taxa in *Plantago* growing on drier sites have higher VD [9]. As a whole, these leaf attributes and stomatal traits reflect a general trend in how land plants adapt when water is limited.

Relationship of Leaf Attributes and Stomatal Traits to Phylogeny

Traits for both leaf anatomy and stomata varied significantly across species, although to a lesser extent for the latter (Table 1, Fig. 1). Several traits, such as SL, SA, UCT and TCT, showed strong phylogenetic signals while SD and SI exhibited a strong convergent evolution. This high level of conservatism demonstrates a distinct evolutionary shift among species [1]. Somewhat contradictory to our findings, Beaulieu *et al.* [43] did not report strong signals in SL ($K = 0.685$) or SD ($K = 0.540$) for 101 angiosperm species. However, Hodgson *et al.* [20] noted that stomatal size was related to both cytological status and phylogeny. The discrepancy between our observations and those of Beaulieu *et al.* are probably related to the choice of plant materials tested. In that earlier study, three growth forms were selected (herb, tree, and shrub), which led to large genetic differences. By contrast, our examination utilized tissues from the same genus, with all plants exposed to the same greenhouse conditions and, consequently, revealing only small genetic differences.

The strong signals for SL, SA, UCT, and TCT indicated that those traits are phylogenetically conserved. However, most traits had weak signals, possibly because of a departure from Brownian motion evolution, such as adaptive evolution, that would not have been correlated with phylogeny. Therefore, this reflected the outcome of selection in heterogeneous environments where species can best acclimate to their current growing conditions [1]. Caruso *et al.* [52] have suggested that any constraints on the development of stomatal traits in *Lobelia cardinalis* primarily arise from a lack of genetic variation. In our study, the correlation between LET and

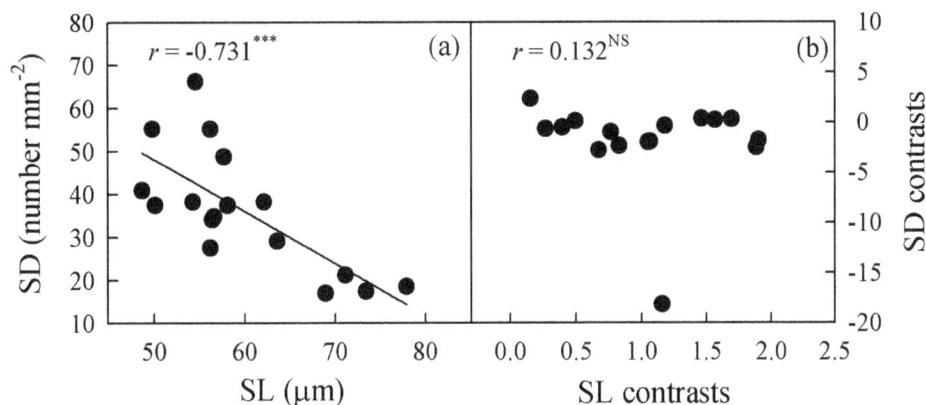

Figure 7. Correlation of stomatal length (SL) with stomatal density (SD). (a) Pearson's regression, and (b) phylogenetically independent contrast correlation.

either SD or SA disappeared when the effect of phylogeny was considered, thus confirming that variations in stomatal traits and leaf attributes are related to that particular influence.

Evolutionary Associations of Stomatal Anatomy with Leaf Traits

Vein density in *Paphiopedilum* was positively correlated with stomatal density, whether or not phylogeny was considered. However, VD was negatively correlated with stomatal size (Fig. 4), indicating that leaf vein has an evolutionary association with stomatal anatomy. This result supports the notion that the development and function of leaf veins and stomata are co-ordinated [11], as the coordinated development of veins and stomata is important for optimizing photosynthetic yield relative to carbon investment in leaf venation [11]. Moreover, coordinated plasticity in veins and stomata is thought to be at least partially related to leaf size; the development of leaf-size plasticity can provide an efficient way for plants to acclimate their hydraulic and stomatal conductance to contrasting transpirational demands under different lighting conditions [11,51]. However, we found that SD and VD for these 17 *Paphiopedilum* species were not affected by leaf size. This was because our experimental materials had been grown in the same environment, and had similar transpirational demands.

We found no evidence for correlated evolution between stomatal traits and leaf thickness or cuticle thickness, which suggests a lack of functional association. Although LET was correlated with stomatal traits when phylogeny was not considered, only two correlations (LET vs SL, UET vs SD) were significant after that correction. The discrepancy between our Pearson's and PIC correlations can be explained in that PICs reflect the historical pattern of diversification among taxa, whereas traditional Pearson's correlations describe present-day relations among taxa [1]. Similar to our results, Beerling and Kelly [8] have suggested that thicker leaves do not necessarily mean more stomata. Nevertheless, previous studies have also shown that species with thick leaves have moderately large stomata [20], and that leaf thickness is negatively correlated with SD along an acidity gradient [18].

The lack of evolutionary correlation of stomatal traits with leaf thickness or cuticle thickness may have several explanations. Selective pressure that drives their development can differ between the two. Evolutionary trends largely depend on the selective force endured in challenging environments [9]. Stomatal density can be influenced by atmospheric CO_2 concentration, heat stress, water status, plant density and light intensity [13,16,17], whereas leaf thickness is affected by light intensity, UV-radiation, rainfall and the supply of soil nutrients [31,35,38]. This inconsistency in evolutionary correlations among functional traits suggests that fundamentally different selective pressures and constraints may be acting [53]. Consequently, for the genus studied here, periodic water shortages and low nutrient availability in karst regions would have contributed to the evolution of leaf anatomy.

The difference in function between leaf cuticle thickness and stomatal traits decreases the coordination between them. In fact,

changes in leaf anatomy do not always reflect adaptations to water availability. For example, leaves of plants growing in habitats with reduced soil nutrients have thicker epidermises than do their relatives in high-nutrient soils [30] because those sclerophyllous tissues develop as a way to protect scarce nutrient investments in leaf material against herbivory and abiotic physical damage [37]. By contrast, in arid environments, a thick cuticle likely has other functions besides that of water barrier, such as preventing physical damage by herbivorous pests [54].

The structural investment toward different leaf traits is largely controlled by an evolutionary trade-off between the antagonistic demands to maximize both photosynthesis and WUE [19,55]. Having a thicker cuticle implies a greater construction cost for the leaf protective structure [28]. If more biomass must be allocated to the same function, the investment is reduced toward other functions. This situation is not cost-efficient to plant survival and competitiveness. Therefore, a correlated evolution among those traits would limit such divergence and adaptive selection [1]. Although many leaf surface characters, e.g., crypts, wax and hairs, can modify the relationship between stomatal size and number, and stomatal function, an evolutionary association between leaf anatomical traits and stomatal traits does not always necessitate water conservation and ecological strategies.

Correlation between Stomatal Density and Size

Stomatal density was significantly correlated with SL, but that association disappeared when phylogeny was considered. The negative correlation found here between SD and SL has been described previously [43,56]. Both stomatal aperture and density are linked to leaf conductance, photosynthetic carbon gain and transpiration [55]. The capacity of plants to fix carbon is constrained by their photosynthetic biochemistry and CO_2 diffusion conductance. When the concentration of atmosphere CO_2 decreases, stomata become denser while the rate of maximum Rubisco carboxylation (V_{cmax}) slows. This co-variation among SL, SD and the V_{cmax} rate reduces the impact that any change in atmospheric CO_2 has on the assimilation of leaf CO_2, resulting in minimum energy cost and reduced nitrogen requirements [3]. A negative correlation between SD and SL also increases plasticity in maximum stomatal conductance to water vapor and CO_2, with minimal alterations in the balance of water loss and epidermal allocations to the stomata [14,56].

In summary, phylogeny has a significant effect on leaf traits and stomatal traits in *Paphiopedilum*. Stomatal length and area and upper cuticle thickness are strongly conserved. We noted a correlated evolution between stomatal traits and vein density in *Paphiopedilum*, but not between stomatal traits and leaf thickness. These findings provide insight into the development of traits related to water economy by orchids under natural selection.

Author Contributions

Conceived and designed the experiments: HH SBZ. Performed the experiments: ZJG MS JJZ. Analyzed the data: ZJG SBZ. Contributed reagents/materials/analysis tools: ZJG JJZ SBZ. Wrote the paper: SBZ KFC HH.

References

1. Ackerly DD, Donoghue MJ (1998) Leaf size, sapling allometry, and Corner's rules: Phylogeny and correlated evolution in maples (*Acer*). The American Naturalist 152: 767–791.

2. Guan ZJ, Zhang SB, Guan KY, Li SY, Hu H (2011) Leaf anatomical structures of *Paphiopedilum* and *Cypripedium* and their adaptive significance. Journal of Plant Research 124: 289–298.

3. Franks PJ, Beerling DJ (2009) CO_2-forced evolution of plant gas exchange capacity and water-use efficiency over the Phanerozoic. Geobiology 7: 227–236.

4. Pittermann J (2010) The evolution of water transport in plants: An integrated approach. *Geobiology* 8: 112–139.

5. Brodribb TJ, Feild TS (2010) Leaf hydraulic evolution led a surge in leaf photosynthetic capacity during early angiosperm diversification. Ecology Letters 13: 175–183.

6. Willson CJ, PS Manos, RB Jackson (2008) Hydraulic traits are influenced by phylogenetic history in the drought-resistant and invasive genus *Juniperus* (Cupressaceae). American Journal of Botany 95: 299–314.

7. Raven JA (2002) Selection pressures on stomatal evolution. New Phytologist 153: 371–386.

8. Beerling DJ, Kelly CK (1996) Evolutionary comparative analyses of the relationship between leaf structure and function. New Phytologist 134: 35–51.

9. Dunbar-Co S, Sporck MJ, Sack L (2009) Leaf trait diversification and design in seven rare taxa of the Hawaiian *Plantago* radiation. International Journal of Plant Sciences 170: 61–75.

10. Woodward FI (1998) Do plants really need stomata? Journal of Experimental Botany 49: 471–480.

11. Brodribb TJ, Jordan GJ (2011) Water supply and demand remain balanced during leaf acclimation of *Nothofagus cunninghamii* trees. New Phytologist 192: 437–448.

12. Edwards D, Kerp H, Hass H (1998) Stomata in early land plants: An anatomical and ecophysiological approach. Journal of Experimental Botany 49: 255–278.

13. Woodward FI (1987) Stomatal numbers are sensitive to increases in CO_2 from pre-industrial levels. Nature 327: 617–618.

14. Franks PJ, Drake PL, Beerling DJ (2009) Plasticity in maximum stomatal conductance constrained by negative correlation between stomatal size and density: An analysis using *Eucalyptus globulus*. Plant, Cell & Environment 32: 1737–1748.

15. Ashton PMS, Berlyn GP (1992) Leaf adaptations of some *Shorea* species to sun and shade. New Phytologist 121: 587–596.

16. Schlüter U, Muschak M, Berger D, Altmann T (2003) Photosynthetic performance of an *Arabidopsis* mutant with elevated stomatal density (*sdd1-1*) under different light regimes. Journal of Experimental Botany 54: 867–874.

17. Xu ZZ, Zhou GS (2008) Responses of leaf stomatal density to water status and its relationship with photosynthesis in a grass. Journal of Experimental Botany 59: 3317–3325.

18. Wang R, Huang W, Chen L, Ma L, Guo C, et al. (2011) Anatomical and physiological plasticity in *Leymus chinensis* (Poaceae) along large-scale longitudinal gradient in Northeast China. PLoS ONE 6: e26209.

19. Aasamaa K, Sõber A, Rahi M (2001) Leaf anatomical characteristics associated with shoot hydraulic conductance, stomatal conductance and stomatal sensitivity to changes of leaf water status in temperate deciduous trees. Australian Journal of Plant Physiology 28: 765–774.

20. Hodgson JG, Sharafi M, Jalili A, Díaz S, Montserrat-Martí G, et al. (2010) Stomatal vs. genome size in angiosperms: The somatic tail wagging the genomic dog? Annals of Botany 105: 573–584.

21. Niklas KJ (1999) A mechanical perspective on foliage leaf form and function. New Phytologist 143: 19–31.

22. Sack L, Frole K (2006) Leaf structural diversity is related to hydraulic capacity in tropical rain forest trees. Ecology 87: 483–491.

23. Brodribb TJ, Feild TS, Jordan GJ (2007) Leaf maximum photosynthetic rate and venation are linked by hydraulics. Plant Physiology 144: 1890–1898.

24. Uhl D, Mosbrugger V (1999) Leaf venation density as a climate and environmental proxy: A critical review and new data. Palaeogeography, Palaeoclimatology, Palaeoecology 149: 15–26.

25. Walls RL (2011) Angiosperm leaf vein patterns are linked to leaf functions in a global-scale data set. American Journal of Botany 98: 244–253.

26. Nardini A, Ramani M, Gortan E, Salleo S (2008) Vein recovery from embolism occurs under negative pressure in leaves of sunflower (*Helianthus annuus*). Physiologia Plantarum 133: 755–764.

27. Scoffoni C, Rawls M, McKown A, Cochard H, Sack L (2011) Decline of leaf hydraulic conductance with dehydration: Relationship to leaf size and venation architecture. Plant Physiology 156: 832–843.

28. Gratani L, Bombelli A (1999) Leaf anatomy, inclination and gas exchange relationships in evergreen sclerophyllous and drought semideciduous shrub species. Photosynthetica 37: 573–585.

29. Watkins JE, Mack MC, Sinclair TR, Mulkey SS (2007) Ecological and evolutionary consequences of desiccation tolerance in tropical fern gametophytes. New Phytologist 176: 708–717.

30. Bucci SJ, Goldstein G, Meinzer FC, Scholz FG, Franco AC, et al. (2004) Functional convergence in hydraulic architecture and water relations of tropical savanna trees: From leaf to whole plant. Tree Physiology 24: 891–899.

31. Cunningham SA, Summerhayes B, Westoby M (1999) Evolutionary divergences in leaf structure and chemistry, comparing rainfall and soil nutrient gradients. Ecological Monograph 69: 569–588.

32. Muchow RC, Sinclair TR (1989) Epidermal conductance, stomatal density and stomatal size among genotypes of *Sorghum bicolour* (L.) Moench. Plant, Cell & Environment 12: 425–431.

33. Helbsing S, Riederer M, Zotz G (2000) Cuticles of vascular epiphytes: Efficient barriers for water loss after stomatal closure? Annals of Botany 86: 765–769.

34. Schreiber L, Riederer M (1996) Ecophysiology of cuticular transpiration: Comparative investigation of cuticular water permeability of plant species from different habitats. Oecologia 107: 426–432.

35. Manetas Y, Petropoulou Y, Stamatakis K, Nikolopoulos D, Levizou E, et al. (1997) Beneficial effects of enhanced UV-B radiation under field conditions: Improvement of needle water relations and survival capacity of *Pinus pinea* L. seedlings during the dry Mediterranean summer. Plant Ecology 128: 101–108.

36. Riederer M, Schreiber L (2001) Protecting against water loss: Analysis of the barrier of plant cuticles. Journal of Experimental Botany 52: 2023–2032.

37. Turner IM (1994) Sclerophylly – primarily protective. Functional Ecology 8: 669–675.

38. Kerstiens G (2006) Water transport in plant cuticles: An update. Journal of Experimental Botany 57: 2493–2499.

39. Cribb P (1998) The Genus *Paphiopedilum* (2nd Edition). Natural History Publications, Kota Kinabalu (Borneo) in association with Royal Botanic Gardens, Kew, UK.

40. Karasawa K, Saito K (1982) A revision of the genus *Paphiopedilum* (Orchidaceae). Bulletin of the Hiroshima Botanical Garden 5: 1–69.

41. Assmann SM, Zeiger E (1985) Stomatal responses to CO_2 in *Paphiopedilum* and *Phragmipedium* – role of the guard cell chloroplast. Plant Physiology 77: 461–464.

42. Zhang S-B, Guan Z-J, Chang W, Hu H, Yin Q, et al. (2011) Slow photosynthetic induction and low photosynthesis in *Paphiopedilum armeniacum* are related to its lack of guard cell chloroplast and peculiar stomatal anatomy. Physiologia Plantarum 142: 118–127.

43. Beaulieu JM, Leitch IJ, Patel S, Pendharkar A, Knight CA (2008) Genome size is a stronger predictor of cell size and stomatal density in angiosperms. New Phytologist 179: 975–986.

44. James SA, Bell DT (2001) Leaf morphological and anatomical characteristics of heteroblastic *Eucalyptus globulus* ssp. *globulus* (Myrtaceae). Australian Journal of Botany 49: 259–269.

45. Blomberg SP, Garland T Jr, Ives AR (2003) Testing for phylogenetic signal in comparative data: Behavioral traits are more labile. Evolution 57: 717–745.

46. Cox AV, Pridgeon AM, Albert VA, Chase MW (1997) Phylogenetics of the slipper orchids (Cypripedioideae, Orchidaceae): Nuclear rDNA ITS sequences. Plant Systematics and Evolution 208: 197–223.

47. R Development Core Team (2011) R: A Language and Environment for Statistical Computing. R Foundation for Statistical Computing, Vienna, Austria. The R Project for Statistical Computing website. Available: http://www.R-project.org. Accessed 2011 Jul 10.

48. Webb CO, Ackerly DD, Kembel SW (2008) PHYLOCOM: Software for the analysis of phylogenetic community structure and trait evolution. Bioinformatics 24: 2098–2100.

49. Felsenstein J (1985) Phylogenies and the comparative method. The American Naturalist 125: 1–15.

50. Jiménez S, Zellnig G, Stabentheiner E, Peters J, Morales D, et al. (2000) Structure and ultrastructure of *Pinus canariensis* needles. Flora 195: 228–235.

51. Murphy MRC, Jordan GJ, Brodribb TJ (2012) Differential leaf expansion can enable hydraulic acclimation to sun and shade. Plant, Cell & Environment doi: 10.1111/j.1365-3040.2012.02498.x.

52. Caruso CM, Maherall H, Mikulyuk A, Carlson K, Jackson RB (2005) Genetic variance and covariance for physiological traits in *Lobelia*: Are there constraints on adaptive evolution? Evolution 59: 826–832.

53. Kembel SW, Cahill Jr JF (2011) Independent evolution of leaf and root traits within and among temperate grassland plant communities. PLoS ONE 6: e19992.

54. Gentry G, Barbosa P (2006) Effects of leaf epicuticular wax on the movement, foraging behavior, and attack efficacy of *Diaeretiella rapae*. Entomologia Experimentalis et Applicata 121: 115–122.

55. Büssis D, von Groll U, Fisahn J, Altmann T (2006) Stomatal aperture can compensate altered stomatal density in *Arabidopsis thaliana* at growth light conditions. Functional Plant Biology 33: 1037–1043.

56. Sack L, Grubb PJ, Marañón T (2003) The functional morphology of juvenile plants tolerant of strong summer drought in shaded forest understories in southern Spain. Plant Ecology 168: 139–163.

11

Some Limitations of Public Sequence Data for Phylogenetic Inference (in Plants)

Cody E. Hinchliff*, Stephen Andrew Smith

Department of Ecology and Evolutionary Biology, University of Michigan. Ann Arbor, Michigan, United States of America

Abstract

The GenBank database contains essentially all of the nucleotide sequence data generated for published molecular systematic studies, but for the majority of taxa these data remain sparse. GenBank has value for phylogenetic methods that leverage data–mining and rapidly improving computational methods, but the limits imposed by the sparse structure of the data are not well understood. Here we present a tree representing 13,093 land plant genera—an estimated 80% of extant plant diversity—to illustrate the potential of public sequence data for broad phylogenetic inference in plants, and we explore the limits to inference imposed by the structure of these data using theoretical foundations from phylogenetic data decisiveness. We find that despite very high levels of missing data (over 96%), the present data retain the potential to inform over 86.3% of all possible phylogenetic relationships. Most of these relationships, however, are informed by small amounts of data—approximately half are informed by fewer than four loci, and more than 99% are informed by fewer than fifteen. We also apply an information theoretic measure of branch support to assess the strength of phylogenetic signal in the data, revealing many poorly supported branches concentrated near the tips of the tree, where data are sparse and the limiting effects of this sparseness are stronger. We argue that limits to phylogenetic inference and signal imposed by low data coverage may pose significant challenges for comprehensive phylogenetic inference at the species level. Computational requirements provide additional limits for large reconstructions, but these may be overcome by methodological advances, whereas insufficient data coverage can only be remedied by additional sampling effort. We conclude that public databases have exceptional value for modern systematics and evolutionary biology, and that a continued emphasis on expanding taxonomic and genomic coverage will play a critical role in developing these resources to their full potential.

Editor: Simon Joly, Montreal Botanical Garden, Canada

Funding: This work was supported by National Science Foundation (http://www.nsf.gov/) award DEB 1207915 to S. A. Smith. The funders had no role in study design, data collection and analysis, decision to publish, or preparation of the manuscript.

Competing Interests: The authors have declared that no competing interests exist.

* Email: cody.hinchliff@gmail.com

Introduction

The GenBank nucleotide database [1] contains more than one hundred million sequences representing more than 275,000 species of life. The successful use of these data to reconstruct comprehensive phylogenies for many large clades has illustrated their potential for phylogenetic inquiry [2,3], with the implication that this potential may extend to the broadest scales, e.g. the tree of life itself [3,4]. By some estimates however, GenBank contains samples of only 3% of Earth's species [5], and studies using sequence data mined from public databases have demonstrated enigmatic results [6–8]. Public sequence data have clear potential for evolutionary biology and hypothesis-testing at very broad scales, but their structure can have significant implications regarding the limits of phylogenetic inference [9,10]. The extent and severity of these limits for existing resources such as GenBank remains largely unexplored (but see [3,4]). Here we demonstrate the potential and the limits to inference of the NCBI GenBank database for comprehensive phylogenetic studies using the land plants, a monophyletic, ancient, and very biodiverse group with over 300,000 extant species [11] as an example.

Results and Discussion

Leveraging public sequence data

We compiled nucleotide sequence data from GenBank for the land plants, including the closely related Charophycean algae [12–14] to facilitate rooting. A total of 128 markers were selected for their relatively broad phylogenetic coverage, including 109 chloroplast, 14 mitochondrial, and 5 nuclear markers including the nuclear ribosomal internal and external transcribed spacers (see Table S1 for a complete list). We chose not to include additional nuclear markers because of challenges of homology assessment at deep phylogenetic scales. We used the program PHLAWD [7] to gather data from GenBank release 185 for these 128 markers, resulting in a data set including 5.1×10^7 genetic sequences and representing over 100,000 plant and algal species. To maximize coverage and reduce computational complexity, we summarized the available sequence data for all of the 13,093 genera for which at least one sequence was available. For each of these genera, we selected the longest available sequence at each of our target loci that had been sequenced for any taxonomic child of that genus, which produced a set of sequence data (usually from multiple species) that was used to represent that genus in the

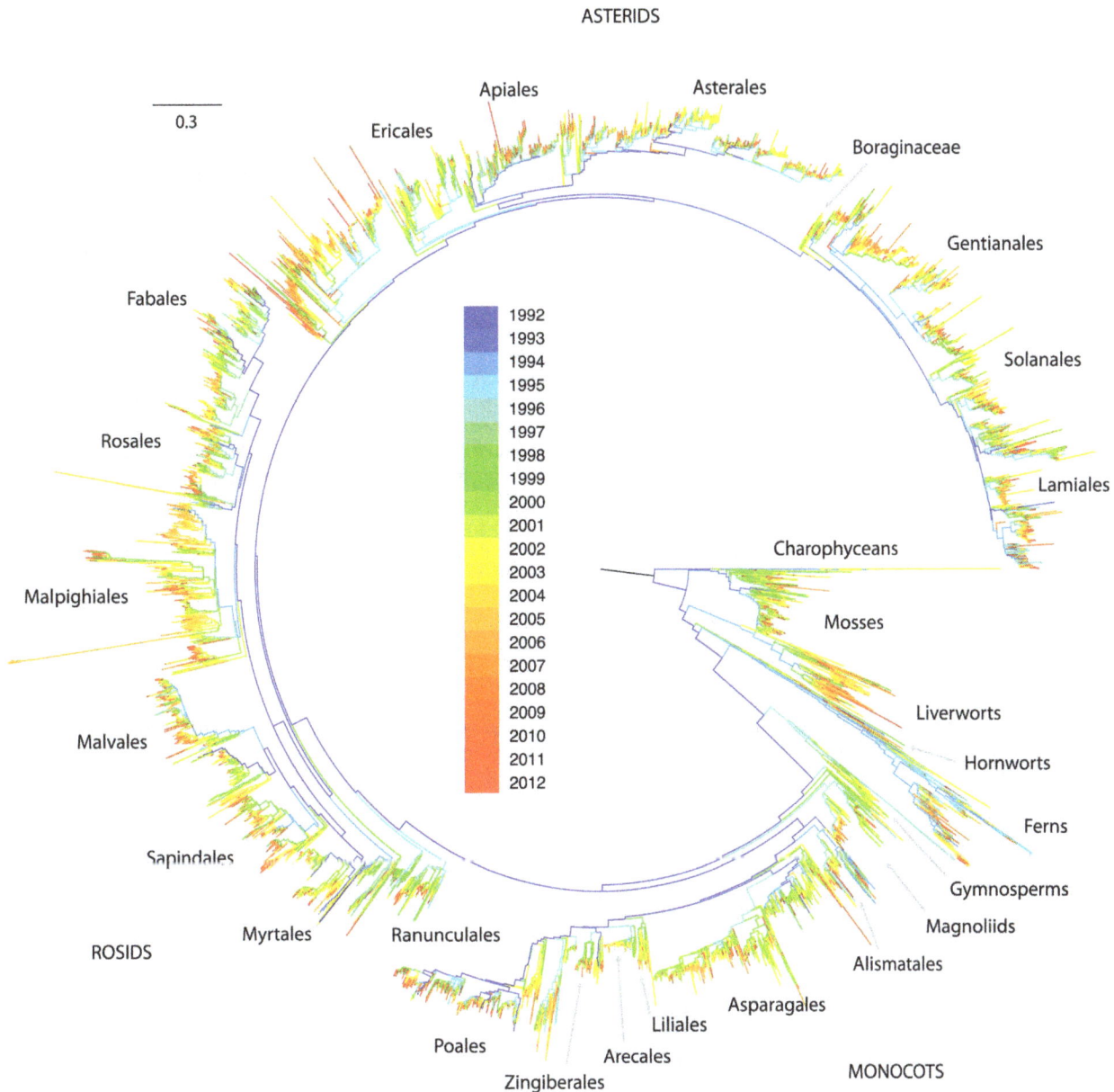

Figure 1. Phylogeny of land plants generated from nucleotide sequence data in GenBank release 185. Tips represent extant genera and branch lengths correspond to substitutions per site. Each branch is labeled according to the age of the oldest exemplar sequence in GenBank for any of its descendant tips, with blue branches representing lineages with older exemplar sequences and red branches showing lineages that have only recently been added. The total number of genera represented in the tree is 13,093. A large version of this figure with legible tip names is presented in File S1.

alignment. Maximum likelihood [15] was used to infer phylogeny using this alignment, yielding the tree topology presented in Fig. 1.

Despite the vast amount of data available on GenBank, only about one third of recognized plant species were represented in GenBank release 185. At deeper taxonomic levels this coverage is considerably better, with about 83% (about 13,400 of 16,167) of recognized land plant genera [16] represented by at least one sequence. The rate of species accumulation on GenBank has stayed relatively constant since the mid 1990's and shows no signs of reaching saturation (Fig. 2, C). Coverage at the generic level however, is approaching the estimated maximum of 16,167 (Fig. 2, B), representing a landmark achievement for plant systematists. Our ability to reconstruct the phylogeny of extant plants has

grown as a function of this increase in lineage representation through time (Fig. 1). Early, often broadly inclusive studies [17–21] resolved many deep divergences (blue branches of Fig. 1), while myriad more detailed studies [22–30] for example) have contributed to resolution near the tips and also increased confidence in deep relationships.

The comprehensive phylogenies made possible by such achievements facilitate inquiry into questions of broad interest to the scientific community. For example, we used the tree topology presented in Fig. 1 to assess the level of monophyly in the land plant classification (as described in the Materials and Methods), finding that monophyly cannot be rejected for 75.6% of the 680 families of Embryophyta in use in the Genbank taxonomy at the

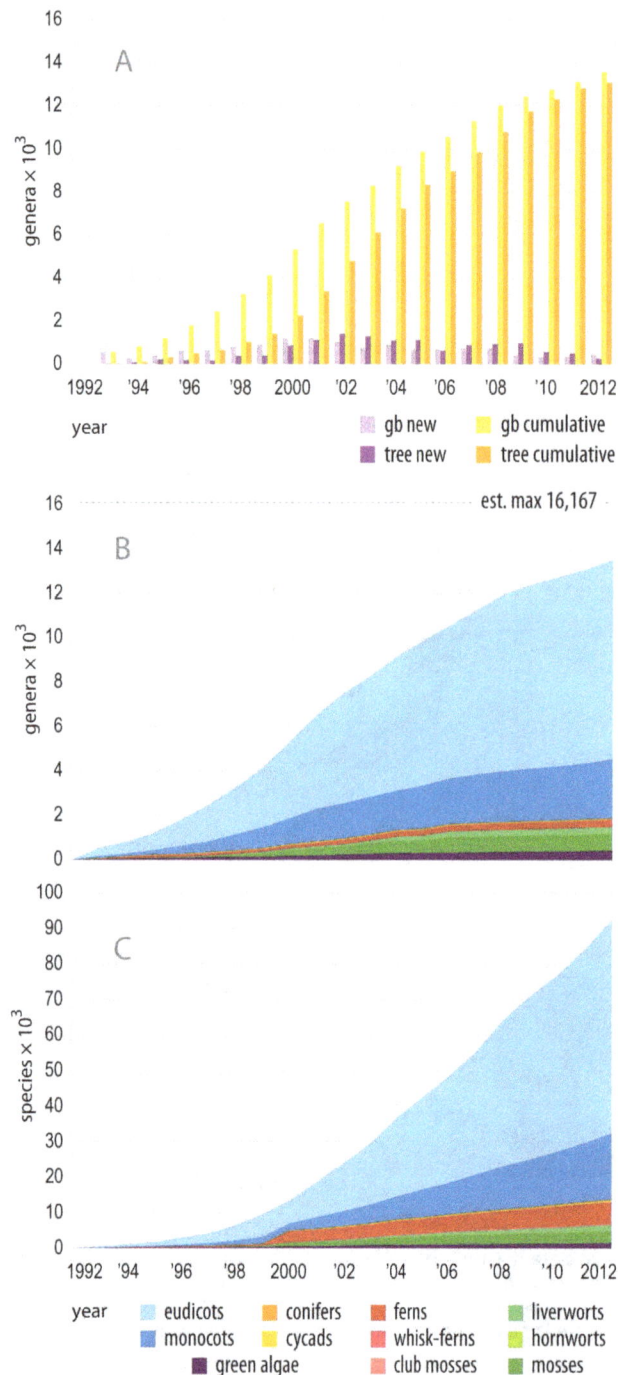

Figure 2. Taxon sampling through time on GenBank and the ages of exemplar sequences. A: purple bars indicate the number of new generic exemplars that were: added each year to GenBank (light purple); and the number of exemplar sequences from that year that were used in the phylogeny presented in Figs. 2 and 5 (dark purple). Yellow bars indicate cumulate values. The lag in the total number of genera represented in the phylogeny is due to the fact that the sequence selection procedures prefer the longest available exemplar sequence, which is often not the oldest. B and C: cumulative numbers of genera (B) and species (C) represented by sequences on GenBank, with colors labeling major groups. Species representation continues to grow at a relatively constant rate, but the rate of new genus addition is slowing.

time of writing. To facilitate additional inquiry into more detailed questions, we also provide a time calibrated version of the Fig. 1 phylogeny using node age constraints from a recent dating analysis of angiosperms [31] in the supplemental materials.

The impact of missing data

It is well-understood that high proportions of missing data (or more specifically, that the distribution and nature of such missing data) can impact phylogenetic inference [3,6,32]. Most large alignments with broad phylogenetic scope compiled from public sequence databases have relatively high proportions of missing data [3,6–8]; for instance, the alignment used to generate the tree of Fig. 1 (13,093 taxa $\times$ 128 loci) contained 96.4% missing data. To measure the limiting effect of these absent data, we used the partial decisiveness metric d [10]. This metric indicates the proportion of all possible edges across all possible trees that are distinguishable given the available data (that is, given the distribution of missing data). d is distributed on interval [0,1] with $d = 0$ indicating a fully uninformative alignment in which no edges may be informed because the available data are not sampled densely enough to permit phylogenetically informative comparisons, and $d = 1$, indicating that all possible edges may be informed (i.e. all possible trees may be inferred). We note that edge *distinguishability*, which is measured by d, is not synonymous with edge *support* (we assess branch support with other measures; see corresponding sections), and that d does not measure phylogenetic signal. An edge is considered distinguishable—that is, the data are decisive for that edge—if the taxonomic sampling is such that the edge *may* be informed by at least some of the present data, but this does not imply that those data *will* support a topology containing that edge [9,10].

The partial decisiveness of our comprehensive alignment was estimated to be $d = 0.863$, indicating that GenBank's sampling is sufficient to inform all but 13.7% of the potential phylogenetic relationships among represented plant genera. This statistic represents a fairly conservative estimate of GenBank's phylogenetic utility in the sense that while d considers all possible edges, only a small subset of those are likely to occur in phylogenetic trees. Accurate phylogenies may be reconstructed while $d < 1$ as long as the data are decisive for the edges present in those tree(s) that actually represent the phylogenetic history of the sampled organisms [6].

Phylogenetic inference relies on the co-sampling of homologous data across lineages; high levels of lineage representation allow the distinguishability of many more edges than low levels simply because lineages for which no data are present can only be arbitrarily placed. Ideally, phylogenetic datasets should contain an adequate number of informative loci (we usually assume that more is better), each sampled for many lineages, thus allowing many edges to be informed by relatively large amounts of data. In Fig. 3, we present patterns of lineage and locus sampling depth across the entire plant chloroplast genome, which comprises the great majority of phylogenetically informative sequence data for plants. In general, chloroplast loci show highly asymmetrical lineage sampling (indicated by dark blue bar plots in the outer ring), and the frequency at which pairs of loci have been sampled for the same lineages is in quite low overall (indicated by the blue ribbons connecting pairs of loci). Only a handful of loci—*atpB*, *matK*, *ndhF*, *psbA-trnH*, *rbcL*, *rps4*, and *trnT-trnL-trnF*—show relatively high levels of lineage representation or taxonomic overlap (primarily with one another). Similar patterns hold for loci in the mitochondrial and nuclear genomes (data not shown), and the problem of sparse data coverage is in fact exacerbated in the case of nuclear genes by challenges associated with homology assessment at deep phylogenetic scales. Only the nuclear ribosomal internal transcribed spacer (ITS) shows levels of lineage sampling similar to the heavily sampled chloroplast loci named above (Fig. 3; Table S3). The partial decisiveness (*d*) of GenBank's nucleotide data, and the great majority of our information about phylogenetic relationships among plants, comes from these fewer than 10 loci (Figs. 3, 4). Many of these best sampled loci, however, are relatively fast evolving (especially ITS and chloroplast intergenic spacers), and contain relatively little phylogenetic signal for resolving deep branches in the tree.

We assessed the extent and potential impact of this sparse data coverage at a more targeted scale by calculating the number of loci in the alignment which had phylogenetically decisive taxon coverage for each branch in the ML tree (see Materials and Methods). These branch-specific patterns of data decisiveness across the land plant genus phylogeny are presented in Fig. 5, and additional figures identifying the individual branches capable of being informed by each locus are presented in the file titled "Supplemental tree figures" that is available in the Data Dryad repository associated with this article. Very deep branches in the tree are informed by many loci (blue hues in Fig. 5), but branches near the tips are informed by relatively few (red hues in Fig. 5). Figure 4 plots sampling depth (in this case the number of loci with decisive taxon sampling) by the proportion of branches in the tree, and demonstrates that the great majority of branches in the tree are informed by relatively few loci.

The patterns shown in Figs. 4 and 5 suggest that GenBank data may lack the sampling depth required to accurately infer comprehensive plant phylogenies. We combined data at the generic level to maximize genetic coverage for the resulting tips, but even this did not entirely overcome sampling issues at shallow phylogenetic depths. This is a significant limitation for a variety of evolutionary analyses that depend on trees—shallow to mid-depth regions of trees encode a far greater amount of phylogenetic information (by sheer number of branches and nodes alone) than deep branches, and have a very high capacity to inform a broad variety of questions regarding evolutionary processes in plants. Thus, we argue that the accurate inference of shallow relationships is critical for many important questions in evolutionary biology (e.g. any analysis involving lineage diversification). At shallower taxonomic depths than genera, however, the limiting effects of sparse data coverage are even stronger than those we present here. Overcoming these limits will require increased sampling to improve both the genetic depth as well as the taxonomic coverage of the data.

Measuring branch support

Estimating branch support is an important mechanism of assessing confidence in topology, but doing so with large, sparse matrices can be challenging. Traditional measures such as bootstraps do not perform well with very high proportions of missing data [6] and replicate tree searches (e.g. Bayesian MCMC, standard bootstrapping, and jackknife) on large alignments can be prohibitively time consuming. In the case of our alignment, a single ML tree search took several weeks using 10×2.4Ghz Intel Xeon processor cores and more than 40 GB of memory; running thousands or more of these is not feasible. We therefore implemented a measure of branch support that is relatively fast to calculate given a single tree and an alignment (even for large alignments), which relies on the information criterion (IC) framework presented by [33]. Specifically, we used the ICA statistic (defined in that paper), which is a branchwise measure of support based on information theory that quantifies, for a given bipartition (e.g. the one implied by a given branch in a tree), the

level of congruence or conflict (with that bipartition) across a set of topologies. The ICA score varies on the interval $[-1,1]$, with 1 indicating perfectly congruent supporting information—the specified bipartition is observed in all of the topologies; -1 indicating perfectly congruent conflicting information—the specified bipartition is never observed, but rather a single conflicting bipartition is observed in all the topologies; and 0 indicating perfectly equivocal information—all observed bipartitions occur at equal frequency. A positive ICA score in general indicates that the specified bipartition is observed at a higher frequency than any single alternative (i.e. conflicting) bipartition, whereas a negative ICA score indicates that some other alternative bipartition occurs at a higher frequency.

Since computational limits prevented us from generating topology replicates for the entire tree, we generated replicates using an approach that we call a localized taxon quartet jackknife, which consisted of selecting tips at random from clades defined by the complete ML tree, and inferring topology for these randomized tip subsamples to generate topology replicates. Each replicate contained a quartet of tips selected to guarantee that any topology inferred for those tips would either be consistent with a given targeted branch in the original ML topology, or would conflict with that branch (see Materials and Methods for a more complete explanation of the subsampling procedure). For each branch in the ML tree, 500 representative ML topologies were generated using these randomly selected quartet replicates, and the resulting topology set was used to calculate the ICA score for that branch. We present this information in Fig. 6, which contains the ML topology colored according to the ICA score estimated for each branch. Blue branches are those with positive ICA scores, that is, the quartet topology consistent with those branches was observed more often than either of the possible conflicting quartet topologies. Branches with negative scores are colored yellow to red, and indicate branches that are not supported by the data—the most frequently resolved quartet topology in these cases was in conflict with the branch. These poorly supported (or controversially resolved) branches are primarily concentrated near the tips of the tree, where data coverage, and by extension the number of loci with decisive taxon coverage, for each branch are low. It is also likely that better resolutions for some of these controversial branches may have been found by running the ML optimization procedure for longer, but such an exhaustive ML search was not feasible, as is often the case for datasets of this size.

To more specifically address the question of whether increased sampling depth (measured here as the number of decisive loci) affects confidence in branch reconstruction, we used a simple linear regression to assess the correlation between the number of decisive loci for a branch, and its ICA score (Fig. 7). Very strong branch support values (i.e. ICA close to 1) are elusive in this dataset, even with high numbers of decisive loci, and the relationship between ICA score and the number decisive loci is correspondingly weak when the entire dataset is analyzed (the "all x" regression line in Fig. 7; $r = 0.0004$, $p = 0.02$). However, very few branches in the tree are informed by more than 25 loci, and the combination of this sparse sampling with the high variance in these data places strong limits on our ability to infer patterns at this scale. It is likely that at least some of the variance in branch support is due to "dirty data" in GenBank, such as misidentified taxa or poor quality sequences, which lead to spurious topology inference when those data are subsampled. Another process that may affect support for deep branches is that even though many loci in the alignment contain decisive taxon sampling for these branches, individual randomized representative taxon quartet replicates for deep branches may often not subsample quartets

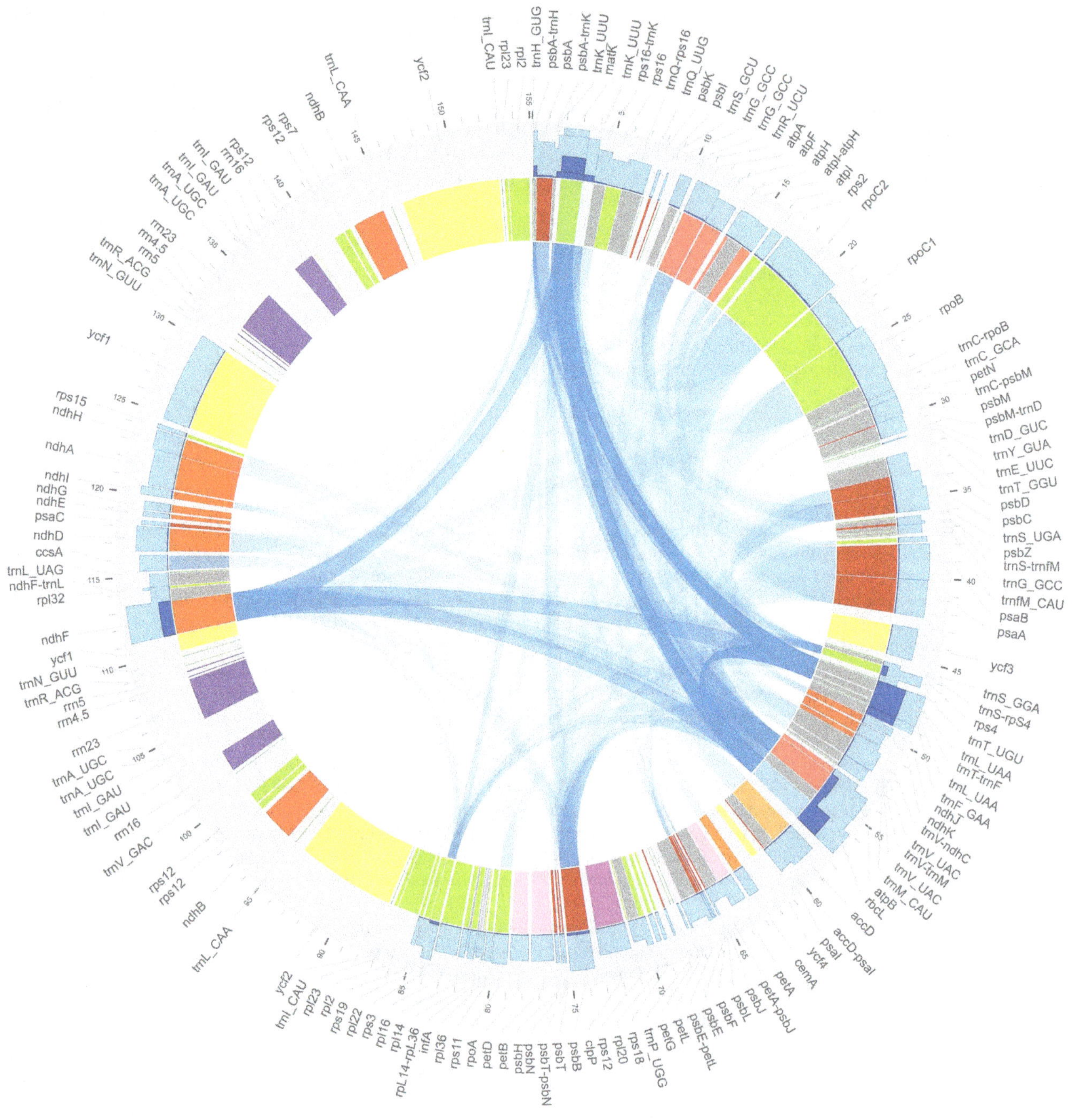

Figure 3. Taxonomic sampling depth and overlap for chloroplast loci in GenBank. Data shown are representation of genera in GenBank release 185 for all chloroplast loci sampled for this study, superimposed on a genome map of the *Coffea arabica* genome. Overall generic sampling depth across the chloroplast is quite low. Dark blue bar plots on the outside of the ring show the number of genera *g* represented by at least one exemplar sequence for each locus, out of $G = 16{,}167$ total genera, while light blue bars show $log(g)/log(G)$, and illustrate relative sampling proportions among loci when absolute proportions are small. Ribbon plots in the center of the figure identify pairs of loci, and indicate the proportion of genera that are represented by exemplar sequences for both loci in the pair. Dark ribbons label locus pairs that are co-represented for many genera; light ribbons label pairs that are co-represented for few. Locus colors correspond to gene groupings by function, and tick marks show linear distance in kilobases. The most well sampled locus in our entire alignment was *trnT–trnL–trnF*, with 55% of genera represented. Mitochondrial and nuclear loci are not shown in the figure, but the most well-sampled nuclear markers were ITS (53%) and ETS (8%; similar to *rps4* in the figure), and for the mitochondrion *atpA* (7%), *rps3* (5%), *matR* (5%), and *atp1* (5%); all other nuclear and mitochondrial markers were sampled for fewer than 4% of genera. Exact counts of genera represented for all markers sampled in this study are available in Table S3.

with decisive sampling, simply because the number of possible quartet combinations for deep branches is large. Near the tips, there is a stronger tendency for subsampled taxa to be represented for the same loci, thus potentially increasing the level of phylogenetic signal in mid-depth replicates. Future studies to more thoroughly characterize the behavior of the ICA statistic, as well as the localized taxon quartet jackknife we present here, would be valuable.

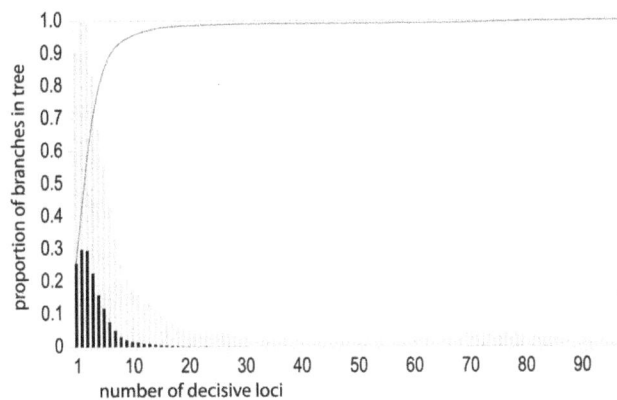

Figure 4. Genetic sampling depth for branches in the ML generic topology of Figs. 1, 5, and 6. Data shown are the proportion of branches for which each number of loci have decisive taxon sampling. Black bars are actual proportions and grey bars are square roots of each proportion, which illustrate relative differences when proportions are very small. The dark grey line is the cumulative proportion of branches, which indicates the proportion of branches in the tree for which the number of loci with decisive taxon sampling is less than or equal to the indicated x value. Exact values used to create this figure are presented in Table S2.

To circumvent the limits imposed by the sparse sampling of branches with high numbers of decisive loci, we performed an additional regression which we limited to a subset of the data. The mean ICA values for branches with relatively low numbers of decisive loci (i.e. fewer than about 25) suggest a positive correlation, though branches informed by very few loci (i.e. fewer than about 5) show a decreasing trend in ICA compared to those with more. We conjecture that the decreasing support values for branches with fewer than about 5 decisive loci may be due to increasing levels of conflicting signal as more data are added, up to the point where the dominant signal becomes strong enough to reverse the trend (other plausible explanations certainly exist). We therefore limited the second regression to branches with greater than 5 but fewer than 25 decisive loci (the "$5 < x < 25$" line in Fig. 7), which we propose constitutes a representative and sufficiently densely sampled subset of the data to accurately quantify any existing trend. In these subsampled data, we find a positive correlation between branch support and the number of loci capable of informing the branch. This relationship is strongly supported ($p < 10^{-9}$), but even so, the correlation itself is weak ($r = 0.016$), suggesting that even with relatively high levels of data coverage (in this case, up to 25 loci, which is higher than many phylogenetic studies being published today), confident inference of at least some phylogenetic relationships may remain challenging. This observation is consistent with studies using very large and densely sampled alignments, which have nonetheless yielded trees containing numerous poorly supported branches [6,13,23,34]. The reasons for this are unclear, but may be related to conflicting phylogenetic signal or in some cases to an overall lack of informative sites even at genome-wide scales.

Concluding remarks

The utility of public sequence databases for phylogenetic inference has reached never before seen levels, facilitating the inference of phylogeny at both broad and deep scales for major groups in the tree of life (Figs. 1 and 2). In plants, we are nearing a threshold where nearly every known land plant genus is represented in GenBank by at least one exemplar sequence,

making phylogeny inference across all lineages of land plants possible at relatively fine scales (Figs. 2 and 5). These comprehensive phylogenies have already shown a unique potential to address broad evolutionary questions that may be difficult to test at less inclusive scales [6,8,32,35,36], and many opportunities exist for researchers who wish to exploit the potential of public sequence databases. Nevertheless, lineage representation remains extremely low for all but a handful of genetic markers (Fig. 3), and resources such as GenBank remain heavily limited by this (Figs. 4, 5, and 6).

The most obvious solution to the problem of low data coverage is simply to increase sampling for informative loci for lineages without it, and indeed one of the implications of the work we present here is that the accurate resolution of comprehensive land plant phylogenies may require the collection of a significant amount of additional sequence data (Figs. 5, 6). This implication is corroborated by the results of previous work by Sanderson [4], which showed that phylogenetically informative sampling is very low for a large number of eukaryotic lineages. In fact, that study found that land plants were among the best sampled major lineages of eukayotic life (after vertebrates), clearly illustrating that despite the relative low sampling depth for land plant genera (Figs 3, 5), the situation is even more extreme in almost all other parts of the tree of life. We suggest that one proactive response to this situation would be to continue to fund and pursue opportunities to improve both taxonomic as well as genetic coverage across the tree of life.

Statistics such as d, ICA, and the measures we have implemented here using these theoretical foundations (Figs. 5 and 6) provide useful tools that can allow the rapid and accurate identification of potential problem areas even in very large phylogenies, and thus may enable efficient and cost-effective collection of targeted data to improve taxonomic and genetic coverage. The concept of data decisiveness in general can be expected to remain a useful theoretical background for data-mining approaches across diverse lineages on the tree of life, and statistics such as d as well as related methods [37,38] may be able to provide a more nuanced assessment of the phylogenetic potential of large databases than previous approaches have allowed [3,4].

The continued accrual of novel sequence data and improvements to taxonomic coverage at the species level will be required to facilitate broad application of these resources at fine evolutionary scales. Improvements to coverage, however, will lead to increasingly larger datasets, which can pose significant technical challenges for analysis methods. An alignment of 100,000 plant species sampled for the same 128 genetic loci as this study resulted in an alignment text file over 6 GB in size, which exceeded the capabilities of available phylogenetic search software. If the relatively stable rate of species accumulation in GenBank remains so for the next four decades, we will reach the estimated (minimum) land plant species diversity of 300,000 around the year 2044, but technical challenges related to computational tractability (given currently available software) would prevent reconstruction of trees with 300,000 plant species, just as those problems prevent the reconstruction of trees with 100,000 plant species today. Encouragingly, methodological advancement in this field is proceeding rapidly [39–43].

In summary, we propose that the continued role of public sequence databases in evolutionary analysis at comprehensive scales will depend critically on advancement in at least three areas: (1) the continued expansion of taxonomic and genetic coverage of these data, (2) ongoing efforts to understand the effects of the complex structure of large phylogenetic datasets, and (3) innovative solutions to the challenges posed by their analysis.

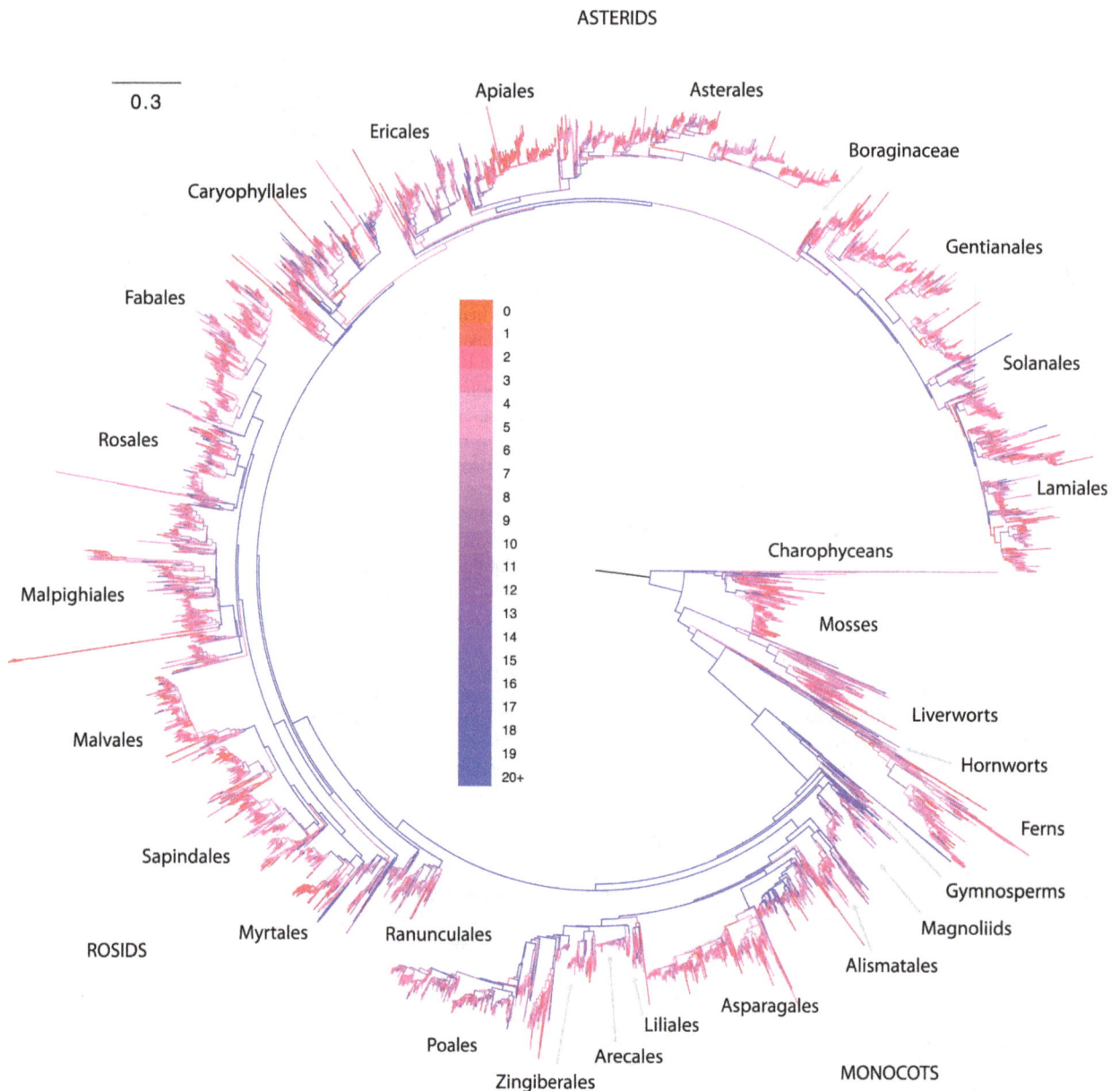

Figure 5. Generic phylogeny of plants with branches colored according to the number of loci with decisive taxon sampling. Branches for which large numbers of loci have phylogenetically decisive taxon sampling are blue, while branches with low numbers are pink to red. Blue branches have a higher capacity than red branches to be informed by the data. Most branches in the tree are able to be informed by data at relatively few loci, but deep branches are generally able to be informed by many. A large version of this figure with legible tip names is presented in File S2.

Analysis

Data collection and phylogeny inference

Data were gathered from GenBank [1] release 185 (accessed in May 2012), using the software PHLAWD [7], which uses an algorithm based on recursive profile alignment (alignment of multiple alignments to one another) to facilitate the alignment of nucleotide sequence data even at relatively deep phylogenetic levels. PHLAWD requires vetted guide sequences to ensure accurate identification of candidate sequences for alignment. High-quality guide sequences for many loci were supplied by Moore et al. [44] and Soltis et al. [22], which we combined with strict coverage and identity requirements to ensure the inclusion of homologous candidate sequences (Table S1, lines with coverage

and identity set at 0.4). For the remaining alignments, we manually selected guide sequences from GenBank, and optimized search parameters to ensure homology and minimize noise. Explicit search terms were used to exclude non-homologous sequences for some loci (Table S1, lines with coverage and identity set to 0), while for others we used coverage/identity cutoffs as well as search terms.

In some cases, loci with uncertain homology were aligned separately for different clades. In these cases, search parameters in Table S1 may appear to indicate that some alignments are taxonomically nested subsets of others, e.g. trnG_intron_bryos defines its search clade as Streptophytina whereas trnG_intron_-tracheophyta appears to use an identical search over Tracheo-phyta only, but because Tracheophyta is nested within Strepto-

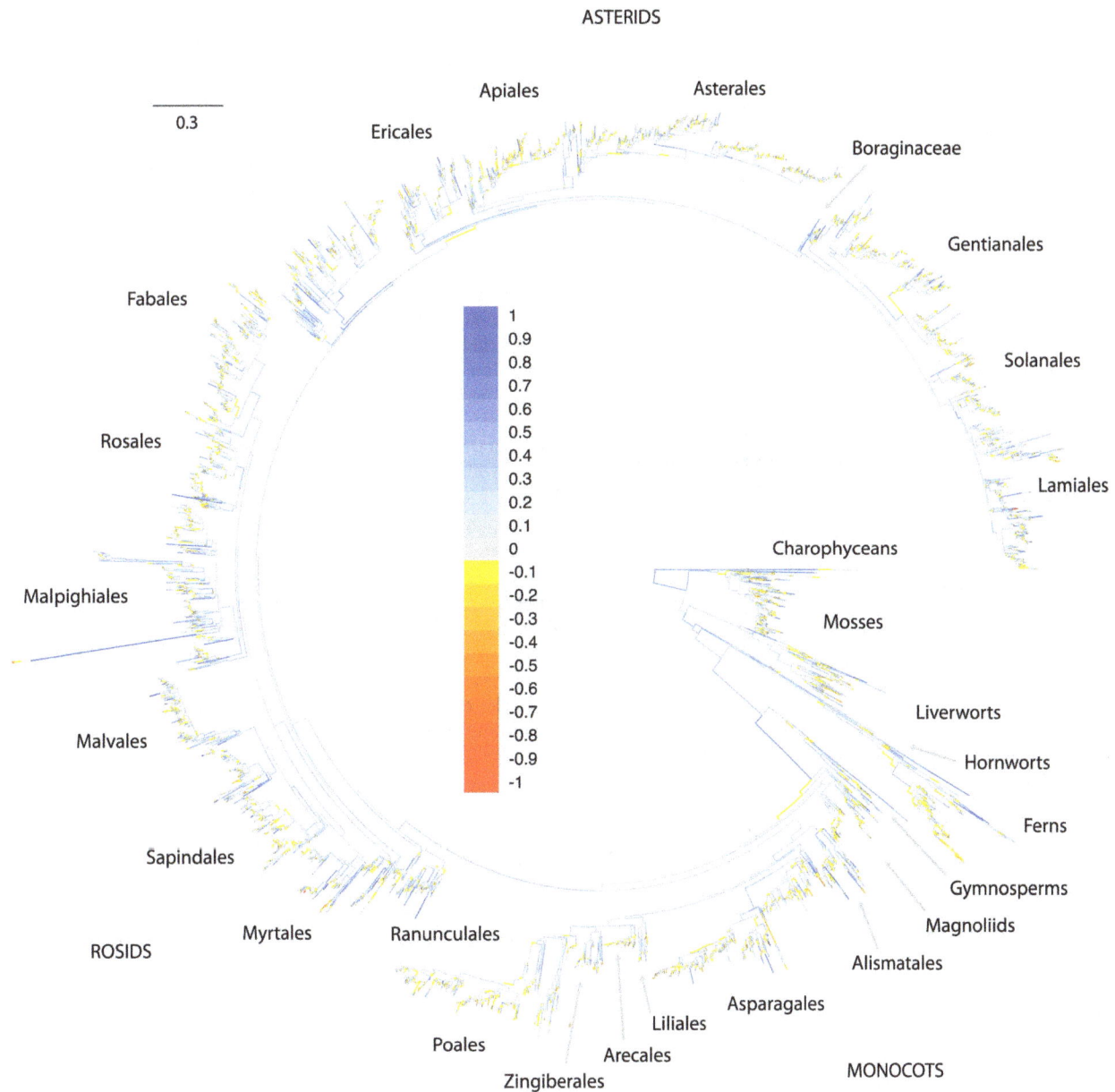

Figure 6. Generic phylogeny of plants with branches colored according to ICA support values. Branches with strong positive values (high support) are light blue, while branches with low positive values (low support) are gray, and branches with negative values (which imply relatively strong conflicting signal in the data) are colored yellow to red. Most branches deep in the tree are moderately well supported, whereas most strongly supported branches occur within smaller clades, and most branches that appear to be in conflict with the signal from the alignment occur near the tips of the tree. Terminal branches (i.e. tips) were pruned from the tree for display purposes, as they do not have meaningful support values. A large version of this figure with legible tip names is presented in File S3.

phytina, it would seem given this nesting that both these alignments would contain Tracheophyta sequences. In this case, however, and in other similar cases, other information such as genetic distance from guide sequences was used to exclude sequences that would otherwise have been represented twice from the the more inclusive alignment.

Each of these PHLAWD alignments was imported into to a SQLite database where sequences were linked on the basis of ncbi taxon id. Using the NCBI taxonomy, we extracted a synthetic concatenated alignment from this database, with each OTU in the alignment corresponding to an NCBI-recognize genus, and each partition corresponding one of the alignments generated with

PHLAWD (i.e. each partition corresponded to a single locus). To populate the alignment, we used exemplar sequences that were chosen on the basis of unaligned length; the longest sequence available for any species in each genus was used to exemplify that genus for each locus. The Python scripts that were used to create the SQLite database and query it are available in the github repository http://github.com/chinchliff/autophy.

The final concatenated alignment file (with empty columns removed) used to create Figs. 1, 5, and 6 consisted of 13,093 tips representing genera, and 128 partitions representing loci. This alignment contained 148,143 total sites encoding 126,121 site patterns, 96.37% missing data, and was 1.9 GB in size. It is

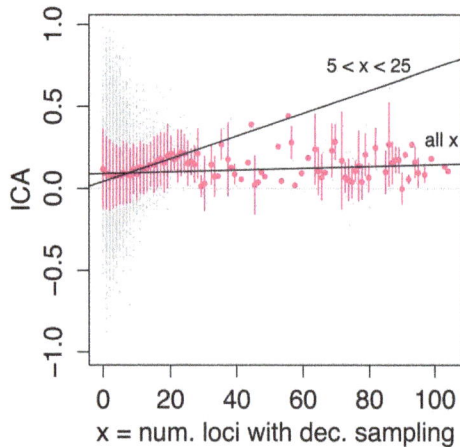

Figure 7. Correlation of branch support (ICA) with locus sampling depth (number of loci with decisive taxon sampling). Original data points correspond to individual internal branches in the ML topology, and are shown in gray. Large colored dots represent mean ICA values for all branches with the corresponding number of loci with decisive sampling, and error bars extend to plus or minus the standard deviation of the data from each mean. Two regression lines are plotted, one for the entire dataset (labeled "all x"; $r^2 < 0.001$, $F = 5.27$, $df = 13,081$, $p = 0.02$) and one for only those branches with greater than five but fewer than 25 loci with decisive sampling (labeled "$5 < x < 25$"; $r^2 = 0.016$, $F = 41.5$, $df = 2,486$, $p < 10^{-9}$).

available in the Data Dryad repository, as are the metadata for all GenBank sequences used in this alignment, including GI numbers. Phylogenetic trees were inferred from this alignment in RAxML 7.3.0 [15], using the command line arguments:

raxmlHPC-PTHREADS-SSE3 -f d -m GTRCAT -p 12345 -q <partitionfile> -s <alignmentfile> -n <name> -j -D -T <numthreads>

The raw topology corresponding to the ML best tree found by RAxML is supplied in Newick format with branch lengths in the Dryad repository. Several extremely long tip branches (potentially representing erroneous phylogenetic chimeras) were manually pruned from this tree for display in Figs. 1, 5, and 6. The pruned topology was used in conjunction with time calibrations from [31] to generate the ultrametric chronogram supplied in the Data Dryad repository, using the program treePL [45].

We used Python scripts (all available in the Dryad repo) to extract sequence age data from GenBank and calculate the ages of the oldest exemplar sequences used to color the tree in Fig. 1. First, the script get_gi_dates_from_gbseq.py was used to extract the date of every land plant sequence from the GenBank release 185 flatfiles. Second, the script calc_age_of_oldest_exemplar_for_nodes.py was used to find the earliest added sequence that could be used to exemplify each tip in the Fig. 1 topology (i.e. the sequence was a sample from a the taxon represented by the tip), and for each internal branch, to identify the age of the oldest exemplar sequence for any of the leaves subtended by that branch.

Assessing the impact of missing data

Partial decisiveness (the d statistic as defined by Sanderson et al. [9]) for the entire dataset was estimated with the software Decisivator (J.W. Brown). We also calculated the number of loci with decisive taxon sampling for each branch (using the Python script calc_branchwise_decisiveness.py, available in the Dryad repo), which is simply a count of the data partitions from the alignment (i.e. loci) for which the four-way partition property of

Steel and Sanderson [10] is satisfied for the given branch. The four-way partition property describes the distinguishability (or not) of an edge based on the presence of minimal data to inform that edge—this property is satisfied for an edge (i.e. the edge is distinguishable) if and only if the alignment contains sufficient taxon by locus sampling that it is *possible* for the given edge to be informed by at least some of the data. For a thorough mathematical exposition of the four-way partition property and its relevance for phylogenetic data decisiveness, we refer readers to the original publications [9,10].

Measuring branch support

We used a localized taxon quartet jackknife approach to subsample the original alignment in order to estimate support values. To explain this, we first define the rationale used for the taxonomic subsampling itself. Let $A = \{a_1, a_2, ..., a_k\}$ represent an alignment of phylogenetic data, where each a_i is a row in the alignment corresponding to a tip t_i in a given rooted, bifurcating tree T with tips $\{t_1, t_2, ..., t_k\}$ and internal edges $\{b_1, b_2, ..., b_{k-3}\}$. In such a tree, each observed internal edge b_j defines four non-overlapping subsets of A: its daughter clades $D_{j1} \subset A$ and $D_{j2} \subset A$, a sister clade $S_j \subset A$, and the rest of the tree $O_j = A - \{D_{j1} \cup D_{j2} \cup S_j\}$. Then, for a given branch b_j, any bifurcating tree topology L_{jm} inferred for any taxon quartet $R_{jm} = \{d_1, d_2, s, o\}$ such that $d_1 \in D_{j1}$, $d_2 \in D_{j2}$, $s \in S_j$, and $o \in O_j$, must either contain an internal edge representing the bipartition $(d_1, d_2 | s, o)$, which is consistent with the tree topology containing b_j, or else L_{jm} will contain an edge implying one of two alternative bipartitions that are inconsistent with b_j (these are $(d_1, s | d_2, o)$ and $(d_1, o | d_2, s)$). We therefore designate any taxon quartet R_{jm} as a *representative taxon quartet replicate* for branch b_j, and L_{jm} as a *representative quartet topology replicate*, for all b_j in T.

We assessed support for each branch in the ML topology inferred using our large generic alignment, by performing 500 random taxon selection procedures to generate representative taxon quartet replicates, inferring the ML topology for each of these quartet replicates using data from the original alignment and RAxML [15], and then calculated the ICA score for the topology $(d_1, d_2 | s, o)$ (consistent with the original ML tree), across all 500 of these representative quartet topology replicates for each branch. ICA is an information theory-based measure of edge support that is calculated across a set of topologies [33]. ICA varies on the interval $[-1, 1]$, with negative values indicating that the targeted edge occurs less frequently across the replicate topologies than some other conflicting edge, positive values indicating that the edge occurs more frequently than any other conflicting edge, and a value of zero indicating that the edge occurs at equal frequency with all alternative (i.e. conflicting) edges. The absolute value of ICA is correlated with the overall frequency of that branch relative to alternative (i.e. conflicting) topologies in the input set. The calculations to yield these ICA scores were performed using software phyx [46].

Colored trees (Figs. 1, 5, and 6, and the "Supplemental tree figures" file in Dryad repository) were generated using the Python script paint_branches.py (available in the dryad repo) and the software FigTree. The paint_branches.py script cross-references node labels in a newick tree file with a CSV file containing values for those nodes, and then assigns branch colors based on those values. It generates a FigTree-formatted tree file with branch color annotations, which can be visualized in FigTree [47] itself and then exported in graphical formats. We used FigTree 1.4.0. The branch-painting script depends on the newick3.py and phylo3.py modules also provided in the Dryad repository.

Additional procedures

Monophyly was assessed for all families in the GenBank taxonomy at the time of writing using the Python script test_monophyly_against_tree.py (available in the Dryad repo). The script accesses a taxonomy through a PHLAWD sequence database; we used a database containing the NCBI taxonomy from GenBank release 185. The test for monophyly involved two parts: first, for a given family, all the tips contained within that family in the taxonomy were identified in the tree. We note that in some cases, not all the taxa defined in the taxonomy were present in the tree. Second, the tree topology was checked to determine if the set of identified tips formed a monophyletic group in the tree, that is, they were all contained in a single clade that contained no other tips. If this condition was met, we inferred that monophyly could not be rejected for the given family. Conversely, if this condition was not met (i.e. the tips associated with a given family did not form a clade in the tree), then monopoly was rejected.

Figure 3 was created using the software Circos 0.62–1 [48]. The karyotype data used to assign genes and named regions were taken from the *Coffea arabica* chloroplast genome on GenBank (accession NC008535). The data for sampling frequencies for loci were extracted from the metadata files generated by the autophy scripts and formatted for Circos using a combination of bash and Python scripting and regular expression search/replace in the text editor Geany [49].

The estimate of time required to reach representation of 300,000 plant species in GenBank was based on an extrapolation using a linear rate estimate of species accumulation (about 6,500 new species/year) between 2000 and 2012. At this rate, it will take 32 years to accumulate the additional 208,000 species required to reach 300,000 from the approximately 92,000 sampled today.

Supporting Information

Table S1   A comma–separated tabular data file describing parameters for all PHLAWD runs.

Table S2   Proportion of branches in tree for which each number of loci contain decisive data.

Table S3   Number of genera represented for each locus in data mined from GenBank 185.

File S1   Full size version of Fig. 1 with tip names.

File S2   Full size version of Fig. 5 with tip names.

File S3   Full size version of Fig. 6 with tip branches and tip names.

Acknowledgments

Thanks to Ya Yang and Joseph Brown for valuable discussions and feedback. Additional supplementary materials are available at the Data Dryad doi 10.5061/dryad.450qq, including the compressed alignment and partition files used with RAxML to generate the tree, a metadata table identifying the GenBank sequence information used to generate the alignment, the ML tree topology found by RAxML, an ultrametric chronogram of this topology and the configuration files used with treePL used to create it (see Materials and Methods), all Python scripts used to generate figures and statistics, and the "Supplemental tree figures" file, which exceeds the size limit set by the journal.

Supplemental tree figures. A collection of figures showing branches on the generic phylogeny for which each locus contains decisive data. This Appendix is a PDF file. View its table of contents to browse the contained figures by locus name. This file exceeds the 100 MB size limit of the journal for hosted supplemental materials, so it is available from the Data Dryad repository instead.

Author Contributions

Conceived and designed the experiments: CEH SAS. Performed the experiments: CEH. Analyzed the data: CEH. Contributed reagents/materials/analysis tools: CEH SAS. Wrote the paper: CEH SAS.

References

1. Benson DA, Karsch-Mizrachi I, Lipman DJ, Ostell J, Wheeler DL (2008) GenBank. Nucleic Acids Research 36: D25–30.

2. National Center for Biotechnology Information (2013) Taxonomy statistics. http://www.ncbi.nlm.nih.gov/Taxonomy/taxonomyhome.html/index.cgi?chapter = STATISTICS&uncultured = hide&unspecified = hide). Accessed 20 November 2013.

3. Driskell AC, Ané C, Burleigh JG, McMahon MM, O'Meara BC, et al. (2004) *Prospects for building the tree of life from large sequence databases*, Science, 306, pp. 1172–1174.

4. Sanderson M (2008) Phylogenetic Signal in the Eukaryotic Tree of Life. Science 321: 121–123.

5. Mora C, Tittensor DP, Adl S, Simpson AGB, Worm B (2011) How Many Species Are There on Earth and in the Ocean? PLoS Biology 9: e1001127.

6. Hinchliff CE, Roalson EH (2013) Using Supermatrices for Phylogenetic Inquiry: An Example Using the Sedges. Systematic Biology 62: 205–219.

7. Smith SA, Beaulieu JM, Donoghue MJ (2009) Mega-phylogeny approach for comparative biology: an alternative to supertree and supermatrix approaches. BMC Evolutionary Biology 9: 37.

8. Smith SA, Donoghue MJ (2008) Rates of Molecular Evolution Are Linked to Life History in Flowering Plants. Science 322: 86–89.

9. Sanderson MJ, McMahon M, Steel M (2010) Phylogenomics with incomplete taxon coverage: the limits to inference. BMC Evolutionary Biology 10: 155.

10. Steel M, Sanderson MJ (2010) Characterizing phylogenetically decisive taxon coverage. Applied Mathematics Letters 23: 82–86.

11. Govaerts R (2003) How many species of seed plants are there?: a response. Taxon 52: 577–582.

12. Lewis LA, McCourt RM (2004) Green algae and the origin of land plants. American Journal of Botany 91: 1535–1556.

13. Timme RE, Bachvaroff TR, Delwiche CF (2012) Broad phylogenomic sampling and the sister lineage of land plants. PLoS ONE 7: e29696.

14. Finet C, Timme RE, Delwiche CF, Marletaz F (2010) Multigene Phylogeny of the Green Lineage Reveals the Origin and Diversification of Land Plants. Current Biology 20: 2217–2222.

15. Stamatakis A (2006) RAxML-VI-HPC: maximum likelihood-based phylogenetic analyses with thousands of taxa and mixed models. Bioinformatics 22: 2688–2690.

16. The Plant List (2013) Edition 1. http://www.theplantlist.org/. Accessed 2013 December 20.

17. Soltis DE, Soltis PS, Nickrent DL, Johnson LA, Hahn WJ, et al. (1997) Angiosperm phylogeny inferred from 18S ribosomal DNA sequences. Annals of the Missouri Botanical Garden 84: 1–49.

18. Qiu Y-L, Lee J, Bernasconi-Quadroni F, Soltis DE, Soltis PS, et al. (1999) The earliest angiosperms: evidence from mitochondrial, plastid and nuclear genomes. Nature 402: 404–407.

19. Chase MW, Soltis DE, Olmstead RG, Morgan D, Les DH, et al. (1993) Phylogenetics of seed plants: an analysis of nucleotide sequences from the plastid gene rbcL. Annals of the Missouri Botanical Garden 80: 528–580.

20. Soltis DE, Soltis PS, Chase MW, Mort ME, Albach DC, et al. (2000) Angiosperm phylogeny inferred from 18S rDNA, rbcL, and atpB sequences. Botanical Journal of the Linnean Society 133: 381–461.

21. Soltis PS, Soltis DE, Chase MW (1999) Angiosperm phylogeny inferred from multiple genes as a tool for comparative biology. Nature 402: 402–404.

22. Soltis DE, Moore MJ, Burleigh JG, Bell CD, Soltis PS (2010) Assembling the angiosperm tree of life: progress and future prospects. Annals of the Missouri Botanical Garden 97: 514–526.

23. Soltis DE, Smith SA, Cellinese N, Wurdack KJ, Tank DC, et al. (2011) Angiosperm phylogeny: 17 genes, 640 taxa, American Journal of Botany 98: 704–730.

24. Chase MW (2004) Monocot relationships: an overview. American Journal of Botany 91: 1645.

25. Olmstead RG, dePamphilis CW, Wolfe AD, Young ND, Elisons WJ, et al. (2001) Disintegration of the Scrophulariaceae, American Journal of Botany 88: 348–361.
26. Hilu KW, Borsch T, Müller K, Soltis DE, Soltis PS, et al. (2003) Angiosperm phylogeny based on matK sequence information. American Journal of Botany 90: 1758–1776.
27. Qiu Y, Li L, Wang B, Chen Z, Dombrovska O, et al. (2007) A nonflowering land plant phylogeny inferred from nucleotide sequences of seven chloroplast, mitochondrial, and nuclear genes. International Journal of Plant Sciences 168: 691–708.
28. Aliscioni S, Bell HL, Besnard G, Christin P, Columbus JT, et al. (2011) New grass phylogeny resolves deep evolutionary relationships and discovers C4 origins. New Phytologist 193: 304–312.
29. Chang Y, Graham SW (2011) Inferring the higher-order phylogeny of mosses (Bryophyta) and relatives using a large, multigene plastid data set. American Journal of Botany 98: 839–849.
30. Moore MJ, Bell CD, Soltis PS, Soltis DE (2007) Using plastid genome-scale data to resolve enigmatic relationships among basal angiosperms. Proceedings of the National Academy of Sciences 104: 19363–19368.
31. Magallon S, Hilu KW, Quandt D (2013) Land plant evolutionary timeline: Gene effects are secondary to fossil constraints in relaxed clock estimation of age and substitution rates. American Journal of Botany 100: 556–573.
32. Smith SA, Beaulieu JM, Stamatakis A, Donoghue MJ (2011) Understanding angiosperm diversification using small and large phylogenetic trees. American Journal of Botany 98: 404–414.
33. Salichos L, Stamatakis A, Rokas A (2014) Novel Information Theory-Based Measures for Quantifying Incongruence among Phylogenetic Trees. Molecular Biology and Evolution, advance access version msu061v2.
34. Thomson RC, Shaffer HB (2010) Sparse Supermatrices for Phylogenetic Inference: Taxonomy, Alignment, Rogue Taxa, and the Phylogeny of Living Turtles. Systematic Biology 59: 42–58.
35. Edwards EJ, Smith SA (2010) Phylogenetic analyses reveal the shady history of C4 grasses. Proceedings of the National Academy of Sciences 107: 2532–2537.
36. Leslie AB, Beaulieu JM, Raic HS, Crane PR, Donoghue MJ, et al. (2012) Hemisphere-scale differences in conifer evolutionary dynamics. Proceedings of the National Academy of Sciences 109: 16217–16221.
37. Misof B, Meyer B, von Reumont MB, Kück P, Misof K, et al. (2013) Selecting informative subsets of sparse supermatrices increases the chance to find correct trees. BMC Bioinformatics 14: 348.
38. Dress AWM, Huber KT, Steel M (2012) 'Lassoing' a phylogenetic tree I: basic properties, shellings, and covers. Mathematical Biology 105: 65–77.
39. Aberer AJ, Krompass D, Stamatakis A (2013) Pruning rogue taxa improves phylogenetic accuracy: an efficient algorithm and webservice. Systematic Biology 62: 162–166.
40. Smith SA, Brown JW, Hinchliff CE (2013) Analyzing and synthesizing phylogenies using tree alignment graphs, PLoS Computational Biology 9: e1003223.
41. Price MN, Dehal PS, Arkin AP (2010) FastTree 2–approximately maximum-likelihood trees for large alignments. PLoS ONE 5: e9490.
42. Stamatakis A, Aberer AJ, Goll C, Smith SA, Berger SA, et al. (2012) RAxML-Light: a tool for computing terabyte phylogenies. Bioinformatics 28: 2064–2066.
43. Chesters D, Vogler AP (2013) Resolving Ambiguity of Species Limits and Concatenation in Multi-locus Sequence Data for the Construction of Phylogenetic Supermatrices. Systematic Biology, advance access.
44. Moore MJ, Soltis PS, Bell CD, Burleigh JG, Soltis DE (2010) Phylogenetic analysis of 83 plastid genes further resolves the early diversification of eudicots. Proceedings of the National Academy of Sciences 107: 4623–4628.
45. Smith SA, BC O'Meara (2012) treePL: divergence time estimation using penalized likelihood for large phylogenies. Bioinformatics 28: 2689–2690.
46. Smith SA (2014) phyx. http://github.com/FePhyFoFum/phyx/. Accessed 25 Mar 25 2014.
47. Rambaut A (2012) FigTree 1.4.
48. Krzywinski M, Birol SJ, Connors J, Gascoyne R, Horsman D, et al. (2009) Circos: an information aesthetic for comparative genomics. Genome Research 19: 1639–1645.
49. Treleaven N, Wendling C, Tröger E, Lanitz F (2012) Geany.

12

The Effect of Phylogeny, Environment and Morphology on Communities of a Lianescent Clade (Bignonieae-Bignoniaceae) in Neotropical Biomes

Suzana Alcantara[1,2*¤], **Richard H. Ree**[2], **Fernando R. Martins**[3], **Lúcia G. Lohmann**[1*]

1 Departamento de Botânica, Instituto de Biociências, Universidade de São Paulo, São Paulo, SP, Brazil, **2** Department of Botany, Field Museum of Natural History, Chicago, Illinois, United States of America, **3** Departamento de Biologia Vegetal, Instituto de Biologia, Universidade Estadual de Campinas – UNICAMP, Campinas, SP, Brazil

Abstract

The influence of ecological traits to the distribution and abundance of species is a prevalent issue in biodiversity science. Most studies of plant community assembly have focused on traits related to abiotic aspects or direct interactions among plants, with less attention paid to ignore indirect interactions, as those mediated by pollinators. Here, we assessed the influence of phylogeny, habitat, and floral morphology on ecological community structure in a clade of Neotropical lianas (tribe Bignonieae, Bignoniaceae). Our investigation was guided by the long-standing hypothesis that habitat specialization has promoted speciation in Bignonieae, while competition for shared pollinators influences species co-occurrence within communities. We analyzed a geo-referenced database for 94 local communities occurring across the Neotropics. The effect of floral morphological traits and abiotic variables on species co-occurrence was investigated, taking into account phylogenetic relationships. Habitat filtering seems to be the main process driving community assembly in Bignonieae, with environmental conditions limiting species distributions. Differing specialization to abiotic conditions might have evolved recently, in contrast to the general pattern of phylogenetic clustering found in communities of other diverse regions. We find no evidence that competition for pollinators affects species co-occurrence; instead, pollinator occurrence seems to have acted as an "environmental filter" in some habitats.

Editor: Shuang-Quan Huang, Central China Normal University, China

Funding: This study has been funded by FAPESP: doctoral grant 2006/59916 0; and CCSD Missouri Botanical Garden. Elizabeth E. Bascom Fellowships for Latin American Female Botanists. The funders had no role in study design, data collection and analysis, decision to publish, or preparation of the manuscript.

Competing Interests: The authors have declared that no competing interests exist.

* E-mail: suzanaalcantara@gmail.com (SA); llohmann@usp.br (LL)

¤ Current address: Department of Integrative Biology, University of California, Berkeley, Berkeley, California, United States of America

Introduction

The importance of species traits for the assembly of communities at local and regional scales is a pervasive topic in ecology [1,2]. In this context, much attention has been paid to two distinct kinds of processes: environmental filtering, i.e., limits imposed by abiotic conditions, and competition, i.e., biotic interactions arising from common use of limited resources [3–5]. While environmental filtering tends to favor co-occurrence of species with similar phenotypes [6–8], competition is thought to create phenotypic "evenness" (overdispersion) of species within communities [5,8,9]. Thus, these processes are expected to exert opposing effects on the phenotypic structure of communities. The dynamics of trait and lineage evolution are thus relevant to community ecology [5,7,10,11], because depending on whether traits are phylogenetically conserved or not, communities can exhibit significant phylogenetic structure [5,9,12,13]. As these assembly processes are not mutually exclusive, the phenotypic and phylogenetic structure of natural communities is expected to reflect their combined effects [10,13,14].

Most studies of plant community assembly have focused on the influences of abiotic aspects or direct interactions among co-occurring plants species [15], although indirect interactions, like those mediated by herbivores or pollinators, have also been shown to be important [15,16]. In particularly, plant-pollinator interactions have been important for the evolution of floral traits and lineages [17], and consequently for the phenotypic and phylogenetic structure of communities [15,18–20]. Pollinator services have been traditionally viewed as a limiting resource, causing plant competition and species phenotypic repulsion on floral traits and flowering patterns [21]. However, two underappreciated processes that cause phenotypic attraction on floral traits in plant communities have increasingly received empirical support: (i) habitat filtering, with environments determining the pollinators and pollination systems that can persist [15], and (ii) facilitative interaction, in which beneficial pollinator sharing by plant species jointly attracts and/or maintains the populations of pollinators [23,24].

Here, we evaluate the role of habitat environmental filtering and competition mediated by pollinators for the structure of communities of a large Neotropical clade of flowering plants, the tribe Bignonieae. Bignonieae includes almost half of the species of the family Bignoniaceae (393 out of 827 species), with most of its taxa occupying a variety of habitats across the Neotropics [24,25]. Most species are lianas, but shrubs are also present in some lineages [26]. The present study owes much inspiration to pioneering research of Bignoniaceae by Gentry [27–29]. He observed that within genera, species of Bignoniaceae tend to have allopatric ranges, narrow habitat preferences, and more divergence in vegetative versus floral traits, suggesting allopatric lineage diversification and adaptation to abiotic conditions at broad spatial scales [27,29]. At local scales, species of Bignoniaceae tend to be self-incompatible, obligatorily outcrossing, and lack natural hybrids, suggesting that competition for pollinators might be an important factor in the assembly of communities [27–29]. Indeed, pollination strategy is supposed to have played a key role in the evolution of the tribe, with changes in floral morphology being associated with shifts in pollinator guilds [29,30]. Up to 20 species of Bignoniaceae have been reported to coexist in natural communities, representing both specialized pollinator guilds (big- to medium-sized bees, bats, hummingbirds, hawkmoths), and more generalist guilds that include butterflies and various smaller insects [27–29]. As a result, it has been suggested that Bignoniaceae communities may be saturated in terms of pollinator use, with individual species being pollinated by a different pollinator group in each community at the same time [29].

A molecular phylogenetic study of Bignonieae [31] has cast these hypotheses in a new light, particularly by showing that most of the traditionally recognized genera are not monophyletic and needed a new circumscription. In addition, floral traits previously considered important for taxonomic delimitation were shown to exhibit considerable homoplasy, the phylogenetic signature of labile or recurrent evolution [30,31]. The objective of the present study is to integrate phylogenetic, environmental, and morphological data with surveys of species co-occurrence to detect the signature of processes driving community assembly in Bignonieae. Specifically, we reformulate Gentry's [29] predictions in an explicit phylogenetic framework, as follows:

Abiotic predictions

Species from communities that are subject to environmental filtering are expected to show phenotypic attraction in the traits associated with habitat specialization. The expectation of phylogenetic structure in such communities (co-occurrence of close versus distant relatives) depends on whether those traits evolve in a labile or conserved manner [13,32]. One potential scenario of labile evolution is that species divergence is frequently driven by habitat specialization in allopatry, in which case we would expect species to have narrow abiotic niches and to infrequently co-occur with close relatives, as proposed formerly for Bignoniaceae [29]. In this case, communities will tend to be assembled from more distant relatives, showing phylogenetic evenness or overdispersion [11]. Alternatively, if niche evolution is phylogenetically conservative, communities assembled through environmental filtering will tend to be composed of close relatives and show phylogenetically clustering [13,32].

Biotic predictions

If competition for pollinators influences species coexistence, communities should exhibit phenotypic repulsion on floral traits, reflecting diversity in pollination strategies [15,20]. If floral traits are phylogenetically conserved, communities can be expected to have overdispersed phylogenetic structure (co-occurrence of distant relatives). On the other hand, labile evolution coupled with competition would create a random pattern of phylogenetic community structure [11,15]. Alternatively, interspecific interactions between co-occurring flowering plants may be facilitative and/or subject to the filtering imposed by the absence of a given pollinator guild [15]. These scenarios would favor phenotypic attraction in plant communities, with the resulting phylogenetic

structure being similar to those mediated by traits involved in habitat filtering (see above). As floral traits of Bignonieae have shown contrasting patterns of evolution, with floral morphologies having evolved in a labile way, while other floral features (i.e., size of attractive parts and allometric pattern) exhibit conserved evolution [30,33,34], it is hard to predict how such floral traits may contribute to the phenotypic and phylogenetic structure of communities of Bignonieae.

In this study, we used a time-calibrated phylogeny of Bignonieae [35] as an evolutionary framework to investigate these predictions. Particularly, we assess the patterns of species co-occurrence and the associated abiotic variables within the context of their phylogenetic structure, in order to test the specific abiotic predictions. We also evaluate the biotic predictions by assessing the phenotypic structure of floral traits within communities and how this phenotypic structure relates to the phylogenetic and distribution patterns of species.

Materials and Methods

Species distribution and communities sampling

We used Alwyn Gentry's transect database as the basis of a dataset of species co-occurrences for Bignonieae (http://www. mobot.org/MOBOT/research/gentry/transect.shtml). In this database, each transect extends 0.1ha, surveyed for the presence and abundance of all plants exceeding 2.5 cm diameter at breast height (dbh). Spatial and environmental variables, such as GIS coordinates and forest physiognomy (i.e., humid or dry forest, savanna), are also recorded. Of the 226 transects available, we restricted our survey to 154 transects located in Central and South America plus Mexico, corresponding to the distribution of Bignonieae (only one species, *Bignonia capreolata*, occurs in the USA). Species of Bignonieae were recorded in 107 transects, of which 18 represented singleton observations and were excluded from further study. Our survey of Gentry's database thus yielded 89 Neotropical transects that contained at least one species of Bignonieae. We supplemented this dataset with additional records of species occurrence and abundance, GIS coordinates, and vegetation physiognomy compiled by one of us (F.R.M.) from floristic inventories. After the exclusion of localities with singletons, this additional dataset yielded five additional sites, substantially improving our sampling of forests in Eastern Brazil (Atlantic rainforest and "Cerrado" areas). A complete account of these 94 localities (hereafter "communities") is provided in Table S1 (see also Fig. S1).

All communities were classified according to their habitat. We based these "habitat" primarily on the WWF biome classification, which are based on a range of abiotic environmental variables that determine the ecological attributes of an area [36], but subdivided the biome "Tropical and Subtropical Moist Broadleaf Forests" into three separate habitats based in its discrete geographic areas: Central American Moist Forests, Amazonian Moist Forests, and Atlantic Moist Forests. The additional biomes represented in our analyses were: "Deserts and Xeric Shrublands," "Tropical and Subtropical Coniferous Forests," "Tropical and Subtropical Dry Forests," and "Tropical and Subtropical Grasslands, Savannas, and Shrublands" (Table S1). Since species distributions on large spatial scales are related to abiotic environmental conditions, assigning biomes generally corroborates the vegetation physiognomy recorded *in situ* for the communities in our dataset. For example, communities classified as occurring in the Moist Broadleaf Forest biome, were generally described as "tropical moist forest vegetation" or "evergreen/semideciduous forests" in Gentry's transects database, while communities classified as occurring in the Tropical Dry Forest biome were described as "dry forest" [37]. In a few cases, we found discrepancies between the physiognomy classification of the plots in our database and the WWF biome classification. In those cases, we favored the *in situ* classification of habitat, since GIS data can be subject to errors associated with coordinate precision and uncertainty in the models used to predict biomes. Thus, our habitat classification corresponds to a biome-based classification with some changes made in agreement with the vegetation physiognomy reported *in situ* (Table S1).

Phylogeny

We based our study on a phylogeny of Bignonieae that was reconstructed from chloroplast and nuclear DNA sequences [31], with branch lengths calibrated to time with fossil constraints [35]. This phylogeny includes 106 species of Bignonieae, selected from the 393 species in the tribe in order to cover the range of their morphological and geographical variation [26,31]. Of the 146 species species encountered in the community dataset, 83 were not included in the molecular phylogeny. To incorporate those additional 83 taxa, we added branches to the tree in polytomous positions corresponding to their most derived morphological synapomorphies [24,31], with lengths assigned according to ultrametric constraints (Fig. S2). This tree was used for all subsequent analyses.

Environmental variables

We extracted data for five abiotic variables from the 94 communities represented in our dataset, using the 2.5 arc-second resolution grid available from the WorldClim database (http://www.worldclim.org) and the GIS software ArcMap 9.1 [38]. Variables were chosen for their power to predict species establishment: mean amplitude of monthly temperature, annual amplitude in mean monthly temperature, mean monthly temperature, annual precipitation, and the distribution of precipitation throughout the year (measured using Walsh's [39] index) (Table S1). We also recorded the biome of each community, based on the WWF world terrestrial ecoregion classification [36].

Floral morphology data

Here, we used the classification of species of Bignonieae according to Gentry's floral morphological "type" [27] derived from an earlier study [30]. In addition, we used quantitative measurements of the 16 floral characters from all four whorls of organs obtained by Alcantara and Lohmann [33]. The morphological dataset used in the present study was complemented with additional information from the species that were found in the plant communities but not sampled in the molecular phylogeny of the group. Floral trait data was recorded as the mean of measurements taken from up to ten specimens per species (see [33] for further details).

Data analyses

We assessed the influence of phylogeny on species co-occurrence from two perspectives, that of the species and that of the community. From a species perspective, we constructed a matrix of pairwise species co-occurrences, measured by Schoener's [40] index of proportional similarity $CI_{ih} = 1 - 0.5 * (\Sigma |p_{ij} - p_{kj}|)$, where p_{ij} is the proportion of plots j with the occurrence of the species i and p_{kj} is the proportion of plots j where the species k occur. We also constructed a corresponding matrix of pairwise phylogenetic (patristic) distances between species pairs. We then tested for correlation between these matrices using a Mantel test

with 9999 permutations [41]. These statistical analyses were carried out using the statistical software R (2004–2008, www.R-project.org). A significant association between these matrices would suggest two opposing scenarios: i) a positive correlation would indicate that distant relatives tend to co-occur, but that closely related species tend not to co-occur, while ii) a negative correlation would indicate the converse.

From a community perspective, we assessed the phylogenetic structure of co-occurring species across sites in order to test whether species in the communities are more or less related than expected by chance. We estimated the net relatedness index (NRI) and the nearest taxon index (NTI) metrics [12] using the software Phylocom ([42]: http://phylodiversity.net/phylocom/). Separate analyses were carried out on site-by-species matrices of presence-absence values and abundance values. The incorporation of species abundance data in the analyses implies that results reflect phylogenetic distances among individuals (abundance-weighted distances) instead of distances among taxa occurring in each sample (see [42] for details). We tested for the significance of NRI and NTI using the null models 0 and 3 available in Phylocom, based on 10,000 randomizations. The null model 0 shuffles the species labels across the phylogeny, randomizing their phylogenetic relationships [42]. The null model 3 uses the independent swap algorithm [43] to create swapped versions of the sample/species matrix, constraining the data to have the same row and column totals of the original matrix. Thus, the number of species per sample and frequency of occurrence of each species across samples are constrained and species co-occurrences are randomized [42]. This null model does not randomize the species abundance values and does not include species from the phylogeny in the randomizations (i.e., the species pool is limited to the species that occur in the matrix). All the analyses were carried out with (i) the whole dataset, which implies that the species pool used to calculate the distributions of null models is formed by all the species present in our sample, and (ii) habitat-specific subsets of samples, where the species pool used to calculate the distributions of null models included only species restricted to the habitat analyzed, in order to detect differences among habitats.

To assess the abiotic variables associated with species occurrences, we carried out a PCA to reduce the five abiotic variables measured for each community to a smaller number of statistically independent variables. For each species, this yielded a set of abiotic PCA scores corresponding to its geographic localities. We quantified the abiotic preferences of a species by calculating the convex hull of points representing its PCA scores. The convex hull is defined as the smallest convex area enclosing a set of points and is a reasonable means of assessing multivariate trait space [6]. This calculation requires at least three points; hence, we excluded the species that only occurred in one or two communities from the dataset. This reduced the number of species from 146 to 76. To test whether species exhibit ecological specialization, i.e., occupy a narrower set of abiotic conditions than expected by chance, we derived a null distribution for the convex hull based on 9999 randomizations of the species-by-locality matrix. These analyses were carried out in the TraitHull program [6], with the total dataset and habitat-specific datasets (i.e., including only the species and communities that occur within a given habitat, see above). During the randomization procedure, we constrained the number of occurrences of each species to be equal to the empirical value. If a species exhibits no abiotic preferences, the convex hull area observed should not fall in the tails of the null distribution; an alternative result would imply that it occupies a smaller or larger region of niche space than expected by chance. We tested this hypothesis with a paired nonparametric two-tailed Wilcoxon signed-ranks test [44].

To evaluate the phylogenetic pattern of abiotic preferences, we calculated the convex hull areas for successively more inclusive clades across the phylogeny. All else being equal, more inclusive clades should have progressively larger convex hulls, owing to cumulative evolutionary divergence of abiotic preferences. If abiotic preferences are phylogenetically conserved (i.e., evolve slowly relative to the rate of cladogenesis), then the convex hulls of closely related species tend to overlap, and the cumulative hull area should be relatively small at recent ancestral nodes. Alternatively, if closely related species are characterized by higher evolutionary divergence in abiotic preferences, the cumulative convex hull area will be relatively larger at recent ancestral nodes. Thus, calculation of convex hull areas for clades of Bignonieae allows us to assess graphically how the disparity in abiotic preferences has accumulated along the phylogeny, without the challenges associated with ancestral state reconstruction.

We also assessed the effect of floral morphology on species co-occurrence from a species perspective and from a community perspective. From a species perspective, we tested for pairwise associations between floral morphology and species co-occurrence using a Mantel test with 9999 permutations. We used the Schoener [40] co-occurrence index to quantify species co-occurrence, and quantified floral differences as the Euclidean distance between species in a multivariate trait space constructed using PCA. From a community perspective, we assessed the intra-community structure of floral morphology, testing whether the floral diversity of species within a community differ from the expectation for communities assembled at random. We calculated the convex hull occupied by co-occurring species, through the PCA scores calculated from floral measurements. As floral morphology and pollinator associations in Bignonieae are also affected by discrete floral traits, we derived scores from Hill-Smith multivariate analyses [45]. All multivariate analyses were carried out in R (2004–2008, www.R-project.org). As floral traits in Bignonieae showed variation in phylogenetic signal [33], we calculated Hill-Smith scores for a series of different trait combinations: (i) all of the 16 continuous traits analyzed; (ii) all the 16 continuous traits analyzed plus the discrete traits "anther position" (included or exserted), "corolla color" (white, red, yellow or magenta), and "nectar guides" (present or absent); (iii) the 16 continuous traits plus the discrete coding of flower morphology; and (iv) separate analyses of the floral trait classes that are evolutionarily conserved and labile, respectively.

To assess how phylogeny is related to floral diversity within communities, we also calculated the phylogenetic diversity [46] of species at each site. To test whether the convex hull of floral traits and the phylogenetic diversity of co-occurring species are different from communities assembled at random, we used the null model implemented in TraitHull [6], which generates a null distribution of 9999 communities with a given number of species, with species sorting from the original species pool. We used a modified version of the TraitHull script that included the estimation of phylogenetic diversity of communities given a tree (available from the authors upon request). Two-tailed Wilcoxon signed-ranks tests were used to test whether observed convex hulls differed from the null distribution [44]. A convex hull in the high tail of the null distribution would indicate that species differ in floral morphology more than expected by chance, while a convex hull in the low tail of the null distribution would indicate that species are more similar than expected [6].

To allow for comparisons among communities with different numbers of species, we ranked the observed values of convex hull

and phylogenetic diversity based on the null distribution generated for each distinct value of community species richness. This ranking was used to compare the pattern of morphological and phylogenetic diversity among communities from different habitats. The correlation between ranked phylogenetic diversity and convex hull values were tested through the Spearman's coefficient of correlation [44]. Estimates of convex hull and phylogenetic diversity for communities located in different habitats using the "habitat species pools" instead of the total species pool were also carried out in order to account for regional differences on species distribution, as might arise if species of Bignonieae are restricted in their distributions by environmental conditions like predicted in the predominance of filtering.

Results

Phylogeny and species distribution

There was no correlation between the paired species co-occurrence index and the paired phylogenetic distance among species (Mantel's test: $r = -0.002$; $p = 0.555$). In general, there was no phylogenetic structure in the communities analyzed, with only a few values of NRI and NTI being statistically significant (Table S1). The same general pattern was observed for both the analyses using the total species pool and using habitat-specific species pools (data not shown); for convenience, we report here only the results for the total species pool (Table S1). The patterns observed by including abundance data did not differ from those obtained with presence/absence data; thus, we report the details of the former. Most NRI and NTI values were negative (NRI: 54 out the 94 communities with null model 0, and 75 communities with the null model 3; NTI: 62 communities with the null model 0, and 53 communities with the null model 3). The communities that showed significant NRI with null model 0 were: B012, C020, C038, M11, and S143 (Table S1). Only B010 showed a positive value of NRI, indicating that the relatedness of individuals within that community was lower than expected. With null model 3, significant NRI were found in the communities C025, C038, C058, R133, T154, and Y166. C058 and R133 showed higher values of NRI than expected, while the others had lower values. NTI were significant for the communities C038, D063, M111, and T155 with the null model 0, being positive only in D063. With the null model 3, only D063 and M111 showed significant values of NTI, which were positive and negative, respectively.

Abiotic preferences and habitat specialization

The two PCA axes used to estimate the abiotic convex hull occupied by species of Bignonieae explained 45.8% and 26.5%, respectively (data not shown), indicating that most variation in the abiotic variables analyzed was included in the convex hull estimates. Species of Bignonieae occupied lower convex hulls (i.e., narrower ranges of abiotic conditions) than expected by chance (Wilcoxon test: $V = 154$; $p = 0.0016$; Fig. 1). These results did not differ from the analyses carried out with habitat-specific subsets (data not shown).

Convex hull calculated for clades in the phylogeny concentrated the most differences amongst species within genera instead of between genera, with lowest divergences in convex hull area occurring in the most inclusive clades (Fig. 2, Fig. S3).

Floral morphology and species co-occurrence

Pairwise floral divergence between species of Bignonieae was not significantly related to co-occurrence (Mantel's test: $r = 0.0025$; $p = 0.452$). In general, floral diversity observed in communities of Bignonieae did not differ from the null expectation that

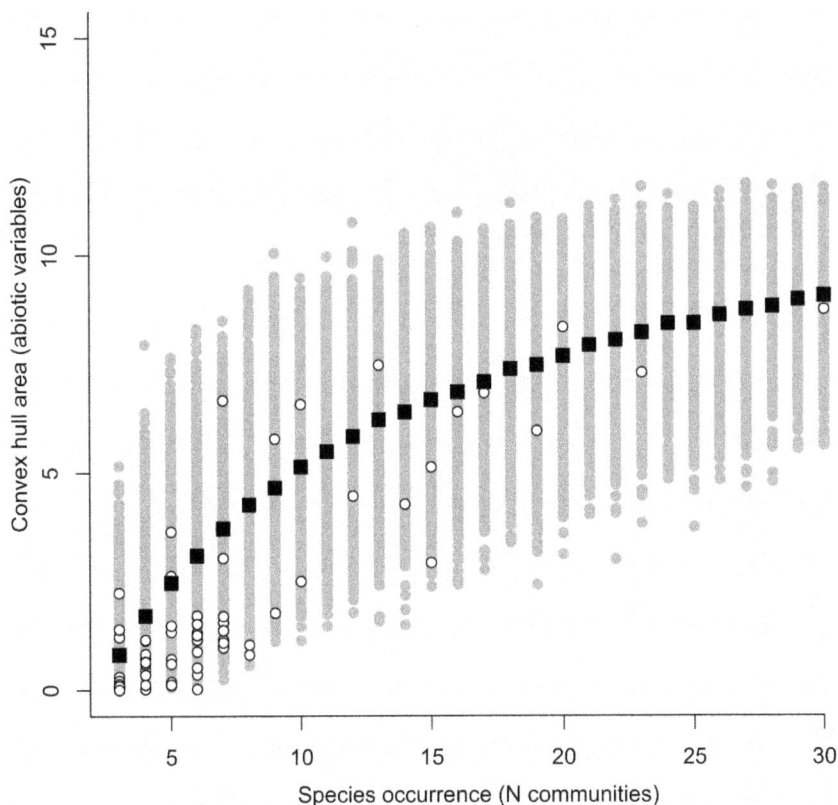

Figure 1. Convex hull area size of 76 species of Bignonieae from their abiotic variables. Convex hull were estimated from the two PC axes scores. Species occurrence indicates the number of communities in which a species was recorded. Open circles represent the observed values of convex hull. Grey circles represent the estimated null distribution of convex hull (see text). Black squares show the mean of the null distribution calculated from each species occurrence number.

communities are assembled randomly (Wilcoxon test: $V = 79$; $p = 0.31$). Analyses carried out with the total species pool and with the habitat species pool did not differ; thus, we only describe the results derived from the former (Fig. S4). Similarly, analyses carried out with different subsets of floral traits also showed similar results; for convenience, we report the results obtained with one combination of traits (the second listed in Materials and Methods). Hill-Smith ordination of this dataset indicated that the two first axes explain 52.1% and 11% of the variation of floral traits, respectively (data not shown).

Rank-based correlation analysis of floral convex hull area and phylogenetic diversity did not reveal general significant associations (Spearman rho = −0.043; p = 0.347; Fig. S5). However, visual inspection of the results revealed notable patterns in three out of the six habitats analyzed. Communities in the Atlantic Moist Forests had higher diversity of floral morphology than the other biomes (Wilcoxon test: $V = 222$; $p = 0.007$), with marginal evidence for a negative correlation with phylogenetic diversity (Spearman rho = −0.612; p = 0.066; Fig. 3A). In contrast, communities from Tropical Dry Forests had relatively low diversity of floral morphology (Wilcoxon test: $V = 811.5$; $p = 0.0034$; Fig. 3B), a pattern also exhibited by the only two communities sampled in the Tropical Savannas (Fig. S5). As far as phylogenetic diversity is concerned, only Tropical Savannas were notably different by presenting lower diversity than the other habitats (Fig. S5).

Discussion

In this paper, we investigated the structure of communities of a Neotropical clade of lianas, bringing phylogeny to bear on questions of how evolutionary patterns of species' traits might influence community assembly. We were particularly motivated by Gentry's [29] predictions that (i) species are specialized to abiotic conditions, and that (ii) communities are saturated in terms of pollination niche. A primary result from our study is that communities of Bignonieae are not phylogenetically structured, i.e., close relatives do not co-occur more than expected by chance or less frequently than expected by chance. This suggests that opposing assembly processes favoring close and distant relatives, respectively, may be at work [14]. This finding is also consistent with the hypothesis that competition among species (e.g., for pollinators) is coupled with labile and presumably adaptive evolution of traits that mediate their competitive interactions [5,11]. The available metrics for characterizing phylogenetic community structure have low power to detect evenness/over-dispersion, i.e., the tendency of distant relatives to co-occur more expected by chance [13]. However, the lack of resolution at the terminals of the phylogeny is not expected to substantially affect detection of phylogenetic structure, but would instead contribute to a signal of random phylogenetic structure [47]. Nevertheless, our results conclusively reject the expectation that tropical communities with large regional species pools exhibit phylogenetic clustering [11,48].

Lack of phylogenetic clustering was persistent in both habitat-specific (regional) and total (continental) species pools. This result

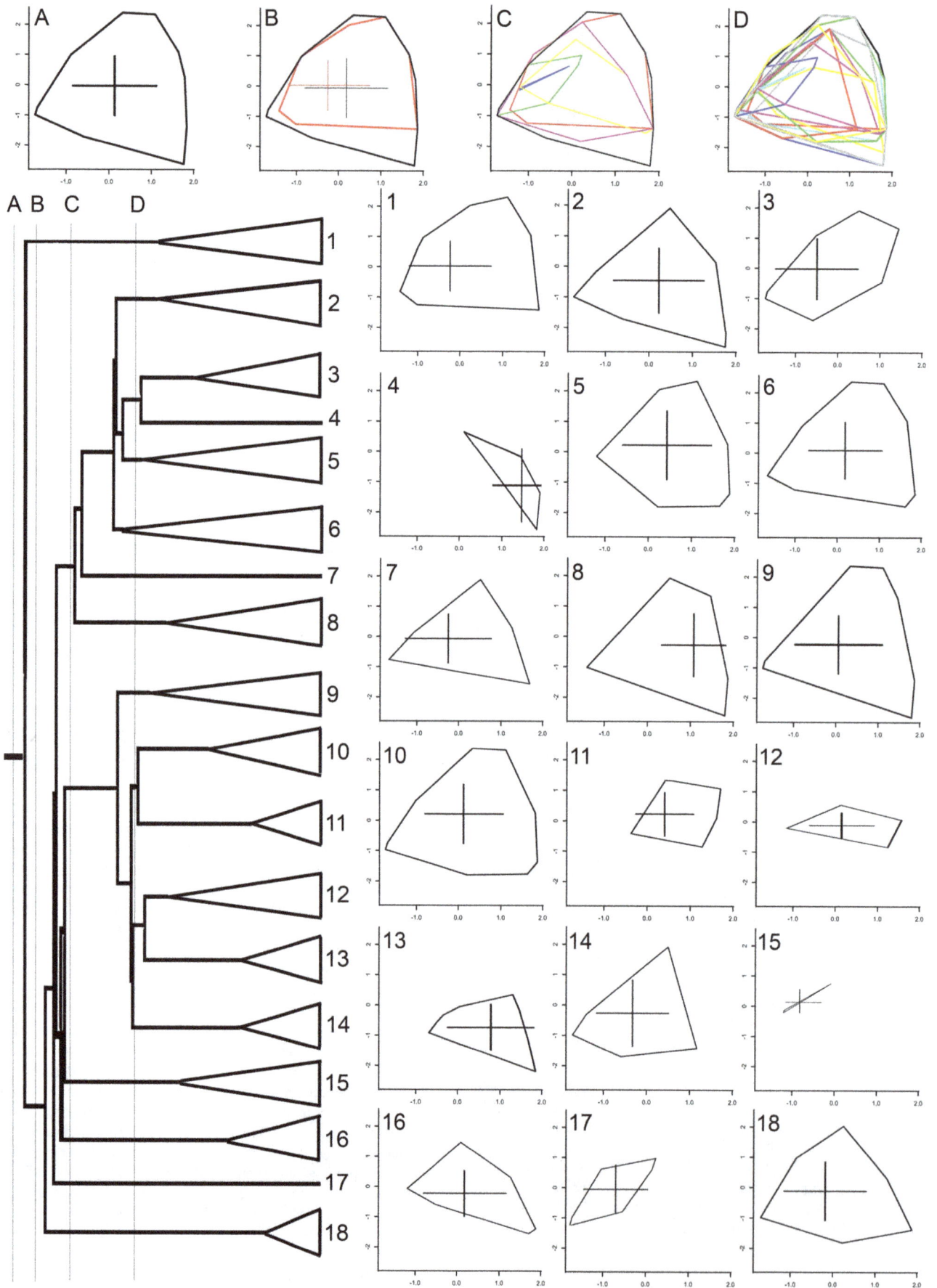

Figure 2. Size of the abiotic variables hyperspace occupied for the most including nodes across the phylogeny of Bignonieae. Graphics indicate the convex hull areas delimited by the abiotic preferences of the species included in each genus (identified by numbers) and more inclusive clades (identified by letters). Total size of convex hulls for individual species and branches of the phylogeny are shown in the Fig. S3. 1. *Adenocalymma.* 2. *Amphilophium,* 3. *Anemopaegma.* 4. *Pyrostegia.* 5. *Mansoa.* 6. *Bignonia.* 7. *Callichlamys.* 8. *Dolichandra.* 9. *Tanaecium.* 10. *Fridericia.* 11. *Xylophragma.* 12. *Cuspidaria.* 13. *Tynanthus.* 14. *Lundia.* 15. *Pachyptera.* 16. *Pleonotoma.* 17. *Martinella.* 18. *Stizophyllum.*

differs from the general trend toward increased phylogenetic clustering at larger geographical scales, or outside the "Darwin-Hutchinson zone" (reviewed in [48]). This increased clustering is expected on continental scales, as a signature of biogeographic processes that reflect dispersal abilities of clades [49]. In Bignonieae, the lack of phylogenetic structure at regional and continental species pools suggests that limited dispersal and/or significant biogeographic barriers have not had major effects on local community structure. In addition, our data set shows that different species from several lineages are broadly distributed and seemingly able to disperse and persist across ecological zones and biomes, suggesting labile evolution of abiotic tolerances [50]. In contrast to this niche-based perspective, the lack of phylogenetic community structure might be attributable to neutral processes of community assembly [51,52]. However, the difficulty in ruling out contrasting niche-based processes that operate on different scales,

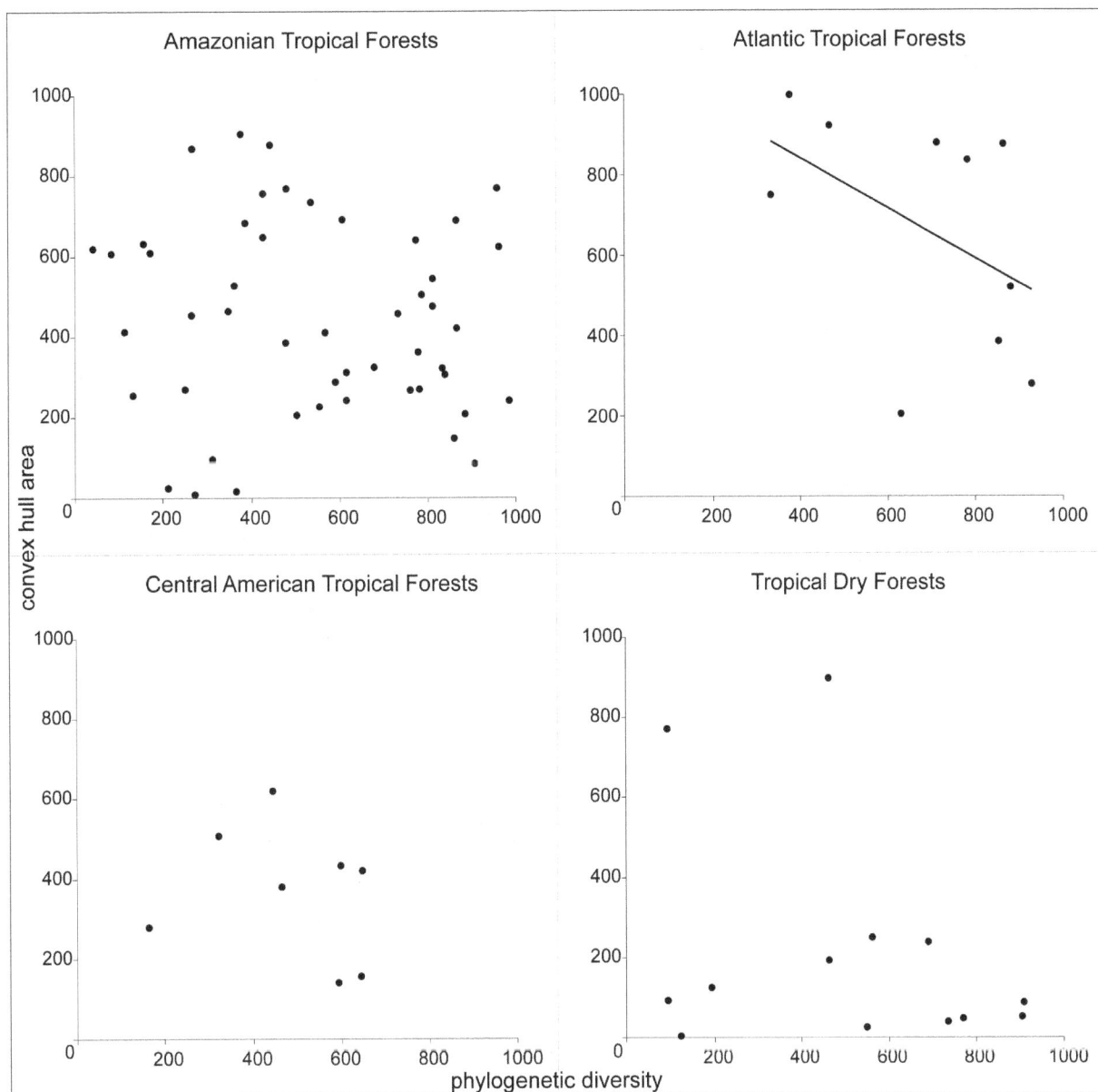

Figure 3. Floral diversity *versus* phylogenetic diversity in communities of Bignonieae in different habitats. Ranked values of (i) convex hull area, representing the morphological floral diversity, and (ii) phylogenetic diversity, calculated as the sum of phylogenetic branch lengths of the species in each community. The recorded points represent each of the communities located at the following habitats: Amazonian Tropical Forests; Atlantic Tropical Forests; Central American Tropical Forests; and Tropical Dry Forests.

and the uncertainty of appropriate null models and species pools, challenge this interpretation [32]. If neutral processes were indeed the prevailing force, one would expect to find no signals concerning habitat preferences. However, we point out that habitat preferences of species and floral traits distribution contribute to the distribution of Bignonieae species across habitats (see below).

Species of Bignonieae tend to occupy a limited portion of the potential convex hull space predicted by their abiotic variables, compared to a null model in which species can occupy any of the communities sampled (Fig. 1). This pattern suggests that, with few exceptions (e.g., *Dolichandra unguis-cati*, *Stizophyllum riparium*, and *Tanaecium pyramidatum*), most species are characterized by specialization to a restricted set of abiotic conditions. Quantification of the convex hull for successively inclusive clades in the phylogeny of Bignonieae shows greater evolutionary divergence at more recent nodes and less divergence at deeper phylogenetic nodes (Fig. 2). Thus, habitat specialization seems to have evolved *within* more recent clades of the phylogeny, like generic or sub-generic clades. It also suggests that similar abiotic preferences have convergently evolved in different clades. The specialization of closely related species to different abiotic conditions corroborates part of Gentry's [29] hypothesis. The second prediction of this hypothesis, that speciation is driven by allopatric specialization, would lead to a negative association among species relatedness and their co-occurrence, which we did not find here. However, most variation in species preference attributes occurs at infra-generic level, for which we have not enough phylogenetic resolution (Fig. S2). Thus, this second prediction still remains to be tested with a phylogeny resolved below the genus level. We did not assess here the effect of potential bias in the geographic locations of communities (i.e., most are located in the Western Amazon; Fig. S1). However, most species of Bignonieae are exclusively Amazonian [24,26]. In addition, we have likely sampled the most common species instead of the rarest ones, suggesting that increased sampling might not change the general pattern of abiotic specialization found here.

The attraction of species possessing traits that enable habitat occupancy characterizes the process of habitat filtering [7,12]. Unfortunately, we have no specific information about the traits in Bignonieae that are functionally associated with the environmental variables we have studied here. In fact, few large-scale tests of coexistence theories in tropical forests have explicitly examined the ecological strategy of co-occurring species [53]. Those studies revealed pervasive habitat specialization affecting species coexistence even in diverse systems [11,14,54]. Moreover, important plant functional traits show evidence of phylogenetic conservatism, as leaf traits [55], wood density [56], and resource allocation patterns [57]. The combination of trait conservatism and environmental filtering has been presumed to account for phylogenetic clustering in many plant communities [48]. Environmental filtering can also cause phylogenetic overdispersion if traits that are important for habitat specialization are labile, with close relatives specializing to different environments [5,9,58]. Further studies focusing on the functional traits coupled with infrageneric phylogenies are needed to evaluate whether this is the case in Bignonieae. This topic is particularly exciting given the relatively rapid and recent evolution suggested for environmental specialization and the increasing changes in natural habitats and global climatic conditions.

There were no effects of floral similarity on species co-occurrence and intracommunity structure, rejecting the hypothesis of saturation by pollinators, to the extent that our measurements of floral morphology accurately reflect pollination mode in Bignonieae [27,30]. This pattern remains even when analyses are carried out with habitat species pools, similar to the pattern found for phylogenetic distance among species. The frequent shifts in floral morphology and the low phylogenetic signal in floral form encountered in Bignonieae were previously interpreted as being indicative of competitive displacement caused by competition for pollinators [33]. Our results imply, however, that the saturation caused by competition by pollinators might have had minor effects on the community assembly of Bignonieae. Nevertheless, there are significant differences in the overall floral diversity of communities located in different habitats. More specifically, less floral diversity than expected by random assemblage found in communities located in the Tropical Dry Forests and Tropical Savannas were detected, while higher diversity was found in Atlantic Tropical Forests.

Similarly to how species' abiotic preferences influences local community structure, habitat specific differences in pollinator pools could directly influence the floral diversity of communities [15]. The local pollinator community can act directly as a biotic filter in an area without suitable pollinators, or indirectly, if the physical environment (i.e., light spectrum, climate, water availability) influences plant-pollinator interactions, determining which pollination systems can persist [15]. Moreover, the occurrence of facilitative interactions between plants that share pollinator guilds has received increased evidence [22,23]. Both pollinator-driven filtering and plant-driven facilitation could create the pattern we found here. Species of Bignonieae are obligate out-crossers and depend on animals for pollination; hence, the absence or rarity of a given pollinator guild in an area would limit species establishment. Finally, correlations among floral morphology and specific vegetative traits associated with abiotic specialization could create differences in floral diversity among habitats, which is also in agreement with the concept of indirect habitat filtering [6,11,59].

Founder-effect colonization of areas by relatively few lineages within Bignonieae may also explain the low morphological diversity in the two communities located in Tropical Savannas, which also showed lower phylogenetic diversity than the other habitats. Evidence indicates that Bignonieae originated in the Atlantic rainforest area and diversified in the Amazon Basin [35], with few species evolving the ability to colonize savannas [26]. The diversification of restricted lineages within Bignonieae in this habitat may not have allowed the accumulation of phylogenetic and morphological diversity compared with that occurred in humid forests.

On the other hand, we did not find any indication of phylogenetic or biogeographic structure in species distributions of Bignonieae species in Tropical Dry Forests. These habitats have already been reported as subject to strong phylogenetic and geographic structure [60]. Instead, our data support the hypotheses that strong environmental filtering may have contributed to the assemblage of Tropical Dry Forests communities, at least in terms of plant-pollinator interactions and their associated morphological traits. Dry areas are known to have the highest levels of bee diversity [61], and most species of Bignonieae have an open-mouthed flower morphology associated with bee pollination [27,30]. This Anemopaegma-type flower is the prevalent morphology within Tropical Dry Forests, and we hypothesize that the predominance of pollination by bees in those areas has limited the occurrence of species with different pollinator vectors. In addition, this floral type was identified as the ancestral morphology of Bignonieae flowers and is widespread between the genera [30]. This would account for the absence of relationship between phylogenetic and morphological diversity in those communities.

The higher morphological diversity found in communities of Atlantic Tropical Forests than in the other habitats has a marginally negative association with their phylogenetic diversity, a trend opposite to the pattern observed in the Tropical Savannas. Evidence indicates that Bignonieae likely originated ca. 50My ago in the same geographical region that is currently occupied by Atlantic Tropical Forests of Brazil [35]. This long-time occupancy and the old age of tropical humid forests would lead to the accumulation of morphological diversity, while the recurrent invasions and diversification at the Amazon Basin would lead to the negative association between phylogeny and morphological diversity observed. Notably, despite the fact that Atlantic Forests are less diverse in their hummingbird fauna than Andean and Amazon Forests, this biome is as diverse as the Andean and Amazonian forests in terms of the number of plant species pollinated by hummingbirds [62]. The morphology associated with hummingbird pollination is the second most common floral form within Bignonieae species, and the most homoplastic one [30]. The suggestion that pollinator faunas have filtered species occurrence across different habitats has important implications for conservation considering the recent worldwide decline of pollinators [15,63,64]. Despite the lack of precise estimates of pollinator diversity on these habitats, these broad patterns represent an intriguing avenue of investigation into the causal relationship between morphology and pollinator diversity in communities of Bignonieae in different habitats.

Conclusion

Our results allowed us to reject the hypothesis that competition for pollinators causes floral saturation and represents a major factor structuring the communities of Bignonieae. Nevertheless, they corroborate Gentry's [27,29] hypothesis that pollination mode may be an important determinant of Bignoniaceae occurrence. We speculate that the specialization to abiotic conditions in this group must have evolved recently, although we did not find the patterns expected by specialization occurred in allopatry, which corroborate only partially the former hypothesis of habitat specialization [29]. Our results differ from the general pattern revealed by most studies of phylogenetic community structure, which report phylogenetic clustering of local communities within larger species pools (reviewed in [48]). Vamosi et al. [48] suggested a common role for habitat filtering coupled with species conserved functional traits, which is opposite to the pattern of evolutionary lability we suggest here. Specialization to abiotic conditions and divergence in floral diversity among habitats suggest a niche-based filtering, concurring with other reports available for tropical forests that suggest that neutral forces may not be sufficient to explain species distributions and the maintenance of diversity in tropical forests [14,53,54].

Supporting Information

Figure S1 Distribution of the 94 communities included in this study. See Table S1 for specific details.

Figure S2 Phylogeny of Bignonieae used in this study, with the manual inclusion of 83 species in 22 polytomies representing genera or infra-generic clades in a time calibrated tree originally containing 106 species, of which 63 species originally included were kept. Branch lengths are represented proportional to time (see text).

Figure S3 Total convex hull size of the 76 species of Bignonieae and of the most inclusive clades of the phylogeny.

Figure S4 Convex hull area estimates for 86 communities of Bignonieae that contain more than 2 species, from the two PC axes obtained from the floral morphology variables included in this study. Species richness indicates the number of species sampled in that community. Open circles represent the observed values of convex hull, and grey circles represent the estimated null distribution of convex hull (see text). Dashed line shows the observed convex hull tendency, while black squares show the mean of the null distribution calculated from each species occurrence number.

Figure S5 Ranked distribution of the observed phylogenetic diversity and convex hull area (calculated from the flower morphological scores) of the species of Bignonieae occurring in the communities studied. Different points represent communities located in different habitats: AMA = Amazonian Moist Forests; ATL = Atlantic Moist Forests; CEN = Central American Moist Forests; DRY = Tropical and Subtropical Dry Forests; DXS = Deserts and Xeric Shrublands; SAV = Tropical and Subtropical Grasslands, Savannas, and Shrublands.

Table S1 Complete list of the 94 communities studied. Location = politic name of locality and country, GEO = geographic coordinate, N = number of species in the community, MMA = annual Mean of Monthly temperature Amplitude (estimated as the average of the values of monthly temperature amplitude), AMMT = annual Amplitude in the Mean monthly Temperature (estimated from the difference between the highest and lowest mean monthly temperature), AMT = annual mean temperature, AP = annual precipitation, Walsh's index = precipitation distribution along the year, Biome = following WWF's classification, Habitat = based on WWF's biomes and on local physiognomy vegetation (see text), Null model 0 = shuffle species in the tips of phylogeny, Null model 3 = independent swap algorithm, NRI (r) and NTI (r) = Net Relatedness Index and Nearest Taxon Index, respectively, with the respective number of randomizations lower than the observed. Significant values (higher than 975 or lower than 25) indicate p<0.05.

Acknowledgments

The authors thank Deren Eaton for help with statistical analyses in R; Dayane Tarabay for the extraction of environmental variables from GIS data; Louis Bernard Klaczko, Luciano Paganucci de Queiroz, Sara Branco and other members of the Flowering Plant Phylogeny and Evolution discussion group at the Field Museum of Natural History for suggestions on earlier versions of this manuscript. This paper is part of the Ph.D. thesis of S.A.

Author Contributions

Conceived and designed the experiments: SA RR FRM LGL. Analyzed the data: SA RR. Wrote the paper: SA RR. Provided community composition data: FRM. Sampled categorical floral traits: LGL. Measured quantitative floral traits: SA. Prepared the scripts used in analysis: RR.

References

1. Diamond JM (1975) Assembly of species communities. In: Cody ML, Diamond JM, editors. Ecology and evolution of communities. Cambridge: Harvard University Press. pp 342–444.
2. Cornwell WK, Ackerly DD (2009) Community assembly and shifts in plant trait distributions across an environmental gradient in coastal California. Ecol Mon 79: 109–126.
3. MacArthur R, Levins R (1967) The limiting similarity, convergence, and divergence of coexisting species. Am Nat 101: 377–385.
4. Weiher E, Clarke GDP, Keddy PA (1998) Community assembly rules, morphological dispersion, and the coexistence of plant species. Oikos 81: 309–322.
5. Cavender-Bares J, Ackerly DD, Baum DA, Bazzaz FA (2004) Phylogenetic overdispersion in Floridian oak communities. Am Nat 163: 823–843.
6. Cornwell WK, Schwilk DW, Ackerly DD (2006) A trait-based test for habitat filtering: Convex hull volume. Ecology 87: 1465–1471.
7. Pausas JG, Verdú M (2008) Fire reduces morphospace occupation in plant communities. Ecology 89: 2181–2186.
8. Pausas JG, Verdú M (2010) The jungle of methods for evaluating phenotypic and phylogenetic structure of communities. BioScience 60: 614–625.
9. Losos JB, Leal M, Glor RE, de Queiroz K, Hertz PE et al. (2003) Niche lability in the evolution of a Caribbean lizard community. Nature 424: 542–545.
10. Ackerly DD (2003) Community assembly, niche conservatism, and adaptive evolution in changing environments. Int J Plant Sci 164: S165–S184.
11. Kraft NJB, Cornwell WK, Webb CO, Ackerly DD (2007) Trait evolution, community assembly, and the phylogenetic structure of ecological communities. Am Nat 170: 271–283.
12. Webb CO, Ackerly DD, Mcpeek MA, Donoghue MJ (2002) Phylogenies and community ecology. Ann Rev Ecol Syst 33: 475–505.
13. Kraft NJB, Ackerly DD (2010) Functional trait and phylogenetic tests of community assembly across spatial scales in an Amazonian forest. Ecol Mon 80: 401–422.
14. Swenson NG, Enquist BJ (2009) Opposing assembly mechanisms in a Neotropical dry forest: implications for phylogenetic and functional community ecology. Ecology 90: 2161–2170.
15. Sargent RD, Ackerly DD (2008) Plant–pollinator interactions and the assembly of plant communities. Tr Ecol Evol 23: 123–130.
16. Becerra JX (2007) The impact of herbivore-plant coevolution on plant community structure Proc Natl Acad Sci U S A 104: 7483–7488.
17. Kay KM, Sargent RD (2009) The role of animal pollination in plant speciation: Integrating ecology, geography, and genetics. Annu Rev Ecol Evol Syst 40: 637–656.
18. Sargent RD, Kembel SW, Emery NC, Forrestel EJ, Ackerly DD (2011) Effect of local community phylogenetic structure on pollen limitation in an obligately insect-pollinated plant. Am J Bot 98: 283–289.
19. Nuismer SL, Jordano P, Bascompte J (2012) Coevolution and the architecture of mutualistic networks. Evolution 67: 338–354.
20. Eaton DAR, Fenster CB, Hereford J, Huang SQ, Ree RH (2013) Floral diversity and community structure in *Pedicularis* (Orobanchaceae). Ecology 93: S182–S194.
21. Pleasants JM (1980) Competition for bumblebee pollinators in rocky mountain plant communities. Ecology 61: 1446–1459.
22. Moeller DA (2004) Facilitative interaction among plants via shared pollinators. Ecology 85: 3289–3301.
23. Hegland SJ, Grytnes JA, Totland O (2009) The relative importance of positive and negative interactions for pollinator attraction in a plant community. Ecol Res 24: 929–936.
24. Lohmann LG, Taylor CM (2014) A new generic classification of Bignonieae (Bignoniaceae). Ann Missouri Bot Gard In press.
25. Olmstead RG, Zjhra ML, Lohmann LG, Grose SO, Eckert AJ (2009) A molecular phylogeny and classification of Bignoniaceae. Am J Bot 96: 1731–1743.
26. Lohmann LG (2003) Phylogeny, classification, morphological diversification and biogeography of Bignonieae (Bignoniaceae, Lamiales). PhD diss. St. Louis: University of Missouri-St.Louis.
27. Gentry AG (1974a) Coevolutionary patterns in Central American Bignoniaceae. Ann Miss Bot Gard 61: 728–759.
28. Gentry AG (1974b) Flowering phenology and diversity in tropical Bignoniaceae. Biotropica 6: 64–68.
29. Gentry AG (1990) Evolutionary patterns in Neotropical Bignonieae. Mem New York Bot Gard 55: 118–129.
30. Alcantara S, Lohmann LG (2010) Evolution of floral morphology and pollination systems in Bignonieae (Bignoniaceae). Am J Bot 97: 782–796.
31. Lohmann LG (2006) Untangling the phylogeny of neotropical lianas (Bignonieae, Bignoniaceae). Am J Bot 93: 304–318.
32. Cavender-Bares J, Kozak K, Fine P, Kembel S (2009) The merging community ecology and phylogenetic biology. Ecol Let 12: 693–715.
33. Alcantara S, Lohmann LG (2011) Contrasting phylogenetic signals and evolutionary rates in floral traits of Neotropical lianas. Biol J Linn Soc 102: 378–390.
34. Alcantara S, Oliveira FB, Lohmann LG (2013) Phenotypic integration in flowers of Neotropical lianas: Diversification of form with stasis of underlying patterns. J Evol Biol 26: 2283–2296.
35. Lohmann LG, Bell C, Calió MF, Winkworth R (2013) Pattern and timing of biogeographic history in neotropical lianas (Bignonieae, Bignoniaceae). Bot J Linn Soc 171: 154–170.
36. Olson DM, Dinerstein E, Wikramanayake ED, Burgess ND, Powell GVN et al. (2001) Terrestrial ecoregions of the world: a new map of life on earth. Bioscience 51: 933–938.
37. Oliver P, Miller J (2002) Global patterns of plant diversity, Alwyn H. Gentry forest transect dataset. St. Louis: Missouri Botanical Garden Press.
38. ESRI. Environmental Systems Research Institute 2005 ArcGIS 9.1. ESRI, Redlands, CA.
39. Walsh RPD (1996) Climate. In: Richards PW, editor. Tropical rain forest. Cambridge: Cambridge University Press. pp 159–205.
40. Schoener TW (1970) Nonsynchronous spatial overlap of lizards in patchy habitats. Ecology 51: 408–418.
41. Manly BFJ (1986) Multivariate statistical methods: A primer. London: Chapman & Hall.
42. Webb CO, Ackerly DD, Kembel SW (2008) Phylocom: software for the analysis of phylogenetic community structure and character evolution. Version 4.0. Available: http://phylodiversity.net/phylocom/. Accessed 2010 May 8.
43. Gotelli N, Entsminger G (2003) Swap algorithms in null model analysis. Ecology 84: 532–535.
44. Sokal RR, Rohlf FJ (1995) Biometry. 3rd edition. New York: Freeman.
45. Hill MO, Smith AJ (1976) Principal component analysis of taxonomic data with multi-state discrete characters. Taxon 25: 249–255.
46. Faith DP 1992 Conservation evaluation and phylogenetic diversity. Biol Cons 61: 1–10.
47. Swenson NG (2009) Phylogenetic resolution and quantifying the phylogenetic diversity and dispersion of communities. PLoS One 4(2): e4390. doi:10.1371/journal.pone.0004390.
48. Vamosi SM, Heard SB, Vamosi JC, Webb CO (2009) Emerging patterns in the comparative analysis of phylogenetic community structure. Mol Ecol 18: 572–592.
49. Wiens JJ, Donoghue MJ (2004) Historical biogeography, ecology and species richness. Tr Ecol Evol 19: 639–644.
50. Ricklefs RE (2006) Evolutionary diversification and the origin of the diversity-environment relationship. Ecology 87: S3–S13.
51. Kembel S, Hubbell SP (2006) The phylogenetic structure of a neotropical forest tree community. Ecology 87: 86–99.
52. Hardy OJ, Senterre B (2007) Characterizing the phylogenetic structure of communities by an additive partitioning of phylogenetic diversity. J Ecol 95: 493–506.
53. Kraft NJB, Valencia R, Ackerly DD (2008) Functional traits and niche-based tree community assembly in an Amazonian forest. Science 322: 580–582.
54. Lebrija-Terros E, Pérez-García EA, Meave JA, Bongers F, Poorter L (2010) Functional traits and environmental filtering drive community assembly in a species-rich tropical system. Ecology 91: 386–398.
55. Ackerly DD, Reich PB (1999) Convergence and correlations among leaf size and function in seed plants: A comparative test using independent contrasts. Am J Bot 86: 1272–1287.
56. Chave J, Muller-Landau HC, Baker TR, Easdale TA, ter Steege H, Webb CO (2006) Regional and phylogenetic variation of wood density across 2456 neotropical tree species. Ecol Appl 16: 2356–2367.
57. McCarthy MC, Enquist BJ, Kerkhoff AJ (2007) Organ partitioning and distribution across the seed plants: assessing the relative importance of phylogeny and function. Int J Plant Sci 168: 751–761.
58. Fine PVA, Daly D, Villa G, Mesones I, Cameron K (2005) The contribution of edaphic heterogeneity to the evolution and diversity of Burseraceae trees in the western Amazon. Evolution 59: 1464–1478.
59. Diaz S, Cabido M, Casanoves F (1998) Plant functional traits and environmental filters at a regional scale. J Veget Sci 9: 113–122.
60. Pennington RT, Lavin M, Oliveira-Filho A (2009) Woody plant diversity, evolution, and ecology in the tropics: Perspectives from seasonally dry tropical forests. Ann Rev Ecol Evol Syst 40: 437–457.
61. Michener CD (1979) Biogeography of the bees. Ann Miss Bot Gard 66: 277–347.
62. Buzato S, Sazima M, Sazima I (2000) Hummingbird-pollinated floras at three Atlantic forest sites. Biotropica 32: 824–841.
63. Biesmeijer JC, Roberts SPM, Reemer M, Oholemuller R, Edwards M, Peeters T, Schaffers AP, Potts SG, Kleukers R, Thomas CD, Settele J, Kunin WE (2006) Parallel declines in pollinators and insect-pollinated plants in Britain and the Netherlands. Science 313: 351–354.
64. Vamosi JC, Knight TM, Steets JA, Mazer SJ, Burd M, Ashman TL (2006) Pollination decays in biodiversity hotspots. Proc Natl Acad Sci USA 103: 956–961.

13

Which Morphological Characteristics Are Most Influenced by the Host Matrix in Downy Mildews? A Case Study in *Pseudoperonospora cubensis*

Fabian Runge[1], Beninweck Ndambi[1,2], Marco Thines[3,4,5]*

1 University of Hohenheim, Institute of Botany, Stuttgart, Germany, **2** University of Hohenheim, Institute of Plant Production and Agroecology in the Tropics and Subtropics, Stuttgart, Germany, **3** Biodiversity and Climate Research Centre (BiK-F), Frankfurt (Main), Germany, **4** Senckenberg Gesellschaft für Naturforschung, Frankfurt (Main), Germany, **5** Johann Wolfgang Goethe University, Department of Biological Sciences, Institute of Ecology, Evolution and Diversity, Frankfurt (Main), Germany

Abstract

Before the advent of molecular phylogenetics, species concepts in the downy mildews, an economically important group of obligate biotrophic oomycete pathogens, have mostly been based upon host range and morphology. While molecular phylogenetic studies have confirmed a narrow host range for many downy mildew species, others, like *Pseudoperonospora cubensis* affect even different genera. Although often morphological differences were found for new, phylogenetically distinct species, uncertainty prevails regarding their host ranges, especially regarding related plants that have been reported as downy mildew hosts, but were not included in the phylogenetic studies. In these cases, the basis for deciding if the divergence in some morphological characters can be deemed sufficient for designation as separate species is uncertain, as observed morphological divergence could be due to different host matrices colonised. The broad host range of *P. cubensis* (ca. 60 host species) renders this pathogen an ideal model organism for the investigation of morphological variations in relation to the host matrix and to evaluate which characteristics are best indicators for conspecificity or distinctiveness. On the basis of twelve morphological characterisitcs and a set of twelve cucurbits from five different Cucurbitaceae tribes, including the two species, *Cyclanthera pedata* and *Thladiantha dubia*, hitherto not reported as hosts of *P. cubensis*, a significant influence of the host matrix on pathogen morphology was found. Given the high intraspecific variation of some characteristics, also their plasticity has to be taken into account. The implications for morphological species determination and the confidence limits of morphological characteristics are discussed. For species delimitations in *Pseudoperonospora* it is shown that the ratio of the height of the first ramification to the sporangiophore length, ratio of the longer to the shorter ultimate branchlet, and especially the length and width of sporangia, as well as, with some reservations, their ratio, are the most suitable characteristics for species delimitation.

Editor: Sung-Hwan Yun, Soonchunhyang University, Republic of Korea

Funding: This study was supported by a grant from the Ministry of Science, Education and the Arts of Baden-Württemberg awarded to FR and by the research funding programme "LOEWE – Landes-Offensive zur Entwicklung Wissenschaftlich ökonomischer Exzellenz" of Hesse's Ministry of Higher Education, Research, and the Arts. The funders had no role in study design, data collection and analysis, decision to publish, or preparation of the manuscript.

Competing Interests: The authors have declared that no competing interests exist.

* E-mail: marco.thines@senckenberg.de

Introduction

The family Peronosporaceae is the largest oomycete family and contains, in addition to smaller groups, the about 100 species of the paraphyletic, hemibiotrophic genus *Phytophthora* and the about 800 species of obligate biotrophic downy mildews [1–4]. Among them are several economically important diseases, like *Phytophthora infestans* (potato late blight), *Bremia lactucae* (lettuce downy mildew), *Plasmopara halstedii* (sunflower downy mildew), *Plasmopara viticola* (grape downy mildew), and *Pseudoperonospora cubensis* (cucurbit downy mildew). As it is often difficult to distinguish downy mildew species on the basis of morphological characters, Yerkes & Shaw [5] favoured a broad species concept for two large groups of downy mildews of Chenopodiaceae and Brassicaceae, respectively. This was contrasting the narrow species concept advocated by Gäumann [6,7], stating that host ranges of downy mildews are often limited to a single host species. Although many plant pathologists have adopted the broad species concept, Gäumann's

concept has mostly been used by taxonomists and has largely been confirmed by molecular studies (e.g. [8–14]). But among obligate biotrophic oomycetes not only *Albugo candida* of the white blister rusts (Albuginales) has been demonstrated to have a broad host range [15–19], but also some downy mildew species, for example in *Hyaloperonospora* [14], *Bremia* [13], and *Pseudoperonospora* [20,21]. *Pseudoperonospora cubensis*, which is one of the most important pathogens of cucurbitaceous crops, is unusual among downy mildews, however, because its reported host range encompasses about 60 species of cucurbitaceous plants in several tribes. The integrity of *P. cubensis* was seldom questioned, although Sawada [22] segregated new species from this species, based on differences in sporangial dimensions associated with some host plants. Although these new species were not considered by subsequent authors [23–26], the potential risk of misinterpreted morphological divergence or similarity is obvious.

Several studies have addressed the morphological variability of downy mildews, mainly regarding ecological conditions. Iwata

[23] and Cohen & Eyal [27] showed that the morphology of sporangiophores of *P. cubensis* varies with different temperatures and intensity of light. Dudka et al. [28] reported humidity as a key environmental factor that has a major impact on sporangial dimensions of *Peronospora alta*. Delanoe [29] pointed out, that even the kind of host tissue from which sporulation takes place may affect the morphology of sporangiophores and sporangia of *Plasmopara halstedii*, as for instance the sporangia produced from roots were more than two times larger than the sporangia produced from leaves. Few studies are available that address the variability of the morphology of different pathogen isolates from the same species on the same host. Kulkarni et al. [30] found that the sporangiophores and also the sporangia of different isolates of *Plasmopara halstedii* varied significantly in their morphology when grown on the same host under controlled conditions. And Salati et al. [31] showed that the size of the sporangiophores of different *Pseudoperonospora cubensis* isolates taken from the same host can vary significantly. Morphology of downy mildews thus seems to be dependent on several different factors, some of which have not been characterized to date. Most of the comparative studies of closely related downy mildews underpin their molecular phylogenetic results with morphological differences of the pathogens, even if there are only very few diagnostic SNPs. Since these pathogens originate from different hosts and data from infection trials for both pathogens on the same host is mostly not available, it often cannot be assessed, whether downy mildews on closely related hosts could be potentially conspecific. However, Runge & Thines [32] demonstrated for the closely related species *P. cubensis* and *P. humuli* that two distinct species may look more dissimilar on different hosts compared to the differences between the two pathogens if the same host was infected. Thus morphologically cryptic, but phylogenetically distinct species might exist on the same host. Conversely, two isolates belonging to the same species could potentially be very dissimilar on two distinct hosts in terms of morphology. The impact of host matrix on sporangial dimensions and a limited number of additional characteristics has already been shown for *P. cubensis* [26,33]. But it is currently unclear, how the broad set of characteristics currently used for species delimitation [8,9,12,34] is influenced by the host matrix, and if increasing phylogenetic distance of the hosts would lead to stronger differences in the morphology of the pathogen, or if other factors are more important. In addition, it is important to assess which characters are most and least influenced by the host matrix to enable the identification of characters by which known species on new hosts could be reliably identified. Some hints regarding the variability of morphological characteristics on different hosts were obtained in our previous work [33], but the limited sampling of hosts among Cucurbitaceae and the few characters investigated were not sufficient to resolve these questions.

Based on twelve hosts and twelve morphological characteristics, the aim of this study was to evaluate, how plastic morphological characteristics used for species delimitation are, on a single host and comparing different host matrices.

Materials and Methods

Infection experiments were carried out using the *Pseudoperonospora cubensis* strain P.C. 26/01 that originally was isolated from *Cucumis sativus* in the Czech Republic. The strain is maintained as a reference strain on *C. sativus* in climate chambers (16°C, 14 h light, and 10 h darkness) since 2007 at the Institute of Botany at the University of Hohenheim. Inoculations for continuous cultivation of the strain and crossinoculations were done using the dab-off technique as described earlier [35]. Uninfected leaves were moistened on the lower leaf surface with deionised water, followed by gently dabbing sporulating leaf parts onto the moist leaf surface. The dab-off technique leads to a range of the spore concentration over the leaf surface from areas without sporangia to areas with very high amounts of sporangia. The advantage is that the optimal inoculum concentrations, which will differ from species to species and even from cultivar to cultivar, need not to be determined, because these will be met through the mosaic pattern in inoculum load throughout the surface of the leaves. Fully mature leaves of *Bryonia dioica*, *Citrullus lanatus*, *Cucumis anguria*, *C. melo*, *C. sativus*, *Cucurbita maxima*, *Cu. moschata*, *Lagenaria siceraria*, and *Luffa cylindrica* were taken from plants in 5- to 10-leaf stage grown in greenhouses at the University of Hohenheim. Additionally leaves of a comparable stage of *Cyclanthera pedata*, *Sicyos angulatus*, and *Thladiantha dubia* were taken from outdoor plants at the Botanical Garden of the University of Hohenheim. After inoculation the leaves were transferred to transparent boxes (approximately 30 cm long, 20 cm wide, 5 cm high) on water-soaked paper towels to ensure that 100% relative humidity (RH) was maintained within the boxes. Crossinoculations were done in three technical replicates. Crossinoculations using *B. dioica*, *Ci. lanatus*, *C. anguria*, *Cu. maxima*, *S. angulatus*, and *T. dubia* were in addition repeated at different time points for testing reproducibility over time. As these tests were successful and did not reveal significant differences, infection trials for the other cucurbitaceous hosts were done only once, in three replicates carried out at the same time. Two days after sporulation was first observed, sporangiophores were picked from the leaf surface with precision tweezers, transferred to a drop of water on a microscopic slide and covered with a coverslip. The morphology of sporangia and sporangiophores were investigated using a Biomed (Leitz, Wetzlar, Germany) light microscope. Pictures of sporangia and sporangiophores were taken using a Canon PowerShot A640 camera (Canon, Tōkyō, Japan). Before each picture series, a picture of a stage micrometer was photographed to calibrate the measurements which were conducted with the AxioVision LE software (Carl Zeiss Imaging Solutions, München, Germany). The characters examined (Figure 1) were the length of the sporangiophores (n = 25), the height of the first ramification (n = 25), the width of the trunk (n = 25) halfway from the base to the first ramification, the number of branching orders (n = 25), the length of the ultimate branchlets (n = 50 each for the longer and the shorter ultimate branchlet), and the length and the width of the sporangia (n = 100, each). For *Ci. lanatus*, *Cu. maxima*, and *Lu. cylindrica* available measurements were included from Runge & Thines [33], and for *B. dioica*, *C. sativus*, and *S. angulatus* available measurements were included from Runge & Thines [32]. For some characters of *P. cubensis* on *B. dioica*, *C. sativus*, and *S. angulatus* measurements of both studies were available. These were combined as no statistically significant differences were apparent between these. This leads to a duplication of the number of measurements taken for the length of the sporangiophores, the height of the first ramification, and the sporangial dimensions in these hosts. In addition the ratio of sporangiophore length to height of the first ramification, the ratio of the longer to the shorter ultimate branchlet, and the ratio of the length to the width of sporangia were calculated. The data were analysed using the STATISTICA '99 software (StatSoft, Tulsa, OK, USA) applying the Mann-Whitney-U-Test [36] to determine the significance of the differences between the species investigated. The STATISTICA 6.1 software (StatSoft, Tulsa, OK, USA) was used for calculating the Spearman's rank correlation [37] at a significance level of $p < 0.004$ ($p < 0.05$ Bonferroni corrected) to determine a potential correlation of the different characters. Values of the Spearman's rank correlation range from 1 to −1 giving the strength of positive or negative correlations. To evaluate the statistically significant differences regarding the usability in species delimitation the plasticity of each character on each host and among all hosts were determined using the range of the standard deviation ((mean+SD)-(mean−SD)) in relation to the total character variation (maximum-minimum). A classification of the plasticity values was done considering a standard deviation interval of about one third of the total variation interval as moderate variation. Thus a low plasticity is presented by a lower value and a high plasticity by a higher value (host related plasticities: 0.28–0.40 (low), 0.41–0.49 (moderate), 0.50–0.70 (high); overall plasticities: 0.20–0.27 (low), 0.28–0.32 (moderate), 0.33–0.38 (high)). This was done for host related plasticity values altogether and separately for the overall plasticity values. Furthermore, a simple method to estimate the maximum of potentially released infective units per sporangiophore was used in conjunction with the sporulation density to obtain an estimate of the suitability of the hosts for pathogen proliferation. Assuming a perfectly ellipsoid shape, the mean volume of the sporangia was calculated ($V = 4/3\pi ab^2$), then divided by the mean volume of one zoospore ($V = 5,814\ \mu m^3$) obtained from a previous study with *P. cubensis* and its closest relative *P. humuli* [38], and finally the mean number of zoospores released by one sporangium was multiplied with the potential maximum of ultimate branchlets according to the number of branch orders on the respective host. To estimate the potential infection pressure giving rise to the next asexual cycle, the sporulation density and the infected leaf area was determined. Therefore symptomatic leaves were photographed together with a scale on a desk lighted from below, in order to increase the visibility of infected leaf areas. Areas with sporulation were confirmed and investigated using an Olympus SK60 stereo microscope (Olympus, Hamburg, Germany) with a KL1500 illumination unit (Schott, Mainz, Germany) and afterwards marked on the pictures. From three different leaves of each host, all sporangiophores of one sporulating area with a diameter of 2 mm ($3.14\ mm^2$) were picked with precision tweezers and counted using a light microscope. Additionally the leaf area and the infected leaf area of the three leaves each were measured using the AxioVision LE software. Afterwards, the potential number of sporangia and zoospores related to the leaf area infected and the total leaf area were calculated.

Results

Four to nine days after inoculation sporulation could be observed (*Cucumis melo, Cyclanthera pedata, Sicyos angulatus, Thladiantha dubia*: 4d; *Cucumis sativus, Cucurbita maxima, Lagenaria siceraria*: 5d; *Cucumis anguria, Luffa cylindrica*: 6d; *Bryonia dioica*: 7d; *Citrullus lanatus*: 8d; *Cucurbita moschata*: 9d). To the best of our knowledge this is the first report of successful inoculation of *Cy. pedata* and *T. dubia* with *Pseudoperonospora cubensis*. The sporulation density on *Cu. moschata* was very low, therefore, only limited measurements could be done (sporangiophores - length, height of first ramification: n = 18; trunk: n = 19; orders of branching: n = 20; ultimate branchlets: n = 37, each for the longer and the shorter ultimate branchlet; sporangia: n = 25).

The morphological characterisation of *P. cubensis* from the different hosts is given in Table 1, the plasticity values of each character of *P. cubensis* from the different hosts and of the hosts altogether and the classification of the plasticity values are given in Table 2. Since twelve hosts of *P. cubensis* were successfully infected 66 pairwise host-host comparisons were available for statistical analysis.

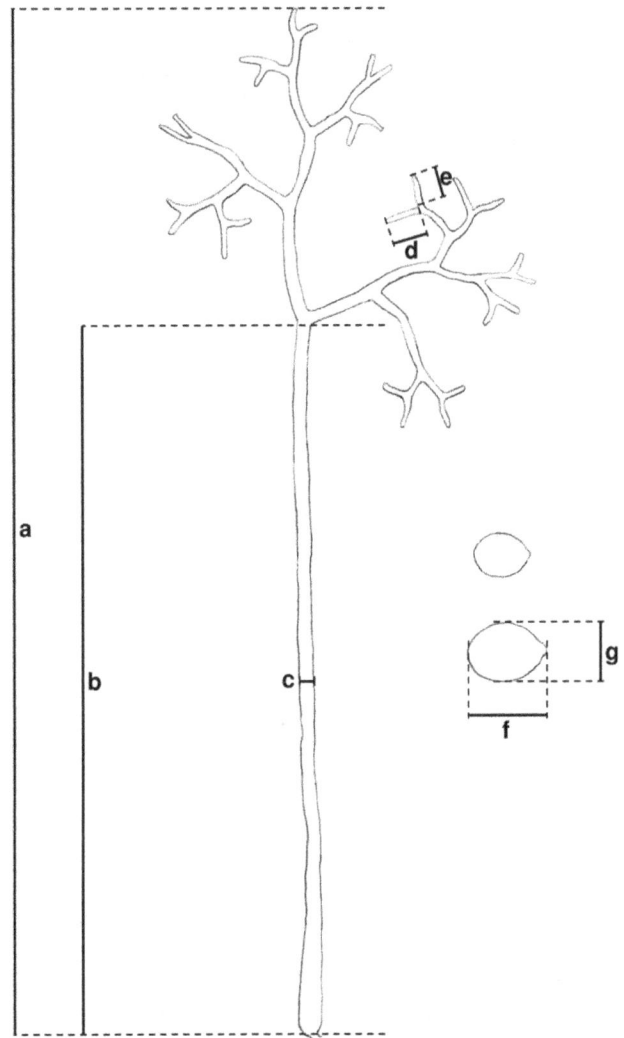

Figure 1. Drawing of a sporangiophore and sporangia of *Pseudoperonospora cubensis* isolated from *Cucumis sativus*. Measurements were taken for the length of the sporangiophore (a), height of the first ramification (b), width of the trunk (c), length of the longer (d) and shorter (e) ultimate branchlets, as well as the length (f) and the width (g) of sporangia.

The sporangiophores varied in their absolute lengths from 153 μm on *Ci. lanatus* to 649 μm on *T. dubia*, in their mean lengths from 266 μm on *Cu. moschata* to 452 μm on *C. melo*, and in their plasticity from 0.43 on *T. dubia* to 0.63 on *C. anguria* (mean 0.52). The height of the first ramification varied in its extremes from 69.4 μm on *Cu. moschata* to 542 μm on *T. dubia*, the mean of the height of the first ramification from 181 μm on *Cu. moschata* to 327 μm on *C. melo*, and its plasticity from 0.39 on *T. dubia* to 0.62 on *Ci. lanatus* (mean 0.49). The lowest ratio of the length of the sporangiophores to the height of the first ramification was observed in *T. dubia* (mean 1.25), the highest in *C. anguria* (mean 1.56, mean of plasticity values 0.43). Moreover the length of the sporangiophores and the height of the first ramification were positively correlated on all hosts with a correlation index of 0.888 (all correlation values significant at a significance level of at least p<0.004). The length of the sporangiophores varied significantly (p<0.05) in 42 of the 66 host-host comparisons (p<0.01: 32; p<0.001: 18; Figure 2A) with the highest overall plasticity value of

Table 1. Morphological characteristics of *Pseudoperonospora cubensis* on various Cucurbitaceae.

Host species	*Thladiantha dubia*	*Bryonia dioica*	*Luffa cylindrica*
Sporangiophores			
Length	(171–)230–333–436(–649) µm	(198–)255–340–424(–530) µm[c]	(253–)328–393–458(–500) µm[a]
Height of first branching	(107–)183–268–353(–542) µm	(130–)186–252–317(–394) µm[c]	(191–)234–284–335(–372) µm
Ratio length/height of first branching	(1.09–)1.14–1.25–1.37(–1.60)	(1.17–)1.23–1.36–1.49(–1.92)[c]	(1.22–)1.29–1.39–1.49(–1.69)
Width of Trunk	(2.1–)2.9–3.8–4.7(–5.9) µm	(3.6–)4.4–5.4–6.4(–7.5) µm[b]	(3.6–)4.4–5.4–6.4(–7.5) µm
Number of branch orders	(2.00–)3.71–4.48–5.25(–6.00)	(3.00–)3.79–4.56–5.33(–6.00)	(4.00–)5.04–5.80–6.56(–7.00)
Ultimate branchlets			
Length	(2.1–)4.0–7.0–10.1(–18.1) µm	(4.9–)6.5–9.5–12.4(–18.0) µm	(2.1–)5.6–8.2–10.7(–14.7) µm[a]
Length of longer ultimate branchlet	(3.0–)4.7–7.8–11.0(–18.1) µm	(5.6–)7.4–10.4–13.4(–18.0) µm[b]	(4.1–)6.6–9.1–11.5(–14.7) µm
Length of shorter ultimate branchlet	(2.1–)3.4–6.3–9.1(–14.2) µm	(4.9–)6.0–8.5–11.0(–14.5) µm[b]	(2.1–)5.0–7.3–9.6(–12.4) µm
Ratio longer/shorter ultimate branchlet	(1.00–)1.03–1.31–1.58(–2.50)	(1.00–)1.06–1.24–1.41(–1.89)[b]	(1.01–)1.01–1.29–1.60(–2.81)
Sporangia			
Length	(12.1–)18.8–21.9–25.1(–28.2) µm	(17.9–)21.1–23.8–26.6(–33.8) µm[c]	(16.3–)19.1–21.7–24.3(–30.1) µm[a]
Width	(10.6–)13.8–16.0–18.2(–23.2) µm	(11.6–)14.1–15.7–17.3(–21.5) µm[c]	(12.2–)14.6–16.5–18.4(–21.4) µm[a]
Ratio length/width	(1.01–)1.21–1.38–1.54(–1.80)	(1.25–)1.40–1.52–1.65(–1.95)[c]	(1.03–)1.20–1.32–1.44(–1.68)[a]

Host species	*Sicyos angulatus*	*Cyclanthera pedata*	*Cucurbita maxima*
Sporangiophores			
Length	(214–)298–392–486(–584) µm[c]	(185–)220–300–380(–497) µm	(205–)236–289–341(–425) µm[a]
Height of first branching	(155–)216–289–363(–495) µm[c]	(125–)151–209–267(–364) µm	(125–)159–208–257(–372) µm
Ratio length/height of first branching	(1.07–)1.24–1.36–1.49(–1.67)[c]	(1.12–)1.31–1.45–1.58(–1.72)	(1.14–)1.21–1.42–1.63(–1.95)
Width of Trunk	(2.9–)3.3–4.0–4.7(–5.3) µm[b]	(3.7–)3.9–4.9–5.8(–6.7) µm	(3.1–)3.8–4.8–5.8(–7.1) µm
Number of branch orders	(4.00–)4.79–5.56–6.33(–7.00)	(4.00–)4.54–5.28–6.02(–7.00)	(3.00–)3.81–4.52–5.23(–6.00)
Ultimate branchlets			
Length	(2.5–)5.3–8.7–12.(–18.7) µm	(3.1–)5.3–7.4–9.6(–13.7) µm	(3.1–)5.8–8.3–10.9(–16.2) µm[a]
Length of longer ultimate branchlet	(4.0–)6.3–9.7–13.2(–18.7) µm[b]	(4.3–)6.1–8.2–10.3(–13.7) µm	(5.0–)6.6–9.0–11.4(–16.2) µm
Length of shorter ultimate branchlet	(2.5–)4.6–7.8–10.9(–17.1) µm[b]	(3.1–)4.8–6.7–8.5(–11.5) µm	(3.1–)5.2–7.7–10.1(–14.3) µm
Ratio longer/shorter ultimate branchlet	(1.01–)1.04–1.30–1.56(–2.11)[b]	(1.01–)1.08–1.26–1.43(–1.76)	(1.00–)1.01–1.21–1.41(–1.89)
Sporangia			
Length	(12.0–)17.9–20.8–23.7(–30.5) µm[c]	(9.3–)17.0–19.6–22.2(–25.5) µm	(13.6–)20.0–23.4–26.8(–32.9) µm[a]
Width	(9.78–)12.5–14.4–16.2(–19.2) µm[c]	(8.6–)12.5–14.1–15.7(–19.6) µm	(11.7–)14.0–16.0–17.9(–23.4) µm[a]
Ratio length/width	(1.01–)1.31–1.45–1.60(–1.80)[c]	(1.05–)1.27–1.39–1.51(–1.67)	(1.01–)1.29–1.47–1.65(–1.84)[a]

Host species	*Cucurbita moschata*	*Citrullus lanatus*	*Lagenaria siceraria*
Sporangiophores			
Length	(173–)195–266–337(–422) µm	(153–)225–301–375(–397) µm[a]	(230–)277–345–413(–498) µm
Height of first branching	(69.4–)130–181–232(–264) µm	(108–)163–228–293(–317) µm	(162–)208–263–318(–372) µm
Ratio length/height of first branching	(1.26–)1.14–1.52–1.90(–2.84)	(1.16–)1.22–1.34–1.45(–1.62)	(1.18–)1.25–1.32–1.39(–1.49)
Width of Trunk	(3.3–)4.1–4.8–5.5(–6.3)µm	(2.9–)3.5–4.4–5.4(–6.1) µm	(3.3–)4.1–4.9–5.7(–6.6) µm
Number of branch orders	(3.00–)3.56–4.45–5.34(–6.00)	(3.00–)3.79–4.56–5.33(–6.00)	(4.00–)3.94–4.64–5.34(–6.00)
Ultimate branchlets			
Length	(3.3–)5.7–9.1–12.5(–17.2) µm	(3.6–)5.5–9.2–12.8(–18.4) µm[a]	(2.2–)5.1–8.1–11.1(–15.7) µm
Length of longer ultimate branchlet	(4.4–)6.2–9.8–13.4(–17.2) µm	(5.1–)6.7–10.5–14.2(–18.4) µm	(5.5–)6.6–9.4–12.2(–15.7) µm
Length of shorter ultimate branchlet	(3.3–)5.4–8.4–11.5(–15.6) µm	(3.6–)4.8–7.9–11.0(–14.9) µm	(2.2–)4.2–6.8–9.3(–12.5) µm
Ratio longer/shorter ultimate branchlet	(1.00–)1.00–1.17–1.34(–1.65)	(1.01–)1.07–1.36–1.65(–2.44)	(1.04–)1.16–1.46–1.75(–2.58)
Sporangia			

Table 1. Cont.

Host species	Cucurbita moschata	Citrullus lanatus	Lagenaria siceraria		
Length	(17.8–)19.8–22.3–24.8(–29.7) μm	(14.3–)19.1–21.6–24.1(–27.1) μm[a]	(18.0–)20.4–23.4–26.5(–38.8) μm		
Width	(12.6–)13.4–14.9–16.4(–19.0) μm	(10.0–)13.4–15.5–17.5(–19.7) μm[a]	(13.9–)15.6–17.7–19.9(–27.5) μm		
Ratio length/width	(1.29–)1.37–1.50–1.63(–1.76)	(1.01–)1.25–1.41–1.57(–1.88)[a]	(1.03–)1.19–1.33–1.46(–1.66)		

Host species	Cucumis anguria	Cucumis sativus	Cucumis melo	Overall mean	Maximum deviation
Sporangiophores					
Length	(251–)273–351–429(–498) μm	(173–)259–354–449(–606) μm[c]	(267–)354–452–550(–642) μm	349	30%
Height of first branching	(123–)175–229–283(–308) μm	(101–)182–257–332(–434) μm[c]	(105–)246–327–409(–472) μm	255	30%
Ratio length/height of first branching	(1.33–)1.30–1.56–1.82(–2.62)	(1.16–)1.25–1.40–1.54(–1.91)[c]	(1.13–)1.17–1.41–1.66(–2.53)	1.39	13%
Width of Trunk	(4.8–)5.6–6.8–8.0(–9.5) μm	(3.1–)4.3–5.3–6.4(–7.2) μm[b]	(3.2–)4.2–4.9–5.6(–6.3) μm	5.0	37%
Number of branch orders	(4.00–)5.22–5.92–6.62(–7.00)	(4.00–)4.60–5.36–6.12(–7.00)	(4.00–)4.35–4.96–5.57(–6.00)	5.0	18%
Ultimate branchlets					
Length	(3.8–)7.0–9.8–12.5(–20.3) μm	(2.9–)5.6–8.4–11.1(–15.2) μm	(3.4–)6.4–9.9–13.4(–21.4) μm	8.6	19%
Length of longer ultimate branchlet	(5.4–)8.1–10.8–13.6(–20.3) μm	(4.0–)6.6–9.4–12.1(–15.2) μm[b]	(4.5–)7.3–10.9–14.5(–21.4) μm	9.6	19%
Length of shorter ultimate branchlet	(3.8–)6.3–8.7–11.1(–14.6) μm	(2.9–)4.9–7.3–9.8(–13.7) μm[b]	(3.4–)5.6–8.9–12.1(–19.5) μm	7.6	19%
Ratio longer/shorter ultimate branchlet	(1.03–)1.09–1.26–1.42(–1.64)	(1.01–)1.08–1.32–1.56(–2.23)[b]	(1.02–)1.06–1.26–1.45(–1.92)	1.29	14%
Sporangia					
Length	(16.9–)20.8–23.6–26.4(–32.2) μm	(14.2–)19.2–22.8–26.4(–36.9) μm[c]	(10.2–)18.7–21.2–23.7(–27.8) μm	22.2	12%
Width	(12.1–)14.7–16.5–18.3(–22.1) μm	(10.3–)14.1–16.3–18.5(–25.1) μm[c]	(9.9–)13.4–14.8–16.1(–18.6) μm	15.7	13%
Ratio length/width	(1.20–)1.33–1.43–1.54(–1.65)	(1.00–)1.26–1.40–1.54(–1.78)[c]	(1.03–)1.31–1.44–1.56(–1.69) μm	1.42	8%

All measurements given in the form (minimum-) standard deviation towards the minimum - mean - standard deviation towards the maximum (-maximum).
[a]Runge & Thines [33].
[b]Runge & Thines [32].
[c]Runge & Thines [32,33] combined.

0.38. In contrast, the ratio of the length to the height of the first ramification varied significantly only in 29 (24, 16) comparisons and had the lowest overall plasticity with 0.20. After eliminating the two most diverging hosts from each of the comparisons, the length of the sporangiophores showed 22 (15, 5) significant differences (*C. melo* and *Lu. cylindrica* were eliminated; Figure 2B), and the ratio of the length to the height of the first ramification showed only 10 (5, 1) significant differences (*T. dubia* and *C. anguria* were eliminated). Although *Ci. lanatus*, *Cu. maxima*, and *T. dubia* showed positive correlations between the length of the sporangiophores and the width of the trunk, no correlation was observed between these characters considering all hosts. The width of the trunk varied from 2.1 μm on *T. dubia* to 9.5 μm on *C. anguria*, the mean values varied from 3.8 μm to 6.8 μm on the same hosts, and the plasticity varied from 0.45 on *C. melo* to 0.61 on *Cy. pedata* (mean 0.52). The trunk showed 33 (30, 24) significant differences among the 66 host-host comparisons with an overall plasticity of 0.32, and after elimination of the two extremes, *T. dubia* and *C. anguria*, only 13 (11, 6) significant differences were found. In the morphology of the sporangiophores including the number of branching orders *B. dioica* had no statistically significant difference to *La. siceraria*, but one difference to *Ci. lanatus* and *C. sativus*, respectively. *Ci. lanatus* was not different from *La. siceraria* and *Cu. maxima*, whereas *Cu. maxima* had no differences compared to *Cu. moschata*. Both were in differing in one character from *Cy. pedata*. Also *Lu. cylindrica* and *S. angulatus* had only one statistically significant difference. The number of branching orders ranged from two to seven with the mean values ranging from 4.45 on *Cu. moschata* to 5.92 on *C. anguria*, and plasticity values from 0.39 on *T. dubia* to 0.70 *La. siceraria* (mean: 0.52). The number of branching

orders showed 41 (35, 26) significant differences and a high overall plasticity of 0.36. After elimination of *C. anguria* and *Lu. cylindrica* there were still 23 (19, 12) significant differences among the remaining 45 host-host comparisons left. Considering all hosts there was a positive correlation between the length of the sporangiophores and the number of branch orders with a correlation index of 0.722.

The overall length of all ultimate branchlets ranged from 2.1 μm on *Lu. cylindrica* and *T. dubia* to 21.4 μm on *C. melo*. The mean values of the ultimate branchlets ranged from 7.0 μm (mean of longer ultimate branchlets 7.8 μm; mean of shorter ultimate branchlets 6.3 μm; mean of plasticity values: 0.47 each) on *T. dubia* to 9.9 μm (10.9 μm, 8.9 μm) on *C. melo*, and the plasticity ranged from 0.33 on *C. anguria* to 0.49 on *Ci. lanatus* and *Cu. moschata* (mean 0.42). The lowest ratio of the longer to the shorter ultimate branchlet had *P. cubensis* on *Cu. moschata* (mean 1.17), whereas on *La. siceraria* it had the highest ratio (mean 1.46). *Pseudoperonospora cubensis* on *Lu. cylindrica* had the lowest plasticity (0.34) and the highest (0.54) was observed on *C. anguria* (mean 0.43).

Considering the longer ultimate branchlets separately 29 (18, 8) host-host comparisons were significantly different, with *T. dubia* and *Cy. pedata* being the most diverging hosts. The shorter ultimate branchlets showed 27 (16, 9) statistically significant differences. Also the ratio of the longer to the shorter ultimate branchlets showed 27 (20, 12) differences, but after *La. siceraria* and *Cu. moschata* were eliminated, only 7 (2, 1) differences were left. The ultimate branchlets had a high overall plasticity of 0.32 (longer 0.34; shorter 0.32). However, the ratio of the longer to the shorter ultimate branchlets had a low overall plasticity of 0.27.

Table 2. Plasticity values of *Pseudoperonospora cubensis* on various Cucurbitaceae.

Host species	Thladiantha dubia	Bryonia dioica	Luffa cylindrica	Sicyos angulatus	Cyclanthera pedata
Sporangiophores					
Length	0.43	0.51	0.53	0.51	0.51
Height of first branching	0.39	0.50	0.56	0.43	0.48
Ratio length/height of first branching	0.45	0.34	0.43	0.41	0.45
Width of Trunk	0.47	0.52	0.50	0.60	0.61
Number of branch orders	0.39	0.51	0.51	0.51	0.49
Ultimate branchlets					
Length	0.38	0.45	0.40	0.42	0.40
Length of longer ultimate branchlet	0.41	0.49	0.46	0.46	0.45
Length of shorter ultimate branchlet	0.47	0.52	0.45	0.43	0.45
Ratio longer/shorter ultimate branchlet	0.37	0.40	0.34	0.47	0.47
Sporangia					
Length	0.39	0.35	0.38	0.32	0.32
Width	0.34	0.31	0.41	0.39	0.29
Ratio length/width	0.41	0.36	0.37	0.36	0.38

Host species	Cucurbita maxima	Cucurbita moschata	Citrullus lanatus	Lagenaria siceraria
Sporangiophores				
Length	0.48	0.57	0.62	0.51
Height of first branching	0.40	0.53	0.62	0.53
Ratio length/height of first branching	0.52	0.48	0.50	0.46
Width of Trunk	0.52	0.48	0.60	0.48
Number of branch orders	0.48	0.59	0.51	0.70
Ultimate branchlets				
Length	0.38	0.49	0.49	0.44
Length of longer ultimate branchlet	0.43	0.56	0.56	0.55
Length of shorter ultimate branchlet	0.44	0.50	0.55	0.50
Ratio longer/shorter ultimate branchlet	0.46	0.52	0.41	0.38
Sporangia				
Length	0.35	0.42	0.39	0.29
Width	0.33	0.47	0.43	0.32
Ratio length/width	0.44	0.53	0.37	0.43

Host species	Cucumis anguria	Cucumis sativus	Cucumis melo	Mean	Overall mean
Sporangiophores					
Length	0.63	0.44	0.52	0.52	0.38
Height of first branching	0.58	0.45	0.44	0.49	0.32
Ratio length/height of first branching	0.40	0.40	0.35	0.43	0.20
Width of Trunk	0.53	0.53	0.45	0.52	0.32
Number of branch orders	0.47	0.50	0.61	0.52	0.36
Ultimate branchlets					
Length	0.33	0.46	0.39	0.42	0.32
Length of longer ultimate branchlet	0.37	0.49	0.42	0.47	0.34
Length of shorter ultimate branchlet	0.45	0.46	0.40	0.47	0.32
Ratio longer/shorter ultimate branchlet	0.54	0.39	0.43	0.43	0.27
Sporangia					
Length	0.37	0.32	0.28	0.35	0.22
Width	0.37	0.30	0.31	0.36	0.23
Ratio length/width	0.46	0.36	0.38	0.40	0.32

Plasticity categories; host related plasticities: 0.28–0.40 (low), 0.41–0.49 (moderate), 0.50–0.70 (high); overall plasticities: 0.20–0.27 (low), 0.28–0.32 (moderate), 0.33–0.38 (high). The classification of the plasticity values was done considering a standard deviation interval of about one third of the total variation interval as moderate variation. A low plasticity is presented by a lower value and a high plasticity by a higher value.

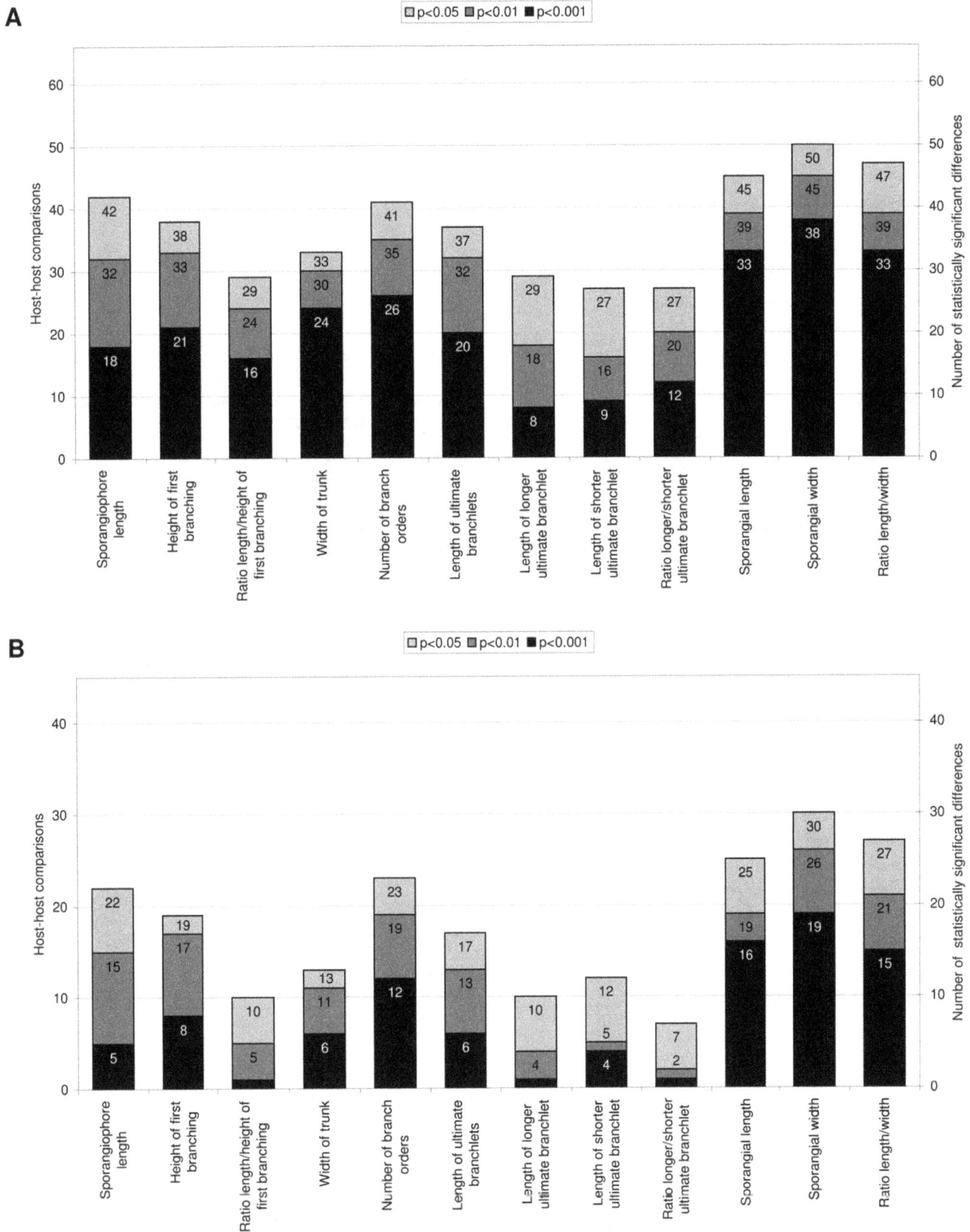

Figure 2. Number of statistically significant differences in morphological characteristics of *Pseudoperonospora cubensis* on a set of twelve cucurbitaceous hosts for all 66 possible host-host comparisons (A) and without the respectively two most deviating hosts, resulting in 45 possible host-host comparisons (B). Excluded in B were the hosts as follows for the characteristics from left to right: 1. *Cucumis melo*, *Luffa cylindrica*; 2. *C. melo*, *Cucurbita moschata*; 3. and 4. *Thladiantha dubia*, *Cucumis anguria*; 5. *C. anguria*, *Lu. cylindrica*; 6. and 7. *T. dubia*, *Cyclanthera pedata*; 8. *T. dubia*, *C. anguria*; 9. *Lagenaria siceraria*, *Cu. moschata*; 10. *Cy. pedata*, *Sicyos angulatus*; 11. *La. siceraria*, *Cy. pedata*; 12. *Lu. cylindrica*, *La. siceraria*.

Table 3. Sporulation density and potential numbers of sporangia and zoospores.

Host species	Thladiantha dubia	Bryonia dioica	Luffa cylindrica	Sicyos angulatus
Leaf area in mm^2	1,897	1,718	6,176	6,000
Sporulating leaf area in mm^2	113	75	62	125
Sporulating leaf area in % of overall leaf area	6.0	4.4	1.0	2.1
Sporangiophores per mm^2 sporulating leaf area	18.0	12.2	5.5	57.3
Sporangiophores per **cm^2** leaf area	107.9	53.4	5.6	119.0
Theoretical approach				
Max. number of sporangia/sporangiophore	22	24	56	47
Volume of sporangia in μm^3 (if perfect ellipsoid)	2,945	3,072	3,111	2,239
Max. number of released zoospores per sporangiophore	113	125	298	182
Max. number of sporangia per mm^2 leaf area	24.1	12.6	3.1	56.2
Max. number of zoospores per mm^2 leaf area	122.3	66.7	16.6	216.8

Host species	Cyclanthera pedata	Cucurbita maxima	Cucurbita moschata	Citrullus lanatus
Leaf area in mm^2	4,030	10,097	8,337	4,316
Sporulating leaf area in mm^2	45	645	1	14
Sporulating leaf area in % of overall leaf area	1.1	6.4	0.0	0.3
Sporangiophores per mm^2 sporulating leaf area	11.9	6.4	20.0	9.5
Sporangiophores per **cm^2** leaf area	13.2	40.6	0.2	3.0
Theoretical approach				
Max. number of sporangia/sporangiophore	39	23	22	24
Volume of sporangia in μm^3 (if perfect ellipsoid)	2,045	3,116	2,604	2,704
Max. number of released zoospores per sporangiophore	137	123	98	110
Max. number of sporangia per mm^2 leaf area	5.1	9.3	0.0	0.7
Max. number of zoospores per mm^2 leaf area	18.0	50.1	0.2	3.3

Host species	Lagenaria siceraria	Cucumis anguria	Cucumis sativus	Cucumis melo
Leaf area in mm^2	9,067	8,181	7,805	7,734
Sporulating leaf area in mm^2	453	462	959	1,754
Sporulating leaf area in % of overall leaf area	5.0	5.6	12.3	22.7
Sporangiophores per mm^2 sporulating leaf area	6.4	8.8	21.4	23.3
Sporangiophores per **cm^2** leaf area	31.8	49.7	263.3	529.4
Theoretical approach				
Max. number of sporangia/sporangiophore	25	61	41	31
Volume of sporangia in μm^3 (if perfect ellipsoid)	3,854	3,350	3,165	2,417
Max. number of released zoospores per sporangiophore	165	349	224	129
Max. number of sporangia per mm^2 leaf area	7.9	30.1	108.2	164.8
Max. number of zoospores per mm^2 leaf area	52.7	173.8	590.1	686.6

On all hosts a positive correlation of the length of the longer to the length of the shorter ultimate branchlet and a negative correlation of the length of the shorter ultimate branchlet to the ratio of the dichotomous ultimate branchlets were observed. No correlation of the length of the longer to the ratio of the dichotomous branchlets could be found, indicating that the shorter ultimate branchlet is the determining factor for the ratio of the length of the longer to the shorter ultimate branchlet. Among all hosts a very strong correlation (0.916) between the lengths of the dichotomous ultimate branchlets was found.

In the characteristics of the ultimate branchlets, *P. cubensis* on *Lu. cylindrica* had no statistically significant differences compared to *Ci. lanatus*, *C. sativus*, *Cu. maxima*, and *S. angulatus*. Furthermore *Cu.*

maxima was not different from *Cu. moschata*, and *Ci. lanatus* was not different from *C. anguria* and *C. melo*. These two were neither different from each other nor from *B. dioica*, which had only one significant difference when compared to *Ci. lanatus*. *Cucurbita moschata* had only one significantly different combination to each of the above mentioned hosts. Furthermore *La. siceraria* had only one significant difference when compared to *Cu. maxima*, *Lu. cylindrica*, and *S. angulatus*.

The sporangia ranged in length from 9.3 μm on *Cy. pedata* to 38.8 μm on *La. siceraria* and in width from 8.6 μm to 27.5 μm on the same hosts, while the mean values ranged from 19.6 μm on *Cy. pedata* to 23.8 μm on *B. dioica* in length and from 14.1 μm on *Cy. pedata* to 17.7 μm on *La. siceraria* in width. The lowest ratio of

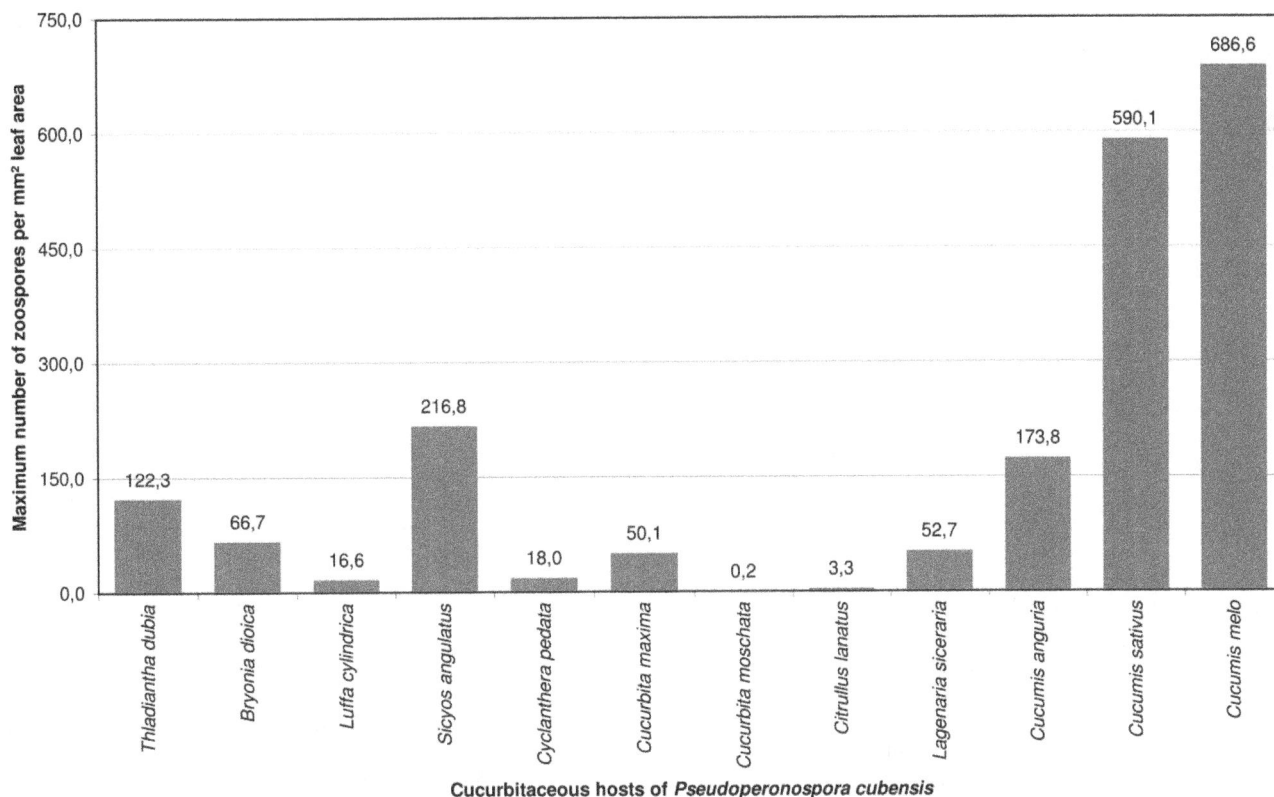

Figure 3. Estimation of the total amount of infective units of *Pseudoperonospora cubensis* per leaf area of the respective cucurbitaceous hosts.

length to width of sporangia was observed for *Lu. cylindrica* (mean 1.32) and the highest ratio had *B. dioica* (mean 1.52). The plasticity of the sporangial length ranged from 0.28 on *C. melo* to 0.42 on *Cu. moschata* (mean 0.35) and the plasticity of the sporangial width ranged from 0.29 on *Cy. pedata* to 0.47 on *Cu. moschata* (mean 0.36), while the plasticity of the ratio of length to width ranged from 0.36 on *B. dioica*, *C. sativus*, and *S. angulatus* to 0.53 on *Cu. moschata* (mean 0.40).

Concerning the length of the sporangia, 45 (39, 33) of the 66 host-host comparisons were statistically significant different. For the width of sporangia, 50 (45, 38) of the combinations were different. For the ratio of sporangial length to width, there were 47 (39, 33) differences found. However, only the overall plasticity of the ratio of sporangial length to width was high with 0.32. The overall plasticity of the sporangial length and the sporangial width was very low, with 0.22 and 0.23, respectively.

On all hosts there was a positive correlation of the length to the width of the sporangia, and also of the length to the ratio of length to width. Furthermore, on almost all hosts, except for *C. melo*, *Cu. moschata*, and *Cy. pedata*, the width of sporangia was correlated negative to the ratio of length to width. Considering all hosts there was a positive correlation (0.664) of the length to the width of the sporangia. *T. dubia* had no significant differences to *C. sativus* and *Ci. lanatus*. In addition *C. melo* was not significantly different from *Cu. moschata*. But all other comparisons had at least one statistically significant difference.

Regarding the sporangial volume, *P. cubensis* on *Cy. pedata* had the smallest sporangia with a mean volume of 2.045 μm^3 (Table 3), and *La. siceraria* had the biggest sporangia with a mean volume of 3.854 μm^3 (*C. sativus*: mean 3,165 μm^3). Thus, considering the

calculated mean of branch orders of the respective host the potential maximum number of released zoospores of one sporangiophore ranges from 98 on *Cu. moschata* to 350 on *C. anguria* (*C. sativus*: 224). The number of sporangiophores covering an infected leaf area of 1 mm^2 ranges in the mean from 5.5 on *Lu. cylindrica* to 57.3 on *S. angulatus* (*C. sativus*: 21.4; for *Cu. moschata* was excluded because of too sparse sporulation). Considering the infected leaf area in relation to the whole leaf surface, the number of sporangiophores per 1 cm^2 ranges from 3.0 on *Ci. lanatus*, where only 0.3% of the surface shows sporulation, to 529.4 on *C. melo*, where 22.7% of the surface shows sporulation (*C. sativus* 263.3, 12.3%). This leads to a potential mean number of 0.7 sporangia and 3.3 zoospores per mm^2 leaf area on *Ci. lanatus* and to a potential number of 164.8 sporangia and 686.6 zoospores on *C. melo* (*C. sativus* 108.2, 590.1). Considering that all sporangiophores were picked from the leaf surface in *Cucurbita moschata*, this host had only 0.2 sporangiophores per 1 cm^2 of leaf area and therefore only a potential maximum number of zoospores of 0.2 per mm^2. A comparison of the potential zoospore discharge is given in Figure 3.

Discussion

Sporulation was first observed on all hosts 4–9 days following inoculation with *Pseudoperonospora cubensis*. An incubation period of 4–12 days has been previously reported, depending on environmental conditions, inoculum load [39], and resistance or susceptibility of the host plant [40]. The variability in the occurrence of first sporulation is in line with previous results from investigations into the genetic variation in resistance to *P. cubensis* among cucurbits [41,42]. The susceptibility of the different cucurbitaceous hosts can be evaluated on the basis of the onset of the sporulation, and also the density of the sporulation. Sporulation on *Cucumis sativus* had the highest density, on *Sicyos angulatus* there were also leaf areas with extremely dense sporulation, but sporulation intensity on *Luffa cylindrica* or *Cyclanthera pedata* was rather low, although in the latter host sporulation occurred already after 4 days. *Cucurbita moschata* was almost resistant to *P. cubensis*, with late onset of sporulation and few emerging sporangiophores. Pathotype specificity was described previously for *Cu. moschata*, but the genetics of host resistance to *P. cubensis* is currently unknown [43]. Nevertheless, if infection experiments with other strains yield similar results, *Cu. moschata* might be a good resource for resistance genes against *P. cubensis*.

The Cucurbitaceae family includes more than 118 genera with 825 species [44]. Of these 20 genera and ca. 60 species are known to be hosts of *P. cubensis* [45,46]. We have found two new potential hosts, *Cy. pedata* and *Thladiantha dubia*, which were susceptible under laboratory conditions. Considering the infection potential of *P. cubensis* and the ongoing discovery of new pathotypes, it seems likely that more cucurbit species will be reported as hosts in the future.

Previous reports of a variety of *Pseudoperonospora* species specific to different cucurbitaceous hosts [22] could not be confirmed in molecular phylogenetic investigations [20,21]. However, two genetically distinct, but morphologically cryptic lineages were found by Runge et al. [21]. Before the onset of molecular phylogenetic investigations, downy mildew species were mainly described on the basis of the host matrix and minor morphological differences on similar hosts, also in *Pseudoperonospora* (e.g. [22]). Although it has been shown that the host matrix has a significant impact on the morphology of some characteristics of *Pseudoperonospora* [33], the extent of this phenomenon could not be fully described previously. Here, we demonstrate that the length of the sporangiophores is a highly variable character depending on the host matrix. It ranges from 153 μm on *Ci. lanatus* to 649 μm on *T. dubia* and has a very high overall plasticity of 0.38. In addition the maximum character deviation of 30% from the overall mean of 349 μm raise doubts if this character is suitable for species delimitation. Although the measured sporangiophores of previous studies [20,23,47–49] were comparatively smaller, they also showed a high variability. Measurements of *P. cubensis* on *La. siceraria* done by Choi & Shin [48] fit perfectly to our results on *La. siceraria* and measurements on *Cu. moschata* made by Choi et al. [20] were also similar to our results on *Cu. moschata*. Minor differences to other studies may be due to the dependence of sporangiophore morphology on temperature as shown by Iwata [23], or on other environmental conditions, especially if specimens were collected in the wild. At comparable temperatures, sporangiophores of *P. cubensis* from *C. sativus* measured by Iwata [23] were smaller than in the present study but sporangiophores taken from *Cu. moschata* were larger than reported here, at comparable temperatures. Although Iwata explained the observed differences with the existence of different biological species of *P. cubensis* with different pathogenicity, these differences are possibly reflecting the intraspecific variability of *P. cubensis* [31]. We thus used both fixed environmental conditions and the same isolate of *P. cubensis* for all the inoculations. Therefore, the observed differences must be induced by the host matrix [33], i.e. differences in the susceptibility and nutrient supply by the hosts [40]. Modification by differences in plant nutrition [50] should be weak, as all plants in the greenhouse were grown in the same soil, as were the plants cut from outdoors. The sporangiophores were very plastic in both their lengths and the height of the first ramification. This led to a lower amount of significant differences. But with a high total plasticity, these characteristics differed significantly in many host-host comparisons and the mean values on the different hosts deviated by 30% in maximum compared to the overall means (349 μm, 255 μm, respectively). Therefore, sporangiophore length and height of the first ramification are not suitable characters for species delimitation. While the height of the first ramification is a character of high variability and high plasticity (0.32), similar to the sporangiophore length, the ratio of these characters is more stable, also shown by the lowest overall plasticity (0.20) and the relatively low amount of statistically significant differences, especially if the most divergent hosts are ignored. Notably the ratio on *C. sativus* matches the value on *C. sativus* in Iwata's [23] analysis, despite the differences in measurements of the respective characteristics of the sporangiophores. Considering the very low overall plasticity (0.20) and a maximum deviation of only 13% from the overall mean (1.39) renders the ratio of sporangiophore length and the height of the first ramification a character potentially useful for species delimitation.

Regarding the width of the trunk, half of all host comparisons were significantly different. In general, the trunk shows rather few statistically supported differences, primarily caused by high to very high plasticity values. The trunk was reported to be slightly thicker in previous studies [23,47–49], but in general the values were comparable. Trunk thickness is a character that is often measured in species descriptions and sometimes is useful for species delimitation [34], but the high overall plasticity in this study and in measurements of *P. humuli* [32,51], the closest relative to *P. cubensis*, and the maximum deviation of 37% from the overall mean (4.97) indicate that for *Pseudoperonospora* this character is not suitable for differentiating between closely related species.

The order of branching is limited in its variability, as it ranges in most cases from three or four to six or seven. However, the frequency distribution of branch orders makes this character highly variable on different hosts with a very high total plasticity of 0.36. Statistically significant differences were observed for order of branching for 41 of 66 host-host comparisons, despite a high plasticity on almost all hosts. Comparison with data from other authors [23,47,48] reveals highly similar figures for the respective hosts, thus suggesting the dependence of the branching orders on the host plant. Thus, although the host related means deviate by a moderate 18% compared to the overall mean (5.02), this character does not seem to be suitable for species delimitation in closely related downy mildews, as the variability of this character is in an overall narrow window. The positive correlation between the length of the sporangiophores and the order of branching indicates a genetic pattern and it seems likely that proper nourishment, caused by adaptation to the host plant, results in longer and more frequently branched sporangiophores carrying more sporangia. Consequently, branching order is also correlated to the amount of possible infections for the next generation.

The ultimate branchlets are a character of moderate variability. Although extremes of 2.1 μm and 21.4 μm were measured, the lengths of the longer and the shorter ultimate branchlets had the fewest statistically significant differences. But the differing plasticity values among the hosts, the high total plasticity and the moderate maximum deviation (19%, each) of the host related means from the overall means (overall 8.6 μm, longer 9.6 μm, shorter 7.6 μm) make it difficult to use these characteristics with confidence in species delimitation. The most stable character is the ratio of the longer to the shorter ultimate branchlet, represented by the fewest statistically significant differences and a low overall plasticity of 0.27 with means deviating by 14% from the overall mean (1.29). This is also expressed by the very strong correlation of the lengths of the longer to the lengths of the shorter ultimate branchlets. In species descriptions, measurements of the ultimate branchlets are often lacking and their ratio has only been considered in a few very recent studies [12,34]. In case of species that are not sister taxa but closely related, like for *Plasmopara angustiterminalis* and *Plasmopara* sp. from *Ambrosia artemisiifolia* and also the case of *Peronospora swinglei* and *P. belbahrii*, the lengths of the ultimate branchlets as well as their ratio support the molecular phylogenetic segregation of the species [12,34]. But for very closely related species, as has been shown by Runge & Thines [32] for *P. humuli* on *S. angulatus*, even moderate deviation of these characters could lead to a breakdown of the significance values. Unfortunately there is no information about ultimate branchlets in the description of any *Pseudoperonospora* species (summarised in [26]) for further comparison of this feature. However, we conclude that at least for species that have already moderately diverged, the lengths of the longer and shorter ultimate branchlets and especially their ratio should be considered and might be useful for species delimitation.

Size and shape of the sporangia of *P. cubensis* are highly affected by the host matrix. The length of the sporangia ranges from 9.3 µm on *Cy. pedata* to 38.8 µm on *La. siceraria* (mean 22.2 µm) and the width from 8.6 µm to 27.5 µm (mean 15.7 µm) on the same respective hosts. The ratio of length to width ranges from 1.00 to 1.95 (mean 1.42). The observed variability is in line with previous observations, as Sawada [22] segregated some species from *P. cubensis* on the basis of host matrix and sporangial dimensions. Due to the known high interspecific variability Waterhouse & Brothers [26] and Gäumann [6,7] also focused their work on sporangial dimensions. Whether genetic background (also due to phylogenetic divergence), host matrix, or environmental conditions could also be responsible for the high variability had never been studied systematically. However, there is an increasing body of evidence from recent studies combining molecular phylogenetics with morphological investigations, e.g. from *Bremia* [9], *Peronospora* [8,12,52,53], and *Hyaloperonospora* [54,55] that sporangial dimensions might be a useful character for species delimitation. In our investigation, there was a positive correlation between the length and the width of sporangia, so the tendency for longer sporangia to also be broader is given. Statistically significant differences were observed for length of sporangia in 45, for width of sporangia in 50, and for the ratio of length to width in 47 of 66 host-host comparisons. In all three characters, at least 33 comparisons had the highest significance level ($p < 0.001$), despite the fact that a single isolate was used and all environmental conditions were controlled. The sporangia were statistically not significantly different only from *T. dubia* and *C. sativus*, *T. dubia* and *Ci. lanatus*, as well as *C. melo* and *Cu. moschata*. The high amount of significant differences is caused by the low plasticity values on the various hosts. Overall, the very low total plasticitiy values of the sporangial length (0.22) and the sporangial width (0.23) make these characteristics very stable and therefore suitable for species delimitations. In contrast, the ratio of these characteristics has a much higher plasticity of 0.32, but under controlled conditions, our study shows that in *P. cubensis* the ratio of the length and the width of the sporangia can vary by about 8%. Given the broad range of sporangial shapes, this character might nonetheless be suitable for species delimitation, decided on a case for case basis, if the intraspecific variation is taken into consideration. For the included characteristics, there was not a single combination without statistically significant differences, highlighting the strong influence of the host matrix on the morphology of downy mildews [33]. Taking into account that the variability under natural environmental conditions will be even high due to the dependence of the morphology on temperature, humidity, and light [23,27,28], care has to be taken when species are delimited solely on the basis of morphology. Considering the

possible variability of the pathogen taken from different parts of the same plant [29] and the influence of the host matrix [26,33], the importance of molecular phylogenetic investigations in addition to morphological investigations becomes obvious. In the absence of molecular data, species delimitation is uncertain, especially if hosts are closely related and morphological differences are subtle. Morphologically indistinguishable pathogens might, however, be different species [4,21], but the same pathogen may also look significantly different depending on the influence of diverse factors given above, especially when considering the potentially strong influence of the host matrix on morphological characteristics.

However, this study reveals that the ratio of the height of the first ramification and sporangiophore length, ratio of the longer to the shorter ultimate branchlets, and especially the length and width of sporangia, as well as, with some reservations, their ratio, are the most suitable characteristics for species delimitations in *Pseudoperonospora*. Additional investigations of intraspecific variation and the influence of host matrix on other downy mildew genera could be useful to investigate if the findings in this study can be generalised for downy mildews.

In a previous study [33], it appeared that the morphological differences of *P. cubensis* might correlate with the phylogenetic position of their hosts in the Cucurbitaceae. In the light of the enlarged dataset in this study and considering the current phylogeny of the Cucurbitaceae [56,57], we found that pathogen morphology on phylogenetically closely related species tended to be similar. For instance *Cu. maxima* and *Cu. moschata* of the Cucurbiteae, *Ci. lanatus* and *La. siceraria* of the Benincaseae, and *Lu. cylindrica* and *S. angulatus* of the Sicyoeae were similar to each other, even in highly variable characters, like the length of the sporangiophores and the number of branch orders, but more dissimilar to other tribes. In addition, on some of the host species (e.g. *Cy. pedata* and *Ci. lanatus*) some characters were similar to the neighbouring tribe, thus occupying an intermediate position between the neighbouring tribe and the tribe of the host species. But, this correlation is not apparent considering the more basal tribes as on *T. dubia* (Thladiantheae) or *B. dioica* (Bryoniae) the pathogen is again more similar to the tribe of origin than to the intermediate tribes, providing evidence that after a certain phylogenetic distance, the correlation between pathogen morphology and phylogenetic position of the host species breaks down. No correlation between sporangial dimension and phylogenetic distance was found.

In summary, the sporangial size and the sporangial shape are of potential value for species delimitation. Sporangiophore length to height of the first ramification and the ratio of the length of the longer to the length of the shorter ultimate branchlet was only weakly affected by host matrix and was invariable and could be of use as well.

The sporangial dimensions and the order of branching were found to be highly dependent on the host matrix. However, these characters impact the infection potential of the next generation, as more often branching sporangiophores bear more sporangia. In addition, it could be assumed that larger sporangia bear more zoospores and can thus produce more potential offspring. *Pseudoperonospora. cubensis* on *Cu. moschata* had only 98 zoospores per sporangiophore, which, in addition to the sparse sporulation (only 0.2 sporangiophores per cm^2 leaf area), reflects the low susceptibility of the host. *Pseudoperonospora cubensis* on *C. sativus* occupied an intermediate position in almost all characters, but regarding sporangial dimensions and branch orders is in the upper third of the observed values. This leads to a possible number of released zoospores per sporangiophore of 224, which is below the maximum of 350 calculated for *C. anguria*, but with 263.3 sporangiophores per cm^2 leaf area markedly beyond the pathogen performance on the other hosts. The calculated result of 108.2 sporangia per mm^2 leaf area is in line with the results of Cohen & Eyal [27], who found a maximum of 102.9 sporangia per mm^2 (sporulation in darkness, leaf temperature 19.4°C) on *C. sativus* in their investigations. Considering this, *P. cubensis* has a possible mean offspring of 590.1 zoospores per mm^2 leaf area on *C. sativus*, whereas on *C. anguria* with 49.7 sporangiophores per cm^2 leaf area it has a mean possible offspring of only 173.8 zoospores per mm^2 leaf area. Although the highest sporulation density was on *S. angulatus* with 57.3 sporangiophores per mm^2 sporulating leaf area (*C. sativus* 21.4, *C. anguria* 8.8), the maximum number of zoospores per mm^2 leaf area is 216.8, due to the relatively low proportion of infected leaf area. In contrast, on *Ci. lanatus P. cubensis* has a mean potential offspring of 3.3 zoospores per mm^2 and on *Lu. cylindrica* a potential offspring of 16.6 zoospores per mm^2. The only host on which *P. cubensis* has a higher offspring than on *C. sativus* is *C. melo* with 686.6 zoospores per mm^2, resulting from sporulation similar in density, but higher in the amount of sporulating leaf area. These findings provide further evidence for the adaptation of the *P. cubensis* strain used in this study to its original host *C. sativus* and the closely related members of *Cucumis*. The estimation of the total amount of infective units per leaf area (Table 3, Figure 3) highlights that there is a huge difference in spore production on the different hosts. It is conceivable that under natural conditions, the infection potential on some hosts is so low that infections would very seldom be found in nature, which might explain the lack of reports of natural infections of *B. dioica* [35], as well as *Cy. pedata* and *T. dubia*.

Oospores could not be investigated in this study, because oospores in *P. cubensis* are usually rare. The only reported occurrence of oospores from *P. cubensis* in Central Europe is in leaves of greenhouse cucumbers in Austria [58]. Despite a field observation of oospores in leaves of cucumber in southern Germany (Runge, unpublished data), the occurrence of oospores seem to be very rare, and the strain used does not readily produce oospores under laboratory conditions and is thus possibly heterothallic. In contrast Cohen et al. [59] have recently reported the massive formation of oospores using two distinct races of *P. cubensis* from Israel. However, when present, oospores could be a useful character in description of species, which have so far not been used frequently for the delimitation of downy mildew species. However, in the genus *Albugo* oospores have been found to be the most important characteristic for species delimitation [15,16,19,60,61] and it could thus be promising to evaluate oospore characteristics for species delimitation in future studies on Peronosporaceae with wider host spectra. Candidates for this research are *Pseudoperonospora humuli*, which is also able to infect a variety of hosts under laboratory conditions [62], as well as *P. celtidis* [63] and *P. urticae*, as oospore production is abundant in these related pathogens. Taking into consideration the findings of Cohen et al. [59] that, provided the right compatible strains or mating types of *P. cubensis* are used, oospores could be formed abundantly, it could be taken into consideration to do further investigations on the relationship of the sister-species *P. humuli* and *P. cubensis* and the other phylogenetic lineages of the *P. cubensis* species cluster [21], based on oospore morphology.

Acknowledgments

Otmar Spring is gratefully acknowledged for providing laboratory and climate chamber space for conducting the experiments, and Aleš Lebeda for providing the laboratory strain of *Pseudoperonospora cubensis*.

Author Contributions

Conceived and designed the experiments: FR MT. Performed the experiments: FR BN. Analyzed the data: FR. Contributed reagents/materials/analysis tools: FR MT. Wrote the paper: FR MT.

References

1. Göker M, Voglmayr H, Riethmüller A, Oberwinkler F (2007) How do obligate parasites evolve? A multi-gene phylogenetic analysis of downy mildews. Fungal Genet Biol 44: 105–122.

2. Voglmayr H (2008) Progress and challenges in systematics of downy mildews and white blister rusts: New insights from genes and morphology. Eur J Plant Pathol 122: 3–18.

3. Thines M, Voglmayr H, Göker M (2009) Taxonomy and phylogeny of the downy mildews (Peronosporaceae). In: Lamour K, Kamoun S, editors. Oomycete Genetics and Genomics: Diversity, Interactions, and Research Tools. Weinheim: Wiley-VCH. pp. 47–75.

4. Runge F, Telle S, Ploch S, Savory E, Day B, et al. (2011) The inclusion of downy mildews in a multi-locus-dataset and its reanalysis reveals high degree of paraphyly in *Phytophthora*. IMA Fungus 2: 163–171.

5. Yerkes WD, Shaw CG (1959) Taxonomy of *Peronospora* species on Cruciferae and Chenopodiaceae. Phytopathology 49: 499–507.

6. Gäumann E (1918) Über die Formen der *Peronospora parasitica* (Pers.) Fries. Beih Biol Zentralblatt 35: 395–533.

7. Gäumann E (1923) Beiträge zu einer Monographie der Gattung *Peronospora* Corda. Beitr Kryptog Flora Schweiz 5: 1–360.

8. Choi YJ, Shin HD, Thines M (2009) Two novel *Peronospora* species are associated with recent reports of downy mildew on sages. Mycol Res 113: 1340–1350.

9. Choi YJ, Thines M, Runge F, Hong SB, Telle S, et al. (2011) Evidence for high degrees of specialisation, evolutionary diversity, and morphological distinctiveness in the genus *Bremia*. Fungal Biol 115: 102–111.

10. García-Blázquez G, Göker M, Voglmayr H, Martín MP, Tellería MT, et al. (2008) Phylogeny of *Peronospora*, parasitic on Fabaceae, based on ITS sequences. Mycol Res 112: 502–512.

11. Thines M (2011) Recent outbreaks of downy mildew on grape ivy (*Parthenocissus tricuspidata*, Vitaceae) in Germany are caused by a new species of *Plasmopara*. Mycol Prog 10: 415–422.

12. Thines M, Telle S, Ploch S, Runge F (2009) Identity of the downy mildew pathogens of basil, coleus, and sage with implications for quarantine measures. Mycol Res 113: 532–540.

13. Thines M, Runge F, Telle S, Voglmayr H (2010) Phylogenetic investigations in the downy mildew genus *Bremia* reveal several distinct lineages and a species with a presumably exceptional wide host range. Eur J Plant Pathol 128: 81–89.

14. Göker M, Voglmayr H, García-Blázquez G, Oberwinkler F (2009) Species delimitation in downy mildews: the case of *Hyaloperonospora* in the light of nuclear ribosomal ITS and LSU sequences. Mycol Res 113: 308–325.

15. Choi YJ, Shin HD, Hong SB, Thines M (2007) Morphological and molecular discrimination among *Albugo candida* materials infecting *Capsella bursa-pastoris* world-wide. Fungal Divers 27: 11–34.

16. Choi YJ, Shin HD, Ploch S, Thines M (2008) Evidence for uncharted biodiversity in the *Albugo candida* complex, with the description of a new species. Mycol Res 112: 1327–1334.

17. Choi YJ, Shin HD, Thines M (2009) The host range of *Albugo candida* extends from Brassicaceae over Cleomaceae to Capparaceae. Mycol Prog 8: 329–335.

18. Thines M, Choi YJ, Kemen E, Ploch S, Holub EB, et al. (2009) A new species of *Albugo* parasitic to *Arabidopsis thaliana* reveals new evolutionary patterns in white blister rusts (Albuginaceae). Persoonia 22: 123–128.

19. Ploch S, Choi YJ, Rost C, Shin HD, Schilling E, et al. (2010) Evolution of diversity in *Albugo* is driven by high host specificity and multiple speciation events on closely related Brassicaceae. Mol Phylogenet Evol 57: 812–820.

20. Choi YJ, Hong SB, Shin HD (2005) A re-consideration of *Pseudoperonospora cubensis* and *P. humuli* based on molecular and morphological data. Mycol Res 109: 841–848.

21. Runge F, Choi YJ, Thines M (2011) Phylogenetic investigations in the genus *Pseudoperonospora* reveal overlooked species and cryptic diversity in the *P. cubensis* species cluster. Eur J Plant Pathol 129: 135–146.

22. Sawada K (1931) Descriptive catalogue of the Formosan Fungi. Report of the Government Research Institute, Department of Agriculture, Formosa 51: 76–78.

23. Iwata Y (1942) Specialization in *Pseudoperonospora cubensis* (Berk. *et* Curt.) Rostov. II. Comparative studies of the morphologies of the fungi from *Cucumis sativus* L. and *Cucurbita moschata* Duchesne. Ann Phytopath Soc Japan 11: 172–185.

24. Iwata Y (1953) Specialization in *Pseudoperonospora cubensis* (Berk. *et* Curt.) Rostov. IV. Studies on the fungus from Oriental pickling melon (*Cucumis melo* var. *conomon* Makino). Bull Fac Agr, Mie Univ 1: 30–35.

25. Palti J, Cohen Y (1980) Downy mildew of cucurbits (*Pseudoperonospora cubensis*): the fungus and its hosts, distribution, epidemiology and control. Phytoparasitica 8: 109–147.

26. Waterhouse GM, Brothers MP (1981) The Taxonomy of *Pseudoperonospora*. Mycol Papers 148: 1–28.

27. Cohen Y, Eyal H (1977) Growth and differentiation of sporangia and sporangiophores of *Pseudoperonospora cubensis* on cucumber cotyledons under various combinations of light and temperature. Physiol Plant Pathol 10: 93–103.

28. Dudka IO, Anishchenko IM, Terent'eva NG (2007) The variability of *Peronospora alta* Fuckel conidia in dependence on the ecological conditions. In: Lebeda A, Spencer-Phillips PTN, editors. Advances in Downy Mildew Research (vol. 3). Kostelec na Hané: Palacký University in Olomouc and JOLA. pp. 39–46.

29. Delanoe D (1972) Biologie et Epidemiologie du mildiou du tournesol. CETIOM Inf Techn 26: 1–61.

30. Kulkarni S, Hegde YR, Kota RV (2009) Pathogenic and morphological variability of *Plasmopara halstedii*, the causal agent of downy mildew in sunflower. Helia 32: 85–90.

31. Salati M, Yun WM, Meon S, Masdek HN (2010) Host range evaluation and morphological characterization of *Pseudoperonospora cubensis*, the causal agent of cucurbit downy mildew in Malaysia. Afr J Biotechnol 9: 4897–4903.

32. Runge F, Thines M (2012) Re-Evaluation of host specificity of the closely related species *Pseudoperonospora humuli* and *P. cubensis*. Plant Dis 96: 55–61.

33. Runge F, Thines M (2011) Host matrix has major impact on the morphology of *Pseudoperonospora cubensis*. Eur J Plant Pathol 129: 147–156.

34. Choi YJ, Kiss L, Vajna L, Shin HD (2009) Characterization of a *Plasmopara* species on *Ambrosia artemisiifolia*, and notes on *P. halstedii*, based on morphology and multiple gene phylogenies. Mycol Res 113: 1127–1136.

35. Runge F, Thines M (2009) A potential perennial host for *Pseudoperonospora cubensis* in temperate regions. Eur J Plant Pathol 123: 483–486.

36. Mann H, Whitney D (1947) On a test of whether one of two random variables is stochastically larger than the other. Ann math Statistics 18: 50–60.

37. Spearman C (1904) The proof and measurement of association between two things. Am J Psychol 15: 72–101.

38. Runge F (2009) Morphologische und molekularbiologische Untersuchungen zum Wirtsspektrum von *Pseudoperonospora cubensis* und *Pseudoperonospora humuli*. Diploma thesis (in German), University of Hohenheim, Stuttgart, Germany.

39. Cohen Y (1977) The combined effects of temperature, leaf wetness, and inoculum concentration on infection of cucumbers with *Pseudoperonospora cubensis*. Can J Bot 55: 1478–1487.

40. Lebeda A, Widrlechner MP (2003) A set of Cucurbitaceae taxa for differentiation of *Pseudoperonospora cubensis* pathotypes. J Plant Dis Prot 110: 337–349.

41. Lebeda A (1991) Resistance in muskmelons to Czechoslovak isolates of *Pseudoperonospora cubensis* from cucumbers. Sci Hortic 45: 255–260.

42. Lebeda A, Widrlechner MP (2004) Response of wild and weedy *Cucurbita* L. to pathotypes of *Pseudoperonospora cubensis* (BERK. & CURT.) ROSTOV. (cucurbit downy mildew). In: Spencer-Phillips PTN, Jeger M, editors. Advances in downey mildew research (vol. 2). Dordrecht: Kluwer. pp. 203–210.

43. Lebeda A, Cohen Y (2011) Cucurbit downy mildew (*Pseudoperonospora cubensis*) – biology, ecology, host-pathogen interaction and control. Eur J Plant Pathol 129: 25–60.

44. Lebeda A, Widrlechner MP, Staub J, Ezura H, Zalapa J, et al. (2007) Cucurbits (Cucurbitaceae; *Cucumis* spp., *Cucurbita* spp., *Citrullus* spp.), Chapter 8. In: Singh R, editor. Genetic resources, chromosome engineering, and crop improvement series, volume 3 – vegetable crops. Boca Raton: CRC. pp. 273–377.

45. Lebeda A (1992) Screening of wild *Cucumis* species against downy mildew (*Pseudoperonospora cubensis*) isolates from cucumbers. Phytoparasitica 20: 203–210.

46. Lebeda A (1999) *Pseudoperonospora cubensis* on *Cucumis* spp. and *Cucurbita* spp. – resistance breeding aspects. Acta Hortic 492: 363–370.

47. Palti J (1975) *Pseudoperonospora cubensis*. CMI Descriptions of Pathogenic Fungi and Bacteria 682: 1–2.

48. Choi YJ, Shin HD (2008) First record of downy mildew caused by *Pseudoperonospora cubensis* on bottle gourd in Korea. Plant Pathol 57: 371.

49. Ko Y, Chen CY, Liu CW, Chen SS, Maruthasalam S, et al. (2008) First report of downy mildew caused by *Pseudoperonospora cubensis* on Chayote (*Sechium edule*) in Taiwan. Plant Dis 92: 1706.

50. Bains SS, Jhooty JS (1978) Relationship between mineral nutrition of muskmelon and development of downy mildew caused by *Pseudoperonospora cubensis*. Plant Soil 49: 85–90.

51. Miyabe K, Takahashi Y (1906) A new disease of hop-vine caused by *Peronoplasmopara humuli* n. sp. Trans Sapporo Nat Hist Soc 1: 149–157.

52. Choi YJ, Constantinescu O, Shin HD (2007) A new downy-mildew of the Rosaceae: *Peronospora oblatispora* sp. nov. (Chromista, Peronosporales). Nova Hedw 85: 93–101.

53. Choi YJ, Denchev CM, Shin HD (2008) Morphological and molecular analyses support the existence of host-specific *Peronospora* species infecting *Chenopodium*. Mycopathologia 165: 155–164.

54. Choi YJ, Shin HD, Voglmayr H (2011) Reclassification of two *Peronospora* species parasitic on *Draba* in *Hyaloperonospora* based on morphological molecular and phylogenetic data. Mycopathologia 171: 151–159.

55. Voglmayr H, Göker M (2011) Morphology and phylogeny of *Hyaloperonospora erophilae* and *H. praecox* sp. nov., two downy mildew species co-occurring on *Draba verna* sensu lato. Mycol Prog 10: 283–292.

56. Kocyan A, Zhang LB, Schaefer H, Renner SS (2007) A multi-locus chloroplast phylogeny for the Cucurbitaceae and its implications for character evolution and classifcation. Mol Phylogenet Evol 44: 553–577.

57. Schaefer H, Renner SS (2011) Phylogenetic relationships in the order Cucurbitales and a new classification of the gourd family (Cucurbitaceae). Taxon 60: 122–138.

58. Bedlan G (1989) Erstmaliger Nachweis von Oosporen von *Pseudoperonospora cubensis* (Berk. et Curt.) Rost. an Gewächshausgurken in Österreich. Pflanzenschutzberichte 3: 119–120.

59. Cohen Y, Rubin AE, Galperin M (2011) Formation and infectivity of oospores of *Pseudoperonospora cubensis*, the causal agent of downy mildew in cucurbits. Plant Dis 95: 874.

60. Thines M, Voglmayr H (2009) An Introduction to the White Blister Rusts (Albuginales). In: Lamour K, Kamoun S, editors. Oomycete Genetics and Genomics: Diversity, Interactions, and Research Tools. Weinheim: Wiley-VCH. pp. 77–92.

61. Thines M (2010) Evolutionary history and diversity of white blister rusts. Polish Bot J 55: 259–264.

62. Hoerner GF (1940) The infection capabilities of hop downy mildew. J Agr Res 61: 331–334.

63. Waite MB (1892) Description of Two New Species of *Peronospora*. J Mycol 7: 105–109.

14

Species Divergence and Phylogenetic Variation of Ecophysiological Traits in Lianas and Trees

Rodrigo S. Rios[1]**, Cristian Salgado-Luarte**[1]**, Ernesto Gianoli**[1,2]*

1 Departamento de Biología, Universidad de La Serena, La Serena, Chile, **2** Departamento de Botánica, Universidad de Concepción, Concepción, Chile

Abstract

The climbing habit is an evolutionary key innovation in plants because it is associated with enhanced clade diversification. We tested whether patterns of species divergence and variation of three ecophysiological traits that are fundamental for plant adaptation to light environments (maximum photosynthetic rate [A_{max}], dark respiration rate [R_d], and specific leaf area [SLA]) are consistent with this key innovation. Using data reported from four tropical forests and three temperate forests, we compared phylogenetic distance among species as well as the evolutionary rate, phylogenetic distance and phylogenetic signal of those traits in lianas and trees. Estimates of evolutionary rates showed that R_d evolved faster in lianas, while SLA evolved faster in trees. The mean phylogenetic distance was 1.2 times greater among liana species than among tree species. Likewise, estimates of phylogenetic distance indicated that lianas were less related than by chance alone (phylogenetic evenness across 63 species), and trees were more related than expected by chance (phylogenetic clustering across 71 species). Lianas showed evenness for R_d, while trees showed phylogenetic clustering for this trait. In contrast, for SLA, lianas exhibited phylogenetic clustering and trees showed phylogenetic evenness. Lianas and trees showed patterns of ecophysiological trait variation among species that were independent of phylogenetic relatedness. We found support for the expected pattern of greater species divergence in lianas, but did not find consistent patterns regarding ecophysiological trait evolution and divergence. R_d followed the species-level pattern, i.e., greater divergence/evolution in lianas compared to trees, while the opposite occurred for SLA and no pattern was detected for A_{max}. R_d may have driven lianas' divergence across forest environments, and might contribute to diversification in climber clades.

Editor: Sylvain Delzon, INRA - University of Bordeaux, France

Funding: This study was funded by the FONDECYT grant 1100585 (http://www.conicyt.cl/fondecyt/). FONDECYT is the Chilean Agency of Research Funding. The funders had no role in study design, data collection and analysis, decision to publish, or preparation of the manuscript.

Competing Interests: The authors have declared that no competing interests exist.

* E-mail: egianoli@userena.cl

Introduction

Climbing plants, in particular woody vines (lianas), are a distinctive component of mature forests in both tropical and temperate regions [1–3]. Data from long-term plots indicate that the dominance of lianas relative to trees is increasing in tropical forests [4,5]. Moreover, liana abundance is negatively associated with tree carbon storage in tropical forests [6,7]. The climbing habit has independently arisen numerous times throughout plant evolution [1,8], and it seems to be a key innovation in angiosperms: climbing plant lineages have greater species richness than their non-climbing sister groups [9]. Thus, evidence from both ecological and macroevolutionary patterns suggests a performance advantage of lianas over trees.

Explanatory factors for the increased abundance and biomass of lianas in tropical forests include increasing forest disturbance, which increases local resource availability, and rising levels of atmospheric CO_2 [4]. Moreover, increased abundance of lianas in seasonal forests during the dry season, as compared to trees, has been related to their increased efficiency in water uptake and transport, and higher photosynthetic rates ([10–12]; but see [13]). Thus, data suggest that lianas are better than trees at exploiting resource pulses. When providing functional arguments for the key innovation of the climbing habit (*sensu* [14]), Gianoli [9] suggested that ecological specialization may arise as a consequence of an hypothetically expanded light niche of lianas in the forest, which would result from the co-occurrence of unsupported (creeping) and supported (climbing) individuals that go up and down the forest canopy. This would maximize interactions with a wide array of antagonistic and mutualistic species [15,16] that, in turn, might promote diversification [17]. It is increasingly recognized that purported evolutionary key innovations may be tested at an ecological time scale [14,18–20].

Ecophysiological traits are fundamental components of plant adaptation to the environment [21,22]. Specifically, A_{max} (maximum photosynthetic rate), R_d (dark respiration rate) and SLA (specific leaf area) play a key role in the phenotypic adjustment to heterogeneous light environments in both lianas and trees [23–26]. Thus, they reflect the balance between carbon gain (A_{max}) and carbon use (R_d), and the allocation of leaf biomass to light interception (SLA), which together determine plant growth and performance across light environments [27,28]. Importantly, variation in plant functional traits observed at the population level is likely to be paralleled by evolutionary divergences under contrasting environments [29]. Moreover, the analysis of the phylogenetic structure of communities can provide insights to our understanding of trait evolution [30]. Recent studies have addressed phylogenetic variation in ecophysiological traits in climbing plants and trees [31–34], but their approach has been either exploratory (aiming to report global patterns) or methodological (testing new analytical tools); to our knowledge, a hypothesis-driven analysis is wanting.

Using data reported for several liana and tree species coexisting in tropical and temperate forests, and focusing on three key ecophysiological traits involved in plant adaptation across forest light gradients: A_{max}, R_d and SLA, we herein compare lianas and trees in terms of trait evolutionary rates, phylogenetic diversity, phylogenetic trait diversity, and the phylogenetic signal. Thus, we compared the rate at which variance in the traits is accumulated among species per unit time at the tips of the phylogenetic tree [35,36]. We also evaluated how similar is the average pair of species of lianas and trees both in terms of mean phylogenetic distance and trait variation [37]. We finally evaluated in lianas and trees the tendency for phylogenetically related species to resemble each other, i.e., the phylogenetic signal [38]. We tested the hypotheses that if the climbing habitat enhances clade diversification [9], and ecological divergence is the process underlying this pattern, then lianas should show higher trait evolutionary rates and greater species and trait divergence than trees under common environmental scenarios.

Materials and Methods

Data collection

We searched the literature for field studies in forest ecosystems where lianas and trees were analyzed for at least one of three ecophysiological traits: A_{max} on an area basis, R_d, and SLA. We only chose those studies carried out in mature forests, where light heterogeneity across microsites is the greatest [3]. We only included native species because they have a long history of adaptation to the environment. We focused on angiosperms because of the availability of tools to reconstruct their phylogenetic history and estimate trait evolution (see below). The final data set included a pool of 63 liana species and 71 tree species belonging to four tropical forests (Gamboa, Panama; San Lorenzo, Panama; Riberalta, Bolivia; Xishuangbanna, China) and three temperate forests (Yakushidake, Japan; Beltsville, USA; Puyehue, Chile). We pooled species from all sites into growth forms, thus we had one liana "super-community" and one tree "super-community". Phylogenetic analyses were conducted on these super-communities (see below). Detailed information, including study species, traits, sites, and data sources, is available in Supporting Information S1.

Phylogeny reconstruction

We produced a phylogeny of all species using a backbone tree based on the angiosperm megatree provided by the Phylodiversity Network in cooperation with the Angiosperm Phylogeny Group (APG; http://www.mobot.org/MOBOT/research/APweb/). Our tree was generated using Phylomatic (http://www.phylodiversity.net/phylomatic/phylomatic.html), a program that returns a working phylogenetic tree after matching the genus and family names of study species to those contained in the angiosperm phylogeny [39]. Comparative inferences require branch lengths for the tree, which were calculated based on the branch length adjustment algorithm (BLADJ) implemented in Phylocom v. 4.2 (www.phylodiversity.net/phylocom) [40]. This algorithm fixes a subset of nodes in the tree to specified ages and evenly distributes the ages to the remaining nodes. Age estimates for major nodes in our tree were taken from [41]. To avoid inaccuracies in tree calibration and to have an updated version of our tree, we corrected the *ages* file with age estimates in [41] included in Phylocom. Corrections followed procedures suggested recently [42]. We also checked and updated age estimates of internal order-level clades according to a net diversification rate estimate of

angiosperms [43]. The few polytomies in the working tree were resolved randomly using the *multi2di* function in R. Values of functional traits of closely related species resulting from such random resolutions were very similar, so results of the final comparative tests were highly robust to topological uncertainty. These and all subsequent analyses were conducted using the R statistical environment version 3.0.2 [44]. Reconstructed phylogenetic trees with associated trait variation are shown in Figures 1–3.

Rate of trait evolution

To assess differences in ecophysiological trait evolution between liana and tree species, we compared estimates of evolutionary rate for A_{max}, R_d, and SLA. To this end, we first used stochastic character mapping, a Bayesian method that uses Monte Carlo simulations to sample the posterior probability distribution of ancestral states and timings of transitions on phylogenetic branches under a Markov process of evolution [45,46]. We built stochastic character-mapped reconstructions for each trait/growth form combination using the *make.simmap* function in the phytools package of R [36]. We thus simulated character history evolution of all three traits in relation to growth form as an initial step, as suggested recently [35,47]. To test the hypothesis that a discrete character state had influenced the rate of a continuous character, one should first stochastically map the discrete trait (e.g., climbing habit), and then test if one state of the discrete character has a different evolutionary rate for the continuous trait of interest (e.g., ecophysiological trait) than the other discrete state [35,47].

The resulting reconstructions of trait states and phylogeny represented a set of phylogenetic topologies, branch lengths and growth forms sampled in proportion to their posterior probabilities. Reconstructions were then used in subsequent analyses as a way of integrating over uncertainty in phylogeny and ancestral states. Finally, we fitted the evolutionary models of character history on the trees to trait data using a likelihood method [35]. This is a maximum likelihood approach that estimates rates of evolution (σ^2). The parameter σ^2 was calculated using the function *brownie.lite* in the *phytools* package [36]. σ^2 is interpreted as the Brownian motion process most likely to have produced the data at the tips of the tree, i.e., the rate at which variance is accumulated among species per unit time. 95% confidence intervals were calculated for each σ^2 to infer differences between lianas and trees in ecophysiological trait evolution.

Trait and phylogenetic diversity

To compare the phylogenetic relatedness among liana species against the phylogenetic relatedness among tree species we used measures of phylogenetic structure. Specifically, we calculated, based on a phylogenetic distance matrix, the mean phylogenetic distance (MPD) and the standardized effect size of the mean phylogenetic distance (SES_{MPD} [48]) between pairs of species for each group. Interspecific phylogenetic distance matrices were obtained from the reconstructed tree of phylogenetic relationships among taxa using the *cophenetic* function in R and unweighted pair-group average (UPGMA) as the clustering method.

Standardized effect sizes describe the difference between average phylogenetic distances in the observed super communities or groups (lianas and trees) compared to null distributions generated for each group with randomization procedures, standardized by the standard deviation of phylogenetic distances in the null data [48]. We compared observed mean distances (branch length) against a null model generated by calculating 999 times the mean phylogenetic distance between 8911 random pairs of species (without replacement) drawn from the matrix of

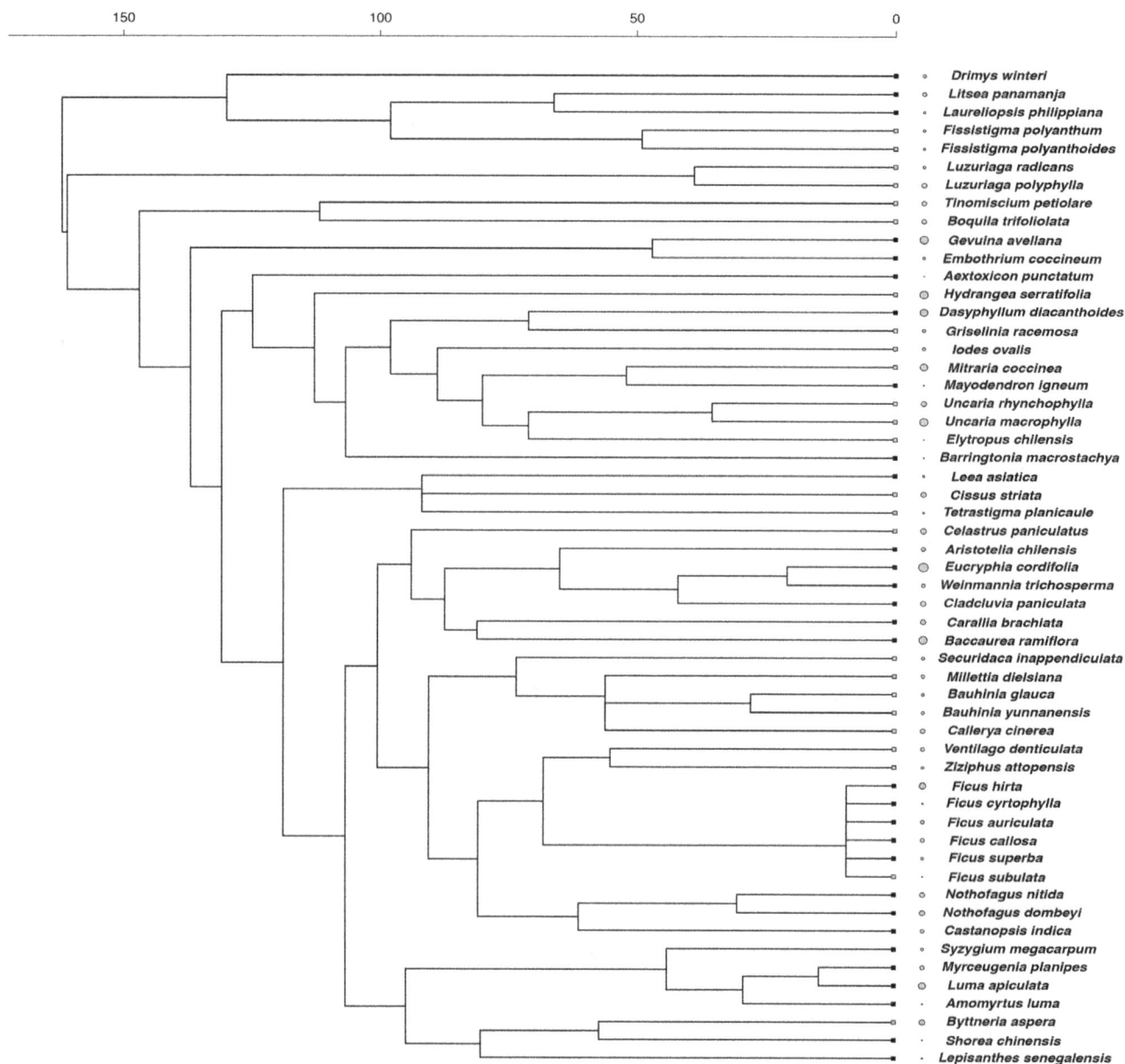

Figure 1. Phylogenetic relationships among tree and liana species and species values of dark respiration rate (R$_d$). Grey circle size represents the proportional magnitude of the trait across species. Square tip symbols represent climbing habit (grey squares = liana, black squares = tree). Timescale is in millions of years before present.

phylogenetic distances between all liana and tree species. In all cases 999 iterations were found to be suitable for our randomization procedures as they were sufficient to attain convergence. The null model was constructed by reshuffling the distance of species labels across the phylogenetic tree using the *ses.mpd* function and the *taxa.labels* algorithm of the *picante* package of R. Positive values of SES$_{MPD}$ (*mpd.obs.z*) and high quantiles (*p*-values >0.95) indicate significant phylogenetic evenness, while negative values of SES$_{MPD}$ and low quantiles (*p*-values <0.05) indicate significant phylogenetic clustering [48]; these outcomes correspond to scenarios where species are more distantly or more closely related than expected by chance, respectively [48,49]. Authors often refer to (weak) evenness or clustering when *p*-values are slightly lower than 0.95 or slightly higher than 0.05, respectively (e.g., [50,51]). Finally, to assess how similar are the

average pair of species within each group in terms of ecophysiological traits; we used the SES$_{MPD}$ as a trait diversity measure. This was done by replacing the phylogenetic distance matrix in the analysis with a trait distance matrix, and proceeding accordingly to calculate standardized values of mean phylogenetic trait distance (SES$_{MTD}$). These results are interpreted in the same way as those of SES$_{MPD}$ with regard to phylogenetic evenness or clustering [48].

Phylogenetic signal

To quantify the degree to which phylogenetic relatedness predicts the similarity of species in functional traits for both trees and lianas, we calculated separately phylogenetic signal for A$_{max}$, R$_d$, and SLA. Phylogenetic signal indicates to what extent phenotypic expression is explained by the lineage to which a

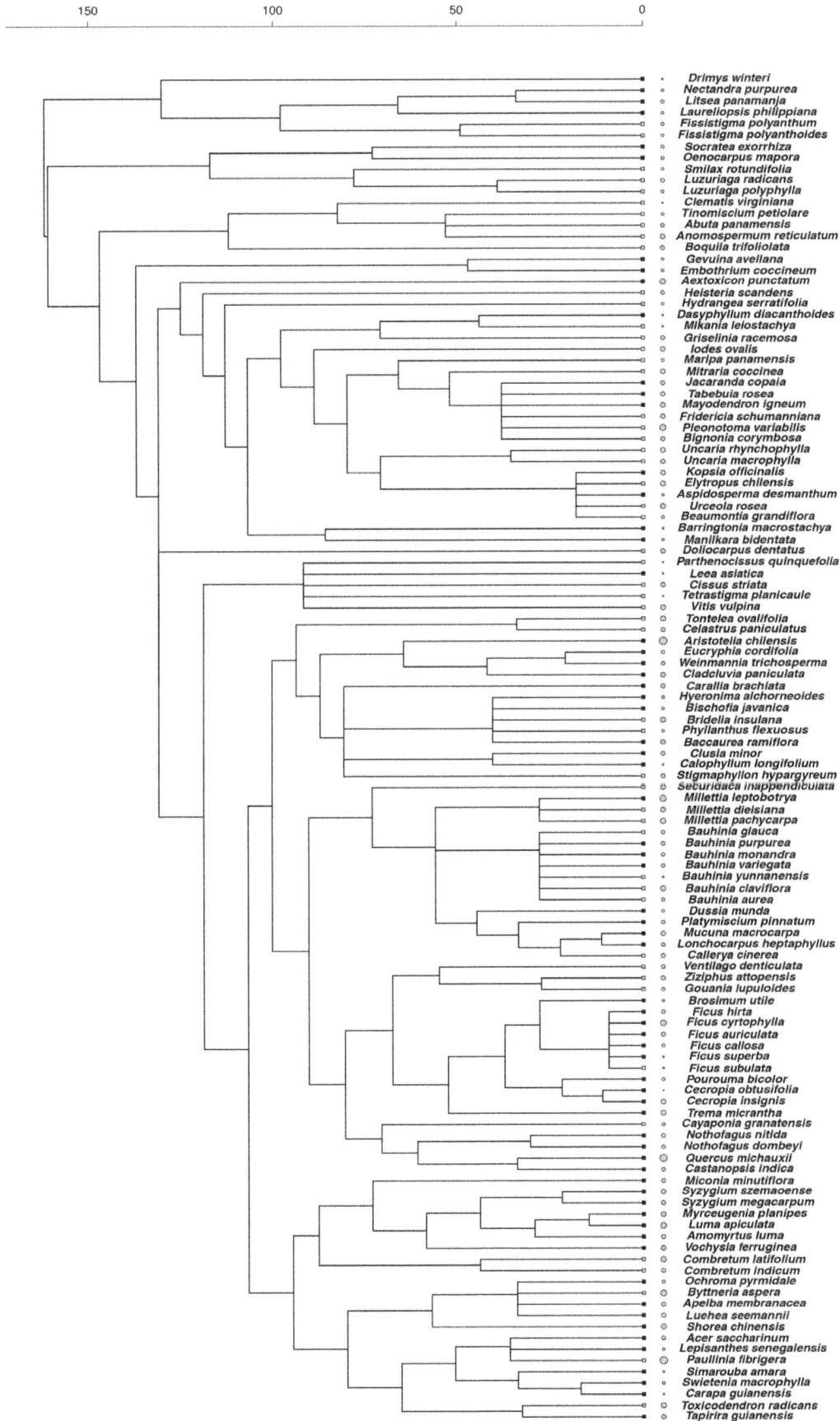

150   100   50   0

Drimys winteri
Nectandra purpurea
Litsea panamanja
Laureliopsis philippiana
Fissistigma polyanthum
Fissistigma polyanthoides
Socratea exorrhiza
Oenocarpus mapora
Smilax rotundifolia
Luzuriaga radicans
Luzuriaga polyphylla
Clematis virginiana
Tinomiscium petiolare
Abuta panamensis
Anomospermum reticulatum
Boquila trifoliolata
Gevuina avellana
Embothrium coccineum
Aextoxicon punctatum
Heisteria scandens
Hydrangea serratifolia
Dasyphyllum diacanthoides
Mikania leiostachya
Griselinia racemosa
Iodes ovalis
Maripa panamensis
Mitraria coccinea
Jacaranda copaia
Tabebuia rosea
Mayodendron igneum
Fridericia schumanniana
Pleonotoma variabilis
Bignonia corymbosa
Uncaria rhynchophylla
Uncaria macrophylla
Kopsia officinalis
Elytropus chilensis
Aspidosperma desmanthum
Urceola rosea
Beaumontia grandiflora
Barringtonia macrostachya
Manilkara bidentata
Doliocarpus dentatus
Parthenocissus quinquefolia
Leea asiatica
Cissus striata
Tetrastigma planicaule
Vitis vulpina
Tontelea ovalifolia
Celastrus paniculatus
Aristotelia chilensis
Eucryphia cordifolia
Weinmannia trichosperma
Cladcluvia paniculata
Carallia brachiata
Hyeronima alchorneoides
Bischofia javanica
Bridelia insulana
Phyllanthus flexuosus
Baccaurea ramiflora
Clusia minor
Calophyllum longifolium
Stigmaphyllon hypargyreum
Securidaca inappendiculata
Millettia leptobotrya
Millettia dielsiana
Millettia pachycarpa
Bauhinia glauca
Bauhinia purpurea
Bauhinia monandra
Bauhinia variegata
Bauhinia yunnanensis
Bauhinia claviflora
Bauhinia aurea
Dussia munda
Platymiscium pinnatum
Mucuna macrocarpa
Lonchocarpus heptaphyllus
Callerya cinerea
Ventilago denticulata
Ziziphus attopensis
Gouania lupuloides
Brosimum utile
Ficus hirta
Ficus cyrtophylla
Ficus auriculata
Ficus callosa
Ficus superba
Ficus subulata
Pourouma bicolor
Cecropia obtusifolia
Cecropia insignis
Trema micrantha
Cayaponia granatensis
Nothofagus nitida
Nothofagus dombeyi
Quercus michauxii
Castanopsis indica
Miconia minutiflora
Syzygium szemaoense
Syzygium megacarpum
Myrceugenia planipes
Luma apiculata
Amomyrtus luma
Vochysia ferruginea
Combretum latifolium
Combretum indicum
Ochroma pyrmidale
Byttneria aspera
Apeiba membranacea
Luehea seemannii
Shorea chinensis
Acer saccharinum
Lepisanthes senegalensis
Paullinia fibrigera
Simarouba amara
Swietenia macrophylla
Carapa guianensis
Toxicodendron radicans
Tapirira guianensis

Figure 2. Phylogenetic relationships among tree and liana species and species values of maximum photosynthetic rate (A_{max}). Grey circle size represents the proportional magnitude of the trait across species. Square tip symbols represent climbing habit (grey squares = liana, black squares = tree). Timescale is in millions of years before present.

species belongs, and it can be compared among clades and among traits [52]. We quantified phylogenetic signal using both Blomberg's K [38] and Pagel's λ [53] statistics for quantitative traits. To calculate these parameters, we first pruned two separate phylogenies, one for the group of lianas and one for the group of trees, using the original tree as a base phylogeny. Then we pruned a tree for each group-trait combination independently, removing taxa for which trait information was not available. The number of species included in each trait/plant growth habit analysis ranged from 26 (R_d/lianas) to 67 (A_{max}/trees), thus meeting the N>20 threshold to achieve good statistical power [38].

Values of $K = 1$ imply that a trait shows exactly the amount of phylogenetic signal expected under a null, stochastic model of character evolution (Brownian motion evolution) [38]. K-values > 1 and <1 imply that close relatives are more similar and less similar, respectively, than expected under a Brownian motion model of trait evolution [38]. If K does not differ from zero it is concluded that the trait has no phylogenetic signal. Statistical significance of K [38] was assessed via permutation tests with 1000 randomizations. The significance of the phylogenetic signal was based on the variance of phylogenetically independent contrasts relative to tip shuffling randomization implemented by the *phylosignal* function of the *picante* package in R [48]. P-values were determined by comparing the variance of standardized independent contrasts for the tip values against variances for randomized data.

The parameter λ scales tree structure in terms of expected variances and covariances in trait change [54]. Thus, λ is a phylogenetic transformation that maximizes the likelihood of the data given a Brownian motion model [54]. When $\lambda = 1$, the trait is consistent with a Brownian motion evolution based on branch lengths represented by the variance-covariance in trait change. Values between 0 and 1 indicate less phylogenetic signal than expected under a Brownian motion model, while values >1 indicate more signal than expected, although λ is not always defined for values greater than one [54]. Values of λ were estimated using the *fitContinuous* function of the *geiger* package. To determine the significance of λ as an indicator of phylogenetic signal, we compared the maximum likelihood estimate of λ against the maximum likelihood of models when $\lambda = 1$ using likelihood ratio tests (LRT).

Results and Discussion

Rate of trait evolution

In general, lianas and trees presented homogenous evolution of ecophysiological traits. In all cases evolutionary rates, as estimated by σ^2, were not significantly different from a single-rate Brownian motion process of evolution (Table 1). Parameter estimate values of σ^2, however, did differ between lianas and trees in two of the three ecophysiological traits considered (Table 1). The evolutionary rate for dark respiration rate (R_d) in lianas was 1.8 times greater than in trees. In the case of the biomass allocation trait (specific leaf area, SLA), the evolutionary rate was 1.2 times greater in trees than in lianas. Evolutionary rates for maximum photosynthetic rate (A_{max}) did not differ between lianas and trees; overall, this trait showed the lowest evolutionary rate among the traits considered (Table 1).

The patterns observed suggest that for all ecophysiological traits a change along any given branch in the phylogeny is independent of both previous changes and changes in other branches of the reconstructed tree. Evolutionary rates in both lianas and trees did not differ from a single-rate Brownian motion model of evolution, which assumes that variance among species in the phylogenetic tree accumulates as function of their time of independent evolution [55]. Thus, it cannot be ruled out that ecophysiological traits evolve at a constant rate over time. A Brownian motion process, however, is not equal to a neutral model of evolution. Brownian motion simply describes the distribution of observed trait changes and may be consistent with adaptive models of evolution [35,56]. Therefore, natural selection could be a plausible force behind the alteration in rate change of traits in relation to growth form (climbers vs. non-climbers).

Evolutionary rates (σ^2) differed between lianas and trees in two of the three ecophysiological traits considered, but in opposite trends. Thus, R_d evolved at a higher rate in lianas, while SLA evolution occurred at a higher rate in trees. This suggests that the outcome of modifications in the selective regime related to the climbing habit depends on the particular plant traits that are under selection (gas-exchange traits vs. biomass allocation traits). Gas-exchange traits have been shown to be of selective value for the exploitation of light availability in mature forests for trees [57], vines [58] and ferns [59]. Our findings suggest that climbers are more evolutionary responsive with regard to R_d than trees. Assuming that (adaptive) ecological speciation is the process behind species divergence in this trait [20], the next step would be to address whether this results from a greater magnitude of selection on R_d or from greater trait heritability [21,60]. Conversely, SLA showed a higher greater evolutionary rate across tree species. This somewhat supports the view of SLA as an essential attribute for tree performance and carbon gain [57,61].

Trait and phylogenetic diversity

We found that mean phylogenetic distance (MPD, non-standardized values) was greater among liana species (259.9 Myr) than among tree species (229.6 Myr). Moreover, there was a clear-cut difference between lianas and trees in the standardized mean phylogenetic distance among species (SES_{MPD}). Whereas lianas showed greater distances between species relative to the null model ($SES_{MPD} = 2.271$; p-value = 0.99), i.e., phylogenetic evenness, trees showed a pattern of phylogenetic clustering ($SES_{MPD} = -3.622$; p-value = 0.006).

Lianas and trees differed in their patterns of trait diversity. For one of the gas-exchange traits (R_d), lianas showed phylogenetic evenness ($SES_{MTD} = 1.266$, p-value = 0.893), which means that trait dissimilarity among liana species was higher than expected by chance, while trees showed phylogenetic clustering ($SES_{MTD} = -1.863$, p-value = 0.039), indicating that tree species were more phenotypically similar than expected by chance (Figure 1). In contrast, for the biomass allocation trait (SLA), lianas exhibited phylogenetic clustering ($SES_{MTD} = -1.194$, p-value = 0.122) and trees showed phylogenetic evenness ($SES_{MTD} = 1.193$, p-value = 0.877) (Figure 2). Finally, the other gas-exchange trait, A_{max}, did not show phylogenetic structure in both lianas ($SES_{MTD} = 0.096$, p-value = 0.536) and trees ($SES_{MTD} = -0.598$, p-value = 0.277) (Figure 3).

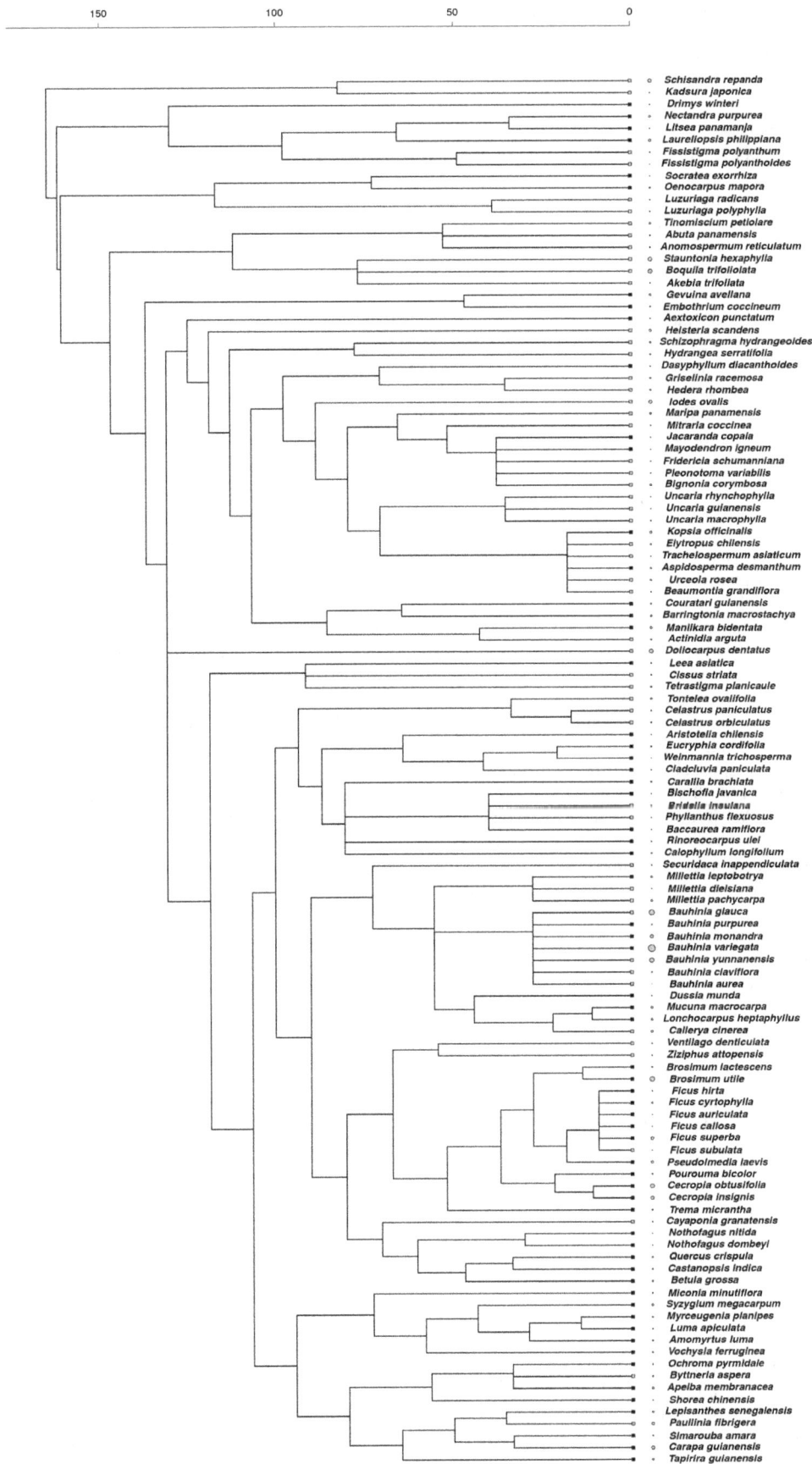

150   100   50   0

Schisandra repanda
Kadsura japonica
Drimys winteri
Nectandra purpurea
Litsea panamanja
Laureliopsis philippiana
Fissistigma polyanthum
Fissistigma polyanthoides
Socratea exorrhiza
Oenocarpus mapora
Luzuriaga radicans
Luzuriaga polyphylla
Tinomiscium petiolare
Abuta panamensis
Anomospermum reticulatum
Stauntonia hexaphylla
Boquila trifoliolata
Akebia trifoliata
Gevuina avellana
Embothrium coccineum
Aextoxicon punctatum
Heisteria scandens
Schizophragma hydrangeoides
Hydrangea serratifolia
Dasyphyllum diacanthoides
Griselinia racemosa
Hedera rhombea
Iodes ovalis
Maripa panamensis
Mitraria coccinea
Jacaranda copaia
Mayodendron igneum
Fridericia schumanniana
Pleonotoma variabilis
Bignonia corymbosa
Uncaria rhynchophylla
Uncaria guianensis
Uncaria macrophylla
Kopsia officinalis
Elytropus chilensis
Trachelospermum asiaticum
Aspidosperma desmanthum
Urceola rosea
Beaumontia grandiflora
Couratari guianensis
Barringtonia macrostachya
Manilkara bidentata
Actinidia arguta
Doliocarpus dentatus
Leea asiatica
Cissus striata
Tetrastigma planicaule
Tontelea ovalifolia
Celastrus paniculatus
Celastrus orbiculatus
Aristotelia chilensis
Eucryphia cordifolia
Weinmannia trichosperma
Cladcluvia paniculata
Carallia brachiata
Bischofia javanica
Bridelia insulana
Phyllanthus flexuosus
Baccaurea ramiflora
Rinoreocarpus ulei
Calophyllum longifolium
Securidaca inappendiculata
Millettia leptobotrya
Millettia dielsiana
Millettia pachycarpa
Bauhinia glauca
Bauhinia purpurea
Bauhinia monandra
Bauhinia variegata
Bauhinia yunnanensis
Bauhinia claviflora
Bauhinia aurea
Dussia munda
Mucuna macrocarpa
Lonchocarpus heptaphyllus
Callerya cinerea
Ventilago denticulata
Ziziphus attopensis
Brosimum lactescens
Brosimum utile
Ficus hirta
Ficus cyrtophylla
Ficus auriculata
Ficus callosa
Ficus superba
Ficus subulata
Pseudolmedia laevis
Pourouma bicolor
Cecropia obtusifolia
Cecropia insignis
Trema micrantha
Cayaponia granatensis
Nothofagus nitida
Nothofagus dombeyi
Quercus crispula
Castanopsis indica
Betula grossa
Miconia minutiflora
Syzygium megacarpum
Myrceugenia planipes
Luma apiculata
Amomyrtus luma
Vochysia ferruginea
Ochroma pyrmidale
Byttneria aspera
Apeiba membranacea
Shorea chinensis
Lepisanthes senegalensis
Paullinia fibrigera
Simarouba amara
Carapa guianensis
Tapirira guianensis

Figure 3. Phylogenetic relationships among tree and liana species and species values of specific leaf area (SLA). Grey circle size represents the proportional magnitude of the trait across species. Square tip symbols represent climbing habit (grey squares = liana, black squares = tree). Timescale is in millions of years before present.

First, in agreement with the study hypotheses, we found greater phylogenetic distance among species within the group of lianas (63 species) than within the group of trees (71 species). This agrees with a recent study in Australian rainforests, where standardized values of mean phylogenetic distance indicated that climbers show weak evenness or no phylogenetic structure, while trees/shrubs show weak to significant phylogenetic clustering [50]. Second, the average phenotypic distance among species for one gas-exchange trait (but not for the other two ecophysiological traits) was greater in the phylogenetic tree of lianas than in that of trees. The environmental gradient experienced by the study species was roughly the same for lianas and trees because data were obtained from sites where trees and lianas coexisted. Therefore, these patterns of (partial) increased phenotypic distance and greater phylogenetic divergence in lianas are consistent with the notion that lianas have a greater differentiation potential than trees [9]. Several plant attributes have been associated with evolutionary rates in angiosperms. For instance, it has been shown that trees and shrubs have lower rates of molecular evolution than herbaceous plants [62,63], and that taller plants have lower rates of molecular evolution [64]; in both cases the outcome is thought to be linked to differences in generation time, which in turn is related to mutation rate. In our study all climbers were woody species (lianas) so herbaceousness should not be a confounded factor. However, there is no available information to reject the possibility that there were longer generation times in the group of trees (see General Conclusions). As to the plant height factor, it is a rather problematic issue, because trees are usually taller than lianas in terms of freestanding height, but if total length is considered, then canopy lianas may be taller. Both issues deserve further scrutiny.

In the realm of community phylogenetics, patterns of phylogenetic evenness in resource-use traits are often interpreted to reflect niche differentiation processes [49]. If trait-based niche differentiation facilitates evolutionary responses to divergent selection, which in turn may lead to incipient speciation [65], then our results of phylogenetic evenness in a liana ecophysiological trait might be linked to the ecological/evolutionary processes that underlie the key innovation of the climbing habit in plants [9]. Whereas results of trait phylogenetic distance suggest that R_d may have played a role as driver of lianas' adaptive divergence, SLA showed greater phenotypic divergence among tree species, as was shown for rates of trait evolution (see above). However, this trait distribution pattern across the phylogenetic tree was not accompanied by an overall greater phylogenetic distance among tree species. This might be interpreted as SLA contributing to tree adaptation to environmental challenges at local scales but do not driving taxonomic divergence across clades.

Phylogenetic signal

Overall, lianas and trees presented mixed but comparable patterns of phylogenetic signal (or lack thereof) in ecophysiological traits (Table 2). In lianas, A_{max} showed no significant phylogenetic signal, but values were lower than expected under a Brownian model of evolution (with $K<1$ and $\lambda<1$). R_d showed mixed results, with significant phylogenetic signal indicated by K, and lower than expected under a Brownian model of evolution, but no significant signal as indicated by λ ($=1$). There was no phylogenetic signal detected for specific leaf area (SLA) using K but strong signal using

λ, and lower than expected under a Brownian model of evolution (Table 2). In trees, whereas no phylogenetic signal was found for both A_{max} and R_d as indicated by K and λ, a significant phylogenetic signal was found for SLA when K is considered (with $K<1$) but no signal was detected by λ, with values lower than expected under a Brownian model of trait evolution ($\lambda\approx0$) (Table 2). Summarizing, in all cases both lianas and trees tended to show patterns of ecophysiological trait variation among species that were independent of phylogenetic relatedness.

Our results are consistent with the general pattern that physiological traits tend to show low values of phylogenetic signal [38]. A global analysis of trait variation in climbing plants reported that SLA showed no phylogenetic signal [34], as found in the present study. Likewise, in agreement with our results, a global-scale study in Angiosperms reported that A_{max} (on an area basis) showed no consistent phylogenetic signal [32]. Conversely, a genus-level study in trees [31] found that A_{max} exhibited significant phylogenetic signal, which seemingly opposes our findings. However, this study used a metric other than Blomberg's K and Pagel's λ, and given that different indices of phylogenetic signal often lead to contrasting outcomes ([68]; and Table 2), these results are not necessarily contrary to those reported here. Another group of ecophysiological traits that could have been studied to seek phylogenetic and evolutionary differences between trees and lianas is that of hydraulic characters. Regarding hydraulic traits, lianas have wider and longer vessels compared to trees, features that enable them to supply a large leaf area with a relatively small allocation to xylem tissue [33,66,67]. However, xylem vessel length did not show significant phylogenetic signal in a recent global analysis including lianas, shrubs and trees [33].

Results indicate that, in both lianas and trees, ecophysiological traits related to light use and carbon economy have undergone evolutionary trajectories different to those expected after phylogenetic relationships, assuming a Brownian motion model of trait evolution [68]. These phylogenetic signal results do not match the patterns of trait divergence and trait evolutionary rates found here. Although under some circumstances (e.g., fluctuating selection in related lineages) a negative association between K and evolutionary rate may be found [52,55], it is generally considered that changes in trait evolutionary rates −and ensuing phenotypic divergence− does not influence phylogenetic signal for continuous characters [52,55].

General conclusions

Lianas and trees differ in a number of anatomical, physiological, morphological and life history traits [69–71]. Among the main differences, trees show a greater allocation of biomass (and carbon) to stems and lianas have lower costs of height gain and larger total leaf area potential. Moreover, compared to shrubs and trees, lianas have lower leaf mass per area (LMA, the inverse of SLA), higher foliar N and higher mass-based photosynthetic rate, which is consistent with the characterization of lianas as fast metabolism/ rapid turnover species [71]. This could be related to hypothetical differences in generation time between lianas and trees that could explain their differential evolutionary rates, as shown here. Nonetheless, when it comes to explain species distribution across the light gradient in forests [72], the life history trade-off between juvenile growth and survival is observed alike in trees and lianas [73].

Table 1. Parameter estimates and 95% confidence intervals (CI) of the evolutionary rate (σ^2) for ecophysiological traits in lianas and trees.

Trait	Lianas				Trees			
	σ^2	95% CI	χ^2	p-value	σ^2	95% CI	χ^2	p-value
A_{max}	0.075	0.072–0.079	−6.8	0.948	0.075	0.072–0.078	−10.6	0.995
R_d	**0.148**	0.132–0.165	−4.1	0.999	**0.083**	0.076–0.091	−13.7	0.991
SLA	**0.093**	0.087–0.097	−11.8	0.999	**0.103**	0.099–0.108	−11.2	0.999

A_{max} = maximum photosynthetic rate; R_d = dark respiration rate; SLA = specific leaf area. Values of σ^2 represent rates at which variance accumulates among species per unit time through a phylogeny with branch lengths in units of millions of years. P-values are for likelihood ratio tests against the chi-square distribution between a single-rate (homogenous) and a heterogeneous Brownian motion process. Bold cells indicate σ^2 values for which 95% CI do not overlap between lianas and trees.

Table 2. Phylogenetic signal, quantified as Blomberg's K and Pagel's λ, for three ecophysiological traits in lianas and trees.

Trait	Lianas				Trees			
	K	p-value	λ	p-value	K	p-value	λ	p-value
A_{max}	0.518	0.157	0.338	0.281	0.308	0.459	0.223	0.503
R_d	**0.874**	0.015	1.000	0.096	0.284	0.573	0.125	0.713
SLA	0.511	0.166	**<0.001**	<0.001	**0.638**	0.001	0.096	0.282

A_{max} = maximum photosynthetic rate; R_d = dark respiration rate; SLA = specific leaf area. Significant phylogenetic signals are shown in bold.

Phylogenetic information is increasingly used to test macroevolutionary hypotheses of trait evolution [74–76]. The study hypotheses, arising from the macroevolutionary pattern of increased taxonomic diversification in lianas [9], received mixed support. Overall, mean phylogenetic distance among liana species was larger than that of trees. Lianas showed a higher evolutionary rate for a gas-exchange trait (R_d), but the biomass allocation trait (SLA) evolved at a higher rate in trees. Likewise, average trait divergence across the phylogenetic tree was greater in lianas for R_d but it was greater in trees for SLA. Therefore, although we have found support for the expected pattern of increased species divergence in lianas compared to trees, we did not find consistent patterns regarding ecophysiological trait evolution and divergence. R_d followed the species-level patterns, i.e., greater divergence/evolution in lianas compared to trees, while the opposite was found for SLA. R_d may have driven lianas' divergence across forest environments and, furthermore, might contribute to the pattern of increased diversification in climber clades.

Acknowledgments

We thank M. Rivadeneira for methodological advice, A. Saldaña for data acquisition, and Gavin Thomas and 3 anonymous reviewers for thoughtful suggestions that substantially improved an earlier version of this manuscript.

Author Contributions

Conceived and designed the experiments: EG CS-L RSR. Performed the experiments: EG CS-L RSR. Analyzed the data: EG CS-L RSR. Contributed reagents/materials/analysis tools: EG CS-L RSR. Wrote the paper: EG CS-L RSR.

References

1. Gentry AH (1991) The distribution and evolution of climbing plants. In: Putz FE, Mooney HA, eds. The biology of vines. Cambridge: Cambridge University Press.pp 3–49.
2. Schnitzer SA, Bongers F (2002) The ecology of lianas and their role in forests. Trends Ecol Evol 17: 223–230.
3. Gianoli E, Saldaña A, Jiménez-Castillo M, Valladares F (2010) Distribution and abundance of vines along the light gradient in a southern temperate rainforest. J Veg Sci 21: 66–73.
4. Schnitzer SA, Bongers F (2011) Increasing liana abundance and biomass in tropical forests: emerging patterns and putative mechanisms. Ecol Lett 14: 397–406.
5. Phillips OL, Martínez RV, Arroyo L, Baker TR, Killeen T, et al. (2002) Increasing dominance of large lianas in Amazonian forests. Nature 418: 770–774.
6. Durán SM, Gianoli E (2013) Carbon stocks in tropical forests decrease with liana density. Biol Lett 9: 20130301.
7. van der Heijden GMF, Schnitzer SA, Power JS, Phillips OL (2013) Liana impacts on carbon cycling, storage and sequestration in tropical forests. Biotropica 45: 682–692.
8. Gianoli E (2014) Evolutionary implications of the climbing habit in plants. In: Schnitzer SA, Bongers F, Burnham RJ, Putz FE, eds. Ecology of lianas. New York: Wiley-Blackwell. in press.
9. Gianoli E (2004) Evolution of a climbing habit promotes diversification in flowering plants. Proc R Soc Lond B 271: 2011–2015.
10. Schnitzer SA (2005) A mechanistic explanation for global patterns of liana abundance and distribution. Am Nat 166: 262–276.
11. Cai Z-Q, Schnitzer SA, Bongers F (2009) Seasonal differences in leaf-level physiology give lianas a competitive advantage over trees in a tropical seasonal forest. Oecologia 161: 25–33.
12. Zhu S-D, Cao K-F (2009) Hydraulic properties and photosynthetic rates in co-occurring lianas and trees in a seasonal tropical rainforest in southwestern China. Plant Ecol 204: 295–304.
13. van der Sande MT, Poorter L, Schnitzer SA, Markesteijn L (2013) Are lianas more drought tolerant than trees? A test for the role of hydraulic architecture and other stem and leaf traits. Oecologia 172: 961–972.
14. Heard SB, Hauser D (1995) Key evolutionary innovations and their ecological mechanisms. Hist Biol 10: 151–173.
15. Gentry AH (1991) Breeding and dispersal systems of lianas. In: Putz FE, Mooney HA, eds. The biology of vines. Cambridge: Cambridge University Press.pp 393–423.
16. Riveros M, Smith-Ramírez C (1995) Patrones de floración y fructificación en bosques del sur de Chile. In: Armesto JJ, Villagrán C, Arroyo MK, eds. Ecología de los bosques nativos de Chile. Santiago: Editorial Universitaria. pp. 235–250.
17. Futuyma DJ, Agrawal AA (2009) Macroevolution and the biological diversity of plants and herbivores. Proc Natl Acad Sci USA 106: 18054–18061.
18. Hunter JP (1998) Key innovations and the ecology of macroevolution. Trends Ecol Evol 13: 31–36.
19. Schluter D (2000) The ecology of adaptive radiation. Oxford: Oxford University Press. 288 p.
20. Funk DJ, Nosil P, Etges WJ (2006) Ecological divergence exhibits consistently positive associations with reproductive isolation across disparate taxa. Proc Natl Acad Sci USA 103: 3209–3213.
21. Ackerly DD, Dudley SA, Sultan SE, Schmitt J, Coleman JS, et al. (2000) The evolution of plant ecophysiological traits: recent advances and future directions. Bioscience 50: 979–995.
22. Wright IJ, Reich PB, Westoby M, Ackerly DD, Baruch Z, et al. (2004) The worldwide leaf economics spectrum. Nature 428: 821–827.
23. Poorter L, Bongers F (2006) Leaf traits are good predictors of plant performance across 53 rain forest species. Ecology 87: 1733–1743
24. Santiago LS, Wright SJ (2007) Leaf functional traits of tropical forest plants in relation to growth form. Functional Ecology 21: 19–27.
25. Valladares F, Niinemets Ü (2008) Shade tolerance, a key plant feature of complex nature and consequences. Annu Rev Ecol Evol Syst 39: 237–257.
26. Gianoli E, Saldaña A, Jiménez-Castillo M (2012) Ecophysiological traits may explain the abundance of climbing plant species across the light gradient in a temperate rainforest. PLoS ONE 7(6): e38831.
27. Lambers H, Chapin FS, Pons TL (1998) Growth and allocation. In: Lambers H, Chapin FS, Pons TL, eds. Plant physiological ecology. New York: Springer. pp. 299–351.
28. Pearcy RW (2007) Responses of plants to heterogeneous light environments. In: Pugnaire FI, Valladares F, eds. Functional plant ecology. Boca Raton: CRC Press. pp. 213–257.
29. Ackerly DD (1999) Comparative plant ecology and the role of phylogenetic information. In: Press MC, Scholes JD, Barker MG, eds. Physiological plant ecology. Oxford: Blackwell Science.pp 391–413.
30. Vamosi SM, Heard SB, Vamosi JC, Webb CO (2009) Emerging patterns in the comparative analysis of phylogenetic community structure. Mol Ecol 18: 572–592.
31. Zheng L, Ives AR, Garland T, Larget BR, Yu Y, et al. (2009) New multivariate tests for phylogenetic signal and trait correlations applied to ecophysiological phenotypes of nine Manglietia species. Funct Ecol 23: 1059–1069.
32. Walls RL (2011) Angiosperm leaf vein patterns are linked to leaf functions in a global-scale data set. Am J Bot 98: 244–253.
33. Jacobsen AL, Pratt RB, Tobin MF, Hacke UG, Ewers FW (2012) A global analysis of xylem vessel length in woody plants. Am J Bot 99: 1583–1591.
34. Gallagher RV, Leishman MR (2012) A global analysis of trait variation and evolution in climbing plants. J Biogeogr 39: 1757–1771.
35. O'Meara BC, Ané C, Sanderson MJ, Wainwright PC (2006) Testing for different rates of continuous trait evolution using likelihood. Evolution 60: 922–933.
36. Revell LJ (2012) phytools: an R package for phylogenetic comparative biology (and other things). Methods Ecol Evol 3: 217–223.
37. Webb CO, Ackerly DD, McPeek MA, Donoghue MJ (2002) Phylogenies and community ecology. Annu Rev Ecol Syst 33: 475–505.
38. Blomberg SP, Garland T, Ives AR (2003) Testing for phylogenetic signal in comparative data: behavioral traits are more labile. Evolution 57: 717–745.
39. Webb CO, Donoghue MJ (2005) Phylomatic: tree assembly for applied phylogenetics. Mol Ecol Notes 5: 181–183.
40. Webb CO, Ackerly DD, Kembel SW (2008) Phylocom: software for the analysis of phylogenetic community structure and trait evolution. Bioinformatics 24: 2098–2100.
41. Wikstrom N, Savolainen V, Chase MW (2001) Evolution of the angiosperms: calibrating the family tree. Proc R Soc Lond B 268: 2211–2220.
42. Gastauer M, Meira-Neto JAA (2013) Avoiding inaccuracies in tree calibration and phylogenetic community analysis using Phylocom 4.2. Ecol Inform 15: 85–90.
43. Magallon M, Castillo S (2009) Angiosperm diversification through time. Am J Bot 96: 349–365.
44. R Core Team (2013) R: A language and environment for statistical computing. R Foundation for Statistical Computing, Vienna, Austria. URL http://www.R-project.org/.
45. Nielsen R (2002) Mapping mutations on phylogenies. Syst Biol 51: 729–739.
46. Huelsenbeck JP, Nielsen R, Bollback JP (2003) Stochastic mapping of morphological characters. Syst Biol 52: 131–158.

47. Revell LJ (2013) A comment on the use of stochastic character maps to estimate evolutionary rate variation in a continuously valued trait. Syst Biol 62: 339–345.

48. Kembel SW, Cowan PD, Helmus MR, Cornwell WK, Morlon H, et al. (2010) Picante: R tools for integrating phylogenies and ecology. Bioinformatics 26: 1463–1464.

49. Kraft NJB, Ackerly DD (2010) Functional trait and phylogenetic tests of community assembly across spatial scales in an Amazonian forest. Ecol Monogr 80: 401–422.

50. Kooyman RM, Rossetto M, Sauquet H, Laffan SW (2013) Landscape patterns in rainforest phylogenetic signal: isolated islands of refugia or structured continental distributions? PLoS ONE 8(12): e80685.

51. Verdú M, Pausas JG (2007) Fire drives phylogenetic clustering in Mediterranean Basin woody plant communities. J Ecol 95: 1316–1323.

52. Ackerly DD (2009) Conservatism and diversification of plant functional traits: evolutionary rates versus phylogenetic signal. Proc Natl Acad Sci USA 106: 19699–19706.

53. Pagel M (1999) Inferring the historical patterns of biological evolution. Nature 401: 877–884.

54. Freckleton RP, Harvey PH, Pagel M (2002) Phylogenetic analysis and comparative data: a test and review of evidence. Am Nat 160: 712–726.

55. Revell LJ, Harmon LJ, Collar DC (2008) Phylogenetic signal, evolutionary process, and rate. Syst Biol 57: 591–601.

56. Hansen TF, Martins EF (1996) Translating between microevolutionary process and macroevolutionary patterns: The correlation structure of interspecific data. Evolution 50: 1404–14017.

57. Salgado-Luarte C, Gianoli E (2012) Herbivores modify selection on plant functional traits in a temperate rainforest understory. Am Nat 180: E42-E53.

58. Gianoli E, Saldaña A (2013) Phenotypic selection on leaf functional traits of two congeneric species in a temperate rainforest is consistent with their shade tolerance. Oecologia 173: 13–21.

59. Saldaña A, Lusk CH, Gonzáles WL, Gianoli E (2007) Natural selection on ecophysiological traits of a fern species in a temperate rainforest. Evol Ecol 21: 651–662.

60. Geber MA, Griffen LR (2003) Inheritance and natural selection on functional traits. Int J Plant Sci 164: S21–S42.

61. Evans JR, Poorter H (2001) Photosynthetic acclimation of plants to growth irradiance: the relative importance of specific leaf area and nitrogen partitioning in maximizing carbon gain. Plant Cell Env 24: 755–767.

62. Dodd ME, Silvertown J, Chase MW (1999) Phylogenetic analysis of trait evolution and species diversity variation among angiosperm families. Evolution 53: 732–744.

63. Smith SA, Donoghue MJ (2008) Rates of molecular evolution are linked to life history in flowering plants. Science 322: 86–89.

64. Lanfear R, Ho SY, Davies TJ, Moles AT, Aarssen L, et al. (2013) Taller plants have lower rates of molecular evolution. Nature Comm 4: 1879.

65. Nosil P, Harmon LJ, Seehausen O (2009) Ecological explanations for (incomplete) speciation. Trends Ecol Evol 24: 145–156.

66. Gartner BL, Bullock SH, Mooney HA, Brown VB, Whitbeck JL (1990) Water transport properties of vine and tree stems in a tropical deciduous forest. Am J Bot 77: 742–749.

67. Ewers FW, Fisher JB (1991) Why vines have narrow stems: histological trends in Bauhinia. Oecologia 88: 233–237.

68. Münkemüller T, Lavergne S, Bzeznik B, Dray S, Jombart T, et al. (2012) How to measure and test phylogenetic signal. Methods Ecol Evol 3: 743–756.

69. Cornelissen JHC, Werger MJA, Castro-Diez P, van Rheenen JWA, Rowland AP (1997) Foliar nutrients in relation to growth, allocation and leaf traits in seedlings of a wide range of woody plant species and types. Oecologia 111: 460–469.

70. Cai ZQ, Schnitzer SA, Bongers F (2009) Seasonal differences in leaf-level physiology give lianas a competitive advantage over trees in a tropical seasonal forest. Oecologia 161: 25–33.

71. Wyka TP, Oleksyn J, Karolewski P, Schnitzer SA (2013) Phenotypic correlates of the lianescent growth form: a review. Ann Bot 112: 1667–1681.

72. Wright SJ (2002) Plant diversity in tropical forests: a review of mechanisms of species coexistence. Oecologia 130: 1–14.

73. Gilbert B, Wright SJ, Muller-Landau HC, Kitajima K, Hernández A (2006) Life history trade-offs in tropical trees and lianas. Ecology 87: 1281–1288.

74. Mooers AØ, Vamosi SM, Schluter D (1999) Using phylogenies to test macroevolutionary hypotheses of trait evolution in cranes (Gruinae). Am Nat 154: 249–259.

75. Adams DC, Berns CM, Kozak KH, Wiens JJ (2009) Are rates of species diversification correlated with rates of morphological evolution? Proc R Soc B 276: 2729–2738.

76. Magnuson-Ford K, Otto SP (2012) Linking the investigations of character evolution and species diversification. Am Nat 180: 225–245.

15

A Molecular Phylogeny for Yponomeutoidea (Insecta, Lepidoptera, Ditrysia) and Its Implications for Classification, Biogeography and the Evolution of Host Plant Use

Jae-Cheon Sohn[1]*, **Jerome C. Regier**[1], **Charles Mitter**[1], **Donald Davis**[2], **Jean-François Landry**[3], **Andreas Zwick**[4], **Michael P. Cummings**[5]

1 Department of Entomology, University of Maryland, College Park, Maryland, United States of America, 2 Department of Entomology, National Museum of Natural History, Smithsonian Institution, Washington DC, United States of America, 3 Agriculture and Agri-Food Canada, Eastern Cereal and Oilseed Research Centre, C.E.F., Ottawa, Canada, 4 Department of Entomology, State Museum of Natural History, Stuttgart, Germany, 5 Laboratory of Molecular Evolution, Center for Bioinformatics and Computational Biology, University of Maryland, College Park, Maryland, United States of America

Abstract

Background: Yponomeutoidea, one of the early-diverging lineages of ditrysian Lepidoptera, comprise about 1,800 species worldwide, including notable pests and insect-plant interaction models. Yponomeutoids were one of the earliest lepidopteran clades to evolve external feeding and to extensively colonize herbaceous angiosperms. Despite the group's economic importance, and its value for tracing early lepidopteran evolution, the biodiversity and phylogeny of Yponomeutoidea have been relatively little studied.

Methodology/Principal Findings: Eight nuclear genes (8 kb) were initially sequenced for 86 putative yponomeutoid species, spanning all previously recognized suprageneric groups, and 53 outgroups representing 22 families and 12 superfamilies. Eleven to 19 additional genes, yielding a total of 14.8 to 18.9 kb, were then sampled for a subset of taxa, including 28 yponomeutoids and 43 outgroups. Maximum likelihood analyses were conducted on data sets differing in numbers of genes, matrix completeness, inclusion/weighting of synonymous substitutions, and inclusion/exclusion of "rogue" taxa. Monophyly for Yponomeutoidea was supported very strongly when the 18 "rogue" taxa were excluded, and moderately otherwise. Results from different analyses are highly congruent and relationships within Yponomeutoidea are well supported overall. There is strong support overall for monophyly of families previously recognized on morphological grounds, including Yponomeutidae, Ypsolophidae, Plutellidae, Glyphipterigidae, Argyresthiidae, Attevidae, Praydidae, Heliodinidae, and Bedelliidae. We also assign family rank to Scythropiinae (Scythropiidae **stat. rev.**), which in our trees are strongly grouped with Bedelliidae, in contrast to all previous proposals. We present a working hypothesis of among-family relationships, and an informal higher classification. Host plant family associations of yponomeutoid subfamilies and families are non-random, but show no trends suggesting parallel phylogenesis. Our analyses suggest that previous characterizations of yponomeutoids as predominantly Holarctic were based on insufficient sampling.

Conclusions/Significance: We provide the first robust molecular phylogeny for Yponomeutoidea, together with a revised classification and new insights into their life history evolution and biogeography.

Editor: Jerome Chave, Centre National de la Recherche Scientifique, France

Funding: Financial support was provided by the U.S. National Science Foundation's Assembling the Tree of Life program, award number 0531769, and the Maryland Agricultural Experiment Station. This is contribution 244 of the Evolution of Terrestrial Ecosystems consortium of the National Museum of Natural History, in Washington, D.C. The funders had no role in study design, data collection and analysis, decision to publish, or preparation of the manuscript.

Competing Interests: The authors have declared that no competing interests exist.

* E-mail: jsohn@umd.edu

Introduction

The Yponomeutoidea constitute one of the early radiations in the so-called ditrysian Lepidoptera, the advanced clade that contains the great majority of lepidopteran species. Yponomeutoids include about 1,800 species worldwide, known heretofore mainly from temperate regions [1,2]. Yponomeutoidea are especially important for tracing the early evolution of Lepidoptera-plant interactions because they are one of the earliest groups

to evolve external feeding [3] and to extensively colonize herbs as well as shrubs and trees [4]. In the modern fauna, those two traits are especially common in the highly diverse lineages of advanced moths, for whose success they may be in part responsible. Some yponomeutoid groups, especially *Yponomeuta*, have served as model systems in studying how insect-plant interactions affect speciation [5]. Yponomeutoidea also include a number of notable pest species. For example, the diamondback moth (*Plutella xylostella*: Plutellidae) is regarded as the most destructive insect pest of cruciferous vegetables, annually causing about a billion US dollars in economic loss [6]. Another notorious pest, the leek moth (*Acrolepiopsis assectella*: Glyphipterigidae), has caused damage to upwards of 70% of leeks and 40–50% of onions in some regions of Europe [7]. Communal larvae of some species sometimes extensively damage local vegetation or even broader landscapes. The small ermine moths (*Yponomeuta* spp.) cause complete defoliation of some trees in northern Europe (e.g. [8,9]) and the U.S. (e.g. the introduced *Y. malinellus* [10]).

Despite their value for tracing the early evolution of Lepidoptera and their importance as pests, the Yponomeutoidea have received relatively little attention from systematists, and their biodiversity remains poorly understood. Especially problematic is the lack of a robust phylogeny, including a synapomorphy-based definition for the superfamily itself. Until the early 20th century, the taxa currently placed in Yponomeutoidea comprised scattered suprageneric groups of Tineina or Tineae, two collective microlepidopteran group names no longer in use (e.g. [11,12,13,14]), or Tineidae (e.g. [15,16]). Although Stephens [17] had already distinguished them from other microlepidopteran groups, it was Fracker [18] who first erected a superfamily for Yponomeutoidea. However, as it lacked unambiguously defining characters, the group remained highly heterogeneous and included many genera that now belong to other superfamilies. A succession of subsequent authors advanced increasingly restrictive re-definitions of Yponomeutoidea (e.g. [14,19,20,21,22,23,24]), but failed to achieve a stable classification because they lacked explicit analyses of phylogenetic relationships (Table 1). Kyrki [25,26], in the first cladistic study, significantly modernized the classification of Yponomeutoidea, in which he included only seven families: Yponomeutidae, Ypsolophidae, Plutellidae, Glyphipterigidae, Heliodinidae, Bedelliidae and Lyonetiidae. However, the lack of robustness of Kyrki's phylogeny hindered acceptance of his classification, leaving other hypotheses, such as those of Moriuti [27] and Heppner [1], still in contention (Fig. 1). Disagreements on the phylogeny of Yponomeutoidea, in turn, have helped to obscure inter-relationships of the basal lepidopteran groups and hindered testing of evolutionary hypotheses bearing on them.

Recent molecular studies of higher phylogeny in Lepidoptera have begun to clarify the phylogenetic position, definition and internal relationships of Yponomeutoidea [28,29,30]. The results of Mutanen et al. [29], who included 23 yponomeutoids in an analysis of 350 lepidopterans sequenced for 8 genes (6.3 kb), were the basis for the revised 10-family classification (Table 1) of van Nieukerken et al. [2]. Here, in the first molecular study aimed specifically at Yponomeutoidea, we greatly expand previous taxon and gene sampling, providing the most comprehensive examination and robust hypothesis to date of phylogeny in this superfamily. We compare our results to all previous classification systems, then trace evolutionary trends in yponomeutoid host associations and biogeography on the new phylogeny.

Materials and Methods

Taxon Sampling

A total of 86 species currently assigned to Yponomeutoidea were included in our analyses. These represent all 17 suprageneric groups recognized by Kyrki [25], and all 10 families recognized by van Nieukerken et al. [2] as well as all subfamilies and tribes therein. The sample collectively spans nearly all zoogeographical regions, including 37 species from the Palearctic, 21 from the Neotropics, 17 from the Nearctic, seven from the Australian region, two from the Oriental region, and two from the Ethiopian region. All yponomeutoid genera for which material could be obtained were included, each represented by a single species except that two or more species were sampled for several broadly distributed, species-rich genera.

The definition of Yponomeutoidea has been considered controversial [31]. For this reason, our putative outgroups, totaling 53 species belonging to 22 families in 12 superfamilies of ditrysian Lepidoptera (see Supplement S1), included all superfamilies that were historically associated with Yponomeutoidea or at least contain genera that were once placed within Yponomeutoidea. Among these are Choreutoidea, Copromorphoidea, Epermenioidea, Galacticoidea, Gelechioidea, Schreckensteinoidea, Urodoidea, and Zygaenoidea. Inclusion of these taxa provides an additional test of the monophyly of Yponomeutoidea in the restricted modern sense. We also included two superfamilies, Tortricoidea and Pterophoroidea, which have never been considered close to yponomeutoids. In contrast to all previous hypotheses, recent molecular studies [28,29,30] have strongly

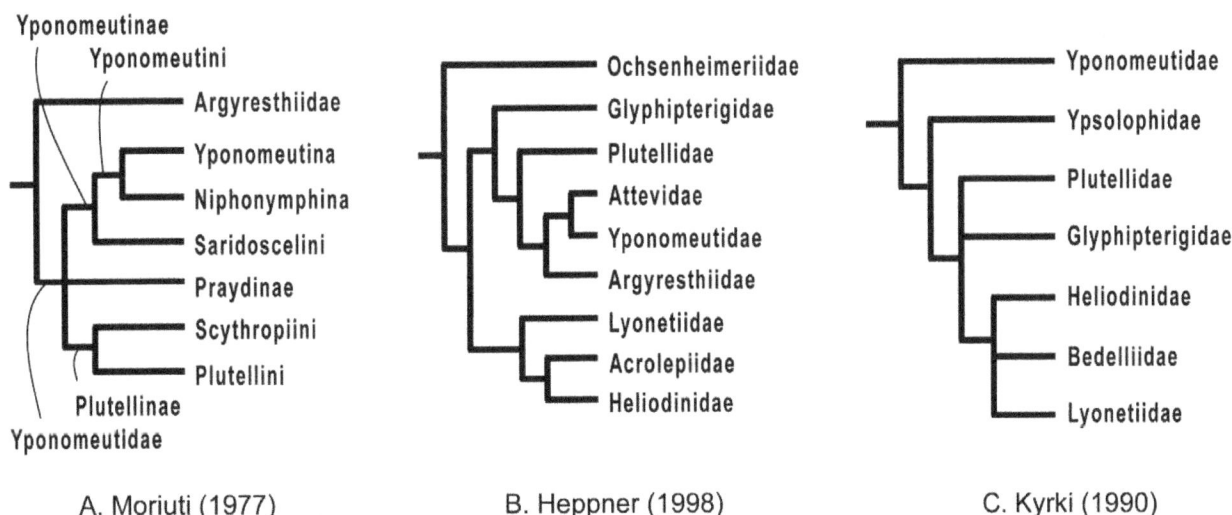

Figure 1. Previous hypotheses of phylogenetic relationships in Yponomeutoidea. A. Moriuti (1977), B. Heppner (1998), C. Kyrki (1990). All figures are redrawn with nomenclature following the original.

Table 1. Previous classifications of Yponomeutoidea.

Common (1970)	Moriuti (1977)	Heppner (1998)	Kyrki (1990)	van Nieukerken et al. (2011)
Yponomeutidae	**Yponomeutidae**	**Yponomeutidae**	**Yponomeutidae**	**Yponomeutidae**
Plutellinae	Yponomeutinae	Yponomeutinae	Yponomeutinae	Yponomeutinae
Yponomeutinae	Yponomeutini	Saridoscelinae	Saridoscelinae	Saridoscelinae
Amphitherinae	Yponomeutina	Cedestinae	Scythropiinae	Scythropiinae
Argyresthinae	Niphonymphina	**Attevidae**	Attevinae	**Attevidae**
Glyphipterigidae	Saridoscelini	**Argyresthiidae**	Praydinae	**Praydidae**
Heliodinidae	Praydinae	**Plutellidae**	Argyresthiinae	**Argyresthiidae**
Aegeriidae	Plutellinae	Ypsolophinae	**Plutellidae**	**Plutellidae**
Douglasiidae	Scythropiini	Plutellinae	Plutellinae	**Ypsolophidae**
Epermeniidae	Plutellini	Scythropiinae	Acrolepiinae	Ypsolophinae
	Argyresthiidae	Praydinae	**Ypsolophidae**	Ochsenheimeriinae
		Acrolepiidae	Ypsolophinae	**Glyphipterigidae**
		Ochsenheimeriidae	Ochsenheimeriinae	Acrolepiinae
		Glyphipterigidae	**Glyphipterigidae**	Orthoteliinae
		Orthoteliinae	Orthoteliinae	Glyphipteriginae
		Glyphipteriginae	Glyphipteriginae	**Heliodinidae**
		Heliodinidae	**Heliodinidae**	**Lyonetiidae**
		Lyonetiidae	**Lyonetiidae**	Cemiostominae
		Cemiostominae	Cemiostominae	Lyonetiinae
		Lyonetiinae	Lyonetiinae	**Bedelliidae**
		Bedelliinae	**Bedelliidae**	

Nomenclature follows the original. Families are indicated in bold.

supported Gracillarioidea as the closest relatives to Yponomeutoidea sensu Kyrki [25,26]. For this reason we sampled gracillarioids especially densely, taking exemplars from most of the known families and subfamilies. We included comparably dense sampling of Tineoidea, which have long been considered, now with increasing molecular evidence ([29] and J. Regier et al., unpublished results), to contain the earliest-branching lineages within the Ditrysia [32]. Finally, to root the entire tree, we added a representative of Tischeriidae, long regarded, also with increasing molecular evidence ([29] and J. Regier et al., unpublished results), to be among the closest relatives to Ditrysia.

Specimen Preparation and Identification

The specimens for this study, obtained by our own collecting as well as from collaborators around the world (see Acknowledgments), are stored in 100% ethanol at −80°C as part of the ATOLep frozen tissue collection at the University of Maryland, College Park, USA (details at http://www.leptree.net/collection). For extraction of nucleic acids we used the legs, head and thorax, or the entire body (always excluding the wings), depending on the size of the specimen. As vouchers we preserved both wings and abdomen for large or medium-sized moths, and wings only for very small ones. Wing voucher images for most of our specimens are available at the Leptree website (http://www.leptree.net/voucher_image_list). Partial COI sequences corresponding to DNA 'barcodes' were generated for each specimen either by the authors or as part of the All-Leps Barcode of Life project (http://www.lepbarcoding.org). Using these sequences, we performed an independent check of the primary identifications of all specimens by searching for matching barcode sequences in the BOLD (Barcode of Life Data system, http://www.boldsystems.org).

Gene Sampling

The sequences initially sampled for this study consisted of eight nuclear genes (Supplment S1), totaling 8,096 bp, for nearly all ingroup taxa (83/86 = 96.5%) and all outgroup taxa. These eight are a subset of the 26 genes sequenced in a study of ditrysian phylogeny by Cho et al. [30], 25 of which were also analyzed in Bombycoidea by Zwick et al. [33]. The eight gene subset was chosen on the basis of its relatively high amplification success rates and phylogenetic utility. The eight genes are: *Gelsolin* (603 bp), *histidyl tRNA synthetase* (447 bp), *AMP deaminase* (768 bp), *glucose phosphate dehydrogenase* (621 bp), *Acetyl-coA carboxylase* (501 bp), *CAD* (2,929 bp), *DDC* (1,281 bp) and *enolase* (1,135 bp). Three species (*Argyresthia austerella*, *Digitivalva hemiglypha*, and *Prays atomocella*), each with close relatives in the eight gene data set, were sequenced for only the five genes (6.6 kb) studied in Ditrysia by Regier et al. [28], namely, *CAD*, *DDC*, *enolase*, *period*, and *wingless* (Figure S1).

Because the initial 8-gene analyses yielded little strong support for deeper nodes, we subsequently added 11–19 more nuclear genes (totaling up to 27 genes and 19,386 bp) for a taxon subset consisting of 28 ingroups and 43 outgroups (Figure S1), amounting to 51% of the total of 139 taxa. The 27 genes include the 26 used by Cho et al. [30], plus one additional gene, *α-spectrin*. All 27 are included in the set of 68 genes studied by Regier et al. [34] across the arthropods. The great majority of taxa (54/65) for which more than eight genes were assayed were sequenced for just the 19 gene set that has recently proven useful in resolving relationships in other superfamilies, including Gracillarioidea [35], Tortricoidea [36] and Pyraloidea [37]. These same studies have also shown that augmentation of the initial gene sample in only a subset of taxa, following Cho et al. [30], is an effective and cost-efficient means for obtaining stronger support at deeper nodes. Partial gene augmentation introduces blocks of nonrandomly missing data that could have adverse effects on phylogeny estimation [38,39]. To test this possibility, we compared the results from the 8+19 gene, deliberately incomplete matrix to those from a 4-gene data set (*glucose phosphate dehydrogenase*, *CAD*, *DDC* and *enolase*) that exhibit a relatively low percentage of missing data (21.5%) among our 139 taxa, due to inadvertent failures of amplification or sequencing.

Gene Extraction, Sequencing and Alignment

A detailed protocol of all laboratory procedures is provided by Regier et al. [34]. Further descriptions, including gene amplification strategies, PCR primer sequences, sequence assembly and alignment methods, can be found in Regier [40] and Regier et al. [28,41]. To summarize, total RNAs were extracted from an excised tissue using the SV Total RNA Isolation System (Promega Co.). The targeted regions of the mRNAs were amplified using Reverse Transcriptase (RT)-PCR, yielding cDNA. Nested PCR for further purification and/or M13 re-amplification for increasing volume were attempted as necessary. Purified amplicons were sequenced on a 3730 DNA Analyzer (Applied Biosystems) at the Center for Biosystems Research at the University of Maryland, College Park. The resulting ABI files and contigs were checked for error manually and then edited and assembled using Geneious Pro 5.3.4 (Biomatters Ltd.). The data were rechecked for error by inspection of the genetic distances among them determined in PAUP* 4.0b8 [42]. The final sequences for each gene were aligned using the "Translation Align" option in Geneious. The final alignments were concatenated with Geneious, separately for the 8-gene and 8-27 gene analyses, and the combined data sets were visually checked. Regions of uncertain alignment, totaling 1,509 characters, were masked and excluded from subsequent analyses. GenBank accession numbers and the percentage sequence completeness for each gene in each taxon are given in Figure S1.

Character Partition and Data Set Design

It is well known that rates of sequence evolution vary among codon positions, reflecting in part different ratios of synonymous versus nonsynonymous substitutions [43,44]. Previous empirical studies (e.g. [28,30,34]) have shown that partitioning data to reflect this variation, or eliminating synonymous change entirely, can reduce or eliminate phylogenetic error due to among-lineage compositional heterogeneity, but at the cost of discarding potentially informative synonymous signal. To gauge the potential effects of differing evolutionary properties between synonymous and non-synonymous substitution on phylogeny inference, we carried out separate analyses using a variety of character coding and/or data partition schemes. These analyses are: (a) "nt123", i.e., all codon positions included and unpartitioned; (b) "degen1" [45,46], i.e., all synonymous differences degenerated, leaving only non-synonymous differences among taxa; (c) "nt123 partitioned" [28], i.e., all codon positions partitioned into mostly non-synonymously evolving ("noLRall1+nt2") versus mostly synonymously- evolving ones ("LRall1+nt3"); and, (d) "codon" analysis [47,48], in which the character states are codons and synonymous and nonsynonymous changes are modeled separately. For the codon analyses (only), a 91 taxon set including only Yponomeutoidea and Gracillarioidea was used, rather than the full 139 taxon data set, to reduce the computational burden. Increased numbers of discrete rate categories in the gamma-distributed rate heterogeneity distribution ('numratecats' in the GARLI configuration) can also dramatically increase computational time. To avoid this problem, we used trial runs to estimate a minimum number of categories beyond which further increase yields no significant improvement in tree likelihood scores. We determined this

number to be three categories. As a third approach to accommodating differences between synonymous and non-synonymous change, we also partitioned the data into first plus second codon positions ("nt12", Figure S3) versus third codon positions ("nt3", Figure S4).

Phylogenetic Analyses

The best substitution model for each data set was determined using jModelTest [49], which in nearly all cases selected GTR+Γ+I, i.e., the General-Time-Reversible model with among-site rate variation accomodated using a gamma distribution plus separate estimation of a proportion of invariable sites. Phylogenetic analyses were conducted with maximum likelihood (ML) methods as implemented in GARLI 2.0 [50], which includes partitioned models. Default settings of the program were used, except that starting tree topology was specified as random; the frequencies with which to log the best score ('logevery') and to save the best tree to file ('saveevery') were set to 100,000 and 100,000 respectively; and, the number of generations without topology improvement required for termination ('genthreshfortopoterm') was set to 5,000. The best tree from 150 independent search replicates was saved, and visualized using FigTree v1.3.1 [51]. To evaluate the robustness of the resulting trees, bootstrap (BP) values were calculated from 1000 pseudoreplicates, each based on 15 heuristic search replicates except that only a single heuristic search replicate was carried out for each pseudoreplicate in the single-gene bootstrap analyses. Because these analyses are so computation-intensive, they were carried out by Grid parallel computing [52], using the Lattice Project [53,54]. For purposes of discussion, we will refer to BP values of 70–79% as "moderate", 80–89% as "strong", and ≥90% as "very strong" support. These conventions, also adopted in previous studies (e.g. [30,35]), are arbitrary and hence serve heuristic purposes only.

Rogue Taxon Analyses

Despite the addition of 11–19 genes to the initial 8-gene data set, some deeper nodes in even our best-supported trees have low bootstrap values. One possible cause of low support is the sensitivity of bootstrap values to taxa of unstable placement [55], termed "rogues" by Wilkinson [56]. Multiple approaches have been suggested for detecting and removing the effects of rogue taxa (reviewed in [57]). We investigated the potential contribution of rogue taxa (Table 2) to low bootstrap values in our data set using the RogueNaRok (RNR) approach of Aberer et al. ([58]; a pun on Ragnarök, the judgement of the gods in Norse mythology). The key feature of RNR is a new optimality criterion for rogue taxon removal, the "Relative Bipartition Information Criterion" (RBIC) [57,59]. The RBIC strikes a balance between improving per-node support in the reduced bootstrap consensus tree (with rogues deleted) and retaining total information by minimizing the loss of bipartitions in the bootstrap consensus tree that results from such deletions. Aberer and Stamatakis [59] compared multiple heuristic approaches to maximizing the RBIC. The best results came from their single-taxon algorithm (STA), which begins by removing taxa one at a time to find the taxon (if any) whose deletion most improves the RBIC. After that taxon is removed, one removes each remaining taxon again, to find the next most "roguish" taxon. The process is repeated until the optimality score stops improving. The RogueNaRok algorithm is a fast generalization of the STA, which allows for "deletion sets" – groups of taxa deleted simultaneously – of varying sizes.

To identify rogue taxa, we used the on-line version of RogueNaRok (RNR) at http://193.197.73.70:8080/rnr/rogue-narok, which is built on RAxML [60]. Bootstrap files were first

Table 2. Rogue taxa identified by the RogueNaRok (RNR) analyses, listed in the order in which they were identified and removed.

Rogue taxon set[*]	Rogue taxon	Code name	SC[**] (%)	Raw Improvement[***]	RBIC
A	Copromorpha sp.	Cmpa	12	0.906667	0.767598
	Xyrosaris lichneuta	Xlic	29.2	0.74	0.773039
	Cycloplasis panicifoliella	Cpan	26.2	0.666667	0.777941
	Hybroma servulella	Hybs	67.0	0.58	0.782206
	Epermenia sinjovi	Esji	30.6	0.26	0.784118
	Philonome clemensella	Pmsa	26.7	0.246667	0.785931
	Opogona thiadelia	Othi	64.1	0.113333	0.786765
	Emmelina monodactyla	Emon	86.9	0.093333	0.787451
	Klimeschia transversella	Ktr	66.4	0.906667	0.794118
	Hemerophila felis	Hfel	90.8	0.186667	0.79549
	Nemapogon cloacella	Nclo	55.1	0.013333	0.795588
B	Narycia duplicella	Nard	34.1	0.373333	0.867413
	Euclemensia bassettella	Cole	81.6	0.146667	0.868587
	Bucculatrix sp.	Bucc	56.9	0.033333	0.868853
C	Homadaula anisocentra	Hani	64.7	0.82	0.870656
	"Wockia" sp.	MX60	19.1	0.2	0.879016
D	*Perileucoptera coffeella*	Leuco	43.2	0.12	0.874545
	Swammerdamia glaucella	Swgl	33.7	0.046667	0.875076

The RBIC (relative bipartition information content) for the reduced consensus tree, after pruning all taxa up to and including any given rogue taxon, is shown in the last column. Ingroup rogue taxa are shown in bold. * Rogue taxon sets = rogue taxa identified on each successive one-at-a-time pass through the taxa. Each such pass, after the first pass, starts from a reduced taxon set from which all previously-identified rogues have been removed. Following the removal of rogue taxon sets A–C, no further rogue taxa could be identified in the entire data set. Rogue taxon set D was identified in an independent analysis of just Yponomeutoidea+Gracillarioidea, excluding other outgroups. A: 139 taxa x 8–27 genes. Initial score = 0.760931, # of partitions in reduced consensus tree = 973. B: 128 taxa (11 rogue taxa deleted from A). Initial score = 0.864427, # of partitions = 443. C: 125 taxa (3 rogue taxa deleted from B). Initial score = 0.870656, # of partitions = 337. D: 91 taxa (Yponomeutoidea+Gracillarioidea). Initial score = 0.873182, # of partitions = 272. ** SC (sequence data completeness) = (# of nucleotides actually sequenced/total # of targeted nucleotides) x 100. ***Raw Improvement: the improvement in support (sum of all bootstrap values) for the reduced consensus tree, if the taxon in question is pruned AND all previously identified rogue taxa are also pruned.

generated and submitted to RNR, which identified possible rogue taxa (i.e. ones whose removal increases the RBIC). The reduced data set was then analyzed with RAxML, and the bootstrap outputs again submitted to RNR. This procedure was repeated until RNR no longer identified any additional rogues. Finally, the putatively rogue-free data sets were subjected to bootstrap analyses using GARLI, to make them directly comparable to the original analyses. This procedure was carried out only for the nt123, 8–27 gene data set, which gave the highest initial bootstrap support overall. In our initial RNR analyses, most of the rogue taxa detected were among the more distant outgroups. This result might stem from increased uncertainty in position due to lower sampling density among these taxa, and might in turn impede detection of more subtle rogue taxon effects within the ingroup, which is what we are most interested in. To circumvent this possibility, we also conducted separate RNR analyses on data sets containing Yponomeutoidea (86 taxa) and Gracillarioidea (11 taxa) only.

Significance Tests of Discord with Previous Hypotheses

Our results appear to contradict a number of prior hypotheses about phylogenetic relationships in Yponomeutoidea, including several depicted in Figure 1. We used the Approximately Unbiased (AU) test of Shimodaira [61] to determine whether our data significantly reject those previous hypotheses, against the alternative that the discrepancy can be explained by sampling error in the sequence data. The test determines whether the best

tree possible under the constraint of monophyly, no matter what its topology may be otherwise, is a significantly worse fit to the data than the best tree without that constraint. Table 3 lists the 12 groups tested for significance of non-monophyly. For each combination of one character set and one apparently non-monophlyetic previous grouping, we performed a GARLI analysis consisting of 150 replicate tree searches, under the constraint of monophyly for the group in question. The constrained tree was then compared to the previously-obtained unconstrained tree. The site likelihoods of the best constrained and unconstrained trees were then estimated with PAUP* [42], and the trees and site likelihoods for all comparisons combined into a single input file for the CONSEL 0.20 package [62,63] with which the Approximately Unbiased test was conducted.

Host Plant Associations and Biogeography

To explore the evolutionary history of Yponomeutoidea with respect to larval host plant associations and biogeography, we compiled data from the literature on these features for all described yponomeutoid species. Given current uncertainty about the limits of the superfamily, we considered only genera whose placements within Yponomeutoidea are secure. Host records were retrieved primarily from the HOSTS website [64]. These data were checked for possible error and supplemented by records from other sources. All suspicious records, possibly representing misidentification of larvae, misidentification of hosts, or confusion with adult-habitat association, were excluded. Individual host

Table 3. Results of Approximately Unbiased (AU) tests for significance of rejection of 12 previous phylogenetic hypotheses.

#	Constraint group	Source	nt123 (p)	degen1 (p)
1	Yponomeutoidea *sensu* Kyrki (Fig. 1)	Kyrki (1990)	**0.001**	**<0.001**
2	Yponomeutoidea *sensu* Heppner (Fig. 1)	Heppner (1998)	**<0.001**	**<0.001**
3	Yponomeutidae s. l. (Fig. 1)	Moriuti (1977)	**<0.001**	**<0.001**
4	Yponomeutidae *sensu* Kyrki (Table 1)	Kyrki (1990)	**<0.001**	**<0.001**
5	Cedestinae	Friese (1960)	**<0.001**	**0.002**
6	Yponomeutidae B1 group	Friese (1960)	**0.001**	**0.001**
7	Plutellidae+Praydidae	Heppner (1998)	**<0.001**	**<0.001**
8	Plutellidae+*Scythropia*	Heppner (1998)	**<0.001**	**0.002**
9	Plutellidae *sensu* Heppner (Table 1)	Heppner (1998)	**<0.001**	**<0.001**
10	#9+*Ochsenheimeria*	Heppner (1998)	**<0.001**	**<0.001**
11	Lyonetiinae+Cemiostominae	Kyrki (1990)	0.259	0.180
12	Lyonetiidae+Bedelliidae	Kuroko (1964)	**0.005**	**0.005**

All analyses are based on the 8–27 gene nt123 and degen1 data sets. P values <0.05 in bold.

records were combined into lists of plant families or higher clades used by each of the 16 major yponomeutoid lineages identified on our molecular phylogeny. Higher classification of host plants follows APG III [65] for angiosperms and Fu et al. [66] for gymnosperms. Host ranges of individual yponomeutoid species were categorized as either oligophagous (feeding on plants in a single order) or polyphagous (feeding on plants in more than one order). The predominant growth form of hosts for each yponomeutoid lineage was categorized as arboreal (trees and shrubs), herbaceous, or scandent (vines and lianas), and alternatively as woody versus herbaceous. We also scored site and mode of feeding. Finally, for each lineage we tabulated the proportions of species and genera for which at least one host plant record is available, using species and generic diversity estimates from van Nieukerken et al. [2] or the first author's unpublished data.

Information on yponomeutoid distributions across major biogeographical regions was assembled from global reviews (e.g. [67,68,69]) and local checklists (e.g. [20,70,71,72,73,74]). Distributions due to human-caused dispersal (accidental or deliberate introduction) were excluded when discernable from non-anthropogenic causes. Data for individual species were compiled into summaries of numbers of species occurring in each region for each major yponomeutoid lineage, as described previously for host plant records. For species occurring in more than one region, each region was counted independently, thus some species were counted more than once. Our compilations are based primarily on described species, but undescribed species were included in several cases where they represent significant expansion of the known distribution of the lineage.

Generalization of host and distribution records by higher taxonomic groups often neglects variation, incompleteness, and bias in such data, introducing errors. For this reason, we did not attempt any formal statistical approach, although we did compute (by hand) parsimony optimizations of predominant feeding mode and host plant growth on a simplified version, reduced to major lineages, of the molecular phylogeny. Our goal was simply to provide a first phylogeny-based summary of evolutionary trends in yponomeutoid host-use evolution and biogeography.

Results

The best-score ML tree found in 150 GARLI searches for the 8–27 gene, 139-taxon nt123 analysis is shown in Figures 2 and 3. Figure 2 shows just the Yponomeutoidea as recovered here (79 taxa), while Figure 3 shows the outgroup region of the tree. Bootstrap values for five different combinations of character coding (nt123, nt123 partitioned, degen1) and gene sample (8 genes only vs. 8+19 genes), plus nt123 with rogue taxa removed, are superimposed on each node of this tree. Overall, the tree is well supported: 65 of the 78 nodes in Figures 2 and 3, or 83%, had strong bootstrap support (≥80%) from at least one analysis. Figure 4 shows the same topology in a phylogram format, with thickened branches denoting bootstrap support of ≥70% from at least one of the bootstrap analyses summarized in Figures 2 and 3.

The most robust phylogenies came from the nt123 analysis of the 8–27 gene deliberately incomplete data set (Fig. 2; Table 4). Within Yponomeutoidea (Fig. 2; 79 taxa) this analysis yielded 59 very strongly supported (BP≥90%), 4 strongly supported (BP = 80–89%) and 3 moderately supported (BP = 70–79%) nodes, for a sum of 66 nodes (of 78 total), or 85%, with BP≥70%. The results for the partitioned nt123 analysis were nearly identical: 58 nodes with BP≥90%, 4 nodes with BP = 80–89% and 3 with BP = 70–79%. The 8–27 gene degen1 analysis yielded 37 nodes with BP≥90%, 6 with BP = 80–89% and 4 with BP = 70–79%, for a total of 47/78 = 60% of nodes with BP≥70%. The codon model results were intermediate between those from nt123 and degen1 but closer to the former, with 54 nodes of BP≥90%, 3 of BP = 80–89% and 2 of BP = 70–79%, for a total of 59/78 = 76% of nodes with BP≥70%. The nt123 unpartitioned and nt123 partitioned trees were nearly identical, disagreeing at only three nodes weakly supported in each. The degen1 tree disagreed with the nt123 tree at 18 nodes, of which 8 were very strongly supported, 2 strongly supported, one moderately supported and 7 poorly supported (BP≤60%) in the nt123 tree. In only two cases, however, was a node strongly supported in the degen1 analysis but not present in the nt123 tree, while in no case was a node strongly supported in one tree and strongly contradicted in the other.

The 8-gene and 8–27 gene nt123 trees were almost entirely congruent, differing in only 2 weakly supported nodes. Of the matching nodes between the two analyses, 12 were better supported in the 8-gene analysis, with a mean difference of

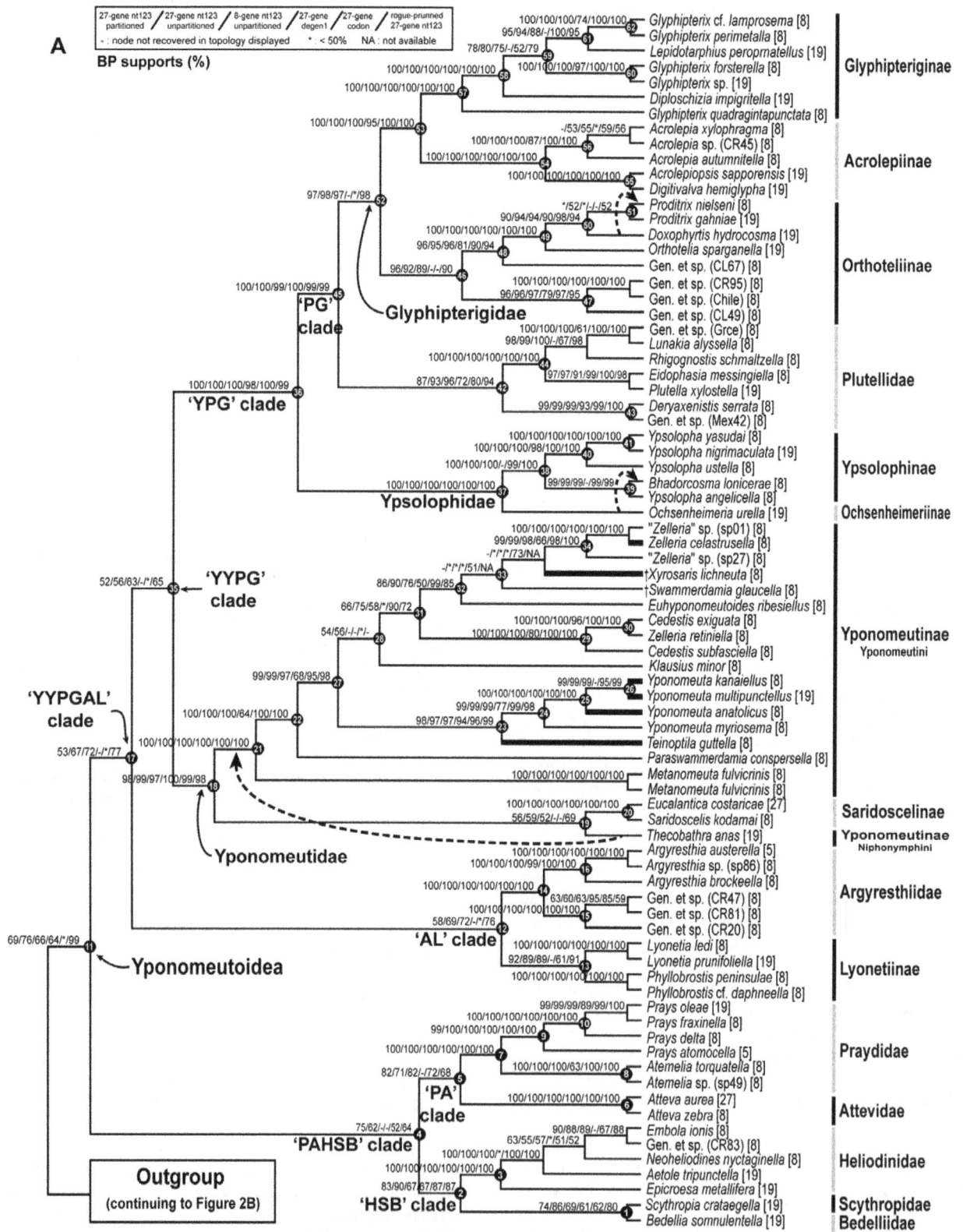

Figure 2. The best ML tree found for nt123 analysis of the deliberately incomplete 8–27 gene, 139-taxon data set, showing Yponomeutoidea only. Bootstrap supports shown above branches: partitioned 8–27 gene nt123/unpartitioned 8–27 gene nt123/8-gene nt123/8–27 gene degen1/8–27 gene codon model/rogue-pruned 8–27 gene nt123 (121 taxa). '−' = node not recovered in the ML tree for that analysis. '*' = bootstrap value <50%. 'NA' = bootstrap value undefined because data were obtained for ≤1 taxon in that clade for that analysis. Dotted lines indicate alternative topologies strongly supported by either degen1 or the codon model. Node numbers for selected nodes (solid circles) are provided to facilitate discussion. Thickened terminal branches denote yponomeutoid species feeding on Celastraceae.

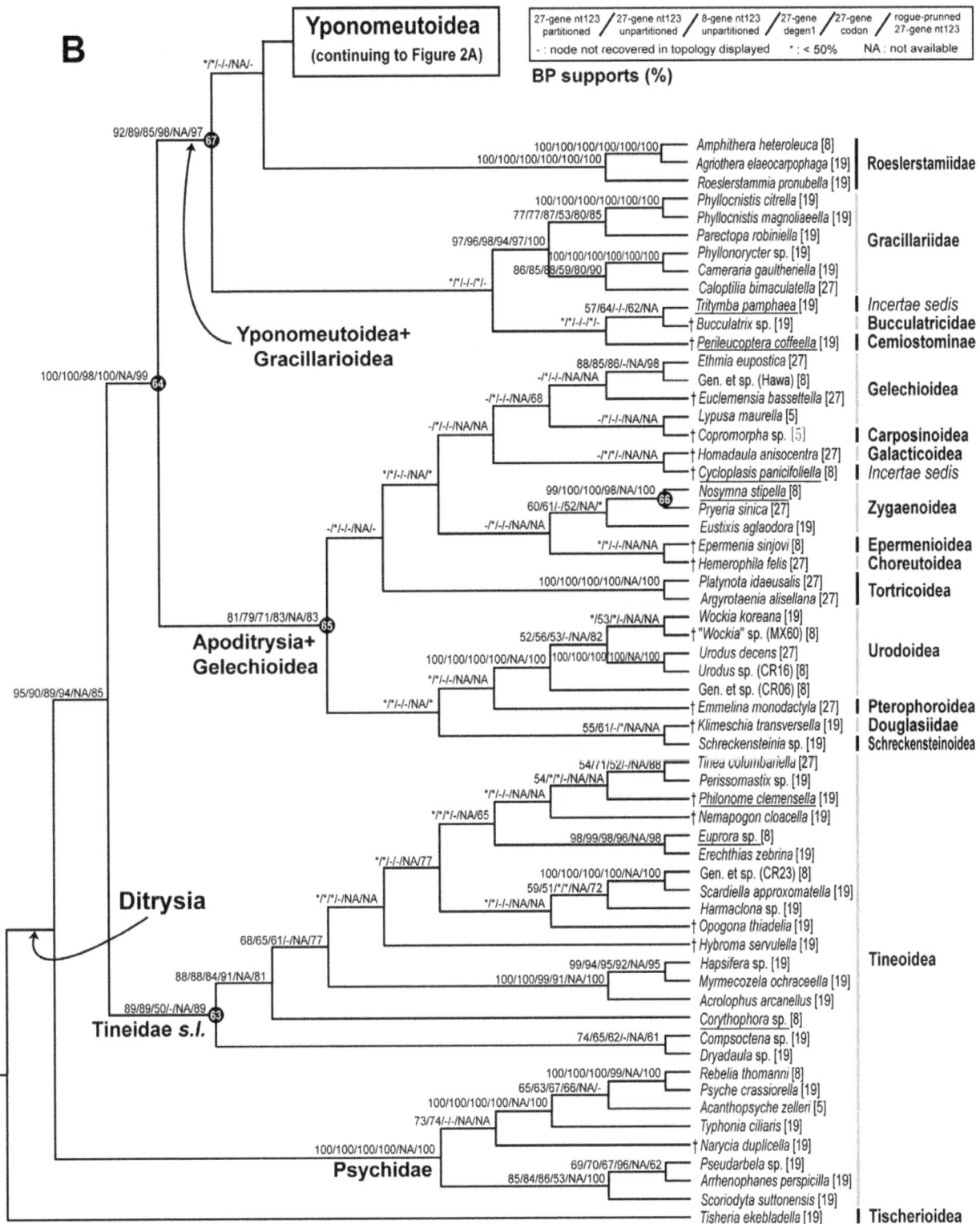

Figure 3. The best ML tree found for nt123 analysis of the deliberately incomplete 8–27 gene, 139-taxon data set (continued from Fig. 2), showing outgroups only. See Figure 2 for notes on bootstrap supports and node numbers. Terminal taxa shown in pink were initially thought to be yponomeutoids.

+3.33% and a range of 1–11%, while the 19+ gene analysis yielded higher support at 16 nodes, with a mean difference of +7.56% and a range of 1–23%. The 8-gene analysis yielded 55 nodes with BP≥90%, 5 with BP = 80–89% and 3 with BP = 70–79%, for a total of 63/78 = 81% of nodes with BP≥70%, only slightly lower than the 19+ gene analysis. However, a few nodes showed

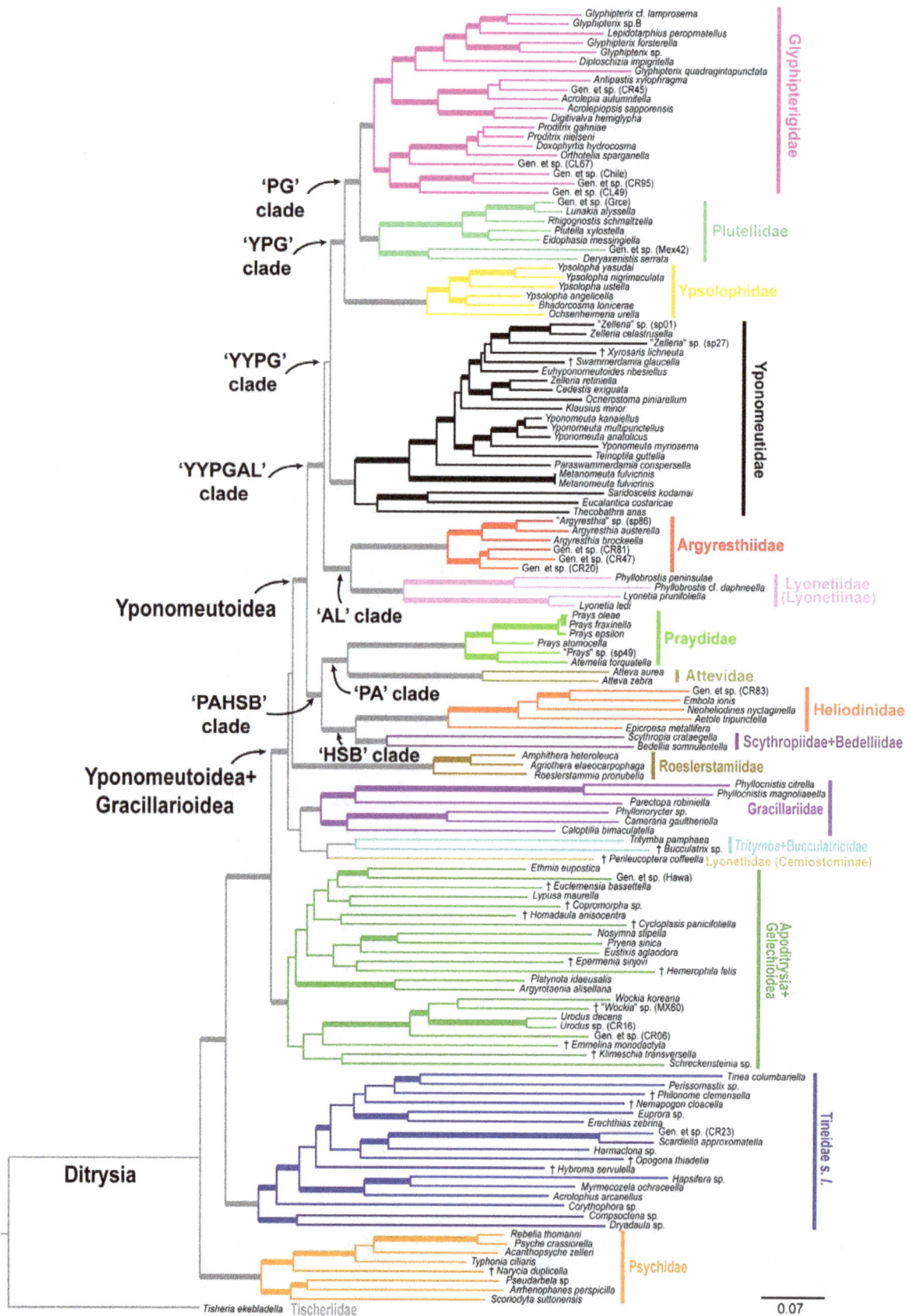

Figure 4. Phylogram representation of ML tree shown in Figures 2 and 3. Branch lengths are proportional to total number of substitutions per site. Thickened branches are supported by ≥70% bootstrap in at least one analysis summarized in Figures 2 and 3.

Table 4. Bootstrap supports for selected clades.

Node #	Selected Clade	4-gene nt123	8-gene nt123	8–27 gene nt123	8–27 gene partition	8–27 gene degen1	8–27 gene Codon	8–27 gene & no-rogue nt123
1	*Bedellia+Scythropia*	<50	69	86	74	61	62	80
2	'H·S·B' clade	56	67	90	83	67	87	87
3	Heliodinidae	100	100	100	100	100	100	100
4	'P·A·H·S·B' clade	–	–	62	75	–	52	64
5	'P·A' clade	96	82	71	82	–	72	68
6	Attevidae	100	100	100	100	100	100	100
7	Praydidae	100	100	100	100	100	100	100
8	*Atemelia*	100	100	100	100	100	100	100
9	*Prays*	89	100	100	99	100	100	100
11	Yponomeutoidea (excl. Cemiostomiinae)	–	66	76	69	64	<50	99
12	'A·L' clade	–	72	69	58	–	<50	76
13	Lyonetiidae (Lyonetiinae)	89	89	89	92	–	61	91
14	Argyresthiidae	100	100	100	100	100	100	100
15	"*Dasycarea*" group	100	100	100	100	100	100	100
16	*Argyresthia*	100	100	100	100	100	100	100
17	'Y·Y·P·G·A·L' clade	–	72	67	53	–	<50	77
18	Yponomeutidae	98	97	99	98	100	99	98
19	Saridoscelinae+*Theco-bathra*	<50	52	59	56	–	–	69
20	Saridoscelinae	100	100	100	100	100	100	100
21a	Yponomeutini	100	100	100	100	100	100	100
21b	Yponomeutini+ *Theco-bathra*	–	–	–	–	82	52	–
23	*Yponomeuta* group	99	97	97	98	94	96	99
29	*Cedestis+Zelleria* (part)	100	100	100	100	80	100	100
31	Node 29+32	–	58	75	66	<50	90	72
32	*Zelleria* (part)+*Xyrosaris*+ *Swammerdamia+Euhypo-nomeutoides*	–	76	90	86	50	99	85
35	'Y·Y·P·G' clade	–	63	56	52	–	<50	65
36	'Y·P·G' clade	96	100	100	100	98	100	99
37	Ypsolophidae	100	100	100	100	100	100	100
38	Ypsolophinae	100	100	100	100	–	99	100
39	*Bhadorcosma+Ypsolopa angelicella*	96	99	99	99	–	99	99
42	Plutellidae	92	96	93	87	72	80	94
43	*Deryaxenistis* group	97	99	99	99	93	99	100
44	Core Plutellidae	100	100	100	100	100	100	100
45	'P·G' clade	95	99	100	100	100	99	99
46	Orthoteliinae	86	89	92	96	–	–	90
47	Neotropical Orthoteliinae	99	96	96	97	79	97	95
48	Core Orthoteliinae	90	96	95	96	81	90	94
51a	*Proditrix*	<50	<50	52	<50	–	–	52
51b	*Doxophytis+Proditrix nielseni*	–	–	–	–	86	56	–
52	Glyphipterigidae	98	97	98	97	–	<50	98
53	Glyphipteriginae+Acro-lepiinae	96	100	100	100	95	100	100
54	Acrolepiinae	100	100	100	100	100	100	100
57	Glyphipteriginae	100	100	100	100	100	100	100
59	*Glyphipterix* (part)+*Lepi-dotarphius*	–	75	80	78	–	52	79

Dashes indicate unrecovered clades. Node numbers corresponding to Figure 2 (a & b for alternative topologies).

substantial increase in support with increased gene sampling. Among these are three that subtend multiple families: Heliodinidae+Bedelliidae+*Scythropia* (Fig. 2, **node 2**; BP = 90/67, 19+ genes/8 genes); Bedelliidae+*Scythropia* (Fig. 2, **node 1**; BP = 86/69); and Yponomeutoidea (Fig. 2, **node 10**; BP = 76/66).

Our rogue taxon analysis using RogueNaRok [58] identified 16 rogue taxa for the 8–27 gene nt123 data set as a whole (Table 2). All but one (Yponomeutidae: *Xylosaris lichineuta*) proved to lie among the outgroups, although several others were thought by some previous authors to belong to Yponomeutoidea (Table 2). Two additional rogue taxa, both yponomeutoids (Lyonetiidae: *Perileucoptera* and Yponomeutidae: *Swammerdamia*), were discovered when only Yponomeutoidea and Gracillarioidea were analyzed. We found no significant correlation between rogue status and sequence data incompleteness (Table 2: SC index). Removal of the 18 rogue taxa resulted in increased bootstrap values for 14 nodes and decreases for 17 nodes in the tree for Yponomeutoidea (Fig. 2). However, 77% of these changes were very small (≤3%). When only changes of >3% are counted, there are just two decreases in support in the rogue-pruned analysis, one of 5% and one of 6%. In contrast, five nodes showed increases, ranging from 7% to 23%. Among the nodes undergoing the strongest improvements in support are Yponomeutoidea (Fig. 2, **node 10**; BP = 99/76, after/before rogue removal); the YPGAL clade (Fig. 2, **node 16**; BP = 77/67); and the AL clade (Fig. 2, **node 11**; BP = 76/69). Half of the increase in bootstrap values across all affected nodes can be explained by deletion of *Perileucoptera coffeella* alone (data not shown).

Discussion

Phylogenetic Signal Sources, Partial Gene Sample Augmentation and Rogue Taxon Analysis

Our results exemplify the ability of combined analyses of multiple genes to produce robust phylogeny estimates even when there is little strong signal from any individual gene [75]; none of the deeper nodes with substantial support (BP≥70) in the concatenated analysis (Fig. 2) were strongly supported by any of the initial 8 genes (Figure S5) or the 11 additional genes sampled for a subset of taxa (data not shown). The utility of concatenated analysis can be undermined when individual gene trees conflict with each other or with the species tree [76]. Our individual gene trees showed little evidence of strong conflict (Figure S5), reinforcing the value of combined analysis for this data set, and implying that the low to modest support for some "backbone" nodes is not in general the result of conflict among gene trees. In a few instances noted below, however, there is indirect evidence that inter-gene conflict may be influencing bootstrap values.

We also see minimal evidence overall of spurious signal resulting from heterogeneity and convergence in base composition. Compositional heterogeneity is especially common at sites undergoing synonymous substitution [75], and our data are no exception; there is highly significant variation in composition across taxa in both nt3 and nt1+nt2, while heterogeneity is minor with synonymous differences removed (the degen1 data set). Conflicting signal due to compositional heterogeneity, in addition to substitutional saturation, may contribute to the inability of nt3 alone (Figure S4) to provide notable support to *any* of the among-family relationships that receive moderate to strong bootstraps from the full data set (nt123), despite providing a great majority of the total evolutionary change inferred from that data set and strongly supporting many individual families and sub-clades thereof. If composition had major effects on phylogenetic inference, however, we might expect to see repeated instances of conflicting moderate to strong bootstrap values between the total data set (nt123), dominated by synonymous change, and non-synonymous change only, as estimated by the degen1 analysis. No such cases were found, although several examples of lesser conflict are pointed out below. Rather than conflicting, the signals from synonymous and non-synonymous change appear to be largely complementary.

Our results provide another instance in which deliberately unequal gene sample augmentation markedly improves support for deeper nodes without introducing any apparent artifacts due to large blocks of non-random missing data. Nt123 analyses of the 8-gene "complete" matrix (27% inadvertently missing data due to sporadic failures of amplification or sequencing) and the deliberately-incomplete 8–27 gene matrix (55% missing data) yielded nearly identical topologies and similar bootstrap values. The 8–27 gene analysis produced higher support overall, however, and markedly increased bootstraps for several deeper nodes, including Yponomeutoidea (Fig. 2, **node 10**). Similar findings have been reported in several recent studies of Lepidoptera [30,33,35].

The potential for even a few "rogue" taxa to substantially reduce bootstrap support, obscuring otherwise strong signal on relationships among the remaining taxa, is now widely recognized [77,78]. Despite multiple proposals, however, it has been unclear how to best identify such taxa and evaluate their effect. We believe that the RogueNaRok procedure of Aberer et al. [58] is an important advance toward solving this problem. It sets out a very reasonable and explicit optimality criterion for deciding which and how many potential rogue taxa should be removed, balancing the increased support gained by deleting those taxa against the information lost through their deletion, and provides well-tested heuristic algorithms for estimating an optimal set of taxa to delete. Application of RogueNaRok following our 8–27 gene, 139-taxon nt123 analysis identified 18 rogue taxa meriting deletion. Removal of these taxa resulted in substantial bootstrap support increases for five nodes, most notably an increase from 76 to 99% for Yponomeutoidea. We predict that RogueNaRok will prove widely useful in phylogenetic studies of large taxon sets.

Monophyly, Composition and Phylogenetic Position of Yponomeutoidea

In this and subsequent sections we evaluate the implications of our molecular results for current understanding of the phylogeny of yponomeutoids, and for their classification. Our exposition proceeds from the base to the tips of the tree in Figure 2, and makes repeated reference to the node numbers labeled on that tree. Representative adult habitus images for nearly all of the 16 families and subfamilies discussed below are provided in Figure 5. The species diversities, geographic distributions and larval feeding habits of these families and subfamilies are summarized in Figures 6 and 7.

All of our molecular analyses support monophyly for Yponomeutoidea (Fig. 2, **node 11**) in approximately the sense of Kyrki [25,26]. Bootstrap support is moderate (BP = 76%, nt123) for the full data set but rises to very strong (BP = 99, nt123) when the 18 rogue taxa are removed. Kyrki [25] initially proposed a single synapomorphy for Yponomeutoidea, the presence of posterior expansions on the 8th abdominal pleuron ("pleural lobes") in males. He later added another possible synapomorphy, a transverse ridge on the second abdominal sternite [26]. On this basis he included seven families: Yponomeutidae, Plutellidae (including Acrolepiidae, later separated by Dugdale et al. [31]), Ypsolophidae, Glyphipterigidae, Heliodinidae, Lyonetiidae, and Bedelliidae. This hypothesis had been questioned because it requires independent losses of the two synapomorphies in some of

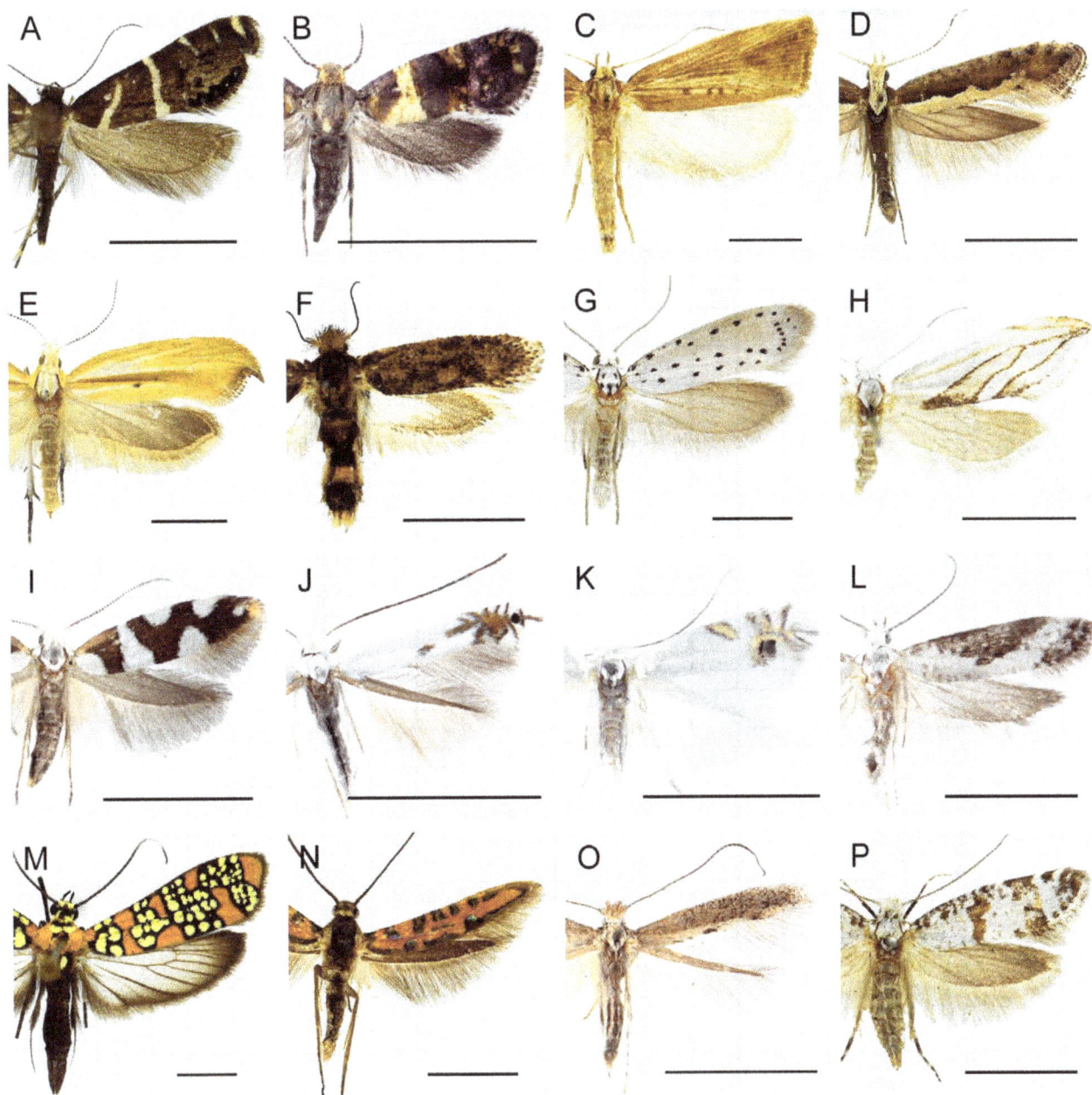

Figure 5. Representative adult habitus images of all yponomeutoid families and subfamilies recognized in this study. Scale bar = 5 mm. A. Glyphipterigidae: Glyphipteriginae, *Glyphipterix bifasciata* (Walsingham); B. Glyphipterigidae: Acrolepiinae, *Acrolepia xylophragma* (Meyrick); C. Glyphipterigidae: Orthoteliinae, *Orthotelia sparganella* (Thunberg); D. Plutellidae, *Plutella xylostella* (Linnaeus); E. Ypsolophidae: Ypsolophinae, *Ypsolopha blandella* (Christoph); F. Ypsolophidae: Ochsenheimeriinae, *Ochsenheimeria vacculella* Fisher von Roeslerstamm; G. Yponomeutidae: Yponomeutinae, *Yponomeuta padellus* Linnaeus; H. Yponomeutidae: Saridoscelinae, *Saridoscelis kodamai* Moriuti; I. Argyresthiidae, *Argyresthia brockeella* (Hübner); J. Lyonetiidae: Lyonetiinae, *Lyonetia ledi* Wocke; K. Lyonetiidae: Cemiostominae, *Leucoptera spartifoliella* (Hübner); L. Praydidae, *Prays fraxinella* (Bjerkander); M. Attevidae, *Atteva aurea* (Fitch); N. Heliodinidae, *Embola ciccella* (Barnes et Busck); O. Bedelliidae, *Bedellia somnulentella* (Zeller); P. Scythropiidae **stat. rev.**, *Scythropia crataegella* (Linnaeus).

the included groups [31]. In our results, the main remaining question about the composition of Yponomeutoidea concerns Lyonetiidae. Our analyses always separate Lyonetiinae from Cemiostominae, placing the former inside Yponomeutoidea but the latter outside, among the gracillarioids. However, the position of *Perileucoptera*, our sole cemiostomine, is exceptionally unstable. It is identified as a rogue taxon by the RNR analysis, and our AU test cannot reject the monophyly of Lyonetiidae (Table 3).

Among the out-groups included in our analyses, Gracillarioidea sensu van Nieukerken et al. [2], i.e. with Douglasiidae excluded, were strongly supported (Fig. 3, **node 67**; BP 85–97, all analyses) as the closest relatives to Yponomeutoidea. This clade has been strongly supported in almost all previous molecular studies (e.g. [28,30,35]). However, the deeper divergences within Yponomeutoidea+Gracillarioidea (the G.B.R.Y. clade of Kawahara et al. [35]) are very weakly supported. Like Kawahara et al. [35], we find no molecular evidence for monophyly of Gracillarioidea.

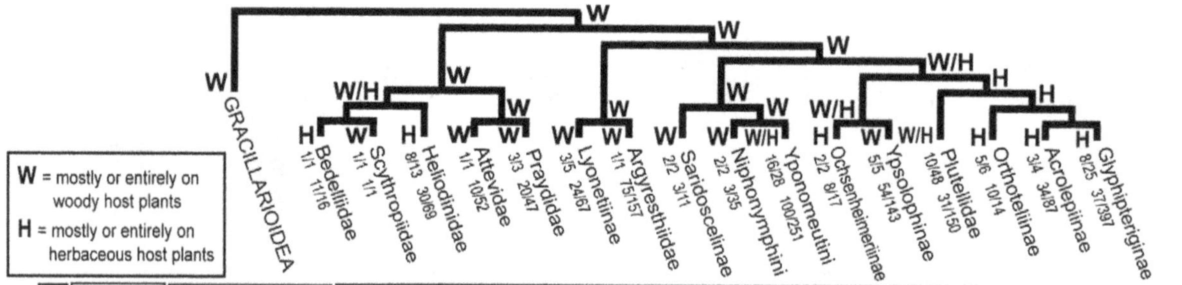

W = mostly or entirely on woody host plants
H = mostly or entirely on herbaceous host plants

Taxon header (with species/genera counts):
- GRACILLARIOIDEA
- Bedelliidae 1/1 11/16
- Scythropidae 1/1 1/1
- Heliodinidae 8/13 30/69
- Attevidae 1/1 10/52
- Praydidae 3/3 20/47
- Lyonetiidae 3/5 24/67
- Argyresthiidae 1/1 75/157
- Saridoscelinae 2/2 3/11
- Niphonymphini 2/2 3/35
- Yponomeutini 16/28 100/251
- Ochsenheimeriinae 2/2 8/17
- Ypsolophinae 5/5 54/143
- Plutellidae 10/48 31/150
- Orthoteliinae 5/6 10/14
- Acrolepiinae 3/4 34/87
- Glyphipteriginae 8/25 37/397

Clade	Order	Family	GRACILLARIOIDEA	Bedelliidae	Scythropidae	Heliodinidae	Attevidae	Praydidae	Lyonetiidae	Argyresthiidae	Saridoscelinae	Niphonymphini	Yponomeutini	Ochsenheimeriinae	Ypsolophinae	Plutellidae	Orthoteliinae	Acrolepiinae	Glyphipteriginae
A	Laurales	Lauraceae								1									
A	Magnoliales	Annonaceae								1									
A	Piperales	Piperaceae				1													1
B	Alismatales	Araceae																	1
B	Asparagales	Amaryllidaceae																3	
B	Asparagales	Asteliaceae															1		
B	Asparagales	Iridaceae															1		
B	Dioscoreales	Dioscoreaceae														1		6	
B	Liliales	Liliaceae															1	5	
B	Liliales	Smilacaceae																3	1
B	Pandanales	Pandanaceae														1			
B	Arecales	Arecaceae														1			
B	Poales	Cyperaceae													2		2	3	12
B	Poales	Juncaceae																	11
B	Poales	Poaceae		2											5		1	3	4
B	Poales	Typhaceae															1		
B	Zingiberales	Zingiberaceae																	1
—	Proteales	Proteaceae								2									
—	Ranunculales	Berberidaceae														1			
—	Saxifragales	Altingiaceae												1					
—	Saxifragales	Crassulaceae											3						2
—	Saxifragales	Grossulariaceae								1			3						
—	Saxifragales	Saxifragaceae											12						
—	Caryophyllales	Aizoaceae				1													
—	Caryophyllales	Amaranthaceae				1													
—	Caryophyllales	Nyctaginaceae				22													
—	Caryophyllales	Phytolaccaceae				1													
—	Caryophyllales	Portulacaceae				3													
—	Santalales	Loranthaceae											3						
—	Santalales	Santalaceae											3						
C	Celastrales	Celastraceae											35			3			
C	Fabales	Fabaceae														1			
C	Fabales	Betulaceae						1		3	4		3			2			
C	Fabales	Fagaceae								3	4		2	1		12			
C	Fabales	Juglandaceae						1											
C	Fabales	Myricaceae								2	1					1			
C	Malpighiales	Salicaceae								3	1		3			8			
C	Rosales	Cannabaceae								1									
C	Rosales	Rhamnaceae								1			2						
C	Rosales	Rosaceae			1					6	14		15			10			
C	Rosales	Ulmaceae						1		1	1					2			
C	Rosales	Urticaceae	1																3
D	Brassicales	Brassicaceae														16			
D	Brassicales	Capparaceae														2			
D	Malvales	Dipterocarpaceae								1									
D	Malvales	Malvaceae														1			
D	Malvales	Thymelaeaceae								5									
D	Myrtales	Myrtaceae								1	1		1						
D	Myrtales	Onagraceae				1													
D	Sapindales	Burseraceae						1											
D	Sapindales	Rutaceae						7											
D	Sapindales	Sapindaceae									1					4			
D	Sapindales	Simaroubaceae					9												
E	Ericales	Ericaceae								2	3	3	4					1	
E	Ericales	Theaceae								2									
E	Gentianales	Rubiaceae											1						
E	Lamiales	Lamiaceae																1	
E	Lamiales	Oleaceae				7							5						
E	Solanales	Convolvulaceae	8																
E	Solanales	Solanaceae																6	1
F	Apiales	Araliaceae				1	1												
F	Asterales	Asteraceae																12	
G	Dipsacales	Caprifoliaceae						3								8	2		
G	Ephedrales	Ephedraceae														10			
H	Cupressales	Cupressaceae								29									
H	Pinales	Podocarpaceae																7	
H	Pinales	Pinaceae								19			8			4			

Left margin clade labels: Monocots (A, B); Rosids (C, D); Asterids (E, F).

Figure 6. Host plant families of 16 major yponomeutoid lineages. The cladogram is simplified from figure 2, annotated with predominant growth form of host plants ('W' for woody plants vs. 'H' for herbaceous plants). Fractions below yponomeutoid taxon names denote host record completeness for genera and species (in that order), calculated from the number of genera or species with host records relative to the total number of known genera or species. Host plant families used by each lineage are denoted by gray cells showing the numbers of species feeding on that plant family. Symbols denote the dominant growth-forms of each plant family: shaded circles = trees and shrubs; open circles = herbs; and shaded stars = veins and lianas. Capital letters next to host plant orders denote membership in clades above the order level: A – magnoliids, B – commelinids, C – fabids, D – malvids, E – lamiids, F – campanulids, G – Gnetophyta, and H – Pinophyta.

Eventually it may be reasonable to merge Gracillarioidea into an Yponomeutoidea *sensu lato*, but such a change is beyond the scope of the present study.

Our results support several earlier morphology-based proposals that excluded a variety of taxa from membership in, or close relatedness to, Yponomeutoidea. Galacticoidea, Urodoidea and Schreckensteinioidea, once placed in Yponomeutoidea [79,80,81,82], are decisively excluded from Yponomeutoidea+−Gracillarioidea, here (Fig. 3, **node 67**) and in all other recent molecular studies. Removal of the putative yponomeutid genus *Nosymna* Walker, 1864 to Zygaenoidea by Heppner [83] is also confirmed by our analyses (Fig. 3, **node 66**), as is the exclusion of *Cycloplasis* Clemens, 1864 from Heliodinidae by Hsu and Powell [84]. Our results place *Cycloplasis* in Apoditrysia+Gelechioidea (Fig. 3, **node 65**; BP = 71–83, all analyses). Two genera previously placed in Lyonetiidae, *Philonome* Chambers, 1872 and *Corythophora* auct Braun, 1915, are here strongly supported as belonging to Tineoidea (Fig. 3, **node 63**; BP = 90, nt123).

Basal Split within Yponomeutoidea

Within Yponomeutoidea (Fig. 2, **node 11**), our results provide moderate to strong support for most nodes above the family level, allowing us to construct a working hypothesis of higher phylogeny across the superfamily. In presenting this hypothesis below, we make repeated use of informal clade names based primarily on the first letters of the names of the included families.

In the tree of Fig. 2, the basal split is between a 'PAHSB clade' (Fig. 2, **node 4**; maximum BP = 75, nt123 partitioned) consisting of Praydidae, Attevidae, Heliodinidae, Bedelliidae and *Scythropia*, and a 'YYPGAL clade' (Fig. 2, **node 17**; maximum BP = 77, rogue-pruned nt123) consisting of Yponomeutidae, Ypsolophidae, Plutellidae, Glyphipterigidae, Argyresthiidae and Lyonetiidae. Because bootstrap support for these clades is modest at best, and they are contradicted, albeit very weakly, by degen1, we regard them as provisional. Neither clade has ever been proposed on the basis of morphology. However, our working hypothesis, including this basal split, fits the molecular data much better than any of the alternative proposals for among-family relationships shown in Figure 1, all of which are decisively rejected (P<0.001) by the AU test (Table 3).

Relationships within the PAHSB Clade

This clade (Fig. 2, **node 4**), for which no morphological synapomorphies are yet known, contains five relatively small yponomeutoid groups. It divides basally into a 'PA clade' (Fig. 2, **node 5**; maximum BP = 82, nt123 partitioned) containing the Praydidae and Attevidae, and an 'HSB clade' (Fig. 2, **node 2**; BP = 90, nt123) consisting of Heliodinidae, Bedelliidae and *Scythropia*. The latter was previously treated as a subfamily of Yponomeutidae.

The PA clade receives moderate to strong support from nearly all of our analyses, except that it is very weakly contradicted by degen1 (BP≤38). The groups based on *Prays* and *Atteva*, here treated as families following van Nieukerken et al. [2], were treated as subfamilies of Yponomeutidae by Kyrki [26], while others have regarded the *Prays* group as closer to Plutellidae than to Yponomeutidae [20,27,85]; Heppner [1] treated it as a subfamily of Plutellidae. All of these hypotheses are strongly contradicted by our results.

While previous ideas about their phylogenetic position receive no support, the molecular data do corroborate Kyrki's [26] assertion of a close relationship between the *Prays* and *Atteva* groups, based on two synapomorphies, the lack of a pecten on the antennal scape and the presence of a larval cranial seta P_1 that lies on or above the line defined by setae Af_2–P_2. A possible additional synapomorphy is the presence of less than four segments in the maxillary palp. Ulenberg [86] also recovered the pairing of the *Prays* and *Atteva* groups within Yponomeutidae, in a parsimony analysis using Kyrki's [26] characters. These putative synapomorphies might be doubted because they are reductions or homoplasious, but the molecular results suggest that they are real. We nonetheless treat these groups as separate families because the molecular evidence is not yet completely incontrovertible.

Monophyly of the Praydidae, here represented by *Prays* and *Atemelia*, is very strongly supported by our data (Fig. 2, **node 7**; BP = 100, all analyses). The members of this group are easily distinguished from other yponomeutoids by an unusually broad male 8th sternum and by female apophyses anteriores lacking a branched costa at the base [20,27]. Our data also strongly resolve the relationships among the four *Prays* species sampled (Fig. 2, **nodes 9, 10**; BP = 89–100, all analyses). Praydidae, comprising 3 genera and 47 species, are a cosmopolitan group that is most diverse in the Old World. The larvae are initially endophagous feeders in leaves, buds or shoots of woody dicots of diverse families; in some species, older larvae feed externally in webs [31].

The two species of *Atteva* included in our sample are likewise strongly grouped (Fig. 2, **node 6**; BP = 100). The Attevidae can be defined by four autapomorphies [25]: the presence of chaetosema; reduction of the hindleg tibia and tarsus, especially in the male; the presence of two subventral setae on the larval meso- and metathorax; and concealment of the labial palps in the pupa. Attevidae are a predominantly pan-tropical group of 52 described species in a single genus *Atteva*, most diverse in the Oriental region. The larvae are communal leaf webbers on woody dicots, with >90% of records from Simaroubaceae [31].

Monophyly of the probable sister group to the PA clade, the HSB clade (Fig. 2, **node 2**; maximum BP = 90, nt123), is supported by all of our analyses. The grouping of Heliodinidae, Bedelliidae and *Scythropia* has not been previously proposed. The closest antecedents are the grouping of Heliodinidae, Bedelliidae and Lyonetiidae by Kyrki [26] and that of Lyonetiidae (including Bedelliinae), Acrolepiidae, and Heliodinidae by Heppner [1]. Kyrki [26] proposed three possible synapomorphies for Heliodinidae+Bedelliidae: larva with a long spinneret; larval seta V_1 not apparent on the thorax; and pupa without a cocoon. It is not known whether *Scythropia* shares any of these traits. The search for morphological synapomorphies of the strongly-supported HSB clade merits further effort.

The molecular data strongly favor monophyly for Heliodinidae as sampled here (Fig. 2, **node 3**; BP = 100, all analyses), corroborating the re-definition of this family by Hsu and Powell

Feeding Modes

■ Internal
▬ External
■▬ Ambiguous

Taxon	# of spp.*	Predominant feeding mode	Predominant diet breadth (%)**	Geographic distribution (%)***
Glyphipterigidae Glyphipteriginae	397	leaf miners or stemborers; herbs, 78% on monocots	Oligophagous (37/37 = 100%)	AO (25) OR (21) NT (19) PA (16) NA (11) ET (8)
Glyphipterigidae Acrolepiinae	87	leaf miners or borers in stems, flower buds, seeds; monocots and asterid herbs	Oligophagous (33/34 = 97%)	PA (50.57) ET (14) NT (9) OR (8) NA (7) AO(5)
Glyphipterigidae Orthoteliinae	14	borers in stems and tiller bases; monocots	Oligophagous (8/10 = 80%)	AO (93) PA (7)NT****
Plutellidae	150	leaf webbers; >50% on Brassicales	Oligophagous (31/31 = 100%)	AO (39) PA (21) OR (14) ET (11) NT (9) NA(6)
Ypsolophidae Ypsolophinae	143	leaf webbers; trees and shrubs, many families	Oligophagous (51/54 = 94%)	PA (66) NA (25) OR (5) NT (3) ET (1)
Ypsolophidae Ochsenheimeriinae	17	leaf miners or stem borers; Poales	Oligophagous (8/8 = 100%)	PA (94) OR (6)
Yponomeutidae Yponomeutini	251	leaf webbers, a few pine needle miners; mostly trees and shrubs, many families	Oligophagous (100/100= 100%)	PA (50) OR (17) AO (15) ET (9) NA (7) NT (2)
Yponomeutidae Niphonymphini	35	leaf webbers; Fabales	Oligophagous (3/3 = 100%)	PA (57) OR (32) ET (5) NA (3) AO (3)
Yponomeutidae Saridoscelinae	11	leaf webbers; shrubs, all Ericaceae	Oligophagous (3/3 = 100%)	NT (41) PA (25) NA (17) OR (17)
Argyresthiidae	157	leaf miners or borers in flower buds, seeds; trees and shrubs, > 60% on conifers	Oligophagous (75/75 = 100%)	PA (49) NA (30) ET (10) OR (6) NT (4) AO (1)
Lyonetiidae Lyonetiinae	67	leaf miners; woody dicots, mostly rosids	Oligophagous (20/24 = 83%)	PA (27) OR (21) AO (18) NT (13) ET (12) NA (9)
Praydidae	47	leaf miners or borers in flower buds or shoots; woody dicots	Oligophagous (19/20 = 95%)	PA (36) OR (20) AO (18) ET (16) NA (6) NT (4)
Attevidae	52	leaf webbers; woody dicots, >90% on Simaroubaceae	Oligophagous (10/10 = 100%)	OR (36) NT (32) AO (28) ET (2) NA (2)
Heliodinidae	69	Leaf miners, some are leaf webbers (ancestral); >85% on Caryophyllales	Oligophagous (30/30 = 100%)	NA (47) NT (42) AO (8) PA (3)
Scythropiidae	1	leaf miners then webbers in later instars; woody Rosaceae	Oligophagous (1/1 = 100%)	PA (100)
Bedelliidae	16	leaf miners; herbs, >70% on Convolvulaceae	Oligophagous (11/11 = 100%)	AO (42) OR (19) PA (19) ET (8) NA (8) NT (4)

* only described species counted.
** the number of oligophagous species/ the number of total species whose hosts are known.
*** AO: Australoceanian, ET: Ethiopian, NA: Nearctic, NT: Neotropical, OR: Oriental, PA: Palearctic.
**** based on the undescribed species included in our analyses

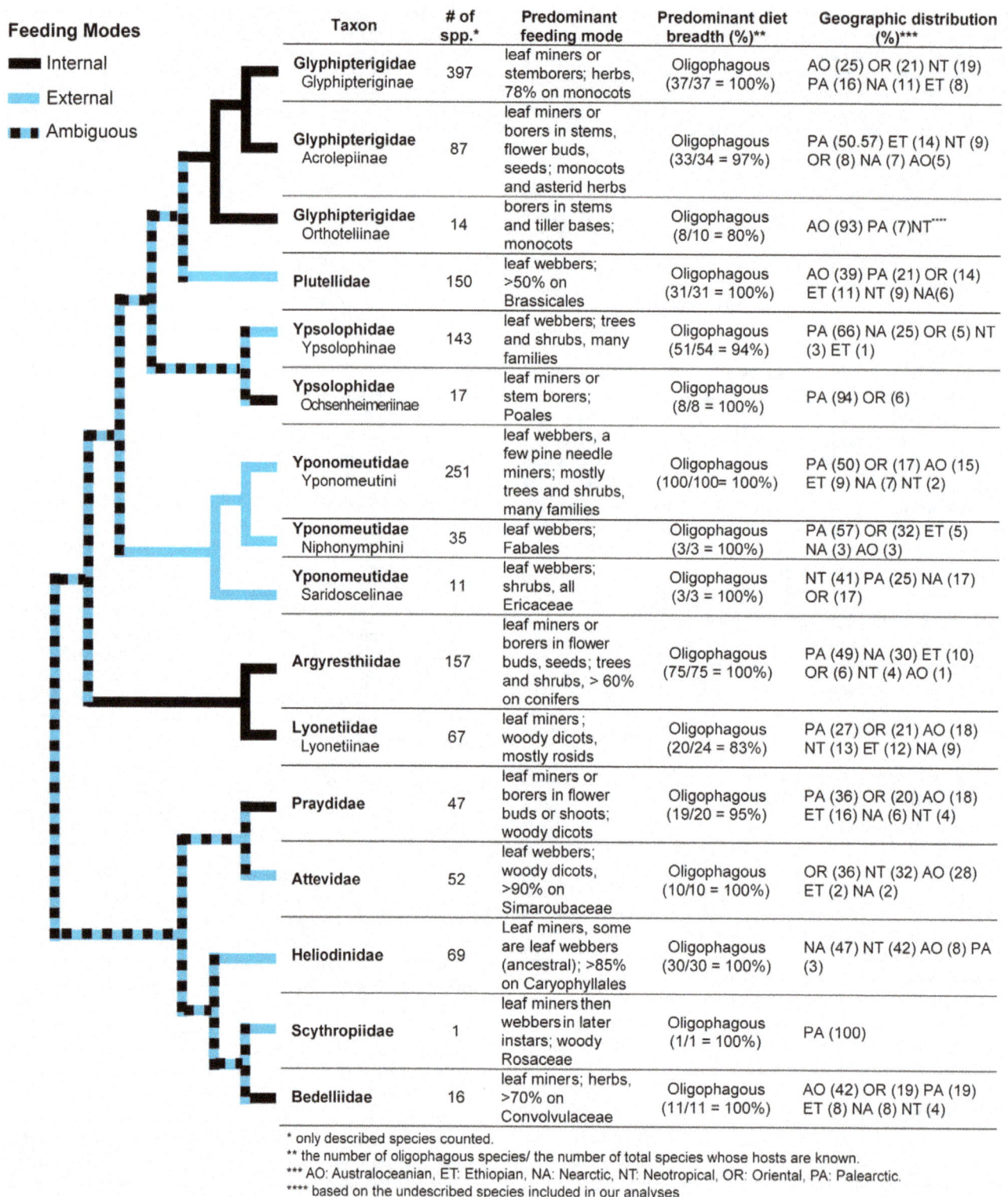

Figure 7. Species diversity, feeding mode, diet breadth and geographic distribution of 16 major yponomeutoid lineages. The tree topology is that of Figure 6. Branch colors indicate predominant feeding modes: black = internal feeding; blue = external feeding; alternating black and blue = state ambiguous under parsimony optimization.

[84]. Kyrki [25] suggested four synapomorphies for heliodinids: in the adult, smooth scaling on the head and absence of the CuP vein in forewing; and in the pupa, strong lateral ridges and stiff, long lateral and dorsal bristles. Only the last trait, however, is limited to the re-defined Heliodinidae. In their cladistic analyses, Hsu and Powell [84] found three additional synapomorphies: female

apophyses anteriores with ventral branches originating from a fused medial sclerite; male tegumen greatly expanded posteriorly, forming a conical or tubular sclerotized sac; and the forewing M vein with two branches. Adult diurnality is another possible synapomorphy [31]. Our data strongly resolve two of the three nodes subtending the five heliodinid genera sampled and yield relationships among these genera that are entirely concordant with the morphological cladistic analysis of Hsu and Powell [84]. Heliodinidae are a widespread but primarily New World group of 13 genera and 69 described species [2]. The larvae are variable in feeding habits, with most species feeding internally in leaves, stems or fruits, while others are externally-feeding leaf webbers, all on herbaceous plants. The great majority of records (>85%) are from Caryophyllales, primarily Nyctaginaceae [84].

The apparent sister group to Heliodinidae is the strongly supported pairing of *Bedellia*+*Scythropia* (Fig. 2, **node 1**), favored in all of our analyses, with bootstraps as high as 86% (8–27 gene nt123). This is an entirely new hypothesis. No morphological synapomorphies are apparent, but a search for these would be worthwhile, given the strength of the molecular evidence. Bedelliidae are often confused with Lyonetiidae or Gracillariidae (see [87] for detailed history). Heppner [88] recently transferred *Philonome* and *Euprora* to Bedelliidae (Bedelliinae auct), but our analyses very strongly place these genera in Tineidae instead (Fig. 3). Kyrki [25,26] maintained separate family status for *Bedellia*. The widespread contrasting view, that *Bedellia* constitutes a subfamily of Lyonetiidae [1,87,89,90], is unsupported by clear morphological synapomorphies and is likewise strongly rejected by our analyses, including the AU test (Table 3, #12). Bedelliidae are a monogeneric, cosmopolitan group of 16 species, most diverse in the Old World [2]. The larvae are leaf miners in herbaceous plants, with 70% of records from Convolvulaceae [31].

The position of *Scythropia* has likewise been controversial. Kyrki [26] suggested that it constitutes the first-diverging subfamily of Yponomeutidae, while others, such as Friese [20], Moriuti [27], and Heppner [1], grouped this genus with Plutellidae. Our results strongly contradict all previous hypotheses about the systematic position of *Scythropia*. We are reluctant to combine it with Bedelliidae, given the current complete absence of morphological support for such a pairing, and therefore hereby elevate Scythropiinae to Scythropiidae **stat. rev.** Larvae of the single, Palearctic species, *Scythropia crataegella*, are initially leaf miners and subsequently feed externally in a communal web, on *Crataegus* and sometimes other woody Rosaceae [31].

Relationships within the YYPGAL Clade

The majority of yponomeutoid species belong to the provisional YYPGAL clade (Fig. 2, **node 17**). This group is monophyletic in all analyses except degen1, where it is only very weakly contradicted (BP<20; tree not shown). However, bootstrap support is moderate at best (BP = 77, rogue-pruned nt123). Limited support for this node may result in part from conflict among gene trees, as suggested by the fact that the bootstrap value for 8–27 genes is lower than that for 8 genes (67 vs. 72%). No grouping like the YYPGAL clade has been proposed previously, and no morphological synapomorphies are apparent.

Within the YYPGAL clade there are three main sub-clades, each with moderate or strong support: an 'AL clade' consisting of Argyresthiidae and Lyonetiidae (Fig. 2, **node 12**; maximum BP = 76, rogue-pruned nt123); Yponomeutidae (Fig. 2, **node 18**; BP≥97, all analyses); and a 'YPG clade' consisting of Ypsolophidae, Plutellidae and Glyphipterigidae (Fig. 2, **node 36**; BP≥97, all analyses). Relationships among these three entities, however, are less clear. All analyses favor grouping of Yponomeutidae plus the

YPG clade to the exclusion of the AL clade (Fig. 2, **node 35**), with the weakly supported exception of degen1. However, bootstrap support for this relationship never exceeds 65%, and is higher for 8 genes than for 8–27 (63 versus 56%), again suggesting the presence of inter-gene conflict.

Relationships within the AL Clade

The AL clade (Fig. 2, **node 12**) comprises Argyresthiidae plus Lyonetiidae: Lyonetiinae. It is monophyletic in all of our analyses except degen1, where it is only very weakly contradicted (BP<20; tree not shown). However, bootstrap support is moderate at best (BP = 77, rogue-pruned nt123). Limited support for this node may result in part from conflict among gene trees, as suggested by the fact that the bootstrap value for 8–27 genes is lower than that for 8 genes (69 vs. 72%). Grouping of these two taxa has never been proposed previously, and no morphological synapomorphies are apparent. In view of all the evidence, we regard this clade as only provisionally established. However, Kyrki's [26] inclusion of Argyresthiidae as a subfamily of Yponomeutidae can be confidently ruled out.

Monophyly for Argyresthiidae as sampled here is very strongly supported (Fig. 2, **node 13**; BP = 100, all analyses). The family had been thought to be monobasic, defined by unique features of the male genitalia including a laterally produced vinculum and sensilla ornaments on the socii [31]. Our results, however, very strongly favor inclusion of a well-supported clade of several Neotropical yponomeutoids (Fig. 2, **node 15**; BP = 100, all analyses) that were originally assigned to, but later excluded from, Acrolepiinae [69]. These species are morphologically divergent from typical *Argyresthia*, which will necessitate a reevaluation of the currently hypothesized argyresthiid synapomorphies. Argyresthiidae are a cosmopolitan group of 157 described species, most species-rich in the Holarctic. The larvae are typically leaf miners or borers in flower buds, seeds or twigs of trees and shrubs [31]. About half of the records are from conifers.

Monophyly of the subfamily Lyonetiinae as sampled here (Fig. 2, **node 13**), comprising two species each of *Lyonetia* and *Phyllobrostis*, is supported by all but one of our analyses, with bootstraps up to 92%, although the two genera are separated by several nodes in the degen1 tree (BP≤64). A close relationship between *Lyonetia* and *Phyllobrostis*, to the exclusion of *Leucoptera* (Cemiostominae), was also supported by a cladistic analysis of morphology [91]. Lyonetiinae are a cosmopolitan group of 5 genera and 67 described species [2]. The larvae are typically leaf miners on woody dicots, of diverse families [31].

The Cemiostominae, in contrast, are one of the most problematic groups in our study. *Perileucoptera*, our sole representative, was identified as a rogue taxon. Cemiostomines differ from Lyonetiinae in many features, e.g. in having shorter antennae, different forewing pattern elements, and spine-like setae on the adult abdomen, leading some authors (e.g. [19,92]) to place them in their own family. Kyrki [26], however, proposed uniting Cemiostominae and Lyonetiinae into a single family, citing as a possible synapomorphy the shared possession of an "eye cap" formed by scales on the antennal scape. Our molecular analyses nearly always separated the two subfamilies, excluding Cemiostominae but not Lyonetiinae from Yponomeutoidea, concordant with the view of Börner [19]. However, bootstrap support for Yponomeutoidea is modest at best except when *Perileucoptera* is excluded from the analysis, and support for alternative positions among the Gracillarioidea for *Perileucoptera* had very low support. Moreover, the four-gene nt123 analysis (Figure S2) grouped Lyonetiinae with Cemiostominae, albeit with very weak support. Finally, our AU test cannot reject the monophyly of Lyonetiinae+Cemiostominae as sampled here (Table 3: # 11). Mutanen et al. [29] also failed to recover Cemiostominae (represented by *Leucoptera*)+Lyonetiinae. Their analysis places *Leucoptera* as sister group to *Atteva* with 76% bootstrap support. Given the weak and conflicting molecular evidence on the placement of *Perileucoptera*, we tentatively retain Cemiostominae as a subfamily of Lyonetiidae pending further investigation. Although the composition of this family remains in doubt, our results do strongly confirm Kyrki's [25] placement of Lyonetiidae in or near Yponomeutoidea: both subfamilies fall within the strongly supported clade Yponomeutoidea+Gracillarioidea (Fig. 3, **node 67**; BP 85–97, all analyses). The Cemiostominae are a cosmopolitan group of about 6 genera and 120 described species; the larvae are typically leaf miners in woody dicots of diverse families [31].

Composition of and Relationships within Yponomeutidae

Different authors have hypothesized very different compositions for Yponomeutidae (Table 1). Our analyses very strongly support a circumscription of this family (Fig. 2, **node 18**; BP = 97–100, all analyses) that corresponds exactly to Yponomeutinae sensu Moriuti [27]. Moriuti [27] proposed two synapomorphies for this group, the presence of spine-like setae on the adult abdominal tergites, and a seta V_1 on the larval head that is as large as a long tactile seta. Kyrki ([26], and see also [86]), in contrast, assigned six subfamilies to Yponomeutidae, three of which are now the separate families Argyresthiidae, Attevidae and Praydidae [2]. Kyrki's hypothesis for Yponomeutidae has gained little support even from other morphological studies [31], and is soundly rejected by our AU test (Table 3: # 4). Yponomeutidae as delimited here are a cosmopolitan group of 32 genera and 297 described species, most diverse in the Palearctic. The larvae are usually communal leaf webbers, although some species of *Zelleria* mine pine needles [31]. A very diverse array of host families is used, mostly woody but some herbaceous.

Within his concept of Yponomeutinae, here treated as a family (Fig. 2, **node 18**), Moriuti [27] recognized two tribes, Yponomeutini and Saridoscelini, which we treat as subfamilies. One of these, here treated as Saridoscelinae, was previously restricted to *Saridoscelis*. The molecular data, however, very strongly indicate that *Saridoscelis* is the sister group to *Eucalantica*, an yponomeutoid genus of previously unsettled position (Fig. 2, **node 20**; BP = 100, all analyses). We therefore hereby re-define Saridoscelinae to include *Eucalantica*. Moriuti [27], followed by Kyrki [26] and Dugdale et al. [31], proposed two synapomorphies for *Saridoscelis*, a unique modification of the male 8th abdominal sternite, and the presence of three branches in the M vein of the hindwing. In *Eucalantica* the condition of the male 8th abdominal sternite is ambiguous; it may or may not share a derived modification with *Saridoscelis*. The number of hindwing M veins is sufficiently homoplasious in Yponomeutoidea that this character too is ambiguous evidence on the grouping of these two genera (J. Sohn, unpublished). Thus, further search is needed for morphological synapomorphies of the Saridoscelinae as here re-defined.

Within his concept of Yponomeutini, here treated as a subfamily, Moriuti [27] recognized two subtribes, here treated as the tribes Yponomeutini and Niphonymphini. The molecular evidence on monophyly of Yponomeutinae as defined here is somewhat complex due to conflicting results regarding the position of our representative of Niphonymphini, *Thecobathra*. In the nt123 and nt123 partitioned analyses, *Thecobathra* groups with Saridoscelinae, but with weak support (Fig. 2, **node 19**; BP 51–59). On the other hand, analyses emphasizing non-synonymous change (degen1 and codon model) place it as sister group to Yponomeutini, with strong support (BP = 82, degen1). Previous morpholog-

ical studies have also supported monophyly for Niphonymphini+Yponomeutini, equivalent to Yponomeutidae sensu Friese [20] and Yponomeutini sensu Moriuti [27]. The 8–27 gene degen1 result, being stronger and concordant with morphology, seems more persuasive than the nt123 placement for *Thecobathra*. We therefore provisionally recognize a subfamily Yponomeutinae composed of Niphonymphini+Yponomeutini.

Our analyses provide robust, consistent evidence on the initial divergences within Yponomeutini as sampled here. *Metanomeuta* branches off first (Fig. 2, **node 21**; BP = 100, nt123), followed by *Paraswammerdamia* (Fig. 2, **node 22**; BP = 99, nt123). *Yponomeuta* is strongly paired with *Teinoptila* (Fig. 2, **node 23**; BP≥94, all analyses), and relationships among the four sampled species of *Yponomeuta* (Fig. 2, **nodes 24, 25, 26**) are also very strongly resolved. The remaining Yponomeutini comprise an assemblage whose monophyly is weakly supported by nt123 (Fig. 2, **node 28**; BP = 56, nt123) and weakly contradicted by degen1, which allies *Klausius* instead with *Teinoptila*+*Yponomeuta* (BP = 57, tree not shown). The remainder of the assemblage (Fig. 2, **node 28**) divides into two strongly supported clades, one consisting of *Cedestis*+*Zelleria retiniella* (Fig. 2, **node 29**; BP = 100, nt123), and the other (Fig. 2, **node 32**; BP = 90, nt123) containing additional species of *Zelleria* plus three other genera, relationships among which are not clearly resolved. These results strongly contradict all previous hypotheses about relationships within Yponomeutini, including Kloet & Hincks [93], Moriuti [27], Heppner [1] and Ulenberg [86]. In addition, our data provide strong evidence for polyphyly of *Zelleria* (Fig. 2, **nodes 30, 34**). Clearly there is much further work to be done on the systematics of Yponomeutini.

Relationships within the YPG Clade

In our analyses, the sister group to Yponomeutidae consists of Ypsolophidae, Plutellidae and Glyphipterigidae. Grouping of the latter three families, the 'YPG clade', is very strongly supported (Fig. 2, **node 36**; BP = 98–100, all analyses). This clade has never been proposed previously, and no morphological synapomorphies are known. The basal split within the YPG clade, also very strongly supported, unites Plutellidae and Glyphipterigidae to the exclusion of Ypsolophidae (Fig. 2, **node 45**; BP≥99, all analyses).

Monophyly of Ypsolophidae including *Ochsenheimeria* is very strongly supported by our data (Fig. 2, **node 37**; BP = 100, all analyses). A similar result was reported by Mutanen et al. [29]. The enigmatic *Ochsenheimeria* group was long assigned to Tineoidea before Kyrki [25] allied it with Yponomeutoidea. Kyrki [26] proposed eight synapomorphies for Ypsolophidae including Ochsenheimeriinae: hindwing veins with Rs and M_1 stalked or coincident; male genitalia with tegumen deeply bilobed at the anterior margin; tuba analis membranous and densely setose; phallus with two cornuti or cornutal zones; female genitalia with long anterior and posterior apophyses; termination of ductus seminalis on ductus bursae close to ostium; signum elongate, band-like, usually with two transverse ridges; and, pupal cremaster without setae. Heppner's [1] placement of Ochsenheimeriinae (raised to the family level) as sister group to all other yponomeutoids (Fig. 1B) is strongly rejected by our data. Our data likewise reject proposals by Moriuti [27] and Heppner [1] to merge Ypsolophidae minus Ochsenheimeriinae into Plutellidae.

Within Ypsolophidae sensu Kyrki, our data provide somewhat contradictory evidence on the basal split. In all analyses that include synonymous change, Ypsolophinae are monophyletic, excluding *Ochsenheimeria*, with very strong support (Fig. 2, **node 38**; BP = 100, nt123). In contrast, under degen1, *Ochsenheimeria* is nested two nodes deep within Ypsolophinae, as sister group to *Bhadorcosma*, with 68% bootstrap support, contradicting two groupings (Fig. 2, **nodes 38, 39**) that have ≥99% bootstrap under nt123. While the signal from nt123 is stronger, we cannot confidently rule out the hypothesis of a paraphyletic Ypsolophinae [31] until this striking conflict is explained. Apart from the position of *Ochsenheimeria*, however, our data provide very strong resolution of all relationships within Ypsolophinae as sampled here (Fig. 2, **nodes 39, 40, 41**; BP = ≥99, nt123). *Ypsolopha* is always paraphyletic in our trees, with respect to either *Bhadorcosma* and *Ochsenheimeria* (degen1) or *Bhadorcosma* alone (all other analyses). Ypsolophidae are a cosmopolitan group of 5 genera and160 described species, most diverse in the Palearctic [2]. The larvae of Ypsolophinae are most often leaf webbers on woody plants, of many different families, while those of Ochsenheimeriinae are leaf miners and borers in Poaceae, Cyperaceae and Juncaceae (Poales).

Relationships within the PG Clade

A sister group relationship between Plutellidae and Glyphipterigidae, very strongly supported by our data (Fig. 2, **node 45**; BP≥99, all analyses), has not been previously proposed. Given the exceptionally robust molecular evidence, a search for morphological synapomorphies seems warranted. Two possible candidates, hypothesized by Kyrki ([26], but see [31]) to unite Plutellinae and Acrolepiinae (now part of Glyphipterigidae), are lamellae postvaginales of the female genitalia consisting of two setose lobes, and loosely meshed cocoons.

Our analyses provide strong and consistent support for monophyly of Plutellidae (Fig. 2, **node 42**; BP = 93, nt123). Like Mutanen et al. [29], we find that the so-called "mega-plutellids" of New Zealand and Tasmania, here represented by *Proditrix* and *Doxophyrtis*, are actually nested within Glyphipterigidae: Orthoteliinae, as sister group to *Orthotelia* (Fig. 2, **node 49**; BP = 100, all analyses). Within Plutellidae sensu stricto [2] as sampled here, our data strongly support a basal split between a North Temperate "core" group consisting of *Plutella* and allies (Fig. 2, **node 44**; BP = 100, all analyses), and a tropical lineage (Fig. 2, **node 43**; BP≥93, all analyses) here represented by the Namibian *Deryaxenistis* and an undescribed genus from Mexico. The plutellid association for *Deryaxenistis*, previously tentative [94,95], is here strongly confirmed. We suspect that this tropical plutellid lineage is greatly under-explored. Its characterization will probably result in a new morphological definition for the family. Kyrki [25] characterized Plutellidae in the restricted sense (*Plutella*-group auct) by male genitalia with curved gnathal processes surrounding the anal tube. This feature, however, is not found in the tropical clade, which may deserve subfamily status. Plutellidae are a cosmopolitan group of 48 genera and150 described species, most diverse in the Australoceanian region [2]. The larvae are typically skeletonizing leaf webbers [31]. More than half of the host records are from Brassicales.

The monophyly of Glyphipterigidae is very strongly supported in all of our analyses (Fig. 2, **node 52**; BP = 98, nt123) except degen1 and the codon model. The conflict concerns a newly-discovered, strongly-supported Neotropical clade of probable Orthoteliinae (Fig. 2, **node 47**; BP = 96, nt123). Under degen1, this clade branches off at the base of the PG clade in the ML tree, but with very weak support; the bootstrap value is actually higher (49%) for glyphipterigid monophyly. Like Mutanen et al. [29], we find Glyphipterigidae to consist of three subfamilies, Glyphipteriginae, Acrolepiinae and Orthoteliinae. Previous hypotheses based on morphology have sometimes included both Glyphipteriginae and Orthoteliinae (Table 1), but never Acrolepiinae, which have been variously treated as a subfamily of Plutellidae [26] or as a family related to Lyonetiidae and Heliodinidae [1]. Morphological synapomorphies for Glyphipterigidae in the new sense [2] have yet

to be discovered. Kyrki & Itämie [96] and Kyrki [26] proposed eight synapomorphies for Glyphipterigidae excluding Acrolepiinae. Three of these – antenna without a pecten, male genitalia without teguminal processes, and larva endophagous – are also common in Acrolepiinae. These traits are also widespread in other lepidopteran lineages, however, leaving their phylogenetic significance uncertain. Within Glyphipterigidae, our data very strongly group Acrolepiinae with Glyphipteriginae to the exclusion of Orthoteliinae (Fig. 2, **node 53**; BP≥95, all analyses). Mutanen et al. [29] reported a similar result.

Our analyses favor a broad concept of the formerly monobasic Orthoteliinae (Fig. 2, **node 46**) that includes both the New Zealand/Tasmanian "mega-plutellids" (Fig. 2, **node 50**), as proposed by Heppner [97] and corroborated also by Mutanen et al. [29], and an assemblage of undescribed genera and species from the Neotropical region. This definition of the subfamily is strongly supported (Fig. 2, **node 46**; 89≤BP≤93) by all analyses except degen1 and the codon model, which, as noted earlier, very weakly place a subclade of Neotropical species (Fig. 2, **node 47**) at the base of either Glyphipterigidae or the PG clade (BP<<50; trees not shown). No morphological synapomorphies are apparent for Orthoteliinae in the new sense.

Within Orthoteliinae, the "mega-plutellids" (Fig. 2, **node 50**) appear closely related to the monobasic Palearctic type genus *Orthotelia* (Fig. 2, **node 49**; BP = 100, all analyses), while the Neotropical fauna may prove to constitute the paraphyletic basal lineages of the subfamily. One undescribed genus from Chile ("CL67") is strongly supported as the nearest relative to the core group that includes *Orthotelia* (Fig. 2, **node 48**; 81≤BP≤96, all analyses), while the remaining Neotropical exemplars form a strongly supported clade (Fig. 2, **node 47**; BP = 96, nt123) that is sister group to all other orthoteliines. Further exploration of the Neotropical biodiversity of Orthoteliinae is clearly desirable. Within the mega-plutellid group (Fig. 2, **node 50**), no analysis yielded strong support for monophyly of *Proditrix* (Fig. 2, **node 51**; BP≤52, all analyses), while degen1 grouped *Doxophyrtis*+*Proditrix nielseni* to the exclusion of *P. gahniae*, with 86% bootstrap (denoted by dotted arrow in Fig. 2). Thus, *Proditrix* may be paraphyletic with respect to *Doxophyrtis*. The Orthoteliinae as here delimited contain 6 genera and 14 described species. The species with known hostplants are typically borers within monocots (>90% of host records).

Monophyly for Acrolepiinae is very strongly supported by our data (Fig. 2, **node 40**; BP = 100, all analyses). Kyrki [25] proposed four synapomorphies for acrolepiines [31]: reduction of the tegumen, teguminal processes, and gnathos; basal widening of the phallus; stalking of hindwing veins M_1+M_2; and stalking of hindwing veins M_3+CuA_1. However, the first of these, involving reduction of the tegumen, is also common in Glyphipteriginae. In addition, stalking of M_3+CuA_1 is found in *Sericostola* (Glyphipteriginae), though not in other glyphipterigine genera for which wing venation is known. Among Acrolepiinae as sampled here, our data strongly favor the grouping of *Acrolepiopsis*+*Digitivalva* (Fig. 2, **node 55**; BP = 100, all analyses) to the exclusion of *Acrolepia* (Fig. 2, **node 56**; BP = 87–100, all analyses). Acrolepiinae are a cosmopolitan group of 4 genera and 87 described species, most diverse in the Palearctic. The larvae are internal feeders in leaves, stems, flower buds and seeds of herbaceous plants, either monocots (*Acrolepiopsis*) or asterids (*Digitivalva, Acrolepia*).

Our analyses very strongly support monophyly for Glyphipteriginae as sampled here (Fig. 2, **node 57**; BP = 100, all analyses). Kyrki & Itämie [96] proposed three possible synapomorphies for Glyphipteriginae [31]: a conical male 8th abdominal segment with an enlarged tergum; a vestigial M-stem and CuP in the forewing venation; and approximation (not stalking) of hindwing veins M_3 and CuA_1. Dugdale et al. [31] note that adult diurnality and a characteristic rhythmic raising and lowering of the wings while at rest may be additional synapomorphies. All divergences within Glyphipteriginae as sampled here are strongly to very strongly supported by nt123 (Fig. 2, **nodes 57–62**; BP 80–100, nt123), and contradicted in only two instances, weakly, by degen1. In our tree, *Glyphipterix quadragintapunctata* is the sister group to a strongly supported clade comprising all remaining Glyphipteriginae including the four other *Glyphipterix* species sampled (Fig. 2, **node 58**; BP = 100, all analyses). The two other genera sampled, *Diploschizia* and *Lepidotarphius*, each have sister groups consisting nearly or entirely of subsets of *Glyphipterix* species, rendering *Glyphipterix* paraphyletic with respect to both. According to Dugdale et al. [31], about two thirds of the species of glyphipterigines are placed in the cosmopolitan type genus, while many of the 20+ other genera are monobasic. Thus, *Glyphipterix* might prove paraphyletic with respect to other genera as well. Glyphipteriginae are a cosmopolitan group of 25 genera and 397 described species, most diverse in the Australoceanian and Oriental regions. The larvae are typically endophagous in the leaves or stems of commelinid monocots.

Host Plant Associations

Previous hypotheses about life history evolution and biogeography of Yponomeutoidea (e.g. [3,4,20,27,86]) have been few, and their evaluation has been hampered by the lack of a robust phylogeny. In this and the next section we review trends in these features in light of our molecular phylogeny, as summarized in Figures 6 and 7.

To characterize the evolution of larval host plant associations, we sought to assess the degree of conservatism with respect to the new ypnomeutoid phylogeny, of mode of feeding, diet breadth (diversity of plant taxa used by individual species), host plant growth form, and host plant taxon membership at the family level and above. We also sought to infer the ancestral conditions and evolutionary directionality of these traits, for Yponomeutoidea as a whole and for subgroups thereof.

Larval feeding mode in the broad sense of internal versus external feeding is strongly conserved at the subfamily level and family level in yponomeutoids (Figure 7). Of the 16 subfamily or family clades identified by our phylogeny, only two show substantial variation in this trait. In Heliodinidae, internal feeding is numerically dominant but several early branching are external feeders, possibly representing the ancestral habit [84]. In Yponomeutidae external feeding is nearly universal, whereas internal feeding, specifically mining in conifer needles, is restricted to several species of the derived genera *Zelleria* and *Cedestis* [20,31]. Despite this stability at the family and subfamily level, however, transitions between internal and external feeding are frequent enough to obscure the deeper-level history of this trait within Yponomeutoidea. For example, parsimony optimization across the entire phylogeny is unable to assign an unambiguous state to any ancestor below the family level (Figure 7). In this frequency of transition between internal and external feeding, Yponomeutoidea contrast strikingly with their nearest relatives, the possibly paraphyletic Gracillarioidea, within which internal feeding is universal.

Although here scored as "external feeding", Scythropiidae (monospecific), as well as some species of Praydidae, Yponomeutidae, Heliodinidae and possibly other families, actually show an intermediate condition, in which initially leaf-mining larvae subsequently switch to become external leaf webbers. Analogous ontogenetic shifts from internal to external feeding are seen in a

number of non-ditrysian groups as well [3], and may represent a pathway by which external feeding arises over evolutionary time as well. External feeding in yponomeutoids, as in most other so-called microlepidopterans, is not fully equivalent to that seen in Macroheterocera (sensu [2]), in that the larvae are not fully exposed, but rather concealed in some way, e.g. by leaf webbing. Nonetheless, given the multiple evolutionary transitions between internal and external feeding now identified, Yponomeutoidea offer promising material for further studies of the causes and consequences of this fundamental feature of evolution in Lepidoptera and other holometabolous insect phytophages [98].

A second aspect of yponomeutoid larval host use that shows striking phylogenetic conservatism is diet breadth. Oligophagy, defined as using plants of a single order, appears to be nearly universal, characterizing >96% of the 448 yponomeutoid species for which we found host records. Moreover, nearly all oligophagous yponomeutoids use only one plant family. We may be under-estimating the incidence of polyphagy, defined as using two or more plant orders, because for many species only a single host record exists. On the other hand, it also is possible that some of the 14 species that have been recorded from two or more plant families represent undetected host-specific sibling species complexes. Whatever the exact incidence of polyphagy in Yponomeutoidea turns out to be, it clearly seems to be dramatically less than that reported for many groups of Apoditrysia, particularly in Macroheterocera [3,99]. Nonetheless, yponomeutoids, like many other insect herbivore clades in which individual species are mostly oligophagous, collectively use an enormous range of host plant families (see below). It may be that models of diversification of insect herbivore species and host associations that depend on plasticity of host use (e.g. [100]) are less applicable to clades of oligophages such as yponomeutoids than to lepidopteran groups with greater mean diet breadth.

A third phylogenetically conserved aspect of yponomeutid host use is growth form of the host plant. Nearly all of the 16 subfamily/family clades supported by our molecular analyses feed on either woody or herbaceous plants, but not both (Fig. 6). The main exceptions are in Plutellidae and Yponomeutini. Most Plutellidae feed on Brassicales or other herbaceous taxa, but eight species of *Chrysorthenches* have been recorded from Podocarpaceae. Most Yponomeutini feed on woody plants, but about 20% feed on herbaceous Saxifragales. Parsimony optimization of herbaceous versus woody plant use on the molecular phylogeny (see Figure 6), when the nearest outgroups, Gracillarioidea, are included, reconstructs an ancestral association with woody plants, followed by relatively few independent origins of herb feeding, in Yponomeutini, the HSB clade and the YPG clade.

Finally, association with particular plant families, orders or more inclusive clades is conserved to a variable but always obviously non-random extent, within and sometimes between the 16 major yponomeutoid clades. There is some suggestion that host-taxon conservatism is stronger among herb feeders than among woody plant feeders, as previously reported for other lepidopterans [99,101]. Most of the taxa with pronounced fidelity to single or closely-related plant families are herb feeders (Fig. 6). For example, Bedelliidae are nearly restricted to Convolvulaceae; Heliodinidae feed almost exclusively on Nyctaginaceae or other Caryophyllales; Ochsenheimeriinae are known only from Poales; and, the great majority of Glyphipteriginae feed on commelinid monocots.

Among woody-plant feeders, the only comparable example is Attevidae, which feed almost exclusively on Simaroubaceae. Larger woody-plant-feeding clades are typically spread across many plant families and orders, with several, most notably

Argyresthiidae, Ypsolophinae and Yponomeutini, using conifers as well as angiosperms as hosts. The Lyonetiinae, for example are recorded from 17 plant families in 10 orders, belonging to major clades [65] including magnoliids, basal eudicot lineages, basal core eudicot lineages, rosids and asterids (Fig. 6). As with other woody-plant-feeding clades, they are most often associated with rosids, particularly Rosales and Fabales, orders that are especially characteristic of north temperate forests. A few woody-plant feeding clades or subclades thereof show unusually frequent association with particular plant clades. The most notable example is *Yponomeuta*, in which 29 of the 42 species with recorded hosts feed on Celastraceae. Several other genera of Yponomeutini also include species feeding on Celastraceae. Our phylogeny, in which the Celastraceae-restricted *Teinoptila* is strongly supported as the sister group to *Yponomeuta*, is consistent with the conclusion of Turner et al. [102] that Celastraceae is the ancestral host for *Yponomeuta*. However, Celastraceae are unlikely to be the ancestral hosts for Yponomeutidae as a whole (contra [86]), as neither Niphonymphini nor Saridoscelinae feed on this family.

Biogeography

Yponomeutoidea have been conventionally considered to be a primarily North Temperate group that is most diverse in the Palearctic region. Tabulation of the zoogeographical composition of the 16 tribe, subfamily and family clades supported by our phylogeny (Figure 7) suggests that this view needs modification. It is indeed the case that in a majority of lineages, nine of 16, species diversity is highest in the Palearctic, equaling or exceeding 50% of total diversity in five of these. However, half of the lineages, eight of 16, are now known to be at least represented in all major zoogeographic regions. Four other yponomeutoid groups have more restricted distributions but are still widespread: Ypsolophinae are nearly absent from the Southern Hemisphere; Ochsenheimeriinae and Niphonymphini are restricted to the Old World; Attevidae are pantropical, extending into the Nearctic Region. Two groups show strongly disjunct distributions. In Saridoscelinae, one of the two genera occurs in the Palearctic and Oriental regions, whereas the other is restricted to the Nearctic and Neotropical regions. Orthoteliinae are found in the Australian region, in Europe, and as demonstrated here for the first time, in the Neotropical region. On-going taxonomic revisions in Ypsolophinae, Yponomeutini, and Argyresthiidae by the first author show that in these groups, Neotropical species diversity has been significantly underestimated. The same may hold true for tropical diversity of yponomeutoids in general.

Summary and Conclusions

Phylogeny and Classification

Our molecular results offer substantial clarification of yponomeutoid relationships at multiple levels of classification:

(1) We find consistent support, rising to very strong (BP = 99%) when rogue taxa are removed, for monophyly of a concept of Yponomeutoidea close to that of Kyrki [25,26].

(2) With one exception, our data are consistent with recognition of all 10 yponomeutoid families included in the classification of van Nieukerken et al. [2], and strongly support monophyly for eight of the nine families for which multiple representatives were sampled. We also find strong support for recognition of an 11th family, Scythropiidae **stat. rev.**, which was previously subordinate within Yponomeutidae.

The chief remaining uncertainty about yponomeutoid family-level classification concerns the subfamily Cemiostominae of Lyonetiidae. Our sole cemiostomine, *Perileucoptera*, is grouped (albeit weakly) with Lyonetiinae in the four-gene nt123 analysis, but is excluded entirely from Yponomeutoidea in all other analyses, suggesting conflict among genes. Such conflict may also underlie the inability of our AU test to reject monophyly for *Perileucoptera*+Lyonetiinae for the full data set, and the identification of *Perileucoptera* as a rogue taxon by RogueNaRok. We leave Cemiostominae in Lyonetiidae until its position is clarified, by further taxon sampling and perhaps gene tree/species tree analysis.

(3) There is strong support for tribal and/or subfamily divisions within the three largest families, and for inter-generic relationships within all families for which two or more genera were sampled (Fig. 2).

(4) We present a new working hypothesis for relationships among yponomeutoid families (Fig. 2) in which 7 of 8 nodes have at least moderate support (BP≥70), and 4 of 8 have strong support (BP≥80), in one or more analyses. It differs markedly from, and fits our data decisively better than, all previous hypotheses.

Our proposed classification and phylogeny are summarized in the following phylogenetically indented list, in which each taxon is taken to be the sister group of all following taxa at the same level of indentation, provided there is no intervening taxon with lesser indentation. Asterisks denote levels of bootstrap support for our proposed supra-familial clades (*, **, *** = BP≥70, 80, 90, respectively, in at least one analysis).

Superfamily Yponomeutoidea.
  'YYPGAL Clade'*.
    'YYPG Clade':
      Family Yponomeutidae.
        Subfamily Yponomeutinae.
          Tribe Yponomeutini.
          Tribe Niphonymphini.
        Subfamily Saridoscelinae.
      'YPG Clade'***:
        Family Ypsolophidae.
          Subfamily Ypsolophinae.
          Subfamily Ochsenheimeriinae.
        'PG Clade'***:
          Family Plutellidae.
          Family Glyphipterigidae.
            Subfamily Orthoteliinae.
            Subfamily Glyphipteriginae.
            Subfamily Acrolepiinae.
  'AL Clade'*:
    Family Argyresthiidae.
    Family Lyonetiidae.
      Subfamily Lyonetiinae.
      Subfamily Cemiostominae.
 'PAHSB Clade'*:
  'PA Clade'**:
    Family Attevidae.
    Family Praydidae.
  'HSB Clade'***.
    Family Heliodinidae.

Family Bedelliidae.
Family Scythropiidae **stat. rev.**

Host Associations

Yponomeutoidea show notable conservatism on the new phylogeny with respect to four aspects of larval host plant use:

(1) **Internal versus external feeding** is strongly conserved at the family level, varying notably only within Heliodinidae and, to a much lesser extent, Yponomeutidae. Parsimony optimization on the molecular phylogeny (Figure 7) points to an internal feeding as the ancestral yponomeutoid condition, with external feeders arising several times independently. This transition may typically pass through an intermediate stage seen in several extant groups, in which larvae mine leaves in the first instar and subsequently switch to external feeding, living in a communal web and skeletonizing leaves.

(2) **Diet breadth** is remarkably conserved across yponomeutoids (Figure 7), with oligophagy, defined as using plants of a single order, characterizing 96% of all species with recorded hosts (albeit uncorrected for singleton records). Moreover, nearly all oligophagous yponomeutoids use only one plant family. It seems therefore possible that at least some of the 14 species that have been recorded from two or more plant families, whose rate of incidence is highest in Lyonetiinae (17%) and Orthoteliinae (20%), will prove to represent undetected host-specific sibling species complexes.

(3) **Growth form of host plants used** is also markedly conserved: with a few exceptions, the 16 family-group taxa supported by our phylogeny feed on either woody plants or herbaceous plants, but not both (Fig. 6). Parsimony optimization of herbaceous versus woody plant use on the molecular phylogeny (Figure 6), when the nearest outgroups, Gracillarioidea, are included, reconstructs an ancestral association with woody plants, followed by several independent origins of herb feeding, in Yponomeutini, the HSB clade and the YPG clade.

(4) **Taxonomic affinity of host plants used**, at the level of plant family, order or more inclusive clade is conserved to a variable but always notable extent within each of the 16 family-group yponomeutoid clades (Figure 7). Most of the clades that are restricted mainly to a single plant family or order are herb feeders; woody plant feeders appear to shift somewhat more readily among plant orders, albeit typically within the rosid plant clade.

Given these strong initial phylogenetic patterns, yponomeutoids appear to provide promising material for future more detailed studies of the evolution and evolutionary consequences of host plant use in early-diverging ditrysian Lepidoptera.

Biogeography

Our tabulation of yponomeutoid distributions in light of the molecular phylogeny shows that Yponomeutoidea are considerably more diverse outside the Palearctic than has previously been appreciated. Half (8) of the 16 family-group clades supported here are now known to occur in all major zoogeographic regions. The known distribution is expanded most markedly by our findings for two groups: Plutellidae, in which the North Temperate "core" group is shown to have a tropical sister lineage; and, the formerly monobasic, exclusively Palearctic Orthoteliinae, which are shown to include both Australoceanic and Neotropical lineages. From these results, in conjunction with recent revisionary studies, it seems likely that tropical and southern continent biodiversity of

Yponomeutoidea, particularly that of the Neotropical Region, has been heretofore considerably under-estimated.

Supporting Information

Figure S1 A spreadsheet showing the included species with annotations of their classification, collecting locality, host plant families, identification check with DNA barcodes, sequence data completeness (fraction of total target sequence actually obtained) and GenBank accession numbers. The eight genes initially sampled are shown to the left of the 11–19 additional genes sampled for a subset of taxa. The genes sampled for the 4-gene nt123 analysis are shown in bold.

Figure S2 The best maximum likelihood tree found in nt123 analysis of the 4-gene, 139-taxon data set. The four genes are listed in Figure S1. The tree is rooted with *Tischeria ekebladella*. Bootstrap values, when >50%, are shown above branches.

Figure S3 The best ML tree found for nt12 (only) analysis of the 8–27 gene, 139-taxon data set, rooted with *Tischeria ekebladella*. Bootstrap values, when >50%, are shown above branches.

Figure S4 The best ML tree for nt3 (only) analysis of the 8–27 gene, 139-taxon data set, rooted with *Tischeria ekebladella*. Bootstrap values, when >50%, are shown above branches.

Figure S5 The best ML cladogram from Figure 2, with bootstrap values for the initial 8 genes (nt123 analysis). Values for 109fin, 205fin, 208fin, and 3007fin are shown above branch, in that order; values for ACC, CAD, DDC and enolase are shown below branches. '−' = node not recovered in the ML tree for that analysis. '*' = bootstrap value <50%. 'NA' = bootstrap value undefined because sequence was obtained for ≤1 taxon for that that gene in that clade. Bootstrap supports for groups with missing taxa are calculated from the remaining taxa.

Acknowledgments

We would like to express our cordial appreciation to two anonymous reviewers for critically editing our manuscript. We are indebted to many colleagues who provided specimens for this study, including Richard Brown (Mississippi Entomological Museum), Soowon Cho (Chungbuk National University), John Dugdale (Landcare Research, Auckland), Ted Edwards (CSIRO Division of Entomology), Terry Harrison (Illinois Natural History Survey), Robert Hoare (Landcare Research, Auckland), Utsugi Jinbo (National Museum of Nature and Science, Tokyo), Lauri Kaila (Finnish Museum of Natural History), Axel Kallies (Walter and Eliza Hall Institute, Victoria), Akito Kawahara (Florida Museum of Natural History, McGuire Center for Lepidoptera and Biodiversity), David Lees (Natural History Museum London), Wolfram Mey (Museum für Naturkinde, Berlin), Marko Mutanen (Zoological Museum, University of Oulu), Kenji Nishida (Universidad de Costa Rica), Ian Sims (Syngenta International Research Centre, Berkshire), Shen-Horn Yen (National Sun Yat-Sen University) and several other contributors to the Leptree frozen tissue collection. The first author especially thanks the museum curators who facilitated his examination of collections in their care, including Axel Hausmann (Bavarian State Collection of Zoology), Martin Lödl (Naturhistorishe Museum Wien), Naomi Pierce (Museum of Comparative Zoology), Jerry Powell (Essig Museum of Entomology), Kevin Tuck

(Natural History Museum London), and Chun-Sheng Wu (Chinese Academy of Sciences, Beijing). This paper was made possible by technical assistance from Kim Mitter, Zaile Du, Hong Zhao, and by the efforts of the entire Leptree team.

Author Contributions

Taxonomy and classification: JCS CM DD JFL. Performed the experiments: JCS JCR. Analyzed the data: JCS JCR CM AZ MPC. Contributed reagents/materials/analysis tools: JCS JCR CM DD JFL AZ MPC. Wrote the paper: JCS JCR CM DD JFL AZ MPC.

References

1. Heppner JB (1998) Classification of Lepidoptera, Part 1. Introduction. Holarctic Lepid 5 (Suppl. 1): 1–148.
2. Nieukerken EJ van, Kaila L, Kitching IJ, Kristensen NP, Lees DC, et al. (2011) Order Lepidoptera Linnaeus, 1758. In: Zhang Z-Q (ed), Animal biodiversity: An outline of higher-level classification and survey of taxonomic richness. Zootaxa 3148: 212–221.
3. Powell JA, Mitter C, Farrell BD (1998) Evolution of larval food preferences in Lepidoptera. In: Kristensen NP (ed), Lepidoptera, moths and butterflies, Vol. 1: Evolution, systematics, and biogeography. Handbook of Zoology 4: 403–422.
4. Grimaldi D, Engel MS (2005) Evolution of the Insects. Cambridge: Cambridge University Press. 755 p.
5. Menken SBJ, Herrebout WM, Wiebes JT (1992) Small ermine moths (*Yponomeuta*): Their host relations and evolution. Annu Rev Entomol 37: 41–66.
6. Talekar NS, Shelton AM (1993) Biology, ecology, and management of the Diamondback moth. Annu Rev Entomol 38: 275–301.
7. Mason PG, Appleby M, Juneja S, Allen J, Landry J-F (2010) Biology and development of *Acrolepiopsis assectella* (Lepidoptera: Acrolepiidae) in eastern Ontario. Can Entomol 142: 393–404.
8. Leather SR (1986) Insects on bird cherry I. The bird cherry ermine moth, *Yponomeuta evonymellus* (L.) (Lepidoptera: Yponomeutidae). Entomol Gaz 37: 209–213.
9. Alonso C, Vuorisalo T, Wilsey B, Honkanen T (2000) *Yponomeuta evonymellus* outbreaks in southern Finland: spatial synchrony but different local magnitudes. Ann Zool Fenn 37: 178–188.
10. Hoebeke ER (1987) *Yponomeuta cagnagella* (Lepidoptera: Yponomeutidae): A Palearctic ermine moth in the United States, with notes on its recognition, seasonal history, and habitats. Ann Entomol Soc Am 80: 462–467.
11. Zeller PC (1839) Versuch einer naturgemassen Eintheilung der Schaben. Isis von Oken 3: 168–220.
12. Bruand CT (1851) Catalogue systématique et synonymique des Lépidoptères du Department du Doubs. Tineides. Mém Soc d'Emul Doubs 3(3): 23–68.
13. Stainton HT (1854) Insecta Britannica. Lepidoptera: Tineina. Vol. 3. London: Reeve and Benham. 331 p.
14. Meyrick E (1928) A Revised Handbook of British Lepidoptera. London: Watkins and Doncaster. 914 p.
15. Staudinger O, Rebel H (1901). Catalog der Lepidopteren des Paläarktischen Faunengebietes. Vol. 3. R. Berlin: Friedländer & Sohn. 368 p.
16. Handlirsch A (1925) Systematische Übersicht. In: Schröder C (ed), Handbuch der Entomologie, Vol. 3. Jena: Gustav Fischer. 377–1143.
17. Stephens JF (1829). A Systematic Catalogue of British Insects. London: Baldwin and Cradock. 388 p.
18. Fracker SB (1915) The classification of lepidopterous larvae. Ill Biol Monogr 2(1): 1–161.
19. Börner C (1939) Die Grundlagen meines Lepidopterensystems. Verh. VII Int Kon Entomol 2: 1372–1424.
20. Friese G (1960) Revision der paläarktischen Yponomeutidae unter besonderer Berücksichtigung der Genitalien (Lepidoptera). Beitr Entomol 10: 1–131.
21. Common IFB (1970) Lepidoptera. In: Mackerras IM (ed), The Insects of Australia. Canberra: CISRO. 765–866.
22. Brock JP (1971) A contribution towards an understanding and phylogeny of the ditrysian Lepidoptera. J Nat Hist 5: 29–102.
23. Heppner JB (1977) The status of the Glyphipterigidae and a reassessment of relationships in Yponomeutoid families and Ditrysian superfamiles. J Lepid Soc 31: 124–134.
24. Kuznetzov VI, Stekolnikov AA (1977) Phylogenetic relationships between the superfamilies Psychoidea, Tineoidea and Yponomeutoidea in the light of the functional morphology of the male genitalia. Part 2. Phylogenetic relationships of the families and subfamilies. Entomol Obozr 56: 19–30.
25. Kyrki J (1984) The Yponomeutoidea: a reassessment of the superfamily and its suprageneric groups (Lepidoptera). Entomol Scand 15: 71–84.
26. Kyrki J (1990) Tentative reclassification of holarctic Yponomeutoidea (Lepidoptera). Nota Lepidop 13(1): 28–42.
27. Moriuti S (1977) Fauna Japonica, Yponomeutidae s. lat. (Insecta, Lepidoptera). Tokyo: Keigaku Publishing Co. 327 p.
28. Regier JC, Zwick A, Cummings MP, Kawahara AY, Cho S, et al. (2009) Toward reconstructing the evolution of advanced moths and butterflies (Lepidoptera: Ditrysia): an initial molecular study. BMC Evol Biol 9: 280.
29. Mutanen M, Wahlberg N, Kaila L (2010) Comprehensive gene and taxon coverage elucidates radiation patterns in moths and butterflies. Proc R Soc B 277: 2839–2848.
30. Cho S, Zwick A, Regier J, Mitter C, Cummings M, et al. (2011) Can deliberately incomplete gene sample augmentation improve a phylogeny estimate for the advanced moths and butterflies(Hexapoda: Lepidoptera)? Syst Biol 60: 782–796.
31. Dugdale JS, Kristensen NP, Robinson GS, Scoble MJ (1998) The Yponomeutidae. In: Kristensen NP (ed), Lepidoptera, Moths and Butterflies. Vol. 1: Evolution, Systematics, and Biogeography. Handbook of Zoology 4: 119–130.
32. Davis DR, Robinson GS (1998) The Tineoidea and Gracillarioidea. In: Kristensen NP (ed). Lepidoptera, Moths and Butterflies. Vol. 1: Evolution, Systematics, and Biogeography. Handbook of Zoology 4: 91–117.
33. Zwick A, Regier JC, Mitter C, Cummings MP (2011) Increased gene sampling yields robust support for higher-level clades within Bombycoidea (Lepidoptera). Syst Entomol 36: 31–43.
34. Regier JC, Shultz JW, Ganley ARD, Hussey A, Shi D, et al. (2008) Resolving Arthropod Phylogeny: Exploring phylogenetic signal within 41 kb of protein-coding nuclear gene sequence. Syst Biol 57 (6): 920–938.
35. Kawahara AY, Ohshima I, Kawakita A, Regier JC, Mitter C, et al. (2011) Increased gene sampling strengthens support for higher-level groups within leaf-mining moths and relatives (Lepidoptera: Gracillariidae). BMC Evol Biol 11: 182.
36. Regier JC, Brown J, Mitter C, Baixeras J, Cho S, et al. (2012) A molecular phylogeny for the leaf-roller moths (Lepidoptera: Tortricidae) and its implications for classification and life history evolution. PLoS One 7 (4): e35574.
37. Regier JC, Mitter C, Solis MA, Hayden JE, Landry B, et al. (2012) A molecular phylogeny for the pyraloid moths (Lepidoptera: Pyraloidea) and its implications for higher-level classification. Syst Entomol 37 (4): 635–656.
38. Lemmon AR, Brown MM, Stanger-Hall K, Lemmon EM (2009) The effect of ambiguous data on phylogenetic estimates obtained by maximum likelihood and Bayesian inference. Syst Biol 58: 501–508.
39. Simmons MP (2012) Radical instability and spurious branch support by likelihood when applied to matrices with non-random distributions of missing data. Mol Phyl Evol 62: 472–484.
40. Regier JC (2008) Protocols, concepts, and reagents for preparing DNA sequencing templates. V. 12/4/08. Available: http://www.umbi.umd.edu/users/jcrlab/PCR_primers.pdf. Accessed 25 January 2010.
41. Regier JC, Cook CP, Mitter C, Hussey A (2008a) A phylogenetic study of the 'bombycoid complex' (Lepidoptera) using five protein-coding nuclear genes, with comments on the problem of macrolepidoteran phylogeny. Syst Entomol, 33, 175–189.
42. Swofford DL (2002) PAUP*. Phylogenetic Analysis Using Parsimony (*and Other Methods). Version 4. Sunderland: Sinauer Associate. Available: http://paup.csit.fsu.edu.
43. Brown WM (1985) The mitochondrial genome of animals. In: MacIntyre RJ (ed), Molecular Evolutionary Genetics. New York: Plenum Publishing Co. 95–130.
44. Griffiths CS (1999) The effect of structure and function of a protein on evolution in protein-coding genes: Problems in retrieving phylogenetic signal. In: Adams NJ, Slotow RH (eds), Proceedings of the 22nd International Ornithological Congress, Durban. Johannesburg: BirdLife South Africa. 754–761.
45. Regier JC, Shultz JW, Zwick A, Hussey A, Ball B, et al. (2010) Arthropod relationships revealed by phylogenomic analysis of nuclear proten-coding sequences. Nature 463: 1079–1083.
46. Zwick A (2010) Degeneracy Coding Web Site. PhyloTools. 10 FEB 2010. Available: http://www.phylotools.com/ptdegen1webservice.htm. Accessed: 1 September 2011.
47. Ren F, Tanaka H, Yang Z (2005) An empirical examination of the utility of codon-substitution models in phylogeny reconstruction. Syst Biol 54 (5): 808–818.
48. Holder MT, Zwickl DJ, Dessimoz C (2008) Evaluating the robustness of phylogenetic methods to among-site variability in substitution processes. Philos Trans R Soc B 363: 4013–4021.
49. Posada D (2008) jModelTest: Phylogenetic Model Averaging. Mol Biol Evol 25: 1253–1256.
50. Zwickl DJ (2011) GARLI 2.0. Available: http://code.google.com/p/garli. Accessed: 3 May 2011.
51. Rambaut A (2009) FigTree v1.3.1. Available at http://tree.bio.ed.ac.uk/software/figtree. Accessed: 25 January 2010.
52. Cummings MP, Huskamp JC (2005) Grid computing. EDUCAUSE review 40: 116–117.
53. Bazinet AL, Cummings MP (2009) The Lattice Project: a Grid research and production environment combining multiple Grid computing models. Distributed & Grid Computing. In: Weber MHW (ed), Science Made Transparent for Everyone. Principles, Applications and Supporting Communities. Marburg: Tectum Publishing House. 2–13.

54. Bazinet AL, Cummings MP (2011) Computing the Tree of Life - Leveraging the power of desktop and service grids. In: 2011 IEEE International Symposium on Parallel and Distributed Processing Workshops and PhD Forum. 1896–1902.

55. Sanderson MJ, Shaffer HB (2002) Troubleshooting molecular phylogenetic analyses. Annu Rev Ecol Syst 33: 49–72.

56. Wilkinson M (1994) Common cladistic information and its consensus representation: Reduced Adams and reduced consensus trees and profiles. Syst Biol 43(3): 343–368.

57. Aberer AJ (2011) RogueNaRok. Available: https://github.com/aberer/RogueNaRok/wiki. Accessed: 10 November 2011.

58. Aberer AJ, Krompaß D, Stamatakis A (2011) RogueNaRok: An efficient and exact algorithm for rogue taxon identification. Heidelberg Institute for Theoretical Studies: Exelixis-RRDR-2011–10, November 2011. Available: http://sco.h-its.org/exelixis/publications.html. Accessed: 10 November 2011.

59. Aberer AJ, Stamatakis A (2011) A simple and accurate method for rogue taxon identification. IEEE BIBM 2011: 118–122.

60. Stamatakis A, Hoover P, Rougemont J (2008) A fast bootstrapping algorithm for the RAxML Web-Servers. Syst Biol 57(5): 758–771.

61. Shimodaira H (2002) An approximately unbiased test of phylogenetic tree selection. Syst Biol 51: 492–508.

62. Shimodaira H, Hasegawa M (2001) CONSEL: for assessing the confidence of phylogenetic tree selection. Bioinformatics 17: 1246–1247.

63. Shimodaira H (2011) CONSEL Home Page. Available at http://www.is.titech.ac.jp/shimo/prog/consel. Accessed: 14 October 2011.

64. Robinson GS, Ackery PR, Kitching IJ, Beccaloni GW, Hernández LM (2010) HOSTS–A Database of the World's Lepidopteran Hostplants. London: Natural History Museum. Available at http://www.nhm.ac.uk/hosts. Accessed: 18 August 2010.

65. APG III (2009) An update of the Angiosperm Phylogeny Group classification for the orders and families of flowering plants: APG III. Bot J Linn Soc 161: 105–121.

66. Fu D-Z, Yang Y, Zhu G-H (2004) A new scheme of classification of living Gymnosperms at family level. Kew Bull 59: 111–116.

67. Meyrick E (1914) Lepidopterorum Catalogus. Vol. 19. Hyponomeutidae, Plutellidae and Amphitheridae. Berlin: W. Junk. 63 p.

68. Gershenson ZS, Ulenberg SA (1998) The Yponomeutinae (Lepidoptera) of the World exclusive of the Americas. K Ned Akad Wet Verh Afd Naturrk 99: 1–202.

69. Gaedike R (1997) Lepidopterorum Catalogus (New Series). Fas. 55 Acrolepiidae. Gainesville: Scientific Publishers. 20 p.

70. Heppner JB, Duckworth WD (1983) Yponomeutoidea. In: Hodges RW (ed), Check List of the Lepidoptera of America North of Mexico. Washington, DC: E. W. Classey Ltd. & London: The Wedge Entomological Research Foundation. 26–28.

71. Heppner JB (1984) Atlas of Neotropical Lepidoptera Checklist. Part I Micropterigoidea–Immoidea. Gainesville: Dr. W. Junk Publishers. 112 p.

72. Karsholt O, Razowski J (1996) The Lepidoptera of Europe, A Distributional Checklist. Stenstrup: Apollo Books. 380 p. & CD-ROM.

73. Edwards ED (1996) Yponomeutidae, Argyresthiidae, Plutellidae, and Glyphipterigidae. In: Nielsen ES, Edwards ED, Rangsi TV (eds), Checklist of the Lepidoptera of Australia. Monographs on Australian Lepidoptera 4: 50–55.

74. Nielsen ES (1996) Heliodinidae and Lyonetiidae. In: Nielsen ES, Edwards ED, Rangsi TV (eds), Checklist of the Lepidoptera of Australia. Monographs on Australian Lepidoptera 4: 56–57.

75. Regier JC, Zwick A (2011) Sources of signal in 62 protein-coding nuclear genes for higher-level phylogenetics of arthropods. PLoS One 6(8): e23408.

76. Liu L, Pearl DK (2007) Species trees from gene trees: reconstructing Bayesian posterior distributions of a species phylogeny using estimated gene tree distributions. Syst Biol 56(3): 504–514.

77. Wilkinson M (1995) More on reduced consensus methods. Syst Biol 44(3): 435–439.

78. Wilkinson M (1996) Majority-rule reduced consensus trees and their use in bootstrapping. Mol Biol Evol 13(3): 437–444.

79. Kyrki J (1988) The systematic position of *Wockia* Heinemann, 1870, and related genera (Lepidoptera: Ditrysia: Yponomeutidae auct.). Nota Lepid 11(1): 45–69.

80. Minet J (1983) Étude morphologique et phylogénétique des organs tympaniques des Pyraloidea. I – généralités et homologies (Lep. Glossata). Ann So Entomol Fr (N S) 19: 175–207.

81. Minet J (1986) Ebauche d'une classification moderne de l'ordre des Lépidoptères. Alexanor 14: 291–313.

82. Dugdale JS, Kristensen NP, Robinson GS, Scoble MJ (1998b) The smaller Microlepidoptera-grade superfamilies. In: Kristensen NP (ed), Lepidoptera, Moths and Butterflies. Vol. 1: Evolution, Systematics, and Biogeography. Handbook of Zoology 4: 217–232.

83. Heppner JB (1995) Lacturidae, new family (Lepidoptera: Zygaenoidea). Tropical Lepid 6(2): 146–148.

84. Hsu Y-F, Powell JA (2005) Phylogenetic relationships within Heliodinidae and systematics of moths formerly assigned to *Heliodines* Stainton (Lepidoptera: Yponomeutoidea). Univ Calif Publ Entomol 124: 1–158.

85. Pierce FN, Metcalfe JW (1935) The Genitalia of the Tineid Families of Lepidoptera of the British Islands. Northants: Oundle. 116+22 p., 68 pls.

86. Ulenberg SA (2009) Phylogeny of the *Yponomeuta* species (Lepidoptera, Yponomeutidae) and the history of their host plant associations. Tijdschr Entomol 152: 187–207.

87. Kuroko H (1964) Revisional studies on the family Lyonetiidae of Japan (Lepidoptera). Esakia 4: 1–61, 17 pls.

88. Heppner JB (2011) Lepidoptera of Florida Checklist. Lepid Novae 4(2–4): 61–193.

89. Kuznetzov VI, Kozlov MV, Seksjaeva SV (1988) To the systematics and phylogeny of mining moths Gracillariidae, Bucculatricidae and Lyonetiidae (Lepidoptera) with consideration of functional and comparative morphology of male genitalia. Proc Zool Inst 176: 52–71.

90. Seksyayeva SV (1994) Comparative morphological analysis of the structure of male genitalia in the family Lyonetiidae (Lepidoptera). Entomol Obozr 73 (3): 716–721.

91. Mey W (2006) Revision of the genus *Phyllobrostis* Staudinger, 1859 (Lepidoptera, Lyonetiidae). Dtsche Entomol Z 53(1): 114–147.

92. Gerasimov AM (1952) Caterpillars, Part 1. Fauna SSSR, Vol. 56, Insects, Lepidoptera (n.s.) 1(2): 1–338 [in Russian].

93. Kloet GS, Hincks WD (1945) A Check List of British insects. Stockport: Kloet and Hincks. 483 p.

94. Mey W (2007) Microlepidoptera: Smaller families. In: Mey W (ed), The Lepidoptera of the Brandberg Massif in Namibia. Esperiana Memoir 4: 9–30.

95. Mey W (2011) Basic pattern of Lepidoptera diversity in southwestern Africa. Esperiana Memoir 6: 7–316.

96. Kyrki J, Itämies J (1986). Immature stages and the systematic position of *Orthotelia sparganella* (Thunberg) (Lepidoptera: Yponomeutoidea). Syst Entomol 11: 93–105.

97. Heppner JB (2005) Primitive sedge moths from New Zealand and Tasmania: Transfer of *Proditrix* and relatives to Orthotelinae (Lepidoptera: Glyphipterigidae). Lepid News 2003 (1–2): 31–42.

98. Winkler IS, Mitter C (2008) The phylogenetic dimension of insect/plant interactions: a review of recent evidence. In: Tillmon K (ed), Specialization, Speciation, and Radiation: The Evolutionary Biology of Herbivorous Insects. Berkeley: University of California Press. 240–263.

99. Menken SBJ, Boomsma JJ, Nieukerken EJ van (2009) Large-scale evolutionary patterns of host plant associations in the Lepidoptera. Evolution 64: 1098–1119.

100. Janz N (2011) Ehrlich and Raven revisited: mechanisms underlying codiversification of plants and enemies. Annu Rev Ecol Evol Syst 42: 71–89.

101. Janz N, Nylin S (1998) Butterflies and plants: a phylogenetic study. Evolution 52: 486–502.

102. Turner H, Lieshout N, Ginkel WE van, Menken SBJ (2010) Molecular phylogeny of the small ermine moth genus *Yponomeuta* (Lepidoptera, Yponomeutidae) in the Palaearctic. PLoS One 5(3): e9933.

16

The Importance of Species Traits for Species Distribution on Oceanic Islands

Kristýna Vazačová[1,2]*, **Zuzana Münzbergová**[1,2]

1 Department of Botany, Faculty of Science, Charles University, Prague, Czech Republic, **2** Institute of Botany, Academy of Sciences of the Czech Republic, Průhonice, Czech Republic

Abstract

Understanding species' ability to colonize new habitats is a key knowledge allowing us to predict species' survival in the changing landscapes. However, most studies exploring this topic observe distribution of species in landscapes which are under strong human influence being fragmented only recently and ignore the fact that the species distribution in these landscapes is far from equilibrium. Oceanic islands seem more appropriate systems for studying the relationship between species traits and its distribution as they are fragmented without human contribution and as they remained unchanged for a long evolutionary time. In our study we compared the values of dispersal as well as persistence traits among 18 species pairs from the Canary Islands differing in their distribution within the archipelago. The data were analyzed both with and without phylogenetic correction. The results demonstrate that no dispersal trait alone can explain the distribution of the species in the system. They, however, also suggest that species with better dispersal compared to their close relatives are better colonizers. Similarly, abundance of species in the archipelago seems to be an important predictor of species colonization ability only when comparing closely related species. This implies that analyses including phylogenetic correction may provide different insights than analyses without such a correction and both types of analyses should be combined to understand the importance of various plant traits for species colonization ability.

Editor: Sylvain Delzon, INRA - University of Bordeaux, France

Funding: This project was supported by GAČR P505/10/0593, GAUK 48807, Mobility Fund of the Charles University in Prague and partly by MSMT and RVO 67985939. The funders had no role in study design, data collection and analysis, decision to publish, or preparation of the manuscript.

Competing Interests: The authors have declared that no competing interests exist.

* Email: vazacova@seznam.cz

Introduction

Species ability to disperse and colonize new habitats is a key prerequisite for their response to ongoing landscape and climate changes [1,2]. Understanding, which are the main traits responsible for this ability, is thus fundamental for prediction of future fates of different species [3,4]. Many recent studies are attempting to understand the importance of species traits for species ability to colonize habitats of different size and isolation (e.g. [5,6]). Most of these studies are done in various fragmented landscapes, predominantly in grasslands and forests. Often these studies demonstrate that species distribution is not only determined by current landscape structure, but is largely a result of landscape structure in the past (e.g. [7,8]).

Strong species response to past landscape structure can be attributed to slow growth dynamics of many perennial species in combination with relatively fast changes in the current landscapes [9,10]. Due to dispersal limitation [11,12] and extinction debt [13,14] the distribution of these species may reflect historical habitat configuration. Species distribution in the landscape may then not reflect species long-term ability to successfully colonize habitats and to survive there. Thus the traits driving species distribution on young habitat fragments in a changing landscape can be different from those in the landscapes fragmented for longer evolutionary time [15,16].

Due to intensive human activity all over the world, it is rather difficult to identify fragmented habitats which remained un-changed for a long time period, on which we could study species ability to colonize new habitats on long time scales. Oceanic islands seem to be suitable candidates of such systems [17,18]. In contrast to continental landscape, oceanic fragments are not a result of human activity and remained almost unchanged in size and number since their origin. Thus the islands are generally thought to be more stable in time as they are fragmented and isolated for much longer time periods. For these reasons they are suitable systems for studying the importance of dispersal traits for species occurrence on isolated patches. Similarly to the studies on habitat fragments on the mainland (e.g. [19,20]), we can predict that species occurring on the youngest and the most isolated islands will have higher dispersal ability than species present on older and more connected islands.

In this study, we analyzed species traits determining distribution of selected native species on the Canary Islands. The Canary archipelago is a suitable model system as it consists of islands differing in their age, size and isolation as well as in species composition. Specifically, we attempted to understand the determinants of species presence on the newest, smallest and most isolated island (El Hierro).

Because closely related species often share a wide range of biological traits, distribution of a species may be related to the traits under study or to other traits correlated with these traits that are characteristic for the whole clade to which the species belongs [21,22]. Comparison of results of analyses with and without phylogenetic correction can help in distinguishing between the traits that are really responsible for a pattern and traits correlated with these within larger species groups. The necessity of phylogenetic correction is a highly debated issue (e.g.[21,22,23]) and it has been suggested that the phylogenetic and ecological explanations for species distribution in a landscape are not mutually exclusive (see also [24]). Separating the phylogenetic and ecological explanations for species distribution is thus difficult. It is, however, generally recognized that both of these types of analyzes should be considered when trying to explain the effect of species traits on species distribution (e.g. [25]).

To consider species phylogeny in this study, we compared dispersal values of 18 pairs of closely related species differing in their presence on El Hierro. In addition, we used the same species to test the relationship between species traits and number of occupied islands in the archipelago. For each species, we assessed the dispersal ability by all possible dispersal vectors acting on islands, i.e. wind, water and animals (anemo-, hydro-, exo- and endozoochory). We also used published sources complemented with our field experience to identify the most likely dispersal mode for each species pair.

Even though nowadays some parts of the islands are quiet heavily inhabited, we suppose that the main dispersal events happened before human's strong influence. Also none of the studied species is purely ruderal. All the species occur in some (semi)-natural habitats such as laurel forests and canary pine woodlands. Such communities obviously suffer from human destructive activities being fragmented and reduced in area, but species extinctions on single islands occur only rarely and were not reported for any of our model species [26].

Although dispersal ability is widely considered as a major determinant of species distribution on islands due to their isolation, other traits, especially those related to species persistence on habitats should not be overlooked as was shown in studies e.g. by Maurer et al. [27] and Saar et al. [8]. For this reason we also tested traits related to species survival and persistence on the islands (i.e. species longevity and woodiness) and traits characterizing species distribution serving as a proxy for amount of seeds available (number of vegetation zones and number of islands occupied by a species). As a number of occupied islands itself can be a function of plant traits, we also explored the life history traits associated with number of occupied islands.

Specifically, we asked the following questions: 1) Which life history traits explain species presence on El Hierro? 2) Which life history traits explain the number of islands occupied by the species? 3) How do the conclusions change when applying phylogenetic correction?

We predict that species occurring on El Hierro will have better dispersal ability and will occupy more islands than species not occurring there. We also expect that species occupying more islands will be more likely r-strategists possessing traits, which enable rapid colonization of free space on islands (i.e. non-woody annuals occupying more vegetation zones).

Materials and Methods

Ethic statement

To test exozoochorous dispersal, we used a pigeon of the King breed, purchased from a local breeder. To minimize subjection to stress during the experiment, the animal was caged in its home aviary ($2 \times 1.5 \times 1$ m) and had free access to commercial diet and water. The bird was not subjected to any invasive intervention which could cause him suffering. As he was tamed since his youth, his manipulation during seed incorporation into feathers did not cause him extreme stress. The manipulation with pigeon was approved by Ministry of Education, youth and sport of the Czech Republic (permission no. 24773/2008-10001) and complied with the relevant legislation of the Czech Republic (article 11, regulation no. 207/2004).

Study site

The Canary Islands are part of the Macaronesian archipelago situated between $27°45'$ and $29°2'$N and between $18°00'$ and $13°37'$W. They consist of 7 main volcanic islands differing in age and size (Figure 1). The age of the islands decreases with increasing distance from the closest mainland (Africa) and from east to west; the easternmost islands are the oldest, while the westernmost are the youngest. Vegetation composition and habitat diversity on islands is highly influenced by altitudinal gradients in combination with predominant north-eastern trade winds [28]. The oldest and most eroded islands Lanzarote and Fuerteventura lack forests, other, steeper and roughed islands (Gran Canaria, Tenerife, La Palma, La Gomera, El Hierro) are covered by thermo-sclerophyllous woodlands, evergreen laurel forests and canary pine woodlands. The highest parts of Tenerife and La Palma host meso-oromediterranean summit broom scrubs [29].

Species selection

We selected 36 species belonging to 22 genera and 15 families, all native to the Canary Islands [30]. The species were grouped into pairs (Table 1). The species within the pair usually belong to the same genera. In three pairs, the two species in the pair represented closely related genera from the same family. Within each pair, the species differed in occurrence on El Hierro, on the youngest Canary Island, but they both were present on the adjacent islands (at least on Tenerife and La Palma or Tenerife and La Gomera). We chose Tenerife as it is considered as a centre of biodiversity of the area and thus can play a key role as a source for species dispersal to the westernmost islands ([31], but see [32]). La Gomera and La Palma were chosen because of their relative proximity to El Hierro and due to their similar size. All the three islands are also similar to El Hierro in the main vegetation zones including *Euphorbia* scrubs, thermo-sclerophyllous woodlands, evergreen laurel forests and canary pine woodlands. Due to these similarities we can suppose that species present on Tenerife and La Palma or Tenerife and La Gomera and not on El Hierro are those which have not been able to reach El Hierro due to dispersal limitation and not due to ecological barriers related to the absence of habitat [31].

We are aware that species presence/absence on El Hierro could be potentially mediated also by human activities. However, this island is less inhabited than the other Canary Islands. While some of the selected genera may occur in ruderal habitats (e.g. the genus of *Reseda*, *Senecio*, *Trifolium*), all of these occur also in some (semi)-natural habitats such as laurel forests and canary pine woodlands and could thus be distributed on the islands prior to increased human activities. We thus suppose that the main dispersal events happened before human's strong influence.

Species selection was further limited to species for which sufficient seed samples could be obtained. For this reason we had to exclude all the previously considered species pairs having fleshy fruits.

Diaspore collection

Diaspores (fruits or seeds representing the most probable dispersal units, see Table 1) for each species were collected in natural populations on the islands except for *Limonium* species. Diaspores of *Limonium* were obtained from the populations in the

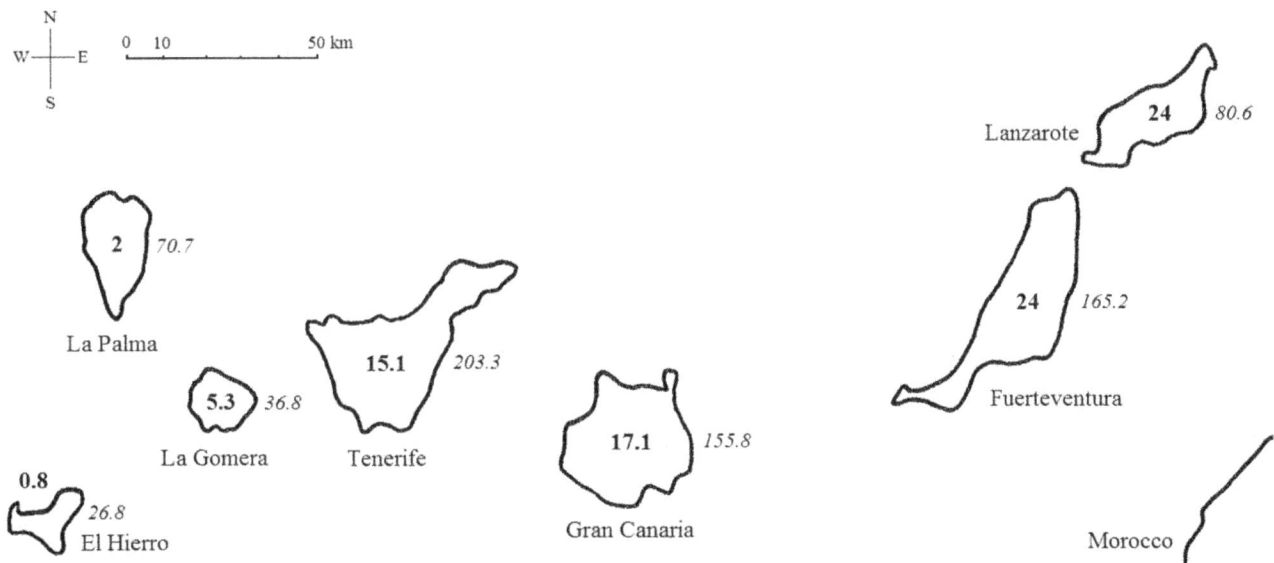

Figure 1. The Canary archipelago. Numbers in bold are island ages (in million years), numbers in italics are island areas (in hectares).

Botanical Garden "Jardín Canario Viera y Clavijo", Gran Canaria. The garden populations originally come from the island populations.

The collection from protected areas was done in cooperation with the Botanical Garden "Jardín Canario Viera y Clavijo", Gran Canaria which obtained appropriate permission for collecting seeds for scientific purposes. The permission was issued by Consejería de Medio Ambiente y Aguas, Islas Canarias. The permission for seed collection from unprotected areas was not required.

In the field we preferably sampled 3 populations per species. For each population, we aimed to collect diaspores from at least 8 individuals. Each population was then tested for dispersal abilities separately. Garden collection was considered as one population and we sampled seeds from 8 individuals in the garden. To have the same number of measurements for the species with seeds collected from the field and from the garden, we had 3 replicates for each dispersal experiment for diaspores collected in the garden.

We used 20 diaspores per species and population for experiments with anemochory, hydrochory and exozoochory and 30 diaspores for testing endozoochory, i.e. 60 and 90 diaspores, respectively. Such number was a compromise between a large amount of species tested and number of seeds used in the literature (c.f. [33]).

For testing other traits related to dispersal (i.e. seed mass and seed viability) we used simple seeds, not fruits. In dispersal modes, where we used fruits as dispersal units, but accounted also for seed viability (i.e. hydrochory and endozoochory), the number of all seeds extracted from the fruits was used as a baseline number of seeds.

Data on all traits used in the study are provided in Supplementary Information (Table S1 and S2).

Traits related to dispersal

Anemochory. The ability of diaspores to disperse by wind was estimated as terminal velocity defined as the maximum rate of seed falling in still air [34]. It was measured as the flight time of a diaspore from predefined height (270 cm [35]). Mean dispersal distance D was expressed as:

$$D = \frac{w \cdot h}{t}$$

where w is the wind speed (being constant for all species), h is the average plant height and t is the terminal velocity. Values of average plant height were obtained from the literature [36,37,38,39,40]

We are aware that our dispersal model is simplified. Nevertheless, it has been successfully used in other studies to characterize mean dispersal distance of diaspores (e.g. [11,6]) and is the easiest way to combine the three key variables affecting wind dispersal. We thus suggest that it is a useful proxy of potential wind dispersal distances for comparison among species.

In the analyses, we used both terminal velocity (m/s) and mean dispersal distance (m). In addition, we tested for the difference in plant height between species present on El Hierro and species absent from El Hierro to see to what extent are the differences in dispersal distance affected by differences in plant height.

Hydrochory. The potential of diaspores to disperse in salt water (buoyancy) was measured as the proportion of diaporess still floating after a defined time period. Diaspores were gently put into beakers filled with salt water having 3.7% salinity (i.e. average salinity of the Atlantic Ocean along the Canary Islands coast). The size of beakers was proportional to the size of diaspores. Sea waves were simulated by continual shaking in electric orbital shaker with frequency of 100 shakes per min. The number of diaspores floating on water surface was checked immediately after putting them into bins and then after 5 minutes of shaking, 1, 2, 6, 24 hours and 7 days of shaking [41]. The experiment was finished after 1 week of diaspores shaking as it is the minimal time a diaspore needs for reaching the Canary islands from mainland when taking into account average speed of water currents in the Atlantic Ocean (60–90 km per week [42]) and the distance between mainland and the closest island (Africa to Fuerteventura, 96 km).

At the end of the experiment, the number of floating and number of sunk diaspores was counted and the two groups of diaspores were then tested for seed viability.

Table 1. List of 18 species pairs used in the study (the first mentioned is species absent from El Hierro).

Species name[1]	Family	Analysed propagule	Most likely dispersal mode
Aeonium sedifolium (Webb ex Bolle) Pit. & Proust	Crassulaceae	Seed	ANEMO
Aeonium spathulatum (Hornem.) Praeger			
Carex perraudieriana Gay ex Bornm.	Cyperaceae	Seed	ANEMO [82]
Carex canariensis Kük.		(with utricle)	
Cistus symphytifolius Lam.	Cistaceae	Seed	ENDO [83]
Cistus monspeliensis L.			
Euphorbia segetalis L.	Euphorbiaceae	Seed	HYDRO [71]*
Euphorbia lamarckii Sweet			ENDO [72]
Hypericum glandulosum Aiton	Hypericaceae	Seed	ANEMO [84]
Hypericum grandifolium Choisy			
Limonium imbricatum (Webb ex Girard) C.F.Hubb.	Plumbaginaceae	Seed	EXO
Limonium pectinatum (Aiton) Kuntze		(with corolla)	
Plantago ovata Forssk.	Plantaginaceae	Seed	EXO [85]
Plantago lagopus L.			
Polycarpaea aristata (Aiton) DC.	Caryophyllaceae	Seed	ANEMO
Polycarpaea nivea (Aiton) Webb			
Reichardia tingitana (L.) Roth	Asteraceae	Achene	ANEMO [82]
Reichardia ligulata (Vent.) G. Kunkel & Sunding		(with pappus)	
Reseda scoparia Brouss. Ex Willd.	Resedaceae	Seed	ANEMO [82]
Reseda luteola L.			
Salvia aegyptiaca L.	Lamiaceae	Seed	EXO [86]
Salvia canariensis L.			
Scrophularia glabrata Aiton	Scrophulariaceae	Seed	ANEMO +BAL [79]
Scrophularia arguta Aiton			
Senecio leucanthemifolius Poir.	Asteraceae	Achene	ANEMO [82]
Senecio glaucus L.		(with pappus)	
Tolpis lagopoda C.Sm. in Buch	Asteraceae	Achene	ANE [85]
Tolpis barbata (L.) Gaertn.		(with pappus)	
Trifolium stellatum L.	Fabaceae	Seed	EXO [87]
Trifolium arvense L.		(with calyx)	
Emex spinosa (L.) Campd.	Polygonaceae	Seed	HYD [88] (Emex), [72] (Rumex)
Rumex bucephalophorus L.		(with spines)	EXO [89] (Emex)*, [90] (Rumex)*
Monanthes laxiflora (DC.) Bolle	Crassulaceae	Seed	ANEMO
Aichryson laxum (Haw.) Bramwell			
Descurainia millefolia (Jacq.) Webb & Berthel.	Brassicaceae	Seed	EXO [87] (Descurainia)
Arabis caucasica Schltdl.			[91] (Arabis)

[1]The species names according to Arechavaleta et al. [30].
*The dispersal mode used as the most likely dispersal mode in the analyses presented (the results do not change when using the other dispersal mode).

In the analyses, we used the proportion of viable seeds which kept floating until the end of the experiment from the total number of viable seeds before the experiment.

The diaspore buoyancy was also expressed as T_{50}, the number of minutes, after which 50 percent of diaspores was still floating. This parameter is commonly used in other studies assessing hydrochory [43,41] however it does not take into account seed viability.

We also used the information on effect of salt water on viability of seeds expressed as the proportion of viable seeds after the experiment (both floating and sunk)/seed viability before the experiment. Viability of seeds was tested by dying the seeds with 0.1% solution of 2,3,5-triphenyl-2H-tetrazolium chloride [44]. In contrast to germination tests, it is not dependent on selection of the right conditions for germination for each individual species and it is thus in fact more reliable for between species comparisons.

Zoochory. Birds are the most important long-distance island dispersers transporting diaspores both externally and internally. The main bird dispersers acting on the Canary Islands are blackbirds (*Turdus merula*), robins (*Erithacus rubecula*), blackcaps (*Sylvia atricapilla* and *S. melanocarpa*, [45]), common ravens (*Corvus corax*, [46]), gulls (*Larus cachinnans*, [47]) and pigeons (*Columba livia*, *C. junoniae* and *C. bolli*).

Bird exozoochory (Epizoochory). Bird exozoochory was tested as diaspore adhesion to bird feathers. As a model species we used a pigeon of the King breed, a utility breed with poor flight ability that is amenable to our experiments.

Although this species is clearly not native to Canary Islands, the functionality of its feathers for diaspore dispersal is readily comparable with native insular pigeon species.

As the seed coat of some species (e.g. *Plantago, Arabis*) contains mucilaginous substances which become sticky when wet, all the diaspores were moistened before the application into pigeon feathers. Moistened diaspores were gently incorporated into feathers on 4 different body parts (on bust, neck and back, under wing). After 1 hour of pigeon free movement in an aviary we checked the numbers of diaspores still attached to feathers. Taking into account the average flight speed of a trained pigeon (80 km/h [48]) and the shortest distance between mainland and the closest island (96 km), diaspores which remained attached to feathers after 1 hour are potentially able to get to the islands by this type of dispersal.

In the analyses we tested the proportion of diaspores which kept attached to feathers after 1 hour (we refer to this value as seed adhesion). This parameter lacks the effect of real bird flight as we do not take into account the air movement around feathers during the flight that can dry out diaspores and cause them to drop earlier than in our simulation. However some behavior of our pigeon during seed testing such as cleaning of feathers was similar to behavior of wild birds. Thus, we still think that our data are sufficient for the purpose to differentiate among diaspores with different ability to disperse by exozoochory.

Bird endozoochory. Bird endozoochory was tested by simulating diaspore gut passage through pigeon digestive tract. Plastic flasks filled up with diaspores were shaken with wet grit (small stones eaten by birds to enhance digestion, commercial mixture for pigeons) for 24 hours in electric orbital shaker (200 shakes per minute [49]). Then diaspores were separated from the grit, rinsed and immersed in 5 ml of 1 M H_2SO_4 (pH$\approx$0.3 [50]) for 4 hours. Intact seeds were retrieved, counted and tested for viability. The proportion of number of viable seeds which survived the simulation to the number of seeds viability before the experiment was used in the analysis. Seed viability after simulation was tested as described above.

Seed mass

Altogether 90 seeds per species were weighted. For this purpose, they were divided to groups by 10 to 30 seeds per group (10 seeds in the group for the largest and 30 for the smallest seeds, to get reasonable size estimates given by the precision of the balance, 0.0001 g). Seed mass is generally recognized as a rough proxy of seed dispersal ability and germination ability (e.g. [51,6]). The same amount of seeds was used for viability testing of intact seeds.

Most likely dispersal mode

For all species pairs the most likely dispersal mode was estimated from available literature (Table 1). Where such data were missing, we estimated the dispersal mode according to our experience with dispersal and diaspore morphology of the species.

Traits related to persistence and distribution

Data on species longevity (short-lived vs. perennial), woodiness (woody vs. not woody) and the number of vegetation zones with species occurrence were gained from Bramwell and Bramwell [38] and Schönfelder and Schönfelder [39,40].

Species distribution was expressed as a number of occupied islands, according to Arechavaleta et al. [30].

Data analysis

To test the importance of life history traits for species presence on El Hierro, we used a generalized linear model with binomial distribution. Species category (present on El Hierro vs. absent from El Hierro) was used as dependent variable and species traits as independent variables. In this analysis the number of islands occupied by a species was counted excluding El Hierro as the effect of El Hierro is already included in the dependent variable.

The importance of traits for species distribution among islands was tested by log-normal regression. Number of islands occupied by a species was used as dependent variable and species traits as independent variables.

The analyses were also performed with phylogenetic correction. Because the exact phylogenetic relationships between the studied species are unknown, we used the simplest version of phylogenetic correction based on comparison of species within the pairs (e.g. [6]). The corrected trait values PC were calculated by applying the formula:

$$PC = \frac{S - MP}{MP}$$

where S is the trait value of a single species (either present or absent from El Hierro) and MP is the mean of the trait value for each species pair. The phylogenetically corrected trait values were used in the tests as described above.

All the tests were done using two different approaches. First, we tested the effect of each trait separately. Afterwards, we combined all the traits in a single model and used forward step wise regression to select an optimal model.

To visualize the similarity between different species in their traits we used principal component analysis (PCA). The data on single species traits were treated as "species", and data on each species represented "samples." The analysis was centered and standardized by "species"; in this way all the traits were expressed in the same, relative, units.

Box plots were done in Statistica 7.0 [52], PCA was processed in CANOCO 4.5 [53]. All the other analyses were done in S-plus 6.2 Professional [54].

Results

Species present on El Hierro and species absent from El Hierro did not differ in any studied dispersal traits. There was, however, marginally significant effect on number of occupied islands (without El Hierro) (Table 2) with species present on El Hierro occupying more islands than species absent from El Hierro (Figure 2A). The results changed dramatically after incorporating phylogenetic correction into analyses. After phylogenetic correction, species presence on El Hierro was significantly influenced by dispersal distance, seed mass, species longevity and by the number of islands occupied by the species (Table 2). Species present on El Hierro dispersed further by wind, had smaller seeds, shorter lifespan and occupied more islands than species absent from El Hierro (Figure 2B–D).

Number of islands occupied by a species was significantly influenced only by species longevity (Table 3). Species occupying more islands were more likely annuals than species occupying fewer islands. This trend remained the same even after phylogenetic correction. All the significant variables also remained in the model after stepwise regression showing that the traits are largely independent of each other (Tables 2 and 3).

Principal component analysis of dispersal traits showed that species within a pair are rather dissimilar in their traits (Figure 3).

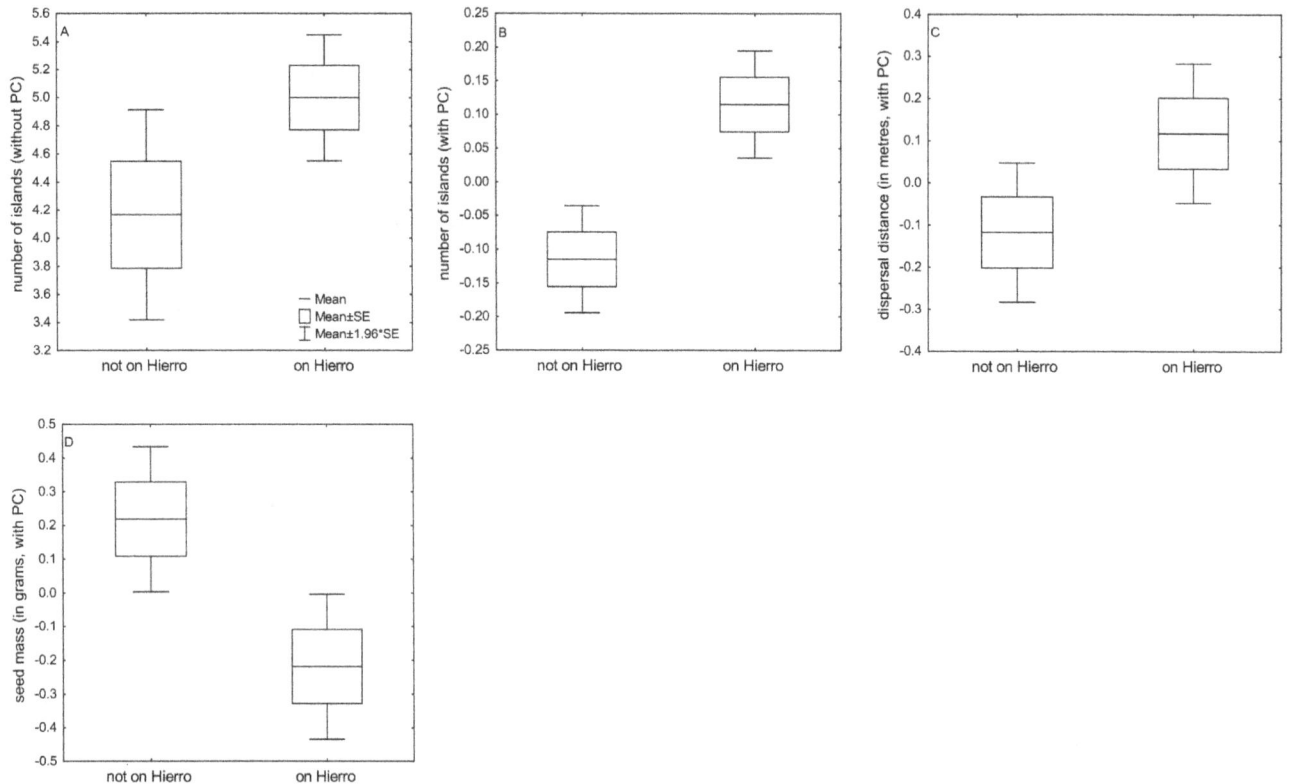

Figure 2. Box plots showing the differences between species present on El Hierro and species absent on El Hierro in number of islands occupied by a species without phylogenetic correction. (PC, A, p = 0.062) and with PC (B, p<0.001), dispersal distance with PC (C, p = 0.05) and seed mass with PC (D, p = 0.007).

As seen in Figure 3, species are partly grouped according to the most likely dispersal mode.

Discussion

The results of the study indicated that species presence on El Hierro, the smallest, youngest and the most remote island is influenced by both dispersal and persistence traits as well as by the number of other Canary Islands occupied by a species. This result was, however, found only after applying phylogenetic corrections. This suggests that the advantage of these traits is relative, and the traits thus play a role only after accounting for other possible differences between closely related species.

Contrasting results with and without phylogenetic correction were found previously also by e.g. Tremlová and Münzbergová [6] for dispersal traits, by Lanta et al. [55] for traits related to plant growth and by Stratton [56] for flower longevity pointing out the necessity for considering phylogenetic information in the analyses. The strong discrepancy between the two types of results is related to the stability of these traits within species phylogenies (e.g. [57,58,59]). The results obtained in this study should thus be interpreted not as the main effects of the given dispersal mode. In contrast, they e.g. suggest that within a given species group (sharing a wide range of biological traits) the species with relatively better dispersal are better colonizers.

Our expectation that species present on El Hierro disperse better than species absent from El Hierro holds only for wind dispersal mode. The importance of anemochory in dispersal among oceanic islands has been mentioned in classical islands studies [60,61]. Regarding the Canary archipelago, seed transport

from the eastern to the western islands (including El Hierro) can be mediated by northeasterly trade winds (which blew during arid Quaternary episodes [62]) as it was reported by e.g. Percy and Cronk [63] or Allan et al.[64]. However, when estimating dispersal distance using simply the data on terminal velocity, plant height and mean wind speed on islands (6.55 m/s [65]) and the nearest distance from El Hierro to neighboring island (La Gomera, 50 km), no species would be able to reach the island by wind. While such simple dispersal model is commonly used to approximate wind dispersal ability of species, such a model is rather simplified [66]. To estimate realistic dispersal distances of species we need to know also other parameters related to wind activity (mainly turbulence and updrafts) and island topography. Considering these types of data in the model is, however, beyond the scope of this study. Another indirect evidence for the importance of wind as an important dispersal mechanism on islands is that species present on El Hierro have smaller seeds (and thus more suitable for flying in the air) than species not present there. Generally, according to Lindborg et al. [25], species with smaller seeds are better dispersers, whereas those with large seeds are better recruiters and tend to have improved establishment in a wider range of habitats [67,68] or when competing with neighbors (see [69]). However, the good competitive ability is not necessarily important for habitats on young volcanic islands arising de novo such as El Hierro. Additionally, the vegetation on El Hierro was repeatedly disturbed by volcanic activities causing extensive landslides further favoring good colonizers over good competitors.

No significant differences in other dispersal traits between species differing in the presence on El Hierro can signify that these species do not disperse by the tested dispersal modes in reality. For

Table 2. Analysis of the relationship between the presence of species on El Hierro and the life history traits.

Traits tested		Without PC				With PC			
		separately		By stepwise		separately		By stepwise	
		P	Dev	P	Dev	P	Dev	P	Dev
Dispersal mode									
ANEMOCHORY	terminal velocity	0.728	0.121			0.326	0.967		
	dispersal distance	0.983	<0.001			0.050	3.847	0.082	3.026
HYDROCHORY	buoyancy	0.927	0.009			0.553	0.353		
	seed survival in salt water	0.624	0.241			0.287	1.135		
	T50	0.686	0.164			0.384	0.759		
EXOZOOCHORY	seed adhesion	0.910	0.013			0.498	0.458		
ENDOZOOCHORY	seed viability after simulation	0.976	0.001			0.474	0.513		
MOST LIKELY DISPERSAL MODE		0.752	0.100			0.654	0.201		
Other traits	seed mass	0.317	1.000			0.007	7.338	0.021	5.340
	seed viability	0.990	<0.001			0.849	0.036		
	plant height	0.540	0.375			0.108	2.585		
Persistence traits	longevity	0.154	2.029			0.001	11.089	0.132	2.263
	woodiness	0.315	1.038			0.103	2.657		
Distribution	no. of vegetation zones	0.331	0.945			0.177	1.824		
	no. of islands	0.062	3.484	0.062	3.484	<0.001	15.14	0.002	9.487

The results are presented with and without phylogenetic correction (PC), Dev indicates deviance explained. DF Error = 35.

Table 3. Analysis of the relationship between the number of occupied islands by a species and the life history traits.

| Traits tested | | Without PC | | | | With PC | | | |
| | | separately | | By stepwise | | separately | | By stepwise | |
		P	Dev	P	Dev	P	Dev	P	Dev
Dispersal mode									
ANEMOCHORY	terminal velocity	0.492	0.472			0.513	0.428		
	dispersal distance	0.967	0.002			0.781	0.077		
HYDROCHORY	buoyancy	0.501	0.453			0.382	0.763		
	seed survival in salt water	0.862	0.030			0.643	0.215		
	T50	0.312	1.021			0.985	<0.001		
EXOZOOCHORY	seed adhesion	0.765	0.089			0.464	0.535		
ENDOZOOCHORY	seed viability after simulation	0.908	0.013			0.967	0.002		
MOST LIKELY DISPERSAL MODE		0.575	0.315			0.979	<0.001		
Other traits	seed mass	0.594	0.284			0.280	1.168		
	seed viability	0.465	0.534			0.425	0.637		
	plant height	0.518	0.418			0.995	<0.001		
Persistence traits	longevity	0.006	7.501	0.006	7.501	0.055	3.693	0.055	3.693
	woodiness	0.038	4.316			0.186	1.75		
Distribution	no. of vegetation zones	0.194	1.687			0.346	0.886		

The results are presented with and without phylogenetic correction (PC), Dev indicates deviance explained. DF Error = 35.

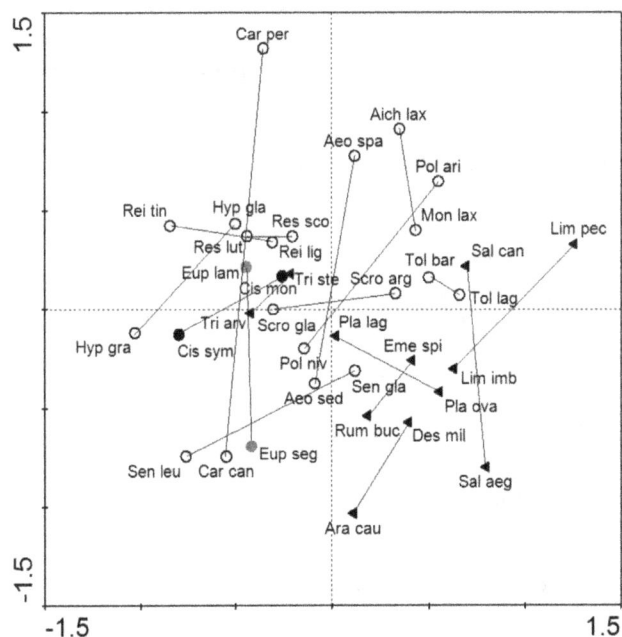

Figure 3. Relationship between individual species determined by principal component analysis (PCA) using trait data as dependent variables. The first axis explained 27.7% of variability, the second axis explained 26.1%. Different symbols indicate species most likely dispersal modes (according to literature): species with solid black circles are most likely dispersed by endozoochory, species with solid grey circles are dispersed by hydrochory, species with opened symbols are dispersed by anemochory and species with solid black triangles are dispersed by exozoochory. Species pairs are connected by lines.

this reason we also tested the most likely dispersal mode, which was based on the selection of the most likely dispersal mode within species pair according to the literature. However, using the most-likely dispersal mode did not show significant differences between species present on El Hierro and absent from El Hierro. The use of such type of dispersal information from a variety of literary sources based on heterogeneous methodology for determining the most likely dispersal mode is questionable, but frequently practiced [70,69]. As a result, the most likely dispersal mode differs according to different authors for some species (e.g. for *Euphorbia* hydrochory in Wald et al. 2005 [71] and endozoochory in Carlquist 1967 [72]). However, even after changing the most likely dispersal mode of some species there were no significant differences between species present and absent on El Hierro in their dispersal ability. Moreover, the only significant wind dispersal in our study was the most frequently chosen most likely dispersal mode. This suggests that the selection of the most likely dispersal mode is not so far from the reality.

According to our results, species present on El Hierro are distributed on more islands (excluding El Hierro) than species absent from El Hierro. This could be due to better wind dispersal ability of species on El Hierro. However, no dispersal trait significantly predicted number of islands occupied by the species. This suggests that better dispersal ability is not generally related to distribution on more islands as we could suppose. No relationship between dispersal and range size was shown e.g. by Kelly and Woodward [70], Goodwin et al. [73] and Lester et al. [74]. Lester et al. [74] assumed that dispersal may only influence species' geographical distributions at certain spatial scales or in particular habitats or environment and/or within certain taxonomic groups,

depending on how the mechanisms by which dispersal and range size are related.

The reason why the number of occupied islands is a good predictor of species' presence on El Hierro could be that the number of occupied islands represents a measure of the amount of available sources (i.e. a proxy of number of plant populations or species abundance) for species' colonization (e.g. [75]). Indeed, to properly measure the amount of available sources we should also know the species local abundances and seed production. Obtaining good information on these two characteristics is, however, rather complicated and such data are not available. Alternatively, number of occupied islands could also be linked to niche width as species with wider distribution range tend to have wider niches and thus more likely occupy a novel habitat (Knappová unpubl.).

Species longevity was another trait influencing species presence on El Hierro.

Species on El Hierro were mainly short-lived (annuals and biennials) showing that short life span enabling rapid production of offspring can be an advantage for colonizing this westernmost island. Due to their ruderal strategy, short-lived species are usually able to grow on newly emerged or disturbed habitats indicating that the island vegetation is still developing. According to Kelly [76] and Kelly and Woodward [70] short-lived species are expected to have smaller ranges than perennials, which is in contrast to our results. We showed the opposite pattern; short-lived species have wider distribution among islands.

There are other possible traits such as seed bank longevity, seed production, pollination mode or detailed characteristics of species habitat requirements (e.g. in the form of indicator values) or species local abundance, which can influence species distribution as was shown e.g. by Pocock [23] and Gabrielová et al. [77]. These studies are mainly done on European species, where most of these data are available as a part of databases [78,33]. No such complete data is, however, available for the whole flora of the Canary Islands.

Possible limitations of the study

Despite the above arguments explaining limited role of dispersal traits in species distribution we cannot exclude the possibility that the importance of species dispersal is undervalued due to our species selection, especially by excluding species with fleshy fruits. We excluded species with fleshy fruits primarily for practical reasons as we were not able to collect sufficient number of fruits due to scarcity and the protection status of some of the potential species (e.g *Sambucus palmensis*, *Pleiomeris canariensis*, *Heberdenia excelsa*). However, as our species list involves mainly anemo- and exozoochorous species, addition of only few pairs of species with fleshy fruits would generate uneven distribution of dispersal modes resulting in few strong outliers. Such data could maybe lead to conclusion that dispersal is more important than we are suggesting based on the current results. On the other hand such a conclusion based on few outliers would not be very robust. We thus suggest that the limited species selection used in this study can also be viewed as an advantage as our study provides relatively robust conclusions for a wide range of anemo- and exozoochorous species.

Another possible critique of our study is that we are working with only 18 pairs of species. Species number was mainly limited due to the approach used to study dispersal, which was dependent on large number of seeds available for each species. Thanks to this approach, we were, however, able to obtain really detailed information on species dispersal by the main dispersal vectors acting among islands. In contrast, other dispersal studies dealing with more species are often based only on categorization of dispersal abilities inferred from the combination of seed visual observation and field experience [79,80] or assessing dispersal by one dispersal mode only ([81]). Such approach enables to cover larger number of species, but species traits are only roughly assessed. As a result, the insights obtained in these studies are more general on one hand, but very rough on the other, not allowing to explore the importance of smaller differences in dispersal ability between different species. We suggest that the results obtained in our study are more likely to indicate possible long-term fates of species in fragmented systems within sets of species of similar growth forms dispersing in similar ways.

Conclusions

The results demonstrated that the relationship between species distribution and species traits depends on the approach we use. Different results were obtained after incorporating phylogenetic relationship between species than when such correction was not used. Thus we suggest to combine both approaches when analyzing closely related species to understand the importance of various plant traits for species distribution.

Supporting Information

Table S1   Values of dispersal traits of 18 species pairs used in the study (the first mentioned is species absent from El Hierro).

Table S2   Values of persistence traits, traits related to distribution and other traits of 18 species pairs used in the study (the first mentioned is species absent from El Hierro).

Checklist S1   ARRIVE Checklist.

Acknowledgments

We thank Dr. D. Bramwell, the director of the Botanical Garden "Jardín Canario Viera y Clavijo" and Dr. J. Caujapé Castells, the Head of Molecular Biodiversity Labs and DNA Bank for enabling collection of seeds from the Botanical Garden and the possibility to participate in seed collection from the field. We also thank other people from these institutes, namely R. Jaén Molina, M. Soto Medina and M. Olangua-Corral for all the kind help. Many thanks belong also to the staff of the Institute of Botany ASCR in Průhonice for helping with all the experiments and William K. Morris and two anonymous reviewers for useful comments to the manuscript.

Author Contributions

Conceived and designed the experiments: KV ZM. Performed the experiments: KV. Analyzed the data: KV ZM. Wrote the paper: KV ZM.

References

1. Schwartz MW, Iverson LR, Prasad AM, Matthews SN, O'Connor RJ (2006) Predicting extinctions as a result of climate change. Ecology 87: 1611–1615.
2. Gallagher RV, Hughes L, Leishman MR (2013) Species loss and gain in communities under future climate change: consequences for functional diversity. Ecography 36: 531–540.
3. Marini L, Bruun HH, Heikkinen RK, Helm A, Honnay O, et al. (2012) Traits related to species persistence and dispersal explain changes in plant communities subjected to habitat loss. Divers Distrib 18: 898–908.
4. Stevens VM, Trochet A, Blanchet S, Moulherat S, Clobert J, et al. (2013) Dispersal syndromes and the use of life-histories to predict dispersal. Evol Appl 6: 630–642.
5. Murray BR, Thrall PH, Gill AM, Nicotra AB (2002) How plant life-history and ecological traits relate to species rarity and commonness at varying spatial scales. Austral Ecol 27: 291–310.
6. Tremlová K, Münzbergová Z (2007) Importance of species traits for species distribution in fragmented landscapes. Ecology 88: 965–977.
7. Chýlová T, Münzbergová Z (2008) Past land use co-determines the present distribution of dry grassland plant species. Preslia 80: 183–198.
8. Saar L, Takkis K, Partel M, Helm A (2012) Which plant traits predict species loss in calcareous grasslands with extinction debt? Divers Distrib 18: 808–817.
9. Lindborg R (2007) Evaluating the distribution of plant life-history traits in relation to current and historical landscape configurations. J Ecol 95: 555–564.
10. Purschke O, Sykes MT, Reitalu T, Poschlod P, Prentice HC (2012) Linking landscape history and dispersal traits in grassland plant communities. Oecologia 168: 773–783.
11. Herben T, Münzbergová Z, Mildén M, Ehrlén J, Cousins SAO, et al. (2006) Long-term spatial dynamics of *Succisa pratensis* in a changing rural landscape: linking dynamical modelling with historical maps. J Ecol 94: 131–143.
12. Münzbergová Z, Cousins SAO, Herben T, Plačková I, Mildén M, et al. (2013) Historical habitat connectivity affects current genetic structure in a grassland species. Plant Biol 15: 195–202.
13. Piqueray J, Cristofoli S, Bisteau E, Palm R, Mahy G (2011) Testing coexistence of extinction debt and colonization credit in fragmented calcareous grasslands with complex historical dynamics. Landscape Ecol 26: 823–836.
14. Hylander K, Ehrlén J (2013) The mechanisms causing extinction debts. Trends Ecol Evol 28: 341–346.
15. Parisod C, Bonvin G (2008) Fine-scale genetic structure and marginal processes in an expanding population of *Biscutella laevigata* L. (Brassicaceae). Heredity 101: 536–542.
16. Parisod C, Christin PA (2008) Genome-wide association to fine-scale ecological heterogeneity within a continuous population of *Biscutella laevigata* (Brassicaceae). New Phytol 178: 436–447.
17. Duarte MC, Rego F, Romeiras MM, Moreira I (2008) Plant species richness in the Cape Verde islands - eco-geographical determinants. Biodivers Conserv 17: 453–466.
18. Hansen DM, Traveset A (2012) An overview and introduction to the special issue on seed dispersal on islands. J Biogeogr 39: 1935–1937.
19. Darling E, Samis KE, Eckert CG (2008) Increased seed dispersal potential towards geographic range limits in a Pacific coast dune plant. New Phytol 178: 424–435.
20. Riba M, Mayol M, Giles BE, Ronce O, Imbert E, et al. (2009) Darwin's wind hypothesis: does it work for plant dispersal in fragmented habitats? New Phytol 183: 667–677.
21. Westoby M, Leishman M, Lord J (1995a) Further remarks on phylogenetic correction. J Ecol 83: 727–729.
22. Westoby M, Leishman M, Lord J (1995b) Issues of interpretation after relating comparative datasets to phylogeny. J Ecol 83: 892–893.
23. Pocock MJO, Hartley S, Telfer MG, Preston CD, Kunin WE (2006) Ecological correlates of range structure in rare and scarce British plants. J Ecol 94: 581–596.
24. Grime JP, Hodgson JG (1987) Botanical contributions to contemporary ecological theory. New Phytol 106: 283–295.
25. Lindborg R, Helm A, Bommarco R, Heikkinen RK, Kuhn I, et al. (2012) Effect of habitat area and isolation on plant trait distribution in European forests and grasslands. Ecography 35: 356–363.
26. Caujapé-Castells J, Tye A, Crawford DJ, Santos-Guerra A, Sakai A, et al. (2010) Conservation of oceanic island floras: present and future global challenges. Perspect Plant Ecol Evol Syst 12: 107–129.
27. Maurer K, Durka W, Stocklin J (2003) Frequency of plant species in remnants of calcareous grassland and their dispersal and persistence characteristics. Basic Appl Ecol 4: 307–316.
28. Fernandéz-Palacios JM, Andersson C (2000) Geographical determinants of the biological richness in the Macaronesian region. Acta Phytogeographica Suecica 85: 41–50.
29. del Arco Aguilar M-J, González-González R, Garzón-Machado V, Pizarro-Hernández B (2010) Actual and potential natural vegetation on the Canary Islands and its conservation status. Biodivers Conserv 19: 3089–3140.
30. Arechavaleta M, Rodríguez S, Zurita N, García A (2010) Lista de especies silvestres de Canarias. Hongos, plantas y animales terrestres. Gobierno de Canarias. 579 p.
31. Sanmartín I, Van der Mark P, Ronquist F (2008) Inferring dispersal: a Bayesian approach to phylogeny-based island biogeography, with special reference to the Canary Islands. J Biogeogr 35: 428–449.
32. Caujapé-Castells J (2011) Jesters, red queens, boomerangs and surfers: a molecular outlook on the diversity of the Canarian endemic flora. In: Bramwell D, Caujapé-Castells J, editors. The biology of island floras. Cambridge University Press London. pp. 284–324.
33. Knevel I, Bekker R, Kunzmann D, Stadler M, Thompson K (2005) The LEDA traitbase collecting and measuring standards of life history traits of the Northwest European flora. Bedum (The Netherlands): Scholma Druk B.V.

34. Thompson K (2005) Terminal velocity. In: Knevel I, Bekker R, Kunzmann D, Stadler M, and Thompson K, editors. The LEDA traitbase collecting and measuring standarts of life history traits of the Northwest European flora. Scholma Druk, B.V., Bedum (The Netherlands). pp. 122–124.

35. Münzbergová Z (2004) Effect of spatial scale on factors limiting species distributions in dry grassland fragments. J Ecol 92: 854–867.

36. Tutin T, Heywood V, Burges N, Moore D, Valentine D, et al. (1964–1980) Flora Europaea. Cambridge University Press.

37. Castroviejo S, Laínz M, López González G, Montserrat P, Muñoz Garmendia F, et al. (1986–2012) Flora Iberica, Plantas Vasculares de la Península Ibérica e Islas Baleares. Real Jardín Botánico, CSIC, Madrid.

38. Bramwell D, Bramwell Z (2001) Wild flowers of the Canary Islands.

39. Schönfelder I, Schönfelder P (2002) Kosmos Atlas Mittelmeer- und Kanarenflora. Kosmos, Stuttgart.

40. Schönfelder P, Schönfelder I (2002) Květena Kanárských ostrovů. Academia, Praha.

41. Römermann C, Tackenberg O, Poschlod P (2005) Buoyancy. In: Knevel I, Bekker R, Kunzmann D, Stadler M, Thompson K, editors. The LEDA traitbase collecting and measuring standards. Scholma Druk, B.V., Bedum (The Netherlands). pp. 124–127.

42. Zhou M, Paduan JD, Niiler PP (2000) Surface currents in the Canary Basin from drifter observations. J Geophys Res 105: 21893–21911.

43. Boedeltje G, Bakker JP, Bekker RM, Van Groenendael JM, Soesbergen M (2003) Plant dispersal in a lowland stream in relation to occurrence and three specific life-history traits of the species in the species pool. J Ecol 91: 855–866.

44. Cottrell H (1947) Tetrazolium salt as a seed germination indicator. Nature 159: 748.

45. Olesen J, Valido Amador A (2004) Lizards and birds as generalized pollinators and seed dispersers of island plants. Ecología insular = Island Ecology: recopilación de las ponencias presentadas en el Symposium de Ecología Insular. Asociación española de ecología terrestre, AEET. pp.229–249.

46. Nogales M, Hernández E, Valdés F (1999) Seed dispersal by common ravens Corvus corax among island habitats (Canarian Archipelago). Ecoscience 6: 56–61.

47. Nogales M, Medina FM, Quilis V, González-Rodríguez M (2001) Ecological and biogeographical implications of Yellow-Legged Gulls (Larus cachinnans Pallas) as seed dispersers of Rubia fruticosa Ait. (Rubiaceae) in the Canary Islands. J Biogeogr 28: 1137–1145.

48. Gessaman JA, Nagy KA (1988) Transmitter loads affect the flight speed and metabolism of homing pigeons. Condor: 662–668.

49. Vazačová K, Münzbergová Z (2013) Simulation of seed digestion by birds: How does it reflect the real passage through a pigeon's gut? Folia Geobot: 1–13.

50. Santamaría L, Charalambidou I, Figuerola J, Green AJ (2002) Effect of passage through duck gut on germination of fennel pondweed seeds. Arch Hydrobiol 156: 11–22.

51. Hewitt N, Kellman M (2002) Tree seed dispersal among forest fragments: II. Dispersal abilities and biogeographical controls. J Biogeogr 29: 351–363.

52. Statsoft I (2013) Statistica 12.0.

53. ter Braak C, Šmilauer P (2002) CANOCO reference manual and CanoDraw for Windows User's Guide: Software for canonical community ordination (version 4.5). Microcomputer Power Ithaca, NY.

54. S-plus 2000 Professional Edition for Windows. Release 2. MathSoft, Inc.

55. Lanta V, Klimešová J, Martincová K, Janeček Š, Doležal J, et al. (2011) A test of the explanatory power of plant functional traits on the individual and population levels. Perspect Plant Ecol 13: 189–199.

56. Stratton DA (1989) Longevity of individual flowers in a Costa Rican cloud forest - ecological correlates and phylogenetic constraints. Biotropica 21: 308–318.

57. Felsenstein J (1985) Phylogenies and the comparative method. Am Nat: 1–15.

58. Harvey PH, Pagel MD (1991) The comparative method in evolutionary biology. Oxford: Oxford University Press

59. Van der Veken S, Bellemare J, Verheyen K, Hermy M (2007) Life-history traits are correlated with geographical distribution patterns of western European forest herb species. J Biogeogr 34: 1723–1735.

60. Carlquist SJ (1965) Island life: A natural history of the islands of the world. New York: Natural History Press.

61. Bramwell D (1979) Plants and islands. London, New York: Academic Press.

62. Ortiz JE, Torres T, Yanes Y, Castillo C, De la Nuez J, et al. (2006) Climatic cycles inferred from the aminostratigraphy and aminochronology of Quaternary dunes and palaeosols from the eastern islands of the Canary Archipelago. J Quaternary Sci 21: 287–306.

63. Percy DM, Cronk QCB (2002) Different fates of island brooms: Contrasting evolution in Adenocarpus, Genista, and Teline (Genisteae, Fabaceae) in the Canary Islands and Madeira. Am J Bot 89: 854–864.

64. Allan G, Francisco-Ortega J, Santos-Guerra A, Boerner E, Zimmer EA (2004) Molecular phylogenetic evidence for the geographic origin and classification of Canary Island Lotus (Fabaceae: Loteae). Mol Phylogen Evol 32: 123–138.

65. Hill G (2003) Wind prospecting on the. Canary Islands. Department of Physical Geography. Goteborg.

66. Tackenberg O, Poschlod P, Bonn S (2003) Assessment of wind dispersal potential in plant species. Ecol Monogr 73: 191–205.

67. Aizen MA, Patterson WA (1990) Acorn size and geographical range in the North-American oaks (Quercus L.). J Biogeogr 17: 327–332.

68. Westoby M, Jurado E, Leishman M (1992) Comparative evolutionary ecology of seed size. Trends Ecol Evol 7: 368–372.

69. Weiher E, van der Werf A, Thompson K, Roderick M, Garnier E, et al. (1999) Challenging Theophrastus: A common core list of plant traits for functional ecology. J Veg Sci 10: 609–620.

70. Kelly CK, Woodward FI (1996) Ecological correlates of plant range size: Taxonomies and phylogenies in the study of plant commonness and rarity in Great Britain. Philos Trans R Soc Lond B Biol Sci 351: 1261–1269.

71. Wald EJ, Kronberg SL, Larson GE, Johnson WC (2005) Dispersal of leafy spurge (Euphorbia esula L.) seeds in the feces of wildlife. Am Midl Nat 154: 342–357.

72. Carlquist S (1967) The biota of long-distance dispersal. V. Plant dispersal to Pacific Islands. Bull Torrey Bot Club: 129–162.

73. Goodwin NB, Dulvy NK, Reynolds JD (2005) Macroecology of live-bearing in fishes: latitudinal and depth range comparisons with egg-laying relatives. Oikos 110: 209–218.

74. Lester SE, Ruttenberg BI, Gaines SD, Kinlan BP (2007) The relationship between dispersal ability and geographic range size. Ecol Lett 10: 745–758.

75. Knappová J, Hemrová L, Münzbergová Z (2012) Colonization of central European abandoned fields by dry grassland species depends on the species richness of the source habitats: a new approach for measuring habitat isolation. Landscape Ecol 27: 97–108.

76. Kelly CK (1996) Identifying plant functional types using floristic data bases: Ecological correlates of plant range size. J Veg Sci 7: 417–424.

77. Gabrielová J, Münzbergová Z, Tackenberg O, Chrtek J (2013) Can we distinguish plant species that are rare and endangered from other plants using their biological traits? Folia Geobot 48: 449–466.

78. Klotz S, Kühn I, Durka W (2002) BIOLFLOR - Eine Datenbank mit biologisch-ökologischen Merkmalen zur Flora von Deutschland. Schriftenreihe für Vegetationskunde

79. Van der Pijl L (1982) Principles of dispersal. Berlin: Springer-Verlag.

80. Vargas P, Nogales M, Jaramillo P, Olesen JM, Traveset A, et al. (2014) Plant colonization across the Galápagos Islands: success of the sea dispersal syndrome. Bot J Linn Soc 174: 349–358.

81. Couvreur M, Vandenberghe B, Verheyen K, Hermy M (2004) An experimental assessment of seed adhesivity on animal furs. Seed Sci Res 14: 147–162.

82. Bonet A, Pausas JG (2004) Species richness and cover along a 60-year chronosequence in old-fields of southeastern Spain. Plant Ecol 174: 257–270.

83. Malo J, Jiménez B, Suarez F (2000) Herbivore dunging and endozoochorous seed deposition in a Mediterranean dehesa. J Range Manage: 322–328.

84. Médail F, Quézel P (1999) The phytogeographical significance of SW Morocco compared to the Canary Islands. Plant Ecol 140: 221–244.

85. Bramwell D (1985) Contribución a la biogeografía de las Islas Canarias. Botan Macaron 14: 3–34.

86. Melendo M, Giménez E, Cano E, Mercado FG, Valle F (2003) The endemic flora in the south of the Iberian Peninsula: taxonomic composition, biological spectrum, pollination, reproductive mode and dispersal. Flora 198: 260–276.

87. Heinken T, Raudnitschka D (2002) Do wild ungulates contribute to the dispersal of vascular plants in central European forests by epizoochory? A case study in NE Germany. Forstwiss Centralbl 121: 179–194.

88. Sadeh A, Guterman H, Gersani M, Ovadia O (2009) Plastic bet-hedging in an amphicarpic annual: an integrated strategy under variable conditions. Evol Ecol 23: 373–388.

89. Evenari M, Kadouri A, Gutterman Y (1977) Eco-physiological investigations on amphicarpy of Emex spinosa (L.) Campd. Flora 166: 223–238.

90. Talavera M, Balao F, Casimiro-Soriguer R, Ortiz MA, Terrab A, et al. (2011) Molecular phylogeny and systematics of the highly polymorphic Rumex bucephalophorus complex (Polygonaceae). Mol Phylogen Evol 61: 659–670.

91. Koch MA, Kiefer C, Ehrich D, Vogel J, Brochmann C, et al. (2006) Three times out of Asia Minor: the phylogeography of Arabis alpina L. (Brassicaceae). Mol Ecol 15: 825–839.

17

Does Land-Use Intensification Decrease Plant Phylogenetic Diversity in Local Grasslands?

Eugen Egorov[1]*, **Daniel Prati**[2], **Walter Durka**[3], **Stefan Michalski**[3], **Markus Fischer**[2], **Barbara Schmitt**[2], **Stefan Blaser**[2], **Martin Brändle**[1]

1 Department of Ecology, Faculty of Biology, Philipps Universität, Marburg, Germany, 2 Institute of Plant Sciences, University of Bern, Bern, Switzerland, 3 Department Community Ecology, Helmholtz Centre for Environmental Research, Halle, Germany

Abstract

Phylogenetic diversity (PD) has been successfully used as a complement to classical measures of biological diversity such as species richness or functional diversity. By considering the phylogenetic history of species, PD broadly summarizes the trait space within a community. This covers amongst others complex physiological or biochemical traits that are often not considered in estimates of functional diversity, but may be important for the understanding of community assembly and the relationship between diversity and ecosystem functions. In this study we analyzed the relationship between PD of plant communities and land-use intensification in 150 local grassland plots in three regions in Germany. Specifically we asked whether PD decreases with land-use intensification and if so, whether the relationship is robust across different regions. Overall, we found that species richness decreased along land-use gradients the results however differed for common and rare species assemblages. PD only weakly decreased with increasing land-use intensity. The strength of the relationship thereby varied among regions and PD metrics used. From our results we suggest that there is no general relationship between PD and land-use intensification probably due to lack of phylogenetic conservatism in land-use sensitive traits. Nevertheless, we suggest that depending on specific regional idiosyncrasies the consideration of PD as a complement to other measures of diversity can be useful.

Editor: Maarja Öpik, University of Tartu, Estonia

Funding: The DFG (German Research Foundation; http://www.dfg.de/en/index.jsp) funded the study in the framework of the Biodiversity Exploratories SSP 1374 "Infrastructure-Biodiversity-Exploratories"(BR 1967/9-1). The funders had no role in study design, data collection and analysis, decision to publish, or preparation of the manuscript.

Competing Interests: The authors have declared that no competing interests exist.

* Email: eugen.egorov@staff.uni-marburg.de

Introduction

Land-use change is one of the primary drivers of biodiversity loss [1,2]. Despite a large amount of studies dealing with the effects of land use on biodiversity, there are still gaps in the understanding of land use – biodiversity relationships. For example, the negative effects of different land-use types on biodiversity can differ in strength or vary in their effects. In addition, regional idiosyncrasies might interact with land use and affect biodiversity responses to land-use intensification, thus impeding general predictions [3]. Recent studies have advocated the consideration of phylogenetic diversity (PD) in ecological analyzes [4–7]. In brief, PD is defined as the total amount of phylogenetic space covered by species in a community. It therefore encapsulates the entire trait space of a community [8] and thus, may serve as a complement to trait diversity if the traits cannot be measured or trait data are not available [4]. Moreover, PD is an important factor for ecosystem function itself. It has been shown that PD can explain more variance in productivity in grasslands than species richness or functional diversity [9]. Plant productivity increased with mycorrhizal PD, which may be caused by niche differentiation, as increasing number of mycorrhizal families provide different advantages to their host plants [10]. Higher Plant PD also increases diversity of higher trophic levels and affects several ecosystem functions and processes [11–13]. That is, higher plant PD reinforces the positive effects of plant species richness on higher trophic levels when species richness is held constant [13]. Finally it has been found that PD promotes ecosystem stability and resilience [12] as well as interacts with plant species richness and alters its effect on herbivory [11]. Despite a consensus that PD is an important factor in understanding biodiversity – ecosystem functions relationships [7] or community assembly rules [14], little effort has been done in analyzing the effects of land-use intensity on PD [15].

In Central Europe managed grasslands are one of the most abundant and species-rich ecosystems [16]. In Germany, about 12% of area is covered by grasslands [17]. Most of these grasslands were established during a long period of low-intensity land-use and a large number of species have adapted to those conditions causing high levels of biodiversity. Land-use intensification in particular during the 20th century posed considerable threats to biodiversity in grasslands, e.g. due to dramatic habitat loss and extinction of less competitive species [18–20]. It is also likely that land-use intensification will be the major driver of biodiversity loss in grasslands during the next decades [1,21]. To attain a compromise between high land-use intensity and biodiversity conservation [1] and to assess the consequences of biodiversity loss a deeper understanding of the relationship between land-use intensification, biodiversity and ecosystem functioning is mandatory.

In general previous studies of plant biodiversity-ecosystem functioning relationships have shown that species richness enhances ecosystem functions [22–24]. Simply counting the number of species, however, is often not sufficient for analyzing the effects of biodiversity on ecosystem functions [25]. More comprehensive approaches consider functional diversity, defined as diversity of traits important for ecosystem level processes [26]. Functional diversity is thought to be the component of biodiversity with the largest effect on ecosystem processes [27–29]. However, implementation of trait data is subject to several limitations. For example, assessment of trait data is time-consuming and the *a priori* choice of specific traits is not always straightforward [26]. To overcome these shortcomings, PD has been proposed as a proxy for functional diversity [5,30]. Recent studies, however, question PD as a proxy and propose it rather as a complement to functional diversity [31]. Despite the current discussion on the use of community phylogenetics in analyzes of assembly processes under several biotic and abiotic conditions [32] the importance of PD to ecosystem processes calls for its implementation into ecological analyzes [4]. While the negative effect of land-use intensification on species richness and functional diversity has been subject to many studies [2,33,34], a relatively small number of studies investigated how increasing land-use intensity affects PD of plant communities, particularly in grasslands. Studies that compared observed phylogenetic community structure of plants with expected patterns [30] revealed shifts in phylogenetic community structure with increasing disturbance and stress [15,35–38]. Similar patterns were also shown within animal communities [39–41]. Changes in phylogenetic community structure may include shifts from overdispersion, where co-occurring species are less phylogenetically related than expected by chance, to clustering, where co-occurring species are phylogenetically more related than expected by chance. Such a shift from overdispersion to clustering is thought to be caused by environmental filtering that selects species with similar ecological traits that are likely to be closely related [15,36; but see 32]. Increasing land use intensity should therefore favor plant species with traits adapted to cope with effects of land-use intensification like fertilization, cattle grazing and frequent mowing. If such traits are phylogenetically conserved and play a major role in the phylogenetic community assembly, communities are likely to become phylogenetically more clustered with increasing land-use intensity. If traits are convergent or show a low phylogenetic signal, plant communities should not exhibit phylogenetic clustering with increasing land-use intensities or even lead to an increase in PD [38].

For conservational purposes the response of rare species to land-use intensification is of great interest. Rare species are in general more vulnerable to land-use intensification than common species [18–20]. Assuming that common species might be better adapted to high land-use intensities, phylogenetic diversity of common species should be less sensitive to land-use intensification than that of rare species. However, to our knowledge there are no studies exploring the response of PD of rare and common species to land-use intensification separately.

Socher et al. [3] showed that strength and direction of the effects of land use on biodiversity can differ between regions. Regional idiosyncrasies can also alter the effect of land use on phylogenetic diversity due to different regional species pools, environmental and geographical variables. It is therefore necessary to compare the effects of land-use intensification on PD among regions. Other limitations of previous research on plant PD are that the majority of studies are either experimental or describe phylogenetic patterns along natural or environmental gradients and are restricted to

certain, often narrow, taxonomic scales [23,42]. Descriptive studies of PD – land-use intensity relationships in human-disturbed systems are still scarce. When analyzing plant PD with respect to man-made disturbance, studies often focus on urban regions [37] or do not encompass the most common agricultural land-use categories such as fertilization, mowing and grazing. Including most common land-use types in descriptive studies of PD – land use relationships in agricultural systems could give new insights on these relationships under "real world" conditions. Previous studies may also suffer from the lack of considering species abundance data. Presence/absence data are highly sensitive to the chance and possible temporary occurrence of a single individual in unusual or unsuitable habitat. Interspecific relationships and interactions between species and ecosystems are based on interactions between individuals, which are cumulative in their effects. Neglecting abundance data may impede to discover important ecological relationships [6].

In this study we use species abundance data to analyze the PD of plant communities in local grasslands (150 sites) across land-use intensification gradients in three regions in Germany. In particular we aimed to answer the following questions:

1) Are there regional differences in the response of phylogenetic diversity to land use?

2) Does land-use intensification decrease phylogenetic diversity of plant communities in grasslands?

3) Does phylogenetic diversity of common and rare species assemblages show different relationships with respect to land-use intensification?

For a better understanding and interpretation of the relationship between PD and land-use intensification, information on the phylogenetic signals in traits relevant for landuse are of interest (i.e. related to a certain ecosystem function or environmental gradient). Thus, we used a set of traits that are likely to be sensitive toland use and tested for phylogenetic signal in those traits.

Materials and Methods

Study area

Our study is part of the Biodiversity Exploratories project, a large German research project to investigate the relationships between land-use, biodiversity and ecosystem functioning (www.biodiversity-exploratories.de). The Biodiversity Exploratories represent three typical regions in Germany covering a south-west – north-east gradient and each region comprises grasslands and forests under a range of land-use types and intensities [43]. The exploratory Schwäbische Alb (hereafter named Alb) is situated in the SW Germany and is part of the UNESCO Biosphere Reserve Schwäbische Alb. The exploratory Hainich-Dün (hereafter named Hainich) is situated in western Thuringia, central Germany. The exploratory Schorfheide-Chorin (hereafter named Schorfheide) is situated in NE Germany and is part of the UNESCO Biosphere Reserve Schorfheide-Chorin. In each region 50 experimental grassland plots representing gradients from semi-natural to intensive land-use were established (overall 150 plots). For more details see [43].

Field work permits were issued by the responsible state environmental offices of Baden-Württemberg, Thüringen, and Brandenburg (according to § 72 BbgNatSchG). The study did not involve protected or endangered species.

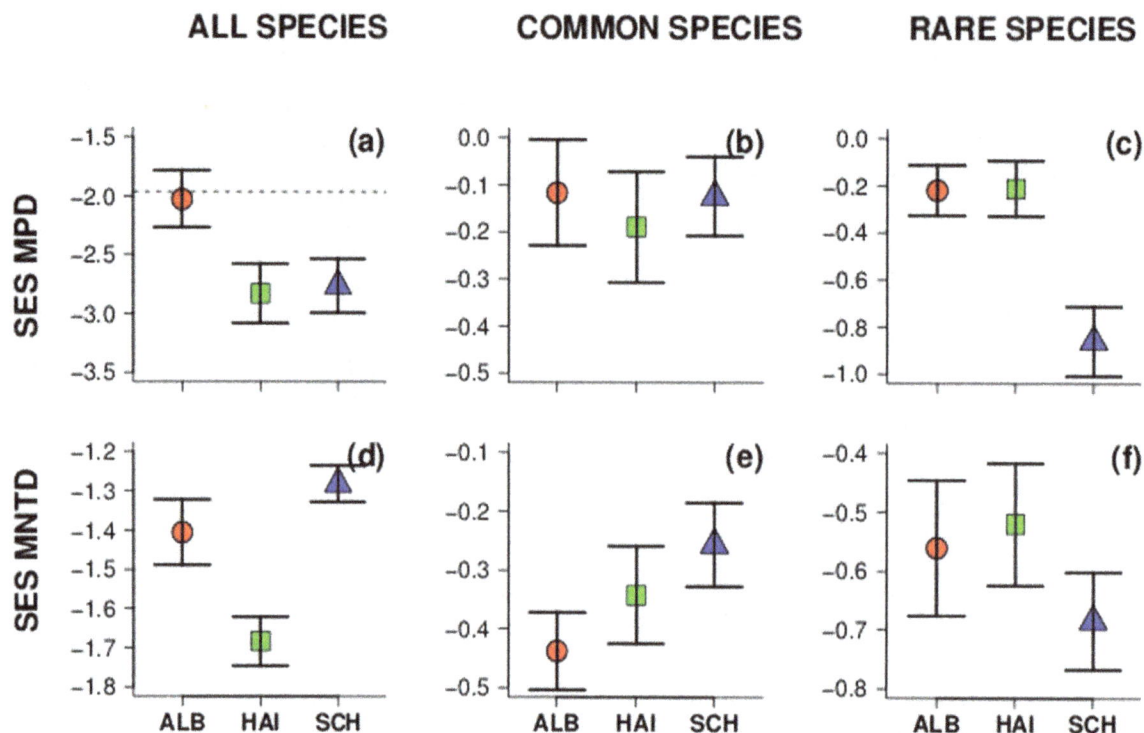

Figure 1. Mean (±SE) values of MPD and MNTD effect sizes for total, common and rare species assemblages in three regions in Germany. (a)–(c) Mean MPD and (d)–(f) mean MNTD for all, common and rare species assemblages in the three regions. Region abbreviations: ALB = SchwäbischeAlb (red circle); HAI = Hainich-Dün (green square); SCH = Schorfheide-Chorin (blue triangle). Error bars indicate ± SE. Points below the dashed line (< −1.96) are significantly clustered. Note different scales of y-axes.

Land-use

Land-use information for each of the 150 grassland plots was obtained by yearly interviews with farmers and land-owners between 2006 and 2010. The acquired information included fertilization level (kg nitrogen ha^{-1} year^{-1}), mowing frequency (number of cuts year^{-1}) and grazing intensity (livestock units×days of grazing ha^{-1} year^{-1}) [43]. The three land-use components were standardized by the respective mean intensity within each region to yield the fertilization, mowing and grazing intensity [44]. For each year the individual components were summed up to a combined quantitative land-use intensity index (LUI). The yearly LUI-values (2006–2010) were averaged for each plot and the obtained means were then used in all our analyses [44].

Vegetation releves and phylogeny

Between 2009 and 2011 we recorded the vegetation on a 4×4 m plot in each of the 150 grasslands three times (2009, 2010 and 2011). For each plot, vascular species richness and their relative abundance in percent cover was estimated. The species were further grouped into common and rare species based on their abundance for each year and region separately, taking into account local (plot) abundance and distribution (number of plots occupied) of each species. Common species were defined as the top 10% in terms of total abundance across plots occupied by a species, while the bottom 90% of the species were defined as rare. Based on these data we calculated the species richness of all, common and rare species as the average richness per plot across the three years. Note that the analyses of plant species richness from our study sites have been already published elsewhere [3,45]. We included these results here only for comparative purposes. Therefore our discussion focuses only on the effects of land-use on

PD. A low number of gymnosperms and ferns with low site incidence were omitted from all analyzes.

Phylogenetic relatedness of species was obtained from a well resolved and dated phylogeny of the Central European flora [46]. In brief, this phylogeny was assembled by manually grafting subtrees on a backbone topology, dating of nodes based on fossil records using the bladj algorithm in PHYLOCOM [47] and calculating an ultrametric tree (for details see [46]). We pruned the overall phylogeny to match the species pool of each of the three regions. As a result we obtained three trees, one for each region, representing the phylogenetic relationships of the respective species pool.

According to the data sharing regulations of the Biodiversity Exploratory Project and in accordance with the rules of the German Science Foundation DFG, the data will be made publicly available no later than five years after collection.

Traits and phylogenetic signal

We compiled functional trait data from different data bases. As traits related to productivity we included the maximal plant height (cm) and specific leaf area (SLA; in cm^2/g). As traits related to reproduction we used data start of flowering (month of the year). Data on the SLA were taken from the LEDA trait data base [48], data on start of flowering and plant height were gathered from BiolFlor data base [49] and from floras [50,51]. Means were calculated when entries differed among the sources, but generally the values were highly consistent across sources. We further compiled performance and persistence traits relevant for agricultural grasslands: (1) soil nutrient indicator value (N, [52], (2) mowing tolerance (M), (3) grazing tolerance (G) and (4) trampling tolerance (T, all according to [53] from [54] and Briemle pers.

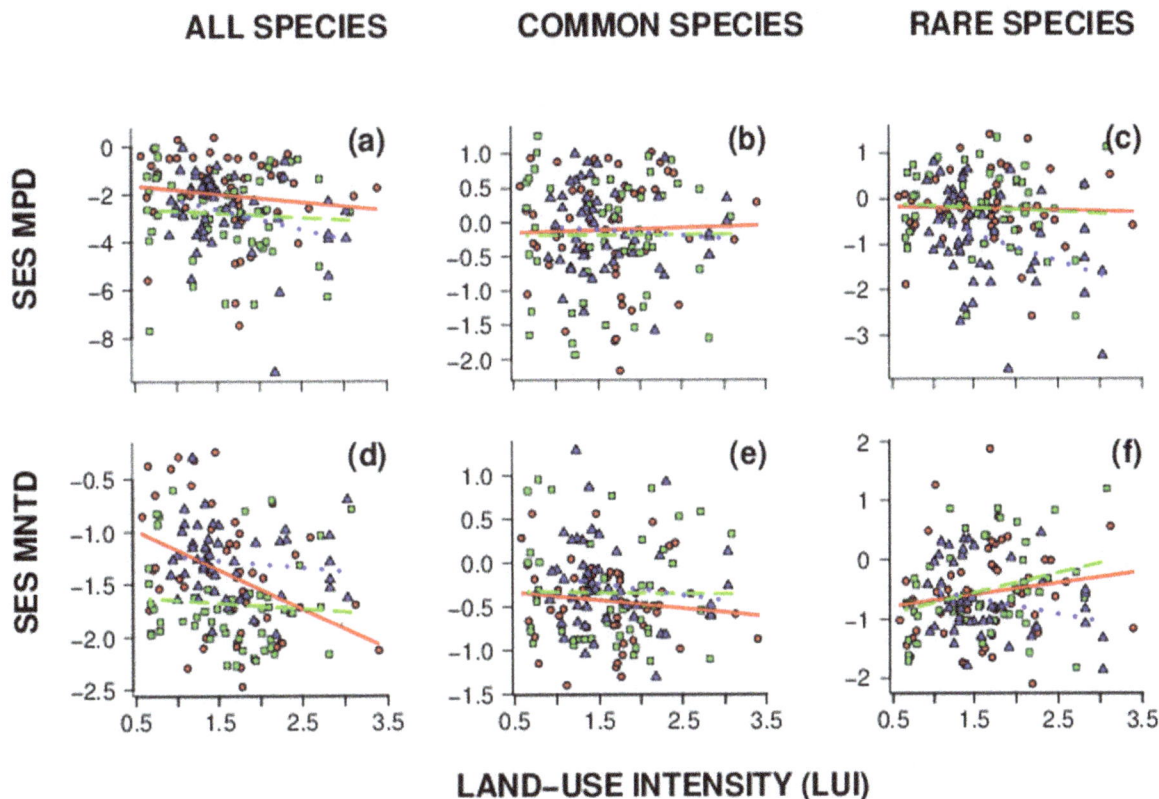

Figure 2. Relationships between mean pairwise distance (effect size MPD), mean nearest taxon distance (effect size MNTD) and land-use intensity (LUI) in three regions in Germany. Linear regression plots showing regression slopes for relationships between (a–c) mean pairwise distance and (d–f) mean nearest taxon distance for total, common and rare species assemblages and land-use intensity (LUI). Color and type code: red solid line/circle = Schwäbische Alb (Alb); green dashed line/square = Hainich-Dün (Hai); blue dotted line/triangle = Schorfheide-Chorin (Sch). Note different scales of y-axes. For significance of regression slopes see Appendix S4.

comm.). For all traits we hypothesized that different agricultural use, in particular fertilization, mowing and grazing selects for species with different traits values. All indicators have numeric values ranging from 1 (low) to 9 (high). Available trait data ranged from 77% (SLA, height and flowering onset) to 86% (G) of the species.

We tested for the strength and significance of phylogenetic signals in traits using Pagel's λ and Blomberg's K implemented in the phytools package [55] in R. We log transformed values for the maximum height to achieve normality. It has been proposed that Pagel's λ is an overall more robust metric than e.g. Blomberg's K [56], however, in general both metrics revealed similar results.

Phylogenetic diversity

Phylogenetic diversity estimates of plots were calculated with the "picante" package in R [57]. We calculated for each year and region separately the mean pairwise distance (MPD) and mean nearest taxon distance (MNTD) [30] weighted by species abundance (estimated % cover) as well as using presence/absence data. Considering % cover as a surrogate for species abundance may only approximate the "true" species abundance distribution within a community. However because of the large number of plots in our study individual counts of species would be very time-consuming and are thus not feasible. Estimates of % cover are at least rough approaches to estimate abundance and we suggest that analyses based on such approaches are more meaningful than considering only presence/absence data, especially in the context of the relative contribution of abundant, subordinate and transient

species [59]. We used a slightly modified calculation of MPD based on abundance data as proposed by Gerhold et al. [60] to reduce effects of species richness. Abundance weighted and presence/absence versions of indices showed moderate correlations (MPD: $r = 0.41$; MNTD: $r = 0.58$). However, results based on the two indices did not differ considerably and therefore we present here only the results of abundance weighted indices (see Appendix S5 and S6 for presence/absence PD results). MPD measures the mean phylogenetic distance between two taxa in a sample and MNTD the mean phylogenetic distance to the nearest taxon in a sample. Hence MPD summarizes all phylogenetic distances including those of very distantly related species (e.g. between species of different orders) while MNTD considers only those between the most closely related species (e.g. between species within a genus). Thus, a stronger relationship of MNTD with land-use intensity compared to MPD would indicate that land-use has a stronger effect on the terminal than on the basal phylogenetic composition of a community. Both metrics depended on species richness and we therefore calculated standardized effect sizes ((observed metric - expected metric)/standard deviation of expected metric). We used a null model that shuffles the tip labels of the phylogeny maintaining all other properties of the sample matrix (i.e. species richness in plots and species prevalence). This null model was chosen since it tests for the null hypothesis, that phylogeny is not an important factor for structuring plants communities. Note that effect sizes of both metrics were calculated for each year and region separately. For each plot we then

Table 1. Effects of region, land-use intensity (LUI) and its interaction with region on effect size of (A) mean pairwise distance (MPD) and (B) mean nearest taxon distance (MNTD) for total, common and rare species assemblages.

A

	df	MPD(total)		MPD(common)		MPD(rare)	
		F	p	F	p	F	p
Region	2	3.44	**0.035**	0.14	0.87	9.04	**0.0002**
LUI	1	3.65	0.06	0.01	0.91	3.71	0.06
Region×LUI	2	0.68	0.51	0.12	0.89	2.15	0.12
Residuals	144						

B

	df	MNTD(total)		MNTD(common)		MNTD(rare)	
		F	p	F	p	F	p
Region	2	10.43	**<0.001**	1.51	0.22	0.73	0.48
LUI	1	7.60	**0.0066**	1.13	0.29	1.33	0.25
Region×LUI	2	3.02	0.052	0.25	0.78	3.51	**0.032**
Residuals	144						

ANOVA table with F and p values of the full models. Significant values in bold.

Table 2. Values for t-statistics and corresponding p values of the linear models with (A) MPD and (B) MNTD as dependent variables and LUI, rarity (two-level factor: common and rare) and their interaction as independent variables.

A

MPD	ALL		ALB		HAI		SCH	
	t	p	t	p	t	p	t	p
Intercept	−0.65	0.52	−0.55	0.58	−0.61	0.54	0.04	0.97
LUI	−0.10	0.92	0.19	0.85	0.03	0.98	−0.42	0.67
Rarity	0.25	0.8	0.04	0.97	0.29	0.77	0.19	0.85
LUI×Rarity	−1.39	0.17	−0.30	0.77	−0.36	0.72	−1.80	0.08

B

MNTD	ALL		ALB		HAI		SCH	
	t	p	t	p	t	p	t	p
Intercept	−1.58	0.12	−1.09	0.28	−1.30	0.20	−0.21	0.83
LUI	−0.87	0.38	−0.61	0.55	−0.08	0.94	−0.98	0.33
Rarity	−2.61	**0.009**	−1.57	0.12	−2.08	**0.041**	−0.66	0.51
LUI×Rarity	1.52	0.13	1.33	0.19	1.69	0.09	−0.72	0.47

Interaction term determines whether rare species PD response differs from that of common species PD. Significant values in bold.

calculated averages across the three years which were further used in all subsequent analyses (see above).

We used simple linear regressions and ANOVAs to analyze the relationships between plant PD and land-use intensification. We considered region (exploratory) as a factor to analyze whether PD differs among regions and whether the relationships between PD and LUI differ among regions (region×LUI interaction). To assess whether rare species assemblages respond more strongly to increasing land use than common species, we compared the slope of the regression lines with an ANCOVA by testing the significance of the LUI×"rarity" interaction. All statistical analyses were conducted in R [58].

Results

A total of 282 vascular plant species were recorded in the three regions from 2009 to 2010 (Appendix S1). We found depending on the considered species pool and the specific traits analyzed varying levels for Pagel's λ and Blomberg's K (Appendix S2). Based on Blomberg's K we found no strong phylogenetic conservatism in analyzed traits (Appendix S2). This suggests that PD cannot be seen as an overall proxy for functional diversity along land-use gradients.

Average total, rare and common species richness differed among regions (Appendix S3). Total and rare species richness decreased with increasing LUI with regional effects modulating the response of. In two regions (Alb, Hainich) total and rare species richness decreased with increasing LUI while in Schorfheide no effect was observed. The relationship between common species and LUI showed very contrasting patterns between regions but there was no overall decrease in species richness (Appendix S3).

Overall, average PD strongly varied among regions. But note that the differences depended on the PD-metric used and whether rare/common species were considered (Fig. 1a–f). When all species were considered, effect size of MPD showed strong significant clustering of communities in two regions (Hainich and Schorfheide) while MNTD estimates showed random patterns in all three regions. Mean phylogenetic community structure was random in respect to phylogeny for common and rare species assemblages in all three regions. After accounting for regional differences, total species MNTD decreased with increasing land-use intensity while MPD showed only a marginally significant decrease with similar relationships in all three exploratories (Table 1). Furthermore, land-use had slightly different effects on MNTD depending on region indicated by a marginally significant region×LUI interaction (Table 1), with a stronger decline of MNTD in one region (Alb: r = −0.39, p<0.01, Appendix S4), in particular. The other two regions showed a non-significant negative trend (Fig. 2). For MPD, only one region (Schorfheide) showed a significant decline with increasing land-use intensity (r = −0.3, p<0.05; Fig. 1, Appendix S4).

In general we found that for both common and rare species PD was not or only weakly affected by increasing land-use intensity. The relationships did not vary among regions except for rare species MPD (Table 1, Fig. 2). Overall, the strength of phylodiversity − land-use intensity relationships did not differ between common and rare species assemblages over three regions as indicated by non-significant LUI×rarity interaction terms in our models (Table 2).

Discussion

Land-use intensification is one of the major threats to global biodiversity in grasslands [21]. However, only a few studies have analyzed the effects of anthropogenic influence on PD of grassland plant communities. Several studies showed that anthropogenic influence can cause a decline in PD of species communities [15,37,61] which possibly may also decrease trait diversity and associated ecological functions [7]. In particular, PD can be important for ecosystem functioning when the ultimate processes, which depend on plant traits and trophic interactions, show a phylogenetic signal [7]. It has been shown that in grasslands PD can act as a better predictor of productivity than species richness or functional diversity [9,62]. Moreover, herbivory was stronger related to phylogenetic relatedness than to plant functional traits [63]. An experimental study by Pellissier et al. [38] revealed an increase in PD after strong fertilization and herbicide application while functional traits showed contrasting relationships presumably by selecting for convergent traits [38]. We found no evidence for strong phylogenetic signal in selected land-use sensitive traits (Appendix S2). Thus, phylogenetic diversity may not capture the relevant functional information leading to a relatively weak response to land-use intensification [31]. On the other side, the significant decrease of PD depending on region and metric used (see below), shows that PD might capture additional information beside the measured traits.

Dinnage [15] showed that the phylogenetic structure of plant communities in disturbed plots of old field sites is more clustered than expected, whereas phylogenetic structure in undisturbed plots does not differ from random expectations. This indicates, that land-use might act similarly to environmental filters and select for (presumably closely related) species with similar traits, which enable species to cope with disturbance. However, Dinnage analyzed the vegetation of an old field system with plowing being the disturbance that affected the phylogenetic diversity. This kind of disturbance mediates phylogenetic succession which can lead to increasing phylogenetic clustering of plant communities [64]. Our study sites are exposed to land-use types completely different to the former study and our results differ in the strength of the PD response to land-use intensification. Although land-use intensification slightly decreased phylogenetic diversity, considering the mean nearest taxon distance (MNTD) in particular, it did not lead to a shift form random to clustered community structures (Table 1, Fig. 2). In general, plant communities exhibited clustered and random phylogenetic structures on plots with both, low as well as high land-use intensities (points <1.96 on y-axis; Fig. 2). There are factors causing clustering of communities, especially when considering the tree-wide patterns (MPD, Fig. 1a) as was shown in several studies [e.g. 5,35]. Whether these factors refer to environmental filters [65,66] or exclusion of weak competitors [32] we cannot distinguish in our study. Land-use intensity, however, seems to play a minor role as determinant of phylogenetic community structure of plants in grasslands. This is contrary to the results of Dinnage [15] but such differences might be caused by different land-use types, with plowing causing a strong disturbance within habitats compared to our land-use types. Note also that in Dinnages study [11] no gradient of land-use intensity was analyzed and the definition of regional species pools was different from our study. Nevertheless, the slight decline of PD in our study may indicate that the influence of factors causing phylogenetic clustering of communities is mediated through or caused by increasing land-use intensity.

Many studies dealing with phylogenetic community structure use only one phylogenetic diversity index like NRI or NTI (equivalent to (−1 * effect size MPD) and (−1 * effect size MNTD), respectively) [e.g. 35,56]. Since the two metrics measure PD at different depths of phylogeny, with MPD (NRI) capturing tree-wide patterns and MNTD (NTI) being more sensitive to the tips of a phylogeny [30], depending on the distribution of traits, results of analyses might differ. However, when both metrics were used, similar results were reported [e.g. 57]. In our study, although the two metrics showed similar relationships with land use, MNTD was more sensitive to increasing land-use intensity. This emphasizes the importance of including different indices into analyzes of PD, as land-use sensitive traits might be conserved within a few relatively young clades (e.g. within families) and thus might be masked when using metrics considering a broader phylogenetic scale (e.g. MPD). Because MNTD shows a stronger response to land-use intensification it is possible that those traits are conserved in the younger nodes of phylogeny. Thus, using MPD might not capture relevant trait information when analyzing the effects of land use on phylogenetic diversity. In fact, as Blomberg's K can be thought of as the partitioning of variance with low values (K<1) indicating variance within clades, this might be the reason for MNTD being more sensitive to land use.

Although common and rare species might differ in several traits [67] or their sensitivity to soil biogeochemical parameters [68] and respond differently to land use and competition [69], we found no significant differences in their response to increasing land-use using analysis of covariance (Fig. 2, Tab 2). This suggests that traits that probably affect the abundance of species are randomly distributed across our plant phylogeny or/and are not affected by land-use. The only trait that was relatively strong conserved in both, common and rare species, was maximum height. Despite a relatively high phylogenetic signal in this trait, it seems that height is not a strong determinant of phylogenetic community structure in both, common and rare species assemblages. Another explanation might be that PD of common and rare species might respond differently to the single LUI components due to different traits not accounted for in our study and combining those to one index might neglect the differences in strength and direction of responses. Likewise, as the effects of land-use on PD did not differ in general between common and rare species communities, but rather showed slightly different patterns on a smaller scale, they should be examined separately if conservation efforts attempt to increase diversity for endangered taxa.

It is well known that regional peculiarities and species pools influence regional phylogenetic diversity [70,71]. For our study regions we found that considering all species Alb had overall high and Hainich overall low PD. Schorfheide showed contrasting patterns depending on the PD-metric used. Low MPD values suggest, that species in communities are closely related when accounting for the whole phylogeny, but high MNTD values indicate, that on lower phylogenetic scales (e.g. within families) species are distantly related. This might be explained by the fact that Schorfheide was more strongly affected by the Pleistocene glaciations than the other regions. One may argue that the plant communities of Schorfheide are still dominated by ecologically similar species belonging to closely related higher clades. Environmental filtering is then likely to cause strong phylogenetic clustering of communities considering the MPD (Fig. 1a). By contrast, within these clades PD might have increased due to limiting similarity [72] causing random community structure (Fig. 1d).

Differences in PD among regions may, to some extent, be also due historical land use rather than current [55] as suggested for species richness or functional diversity [70,73]. Such regional differences call for a careful consideration of regional particularities when providing management strategies to maintain or increase phylogenetic diversity of grassland plant communities under "real world" conditions.

The theory behind phylogenetic patterns along disturbance gradients relies on several hypotheses about distribution of ecological traits across phylogenetic trees [9,30,32,62,74]. We showed that although potentially land-use relevant traits show some levels of phylogenetic conservatism, PD still can provide additional information. The consideration of PD is therefore in particular importantin situations when functional traits of species are not available. Phylogenetic methods can complement ecological analyzes, but it must be pointed out that PD cannot be seen as a surrogate for other biodiversity metrics, functional diversity in particular.

Supporting Information

Appendix S1 Phylogenetic tree of plants in three regions in Germany used in this study.

Appendix S2 Phylogenetic signal in 7 traits considered as sensitive to land use for all, common and rare species in the three regions (ALB: Schwäbische Alb, HAI: Hainich-Dün and SCH: Schorfheide-Chorin) and in all regions combined. Significant values are in bold.

Appendix S3 Mean (±SE) values and regression slopes of species richness for total, common and rare species assemblages in three regions in Germany.

Appendix S4 Correlation coefficients and significance of the respective regression slopes from Fig. 1 and Appendix S3 and for all regions combined.

Appendix S5 Relationships between presence/absence based mean pairwise distance (effect size MPD), mean nearest taxon distance (effect size MNTD) and land-use intensity (LUI) in three regions in Germany.

Appendix S6 Mean (±SE) values of presence/absence based MPD and MNTD effect sizes for total, common and rare species assemblages in three regions in Germany.

Acknowledgments

We thank the managers of the three Exploratories, Swen Renner, Sonja Gockel, Kerstin Wiesner, and Martin Gorke for their work in maintaining the plot and project infrastructure; Simone Pfeiffer and Christiane Fischer giving support through the central office, Michael Owonibi for managing the central data base, and Eduard Linsenmair, Dominik Hessenmöller, Jens Nieschulze, Ingo Schöning, François Buscot, Ernst-Detlef Schulze, Wolfgang W. Weisser and the late Elisabeth Kalko for their role in setting up the Biodiversity Exploratories project. We also want to thank Steffen Boch, Stephanie Socher and Jörg Müller for providing vegetation data for our analyses.

Author Contributions

Conceived and designed the experiments: MB MF. Performed the experiments: EE. Analyzed the data: EE. Contributed reagents/materials/analysis tools: WD SM BS SB DP. Wrote the paper: EE. Provided editorial advice: DP WD MB.

References

1. Foley JA, DeFries R, Asner GP (2005) Global Consequences of Land Use. Science 309: 570–574. doi:10.1126/science.1111772.
2. Flynn DFB, Gogol-Prokurat M, Nogeire T, Molinari N, Richers BT, et al. (2009) Loss of functional diversity under land use intensification across multiple taxa. Ecol Lett 12: 22–33. doi:10.1111/j.1461-0248.2008.01255.x.
3. Socher SA, Prati D, Boch S, Müller J, Klaus VH, et al. (2012) Direct and productivity-mediated indirect effects of fertilization, mowing and grazing on grassland species richness. J Ecol 100: 1391–1399. doi:10.1111/j.1365-2745.2012.02020.x.
4. Cadotte MW, Cardinale BJ, Oakley TH (2008) Evolutionary history and the effect of biodiversity on plant productivity. Proc Natl Acad Sci 105: 17012–17017. doi:10.1073/pnas.0805962105.
5. Cavender-Bares J, Kozak KH, Fine PVA, Kembel SW (2009) The merging of community ecology and phylogenetic biology. Ecol Lett 12: 693–715.
6. Vamosi SM, Heard SB, Vamosi JC, Webb CO (2009) Emerging patterns in the comparative analysis of phylogenetic community structure. Mol Ecol 18: 572–592.
7. Srivastava DS, Cadotte MW, MacDonald AAM, Marushia RG, Mirotchnick N (2012) Phylogenetic diversity and the functioning of ecosystems. Ecol Lett: n/a–n/a. doi:10.1111/j.1461-0248.2012.01795.x.
8. Wiens JJ, Ackerly DD, Allen AP, Anacker BL, Buckley LB, et al. (2010) Niche conservatism as an emerging principle in ecology and conservation biology. Ecol Lett 13: 1310–1324. doi:10.1111/j.1461-0248.2010.01515.x.
9. Flynn DFB, Mirotchnick N, Jain M, Palmer MI, Naeem S (2011) Functional and phylogenetic diversity as predictors of biodiversity–ecosystem-function relationships. Ecology 92: 1573–1581. doi:10.1890/10-1245.1.
10. Maherali H, Klironomos JN (2007) Influence of Phylogeny on Fungal Community Assembly and Ecosystem Functioning. Science 316: 1746–1748. doi:10.1126/science.1143082.
11. Dinnage R (2013) Phylogenetic diversity of plants alters the effect of species richness on invertebrate herbivory. PeerJ 1: e93. Available: https://peerj.com/articles/93/. Accessed 14 January 2014.
12. Cadotte MW, Dinnage R, Tilman D (2012) Phylogenetic diversity promotes ecosystem stability. Ecology 93: S223–S233. doi:10.1890/11-0426.1.
13. Dinnage R, Cadotte MW, Haddad NM, Crutsinger GM, Tilman D (2012) Diversity of plant evolutionary lineages promotes arthropod diversity - Ecology Letters - Wiley Online Library. Available: http://onlinelibrary.wiley.com/doi/10.1111/j.1461-0248.2012.01854.x/full. Accessed 14 January 2014.
14. Mouquet N, Devictor V, Meynard CN, Munoz F, Bersier L-F, et al. (2012) Ecophylogenetics: advances and perspectives. Biol Rev 87: 769–785. Hoiss B, Krauss J, Potts SG, Roberts S, Steffan-Dewenter I. (2012).X.2012.00224.x.
15. Dinnage R (2009) Disturbance Alters the Phylogenetic Composition and Structure of Plant Communities in an Old Field System. PLoS ONE 4: e7071. doi:10.1371/journal.pone.0007071.
16. Pärtel M, Bruun HH, Sammul M (2005) Biodiversity in temperate European grasslands: origin and conservation European Grassland Federation; Distributed by British Grassland Society.
17. Statistisches Bundesamt (2012) Statistisches Jahrbuch 2012. Wiesbaden.
18. Suding KN, Collins SL, Gough L, Clark C, Cleland EE, et al. (2005) Functional- and abundance-based mechanisms explain diversity loss due to N fertilization. Proc Natl Acad Sci 102: 4387–4392. doi:10.1073/pnas.0408648102.
19. Kleijn D, Kohler F, Báldi A, Batáry P, Concepción ED, et al. (2009) On the relationship between farmland biodiversity and land-use intensity in Europe. Proc R Soc B Biol Sci 276: 903–909. doi:10.1098/rspb.2008.1509.
20. Storkey J, Meyer S, Still KS, Leuschner C (2011) The impact of agricultural intensification and land-use change on the European arable flora. Proc R Soc B Biol Sci 279: 1421–1429. doi:10.1098/rspb.2011.1686.
21. Sala OE, Chaplin FS, Armesto JJ, Berlow E, Bloomfield J, et al. (2000) Global Biodiversity Scenarios for the Year 2100 Science 287: 1770–1774. doi:10.1126/science.287.5459.1770.
22. Tilman D, Wedin D, Knops J (1996) Productivity and sustainability influenced by biodiversity in grassland ecosystems. Nature 379: 718–720. doi:10.1038/379718a0.
23. Hector A, Schmid B, Beierkuhnlein C, Caldeira M, Diemer M, et al. (1999) Plant diversity and productivity experiments in European grasslands. Science 286: 1123.
24. Cardinale BJ, Wright JP, Cadotte MW, Carroll IT, Hector A, et al. (2007) Impacts of plant diversity on biomass production increase through time because of species complementarity. Proc Natl Acad Sci 104: 18123–18128. doi:10.1073/pnas.0709069104.
25. Cardinale BJ, Srivastava DS, Emmett Duffy J, Wright JP, Downing AL, et al. (2006) Effects of biodiversity on the functioning of trophic groups and ecosystems. Nature 443: 989–992. doi:10.1038/nature05202.
26. Petchey OL, Gaston KJ (2006) Functional diversity: back to basics and looking forward. Ecol Lett 9: 741–758. doi:10.1111/j.1461-0248.2006.00924.x.
27. Tilman D, Knops J, Wedin D, Reich P, Ritchie M, et al. (1997) The Influence of Functional Diversity and Composition on Ecosystem Processes. Science 277: 1300–1302. doi:10.1126/science.277.5330.1300.
28. Chapin III FS, Zavaleta ES, Eviner VT, Naylor RL, Vitousek PM, et al. (2000) Consequences of changing biodiversity. Nature 405: 234–242. doi:10.1038/35012241.
29. Loreau M (2000) Biodiversity and ecosystem functioning: recent theoretical advances. Oikos 91: 3–17. doi:10.1034/j.1600-0706.2000.910101.x.
30. Webb CO, Ackerly DD, McPeek MA, Donoghue MJ (2002) Phylogenies and Community Ecology. Annu Rev Ecol Syst 33: 475–505. doi:10.1146/annurev.ecolsys.33.010802.150448.
31. Bernard-Verdier M, Flores O, Navas M-L, Garnier E (2013) Partitioning phylogenetic and functional diversity into alpha and beta components along an environmental gradient in a Mediterranean rangeland. J Veg Sci 24: 877–889. doi:10.1111/jvs.12048.
32. Mayfield MM, Levine JM (2010) Opposing effects of competitive exclusion on the phylogenetic structure of communities: Phylogeny and coexistence. Ecol Lett 13: 1085–1093. doi:10.1111/j.1461-0248.2010.01509.x.
33. Díaz S, Cabido M, Zak M, Martínez Carretero E, Aranibar J (1999) Plant functional traits, ecosystem structure and land-use history along a climatic gradient in central-western Argentina. J Veg Sci 10: 651–660. doi:10.2307/3237080.
34. Stevens CJ, Dise NB, Mountford JO, Gowing DJ (2004) Impact of Nitrogen Deposition on the Species Richness of Grasslands. Science 303: 1876–1879. doi:10.1126/science.1094678.
35. Kluge J, Kessler M (2011) Phylogenetic diversity, trait diversity and niches: species assembly of ferns along a tropical elevational gradient. J Biogeogr 38: 394–405. doi:10.1111/j.1365-2699.2010.02433.x.
36. Brunbjerg AK, Borchsenius F, Eiserhardt WL, Ejrnaes R, Svenning J-C (2012) Disturbance drives phylogenetic community structure in coastal dune vegetation. J Veg Sci 23: 1082–1094. doi:10.1111/j.1654-1103.2012.01433.x.
37. Knapp S, Kühn I, Schweiger O, Klotz S (2008) Challenging urban species diversity: contrasting phylogenetic patterns across plant functional groups in Germany. Ecol Lett 11: 1054–1064.
38. Pellissier L, Wisz MS, Strandberg B, Damgaard C (2014) Herbicide and fertilizers promote analogous phylogenetic responses but opposite functional responses in plant communities. Environ Res Lett 9: 024016. doi:10.1088/1748-9326/9/2/024016.
39. Graham CH, Parra JL, Rahbek C, McGuire JA (2009) Colloquium Papers: Phylogenetic structure in tropical hummingbird communities. Proc Natl Acad Sci 106: 19673–19678. doi:10.1073/pnas.0901649106.
40. Machac A, Janda M, Dunn RR, Sanders NJ (2011) Elevational gradients in phylogenetic structure of ant communities reveal the interplay of biotic and abiotic constraints on diversity. Ecography 34: 364–371. doi:10.1111/j.1600-0587.2010.06629.x.
41. Hoiss B, Krauss J, Potts SG, Roberts S, Steffan-Dewenter I (2012) Altitude acts as an environmental filter on phylogenetic composition, traits and diversity in bee communities. Proc R Soc B Biol Sci. Available: http://rspb.royalsocietypublishing.org/cgi/doi/10.1098/rspb.2012.1581. Accessed 26 September 2012.
42. Cavender-Bares J, Ackerly DD, Baum DA, Bazzaz FA (2004) Phylogenetic Overdispersion in Floridian Oak Communities. Am Nat 163: 823–843. doi:10.1086/386375.
43. Fischer M, Bossdorf O, Gockel S, Hänsel F, Hemp A, et al. (2010) Implementing large-scale and long-term functional biodiversity research: The Biodiversity Exploratories. Basic Appl Ecol 11: 473–485. doi:10.1016/j.baae.2010.07.009.
44. Blüthgen N, Dormann CF, Prati D, Klaus VH, Kleinebecker T, et al. (2012) A quantitative index of land-use intensity in grasslands: Integrating mowing, grazing and fertilization. Basic Appl Ecol 13: 207–220. doi:10.1016/j.baae.2012.04.001.
45. Allan E, Bossdorf O, Dormann CF, Prati D, Gossner MM, et al. (in press) Inter-annual variation in land-use intensity enhances grassland multidiversity. PNAS.
46. Durka W, Michalski SG (2012) Daphne: a dated phylogeny of a large European flora for phylogenetically informed ecological analyses. Ecology 93: 2297–2297. doi:10.1890/12-0743.1.
47. Webb CO, Ackerly DD, Kembel SW (2008) Phylocom: software for the analysis of phylogenetic community structure and trait evolution. Bioinformatics 24: 2098–2100. doi:10.1093/bioinformatics/btn358.
48. Kleyer M, Bekker RM, Knevel IC, Bakker JP, Thompson K, et al. (2008) The LEDA Traitbase: a database of life-history traits of the Northwest European flora. J Ecol 96: 1266–1274. doi:10.1111/j.1365-2745.2008.01430.x.
49. Klotz S, Kühn I, Durka W (2002) BIOLFLOR - Eine Datenbank zu biologisch-ökologischen Merkmalen der Gefäßpflanzen in Deutschland. Bonn: Bundesamt für Naturschutz.
50. Binz A, Heitz C (1990) Schul- und Exkursionsflora fur die Schweiz: mit Berucksichtigung der Grenzgebiete: Bestimmungsbuch fur die wildwachsenden Gefasspflanzen. 19. Aufl. Basel: Schwabe. 659 p.
51. Jäger EJ, Werner K (2005) Exkursionsflora von Deutschland. 10th ed. Heidelberg: Spektrum Akademischer Verlag.
52. Ellenberg H, Weber HE, Düll R, Wirth V, Werner W, et al. (1992) Zeigerwerte von Pflanzen in Mitteleuropa. Göttingen: Verlag Erich Goltze. 1–248.
53. Briemle G, Ellenberg H (1994) Zur Mahdverträglichkeit von Grünlandpflanzen. Nat Landschaft 69: 139–147.
54. Briemle G, Nitsche S, Nitsche L. (2002) Nutzungswertzahlen für Gefäßpflanzen des Grünlandes.

55. Revell LJ (2012) phytools: an R package for phylogenetic comparative biology (and other things): phytools: R package. Methods Ecol Evol 3: 217–223. doi:10.1111/j.2041-210X.2011.00169.x.

56. Münkemüller T, Lavergne S, Bzeznik B, Dray S, Jombart T, et al. (2012) How to measure and test phylogenetic signal: How to measure and test phylogenetic signal. Methods Ecol Evol 3: 743–756. doi:10.1111/j.2041-210X.2012.00196.x.

57. Kembel SW, Cowan PD, Helmus MR, Cornwell WK, Morlon H, et al. (2010) Picante: R tools for integrating phylogenies and ecology. Bioinformatics 26: 1463–1464. doi:10.1093/bioinformatics/btq166.

58. R Core Team (2012) R: A language and environment for statistical computing. Vienna, Austria: R Foundation for Statistical Computing.

59. Grime JP (1998) Benefits of plant diversity to ecosystems: immediate, filter and founder effects. J Ecol 86: 902–910. doi:10.1046/j.1365-2745.1998.00306.x.

60. Gerhold P, Price JN, Püssa K, Kalamees R, Aher K, et al. (2013) Functional and phylogenetic community assembly linked to changes in species diversity in a long-term resource manipulation experiment. J Veg Sci 24: 843–852. doi:10.1111/jvs.12052.

61. Helmus MR, Keller W (Bill), Paterson MJ, Yan ND, Cannon CH, et al. (2010) Communities contain closely related species during ecosystem disturbance. Ecol Lett 13: 162–174. doi:10.1111/j.1461-0248.2009.01411.x.

62. Cadotte MW, Cavender-Bares J, Tilman D, Oakley TH (2009) Using Phylogenetic, Functional and Trait Diversity to Understand Patterns of Plant Community Productivity. PLoS ONE 4: e5695. doi:10.1371/journal.pone.0005695.

63. Paine CET, Norden N, Chave J, Forget P-M, Fortunel C, et al. (2012) Phylogenetic density dependence and environmental filtering predict seedling mortality in a tropical forest: Neighbourhood similarity predicts seedling mortality. Ecol Lett 15: 34–41. doi:10.1111/j.1461-0248.2011.01705.x.

64. Valiente-Banuet A, Verdú M (2007) Facilitation can increase the phylogenetic diversity of plant communities. Ecol Lett 10: 1029–1036. doi:10.1111/j.1461-0248.2007.01100.x.

65. Butterfield BJ, Cavieres LA, Callaway RM, Cook BJ, Kikvidze Z, et al. (2013) Alpine cushion plants inhibit the loss of phylogenetic diversity in severe environments. Ecol Lett 16: 478–486. doi:10.1111/ele.12070.

66. Culmsee H, Leuschner C (2013) Consistent patterns of elevational change in tree taxonomic and phylogenetic diversity across Malesian mountain forests. J Biogeogr: n/a–n/a. doi:10.1111/jbi.12138.

67. Farnsworth EJ (2007) Plant life history traits of rare versus frequent plant taxa of sandplains: Implications for research and management trials. Biol Conserv 136: 44–52. doi:10.1016/j.biocon.2006.10.045.

68. Kleijn D, Bekker RM, Bobbink R, De Graaf MCC, Roelofs JGM (2008) In search for key biogeochemical factors affecting plant species persistence in heathland and acidic grasslands: a comparison of common and rare species. J Appl Ecol 45: 680–687. doi:10.1111/j.1365-2664.2007.01444.x.

69. Dawson W, Fischer M, van Kleunen M (2012) Common and rare plant species respond differently to fertilisation and competition, whether they are alien or native. Ecol Lett 15: 873–880. doi:10.1111/j.1461-0248.2012.01811.x.

70. Anacker BL, Harrison SP (2012) Historical and Ecological Controls on Phylogenetic Diversity in Californian Plant Communities. Am Nat 180: 257–269. doi:10.1086/666650.

71. Blanchet S, Helmus MR, Brosse S, Grenouillet G (2013) Regional *vs* local drivers of phylogenetic and species diversity in stream fish communities. Freshw Biol: n/a–n/a. doi:10.1111/fwb.12277.

72. MacArthur R, Levins R (1967) The Limiting Similarity, Convergence and Divergence of Coexisting Species. Am Nat 101: 377–385.

73. Klaus VH, Hölzel N, Boch S, Müller J, Socher SA, et al. (n.d.) Direct and indirect associations between plant species richness and productivity in grasslands: regional differences preclude simple generalization of productivity-biodiversity relationships. Available: http://www.preslia.cz/P132Klaus.pdf. Accessed 19 December 2013.

74. Losos JB (2008) Phylogenetic niche conservatism, phylogenetic signal and the relationship between phylogenetic relatedness and ecological similarity among species. Ecol Lett 11: 995–1003. doi:10.1111/j.1461-0248.2008.01229.x.

18

A Well-Resolved Phylogeny of the Trees of Puerto Rico Based on DNA Barcode Sequence Data

Robert Muscarella[1]*, María Uriarte[1], David L. Erickson[2], Nathan G. Swenson[3], Jess K. Zimmerman[4], W. John Kress[2]

1 Department of Ecology, Evolution and Environmental Biology, Columbia University, New York, New York 10027, United States of America, 2 Department of Botany, MRC-166, National Museum of Natural History Smithsonian Institution, P.O. Box 37012, Washington, D. C., 20013, United States of America, 3 Department of Plant Biology, Michigan State University, East Lansing, Michigan 48824, United States of America, 4 Department of Environmental Science, University of Puerto Rico, San Juan, Puerto Rico 00925, United States of America

Abstract

Background: The use of phylogenetic information in community ecology and conservation has grown in recent years. Two key issues for community phylogenetics studies, however, are (i) low terminal phylogenetic resolution and (ii) arbitrarily defined species pools.

Methodology/principal findings: We used three DNA barcodes (plastid DNA regions *rbcL*, *matK*, and *trnH-psbA*) to infer a phylogeny for 527 native and naturalized trees of Puerto Rico, representing the vast majority of the entire tree flora of the island (89%). We used a maximum likelihood (ML) approach with and without a constraint tree that enforced monophyly of recognized plant orders. Based on 50% consensus trees, the ML analyses improved phylogenetic resolution relative to a comparable phylogeny generated with PHYLOMATIC (proportion of internal nodes resolved: constrained ML = 74%, unconstrained ML = 68%, PHYLOMATIC = 52%). We quantified the phylogenetic composition of 15 protected forests in Puerto Rico using the constrained ML and PHYLOMATIC phylogenies. We found some evidence that tree communities in areas of high water stress were relatively phylogenetically clustered. Reducing the scale at which the species pool was defined (from island to soil types) changed some of our results depending on which phylogeny (ML vs. PHYLOMATIC) was used. Overall, the increased terminal resolution provided by the ML phylogeny revealed additional patterns that were not observed with a less-resolved phylogeny.

Conclusions/significance: With the DNA barcode phylogeny presented here (based on an island-wide species pool), we show that a more fully resolved phylogeny increases power to detect nonrandom patterns of community composition in several Puerto Rican tree communities. Especially if combined with additional information on species functional traits and geographic distributions, this phylogeny will (i) facilitate stronger inferences about the role of historical processes in governing the assembly and composition of Puerto Rican forests, (ii) provide insight into Caribbean biogeography, and (iii) aid in incorporating evolutionary history into conservation planning.

Editor: Damon P. Little, The New York Botanical Garden, United States of America

Funding: This work was supported by NSF DEB 1050957 to MU, NSF DEB 1311367 to MU and RM, and the Smithsonian Institution. The funders had no role in study design, data collection and analysis, decision to publish, or preparation of the manuscript.

Competing Interests: The authors have declared that no competing interests exist.

* Email: bob.muscarella@gmail.com

Introduction

The use of phylogenetic information in community ecology and conservation has grown dramatically in recent years [1,2,3]. This body of research has been largely stimulated by the idea that evolutionary relationships can provide insights into the historical processes governing assembly of local communities [4,5,6]. From a conservation perspective, phylogenies may reveal aspects of biodiversity that are not observable from traditional metrics of species diversity [7,8,9,10,11]. By providing a historical context, phylogenies help merge our understanding of ecological, evolutionary, and biogeographic drivers of community composition [12].

One key issue for research in community phylogenetics is how to best estimate phylogenetic relationships among species in diverse communities (*e.g.*, tropical forests). To date, the program PHYLOMATIC [13] has become a primary method by which ecologists integrate phylogenetic information with analyses of community patterns (*e.g.*, [14,15,16]). For plants, PHYLOMATIC generates community phylogenies by pruning a megatree of angiosperms given a user-defined species list. This approach offers a repeatable and accessible way to obtain phylogenies using existing data (*also see* [17]), however, PHYLOMATIC phylogenies typically have low or no taxonomic resolution among closely related species (*e.g.*, within plant families or genera). Low taxonomic resolution can reduce statistical power for detecting nonrandom patterns of community structure [18,19] and can bias estimates of phylogenetic signal [20]. Furthermore, because single genera often contain numerous species with diverse life-history characteristics (*e.g.*, [21,22]), resolving evolutionary relationships among congeners is critical for interpreting the link between patterns of phylogenetic community composition and the history of trait evolution. Finally, low taxonomic resolution can preclude inferences about biogeographic influences on local assemblages. The issue is particularly acute with respect to relatively recent evolutionary history (i.e., speciation events), which arguably represent a key connection between local and regional processes (*see* [23] *and references therein*).

In contrast to megatree approaches such as PHYLOMATIC, phylogenies based on genetic data typically provide comparatively high taxonomic resolution. Generating molecular phylogenies, however, requires a significant investment of resources and expert knowledge. Additionally, determining how to estimate phylogenies among the very distantly related species that are typical of community-based phylogenies (as opposed to clade-based phylogenies) remains an active area of research. One potentially promising approach is to integrate existing information on evolutionary relationships in the form of a constraint tree [24]. More research is required, however, to determine the influence of constraint trees on phylogenetic reconstruction and downstream analyses of community phylogenetic patterns.

Another characteristic of many existing studies of community phylogenetic structure lies in the lack of consistent methodology in defining species pools when testing hypotheses about mechanisms driving community assembly (*e.g.*, competition versus environmental filtering) [4]. Generally, these analyses are based on null models that compare an observed metric of phylogenetic composition (*e.g.*, NRI-, the net relatedness index) with a random expectation based on assemblages drawn from a regional species pool [16]. In practice, studies often delimit the 'regional pool' as the set of species encountered in the study, regardless of the ecological significance of the study area boundaries (*e.g.*, forest dynamics plots). Examining species assemblages within such arbitrarily defined regions can provide information on processes occurring at certain scales (*e.g.*, [16,25]). However, varying the spatial scale at which species pools are defined can provide important opportunities to evaluate the relative strength of local assembly processes (*e.g.*, interactions that occur among neighboring trees) versus processes that occur over larger spatial and temporal scales (*e.g.*, evolution and biogeography) and across broader environmental gradients (*e.g.*, [26,27,28,29,30,31]). For example, numerous studies in phylogenetic community ecology have shown that as the spatial (and taxonomic) extent of the species pool increases, the phylogenetic composition of local communities tends to appear increasingly 'clustered' (i.e., co-occurring species are more closely related than expected by random chance). Other studies have shown more mixed results (*see references in* [31], [32]), which may emerge, for example, if a larger species pool includes sister taxa absent from the smaller pool. In any case, scale-dependency of community patterns likely reflects the scales at which different assembly processes influence community structure [5,33,34,35]. As such, we can gain valuable insights on community assembly by adjusting species pools to suit particular hypotheses about the scales at which different assembly processes act [6,28,29,31,36,37,38,39].

In this study, we used DNA sequence data to generate an island-wide phylogeny for nearly all of the native and naturalized tree species of Puerto Rico. Specifically, we used sequence data from three regions of plastid DNA which are commonly used as plant DNA barcodes (*rbcL, matK, trnH-psbA*; [24]) to resolve evolutionary relationships among 527 recognized species with a maximum likelihood (ML) approach. We compare phylogenetic resolution of two ML phylogenies (built with and without the use of an ordinal-level constraint tree) and a comparable phylogeny derived from PHYLOMATIC. We then explore the implications of these different methods in a case study where we examined the phylogenetic structure of tree communities in 15 protected forests in Puerto Rico. These 15 forests span a wide variation in environmental conditions, providing an ideal template for evaluating the effects of local environmental variation on phylogenetic community structure within the island of Puerto Rico (Table 1). We addressed the following specific questions:

1. How does the use of a constraint tree influence (i) the level of bootstrap support in a DNA barcode phylogeny of Puerto Rican trees, and (ii) the degree to which a molecular phylogeny corresponds with currently recognized taxonomic groups? We predicted that the constraint tree would provide higher levels of bootstrap support among unconstrained nodes and increase concordance with current taxonomy relative to the unconstrained analysis.

2. How do patterns of community phylogenetic structure in Puerto Rican forests differ when based on a DNA barcode phylogeny versus a PHYLOMATIC phylogeny? We predicted that an increase in statistical power provided by the higher resolution of a molecular phylogeny would lead to a stronger signal of non-random phylogenetic structure.

3. How does phylogenetic structure in Puerto Rican forests change with respect to different species pool definitions? We predicted co-occurring species would tend to appear relatively phylogenetically clustered with respect to the full island species pool because of a strong role for environmental filtering across broad environmental gradients. We predicted that a more restricted species pool definition would reduce the level of phylogenetic clustering if niche differentiation (competitive exclusion) becomes more apparent at small spatial scales.

Materials and Methods

All necessary permits were obtained for the described study, which complied with all relevant regulations. Specifically, the Departmento de Recursos Naturales y Ambientales (DRNA) of Puerto Rico granted permit #2011-IC-046 to collect plant specimens in the state forests of Puerto Rico. Herbaria staff at the University of Puerto Rico, Rio Piedras and the US National Herbarium provided permission to sample tissue from their collections.

Study area and species

The island of Puerto Rico encompasses six Holdridge life zones [40] ranging from subtropical dry forest to subtropical rainforest in an area of 8,740 km^2 [41]. Mean annual precipitation ranges drastically, from ca. 700–4,500 mm yr^{-1} [42]. The island's complex geologic history is reflected in its rugged topography (0–1,338 m a.s.l.) and diverse parent soil materials, which include volcanic, limestone, alluvial, and ultramafic materials [43]. Substantial existing data on the flora (*e.g.*, [24,44,45]) provide a strong foundation for our work.

We created an initial list of Puerto Rican trees with the species list from the USFS Forest Inventory and Analysis (FIA) Caribbean field guide [46]. With guidance from local experts (P. Acevedo-Rodríguez, F. Areces, F. Axelrod, M. Caraballo, J. Sustache, and P. Vives, *personal communication*), we modified this list by (1) updating nomenclature to be consistent with Acevedo-Rodríguez and Strong [45], (2) removing species occurring only under cultivation and (3) adding native and naturalized tree species

Table 1. Environmental characteristics and generalized results of 'nodesig' analysis (*i.e.*, over and underrepresented lineages) for 15 protected forests in Puerto Rico.

Forest	Area (ha)[1]	Holdridge Life Zone(s)[2]	Elevation Range (mean) (m)[3]	Mean Annual Precipitation (mm yr⁻¹)[4]	Primary Geologic Substrate[5]	Species Richness[6,7]	Overrepresented Groups	Underrepresented Groups
Aguirre	432	df-S	0–4 (1)	953	Unconsolidated	33	Combretaceae, Fabaceae	Combretaceae, Fabaceae
Boquerón	623	df-S	0–5 (1)	786	Unconsolidated	19	Combretaceae	Combretaceae
Cambalache	649	mf-S	31–188 (157)	1,593	Limestone	152	Arecaceae, Burseraceae, Anacardiaceae, Celastraceae	Melastomataceae
Carite	2,699	wf-S, wf-LM	296–839 (657)	2,018	Volcanic	146	Lauraceae, Solanaceae, *Psychotria* (Rubiaceae), *Myrcia* (Myrtaceae), *Clusia* (Clusiaceae), Meliaceae	*Exostema*, *Guettarda*, and *Stenostomum* (Rubiaceae), Fabaceae (Mimosoidae)
Ceiba	237	df-S	0–11 (4)	1,408	Unconsolidated	5	Solanaceae, Combretaceae, Rhizophoraceae	
Guajataca	955	mf-S	192–310 (249)	1,981	Limestone	197	Nyctaginaceae, Sapindales, Meliaceae	
Guánica	3,831	df-S	0–210 (81)	876	Limestone	133	*Coccoloba* (Polygonaceae), *Crescentia* (Bignoniaceae), Capparaceae, Sapindales, Fabales	Melastomataceae, Laurales, Ericales
Monte Guilarte	1,705	wf-S, wf-LM	629–1079 (909)	2,156	Volcanic	87	*Piper* (Piperaceae), *Miconia* (Melastomataceae), Meliaceae, *Inga* (Fabaceae)	
El Yunque	11,429	mf-S, wf-S, wf-LM, rf-S, rf-LM	87–1011 (570)	3,758	Volcanic	215	Solanaceae, Melastomataceae, Meliaceae, Laurales	Rutaceae, Fabaceae, Celastraceae
Maricao	4,168	mf-S, wf-S, wf-LM	130–871 (511)	2,126	Serpentine	212	Araliaceae, Aquifoliaceae, Meliaceae	Lamiales
Río Abajo	2,284	mf-S, wf-S	209–380 (313)	2,079	Limestone	175	Meliaceae	
Piñones	732	mf-S	0–2 (1)	1,398	Unconsolidated	31	Combretaceae, Malvaceae, *Pterocarpus* (Fabaceae)	
Susúa	1,298	mf-S	107–501 (264)	1,395	Serpentine	180	Rubiaceae	Melastomataceae
Toro Negro	2,763	wf-S, wf-LM	486–1284 (988)	2,248	Volcanic	133	Aquifoliaceae, Primulaceae, Meliaceae, *Piper* (Piperaceae), Araliaceae, Laurales, Solanaceae, Melastomataceae	*Guettarda* and *Stenostomum* (Rubiaceae), Fabaceae
Vega	482	mf-S	27–110 (67)	1,668	Limestone	86	Arecaceae, Meliaceae, Celastraceae, Moraceae, Urticaceae *Drypetes* (Putrajivaceae)	

Environmental and occurrence data are from [1]Gould et al. [90], [2]Ewel & Whitmore [41], [3]Gesch [91], [4]Daly et al. [42], [5][43], [6]Little & Wadsworth [63], and [7]Little et al. [64]. Forest life zones are coded as: subtropical dry (df-S), subtropical moist (mf-S), subtropical wet (wf-S), lower montane wet (wf-LM), subtropical rainforest (rf-S), lower montane rainforest (rf-LM).

known to occur in Puerto Rico but absent from the FIA list. Our final list of target species contained 594 species of seed plants representing 33 orders, 86 families, and 304 genera (Table S1). Of these, we were able to compile DNA sequence data for 523 (89%) species representing all 32 orders, 85 families (99%), and 287 genera (94%). The single excluded family (Cunoniaceae) is represented in Puerto Rico by a single rare species of shrub and most of the other species missing from our dataset are relatively uncommon and distributed widely throughout taxonomic groups. As a result, we do not expect the missing species to influence overall results of community phylogenetic analyses. However, it will be enlightening to include these species when sequence data become available in order to better understand the contributions of rare species to phylogenetic diversity [47].

Tissue collection and lab procedures

We acquired DNA sequence data from a variety of sources. Primarily, we obtained leaf tissue either from freshly collected specimens or existing herbarium sheets. For fresh specimens, we dried leaf tissue in silica gel prior to DNA extraction. Prior to depositing voucher specimens at the US National Herbarium (US), we verified species identifications by referring to the herbarium at the University of Puerto Rico, Río Piedras (UPRRP) and through consultation with local experts (F. Areces, F. Axelrod, P. Vives, *personal communication*). For 95 species, we collected leaf tissue from dry material sampled from herbarium specimens at UPRRP or US. DNA extraction, amplification and sequencing protocols followed Kress *et al.* [24]. Specifically, we used the following lab procedures for fresh and dried leaf tissue. After disrupting tissue with a Tissuelyzer (Qiagen Cat. #85210), we incubated samples overnight at 55°C in a CTAB-based extraction buffer (AutoGen, Holliston, MA). Following incubation, we removed the supernatant and placed it in clean, 2 ml 96-well plate for submission to a DNA extraction robot (AutoGen 960, Holliston, MA). We hydrated DNA extractions in 100 mM Tris-HCl (pH 8.0) and then transferred them to Matrix barcode tubes (MatrixTechnologies Cat. # 3735) and stored them at −80°C. Working stocks of DNA were transferred to a microtiter plate and diluted 5× with water prior to PCR. We used routine PCR, with no more than three attempts per sample to recover PCR amplicons for each sample. The PCR cycling conditions were exactly the same for *rbcL* and *trnH-psbA* (95°C 3 min, [94°C 30 sec, 55°C 30 sec, 72°C 1 min]×35 cycles, 72°C 10 min) following procedures outlined in Kress and Erickson [48]. The PCR cycling conditions for *matK* required lower annealing temperatures and more cycles (95°C 3 min [94°C 30 sec, 49°C 30 sec, 72°C 1 min]×40 cycles, 72°C 10 min) following Fazekas *et al.* [49] and included DMSO at a final concentration of 5%. We purified successful PCR reactions with a 56 diluted mixture of ExoSap (USB, Cat. # 78201). For sequencing, 2–4 ul of the purified PCR was used in a 12 ul reaction (0.8 ul BigDye terminator sequencing mixture (V3.1; ABI, Cat. 4337457), 2.0 ul of a 56 buffer (400 u Molar Tris-HCL pH 8.0), 1 ul of 1 uMolar primer and distilled water to volume). Sequencing of *matK* PCR products included DMSO to a final concentration of 4% in the reaction mixture. Cycling sequencing protocols were the same for all markers, (95°C 15 sec [94°C 15 sec, 50°C 15 sec, 60°C 4 min]×30 cycles). Following cycle sequencing, products were purified on a column of sephadex and sequence reactions were read on an ABI 3730 (Applied Biosystems).

We also incorporated existing sequence data for 143 species previously sequenced from the Luquillo Forest Dynamics Plot [24] and for 25 species from GenBank [50]. We excluded 67 species from analyses for which we were unable to acquire reliable

sequence data either because tissue was not available or because of failure during DNA sequencing (Table S1).

Sequence editing, alignment, and assembly

We used GENEIOUS (R6, version 2.4.1; Biomatters Ltd.) to trim and assemble trace files for each marker into bidirectional contigs. Separately for each marker, we aligned sequences using SATé [51]. SATé is an iterative algorithm that divides the original sequence data set using a tree-based decomposition; we aligned these smaller sets of sequences using MAFFT [52] and merged these sub-alignments into a global alignment without disrupting the individual sub-alignments using MUSCLE [53]. SATé is particularly effective for conducting multiple sequence alignment among very distantly related taxa through the use of merging sub-alignments among related sequences, and has been widely applied for studies of very broad phylogenetic application [54,55]. We then concatenated the three separate marker alignments to produce an aligned three-gene matrix. Gaps were not coded and were treated as missing data in phylogenetic reconstruction.

Phylogenetic reconstruction

We generated a phylogeny using maximum likelihood (ML) methods, implemented in RAxML (Stamatakis et al. 2005) via the CIPRES Science Gateway [56]. Based on jModelTest2 [57], we modeled nucleotide substitution using a GTR+GAMMA model, with substitution rates estimated independently for each gene. We evaluated node support for the topology with the highest likelihood using 100 bootstrap runs. In addition, we trimmed Phylomatic reference tree R20120829 [58] to use for comparative purposes. While other methods for phylogenetic reconstruction are available (*e.g.*, parsimony), we focus here on a comparison between ML methods and a very commonly used method of generating phylogenies for community ecology (Phylomatic).

Rather than including densely sampled small taxonomic units, community phylogenies often contain smaller numbers of more distantly related species (*e.g.*, 32 orders represented in our dataset, represented by 18 species, on average). Resolving both shallow and deep relationships requires distinct molecular data sets that are difficult to assemble. When strong prior information is available, one approach to confront this issue is to enforce some relationships through the use of a constraint tree (*see for example* [59]). In the case of our study, the Angiosperm Phylogeny Group III [60] represents the authoritative standard for current relationships up to the family level in angiosperms. However, within the AGP III phylogeny, relationships between species are generally not resolved beyond the family level, thus providing an ideal opportunity to use DNA barcodes to resolve these finer-scale relationships. To test the ability of a constraint tree to improve phylogenetic resolution among distantly related taxa, we repeated the ML analysis detailed above using the APG III phylogeny [60] to constrain the topology of ordinal and deeper nodes. This approach allowed the topology within each order to be resolved with DNA barcode sequence data while ordinal and deeper nodes were enforced a priori. We dated both the constrained and unconstrained ML phylogenies using PATHd8 [61] with age constraints based on fossil records provided in the Appendix of Magallón & Castillo [62] (input files for our analyses are provided in Appendix S1 and S2). The constraints we used included one fixed age estimate for the angiosperm crown group and 35 minimal age estimates for other clades represented in our phylogeny ([62]; Appendix S1 and S2). We used this approach because dated ultrametric trees are the standard for community phylogenetics studies; however, we also provide the undated, non-ultrametric trees in Appendix S3. To explore the distribution of

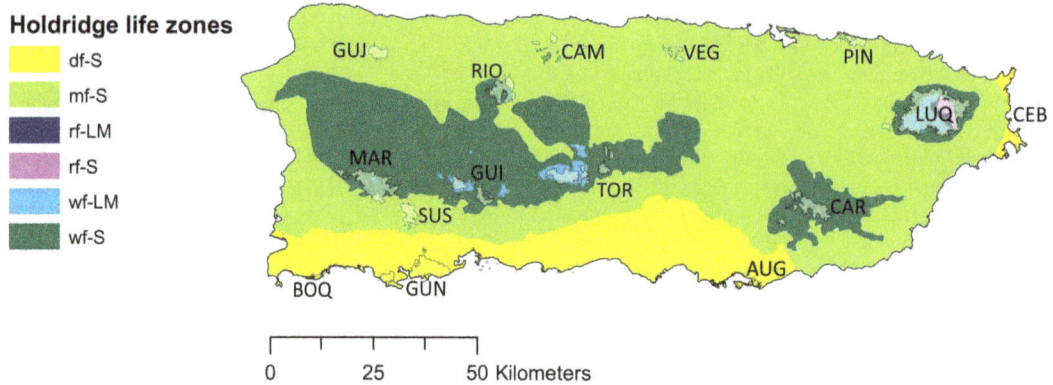

Figure 1. A map of Puerto Rico including the 15 state forests used in this study [90]. Forest life zones are coded as: subtropical dry (df-S), subtropical moist (mf-S), subtropical wet (wf-S), lower montane wet (wf-LM), subtropical rainforest (rf-S), lower montane rainforest (rf-LM). Refer to Table 1 for forest codes.

uncertainty across the phylogeny, we calculated the proportion of recognized taxonomic groups (orders, families, and genera) that were found to be monophyletic in each analysis and the proportion of resolved nodes within each of these groups.

Case study: Phylogenetic composition of Puerto Rican forests

We measured the phylogenetic composition of 15 protected forests in Puerto Rico based on species occurrence data (presence/absence) from Little & Wadsworth [63] and Little *et al.* [64]. As a synthesis of observations made by local experts, these volumes are the most commonly used references to describe tree composition of Puerto Rico's protected forests. The 15 forests examined here span a wide range of environmental conditions (precipitation range: ca. 800–3,800 mm yr^{-1}, elevation range: ca. 0–1,300 m a.s.l.) and occur across four main soil parent materials: unconsolidated, limestone, volcanic, and serpentine (Table 1, Fig. 1). We excluded taxa not included in our phylogeny – these accounted for only 2% of the total observations in the community dataset. With the remaining data, we quantified phylogenetic composition of each forest using the net relatedness index (NRI) and nearest taxon index (NTI) [4]. These indices describe whether sets of co-occurring taxa are more or less closely related than random assemblages of equal species richness drawn from a pool of species. Specifically, NRI measures the average degree of relatedness among all members of the community and thus emphasizes deeper branches of the phylogenetic tree. In contrast, NTI is based on the average distance between closest relatives in each assemblage and thus emphasizes compositional patterns at the tips of the phylogeny [4]. These metrics are calculated as: $NRI = -(r_{obs} - mean(r_{rand}))/sd(r_{rand})$, where r is either the co-occurring taxa (for NRI) or mean phylogenetic branch length separating nearest neighbors (for NTI). The observed value is r_{obs} and r_{rand} is a distribution of values based on assemblages drawn from a species pool. We calculated NRI and NTI for each forest using two different species pools: the full list of species in our dataset (the 'island pool'), and the list of species recorded from forests on the same soil parent material (the 'soil pool'). For example, for Guánica forest (limestone soil), we calculated two values of NRI: one value (NRI_{ISLAND}) based on null assemblages drawn from the entire species list and another value (NRI_{SOIL}) based on the list of species recorded from all forests on limestone soil (the soil pool).

We computed NRI and NTI using the ses.mpd and ses.mntd functions of the 'picante' package [65] for R v 3.0.0 [66]. We ran the analyses for 999 iterations and used the 'taxa-labels' null model. We chose this null model to control for the observed species occupancy rates and species richness of each forest. Positive values of NRI and NTI indicate phylogenetic clustering whereas negative values indicate phylogenetic evenness. We performed these analyses using the constrained ML 50% consensus tree and the PHYLOMATIC phylogeny. We based these analyses on the constrained ML 50% consensus tree because it reflects the uncertainty of our phylogenetic hypothesis given our data, while also incorporating the strong evidence resolving deep relationships provided by the APG III constraint tree.

We quantified shifts in NRI and NTI values between the two species pool definitions using paired t-tests and we quantified the similarity of these values between phylogenies with Pearson's correlation coefficient. In addition to overall patterns of community phylogenetic composition, we used the 'nodesig' algorithm in PHYLOCOM v 4.2 [67] to determine the particular clades that contribute significantly more or fewer species than expected to the composition of each forest.

Results

DNA barcode sequences

From fresh tissue, we successfully recovered sequence data from 85%, 75%, and 94% of samples for *rbcL*, *matK*, *trnH-psbA*, respectively. The final three-gene alignment comprised 3,366 base pairs (549 bp for *rbcL*, 1,070 bp for *matK*, and 1,747 bp for *trnH-psbA*). The data matrix had 62.2% missing data (including gaps coded as missing data and species for which we did not recover sequence data). This amount is far more compact than previous alignments of the same three regions that used a nested partitioning of the *trnH-psbA* alignment, resulting in >95% missing data [24]. Considering each region separately, the amount of missing data was 23.1%, 49.2%, and 82.1% for *rbcL*, *matK*, and *trnH-psbA*, respectively.

Phylogenetic analyses

We provide the constrained and unconstrained ML trees, with bootstrap support, as well as the PHYLOMATIC phylogeny used in our analyses in Appendix S3. Overall, we found relatively strong support for the majority of nodes in the both the constrained and unconstrained ML trees (Fig. 2). Across all nodes, 74% of nodes in the constrained ML tree received ≥50% bootstrap support and 52% received ≥80% bootstrap support. Considering only the 468

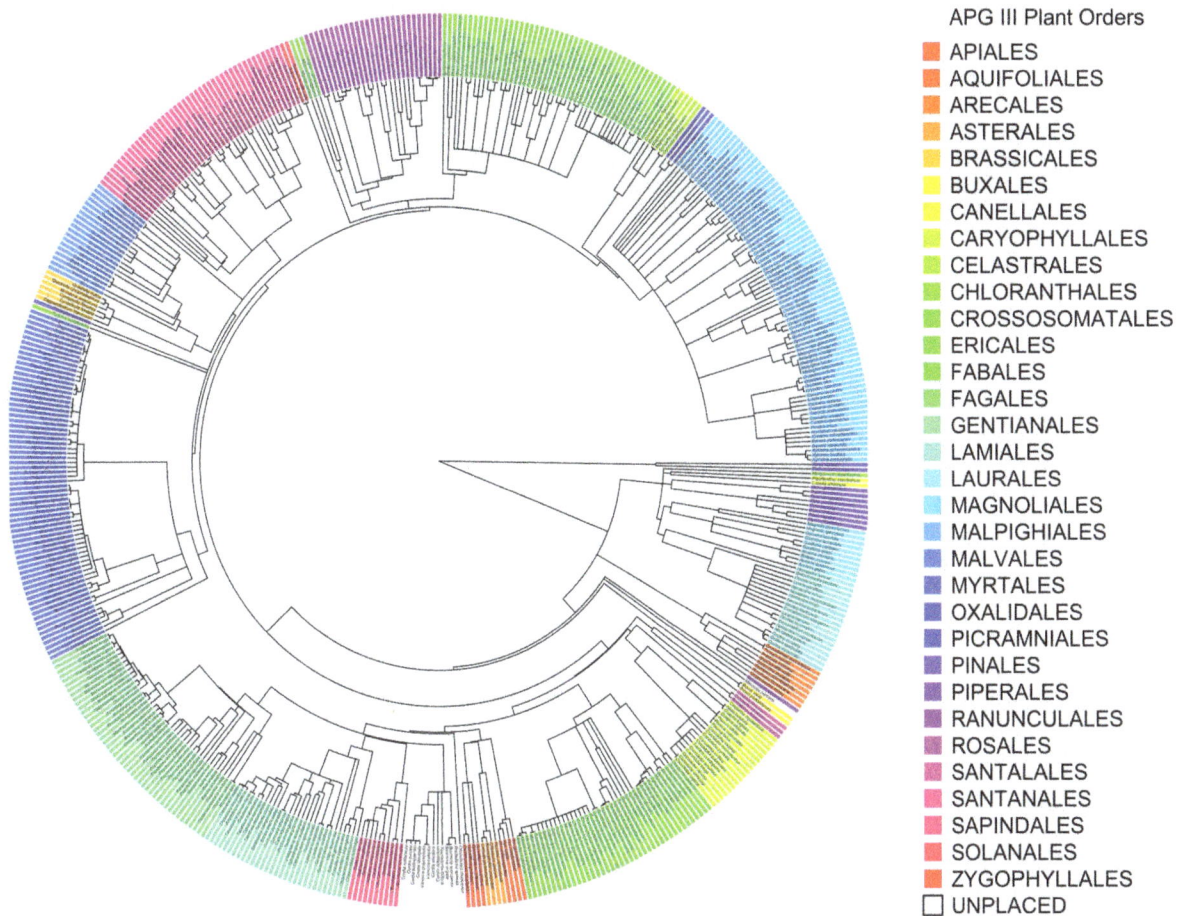

APG III Plant Orders
- APIALES
- AQUIFOLIALES
- ARECALES
- ASTERALES
- BRASSICALES
- BUXALES
- CANELLALES
- CARYOPHYLLALES
- CELASTRALES
- CHLORANTHALES
- CROSSOSOMATALES
- ERICALES
- FABALES
- FAGALES
- GENTIANALES
- LAMIALES
- LAURALES
- MAGNOLIALES
- MALPIGHIALES
- MALVALES
- MYRTALES
- OXALIDALES
- PICRAMNIALES
- PINALES
- PIPERALES
- RANUNCULALES
- ROSALES
- SANTALALES
- SANTANALES
- SAPINDALES
- SOLANALES
- ZYGOPHYLLALES
- □ UNPLACED

Figure 2. A maximum likelihood phylogeny constrained at the ordinal level representing 526 native and naturalized tree species of Puerto Rico (the single tree fern in the phylogeny is excluded to aid visualization). Ordinal placement according to APG III [60] is color coded.

unconstrained nodes, 71% received ≥50% bootstrap support and 46% received ≥80% bootstrap support. The unconstrained ML tree had slightly lower levels of support with 68% of nodes receiving ≥50% support and 43% of nodes receiving ≥80% support. Both the constrained and unconstrained ML trees had higher resolution than the PHYLOMATIC tree, in which only 52% of internal nodes were resolved. For the constrained ML tree, monophyly was supported for 91% of families and 87% of genera (monophyly of orders was constrained). In comparison, monophyly was supported for 72% of orders, 85% of families, and 87% of genera in the unconstrained ML tree. In both cases, the non-monophyly of currently recognized families related to the placement of taxa for which we did not have sequence data for all three barcode regions. For the constrained ML tree, the average proportion of nodes within orders, families, and genera with ≥50% bootstrap support was 0.81 (± SD 0.20), 0.87 (± SD 0.20), and 94% (± SD 0.19), respectively. For the unconstrained ML tree, the average proportion of nodes within orders, families, and genera with ≥50% bootstrap support was 0.92 (± SD 0.14), 0.89 (± SD 0.18), and 92% (± SD 0.20), respectively.

Case study: Phylogenetic composition of Puerto Rican forests

Some patterns of phylogenetic community structure varied with respect to the phylogeny and species pool used in analyses (Fig. 3).

For NRI, which emphasizes tree-wide patterns, Guánica dry forest was significantly clustered (i.e., taxa were more closely related than expected) based on the full island species pool for both the ML and PHYLOMATIC phylogenies (Fig. 3A). None of the other 14 forests departed from random expectations for NRI when based on the island pool. When considering the (reduced) soil species pools, the composition of the two wettest forests (Toro Negro and El Yunque, both located on volcanic soils) were significantly over-dispersed (i.e., taxa were less closely related than expected), although the NRI_{SOIL} value for Toro Negro was only significant with respect to the ML phylogeny (Fig. 3B). For NTI, which emphasizes compositional patterns at the tips of the phylogeny, Cambalache forest was significantly clustered with respect to the full island species pool but only for the ML phylogeny (Fig. 3C). None of the forests had significantly nonrandom NTI values when the analyses were based on the (reduced) soil species pools, regardless of which phylogeny was used (Fig. 3D).

None of the forests shifted from significantly clustered to significantly even when comparing NRI or NTI values based on the two different species pools. However, as we predicted, the (reduced) soil species pools caused both of these metrics to become more negative (i.e., decreased the signal of phylogenetic clustering) when calculated with the ML phylogeny (paired t-test: NRI: $t = 2.79$, $df = 14$, $p < 0.01$; NTI: $t = 4.34$, $df = 14$, $p < 0.001$). In contrast, these species pool definitions did not significantly change

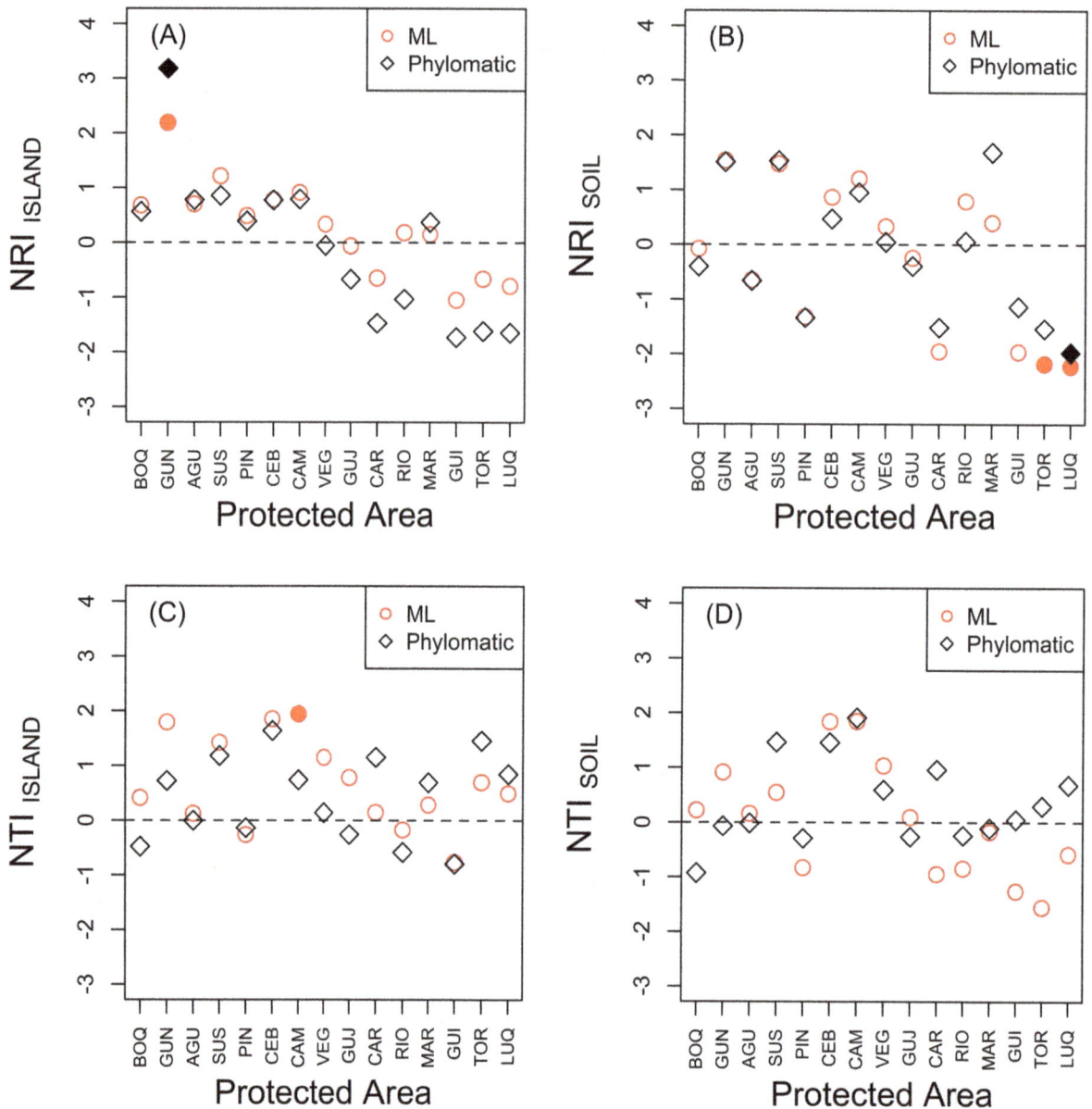

Figure 3. The net relatedness index (NRI) (A, B) and nearest taxon index (NTI; C, D) based on species occurrence records from Little & Wadworth [63] and Little *et al.* [64] versus reserve area [90] for 15 state forests in Puerto Rico. Leftmost panels are based on a null model using the full island species pool; right panels (B, D) are based on species pools restricted to primary soil types. Forests are sorted from left to right in order of their mean annual precipitation. Positive values indicate phylogenetic clustering and negative values indicate phylogenetic evenness. Filled symbols indicate values that are significantly different from a null model. Refer to Table 1 for forest codes.

NRI or NTI when calculated with the PHYLOMATIC phylogeny (paired t-test: NRI: t = 0.39, df = 14, p = 0.35; NTI: t = 0.28, df = 14, p = 0.39). Values of NRI calculated with each phylogeny were strongly correlated for both species pool definitions (island pool: Pearson's r = 0.96, p<0.001; soil pool: Pearson's r = 0.92, p<0.001) but values of NTI were less strongly correlated between these two phylogenies, and were not significantly correlated when based on the soil species pool (island pool: Pearson's r = 0.60, p = 0.02; soil pool: Pearson's r = 0.48, p = 0.06).

The node-based analysis identified particular clades that were relatively over- and under represented in each forest compared with a random expectation (Table 1) and, overall, the ML and PHYLOMATIC phylogenies produced largely congruent results (Appendix S4). One of the more consistent results was that species

belonging to Melastomataceae tended to be significantly under-represented in relatively dry forests on limestone and serpentine soils (*i.e.*, Guánica, Cambalache, Maricao, and Susúa) and relatively overrepresented in three relatively wet forests on volcanic soils (Guilarte, Luquillo and Toro Negro). Also, phylogenetic clustering of Guánica forest appears to be primarily driven by an overrepresentation of Fabaceae and Capparaceae, together with an underrepresentation of magnoliids, Ericales, and Melastomataceae (Appendix S4).

Discussion

The island-wide phylogeny for Puerto Rican trees presented here represents the community phylogenetics approach applied at a regional scale with the use of DNA sequence data. Both the constrained and unconstrained ML phylogenies provided increased phylogenetic resolution in comparison with a corresponding PHYLOMATIC tree, a predominant tree-building approach used in studies of community phylogenetics. In this study, the use of an ordinal-level constraint tree provided slightly higher phylogenetic resolution compared to the unconstrained analysis. In our case study, we uncovered patterns of nonrandom phylogenetic structure in Puerto Rican forests that depended on the phylogeny used as well as the scale at which the regional species pool was defined. Considering the rapidly increasing availability of DNA sequence data, future regional scale work in community phylogenetics will benefit from highly resolved phylogenies that include many taxa sampled across large areas and broad environmental gradients [39,68,69].

Comparison between phylogenies and taxonomic resolution

Although the ML phylogenies generated in this study were not completely resolved, the constrained 50% consensus tree did increase tip resolution by 22% in comparison with the PHYLO-MATIC tree. This relatively high degree of phylogenetic resolution has a number of important implications for community phylogenetic analyses [18,19,20]. First, poorly resolved phylogenies tend to reduce statistical power for detecting nonrandom patterns of community structure (e.g., with NRI and NTI), an issue that appears to be more severe with larger phylogenies [18]. Swenson [18] found that statistical power was most strongly reduced, however, when deeper nodes were unresolved (i.e. among orders and families) as opposed to more recent nodes (i.e. among species). As a result, we expect that the remaining unresolved nodes in our ML tree have a relatively small effect on analyses of phylogenetic structure for Puerto Rican tree communities because our constraint tree fixed the resolution of the deeper nodes. At the same time, the relatively deep nodes of the PHYLOMATIC phylogeny are also resolved, suggesting that a reduction in statistical power for detecting nonrandom patterns between our ML tree and the PHYLOMATIC tree may be most pronounced for metrics that focus on phylogenetic patterns among close relatives (e.g., NTI).

A second issue related to poorly resolved phylogenies is an upward bias when estimating phylogenetic signal [20]. In other words, the tendency for close relatives to have similar functional traits tends to be overestimated when phylogenies are poorly resolved. This bias is of particular concern when examining patterns of phylogenetic community composition given the central role of phylogenetic signal of traits relevant for species co-occurrence [70]. In general, the relatively high degree of tip resolution afforded by molecular data can strengthen inferences that rely on linking phylogenetic and functional patterns of community composition.

A major challenge in generating large-scale community phylogenies (and systematic biology, in general) is how to recover accurate phylogenetic relationships given limited data. Researchers have long debated the relative benefits of increasing sequence length versus increasing taxon sampling to improve the accuracy of phylogenetic reconstruction (e.g., [71,72,73,74,75]). This issue, however, has rarely been discussed in the context of community phylogenetics even though community-based analyses typically have relatively sparse taxon sampling compared to clade-based analyses. One implication of sparse taxon sampling is that long-branch attraction can reduce the accuracy of inferred topologies [74,76]. We confronted this potential issue by using a constraint tree to leverage strong prior information on deep phylogenetic

relationships. In our case, the Angiosperm Phylogeny Group [60] provides a synthesis of well-supported relationships among the plant orders. Overall, bootstrap support for the constrained ML tree was higher than for the unconstrained tree although we had originally expected a stronger effect of using the constraint tree. The fairly high success of recovering recognized orders in the unconstrained analysis likely derives from the large sample size included in this study and the particular genes used; they were chosen, in part, for their high performance in phylogenetic analyses [48,77].

While this study used a less sparse data matrix than previous work [24], the alignment procedure we use still resulted in a relatively sparse data matrix, particularly for the trnH-psbA region. The reason for this is that the SATé alignment algorithm knits together small alignments and introduces gaps when making a consensus alignment [51]. Evidence suggests that introducing gaps does not affect the overall phylogenetic results as seen with the success of phylogenetic reconstructions using super matrix methods that produce extremely sparse alignments [78] and studies that successfully align non-coding ITS and chloroplast intergenic spacer data for very large phylogenetic assemblages [79]. These studies suggest that missing data is not critical, particularly if one gene is shared among all taxa. Furthermore, while the effects of missing data on phylogenetic analyses are complex [80], several studies suggest that even taxa with large amounts of missing data can be accurately placed in phylogenies as long as the total number of characters sampled is large (e.g., [81,82]). In addition, Wiens [80] showed that, in some cases, taxa with large amounts of missing data can improve overall phylogenetic accuracy, particularly with model-based phylogenetic methods (e.g., likelihood) [83]; but see [84]. In our case, some instances of non-monophyly of recognized taxonomic groups were caused by individual taxa for which we did not have the full complement of three gene regions. Continued investigation of the influence of missing data on large phylogenetic analyses will help clarify the conditions under which missing data may decrease phylogenetic accuracy.

Case study: Phylogenetic composition of Puerto Rican forests

Our analysis of Puerto Rican tree communities provides an initial look at broad patterns of phylogenetic structure at a regional scale. For the most part, the ML and PHYLOMATIC phylogenies provided congruent results in terms of NRI, which is a tree-wide metric of phylogenetic composition. In contrast, NTI values, which are more sensitive to variation at the tips of a phylogeny, were not surprisingly, more variable between the two phylogenies. Another difference between the two phylogenies was how the species pool influenced the results. Reducing the scale at which the regional species pool was defined (i.e., from the island to pools in each soil type) caused a decrease for both NRI and NTI when based on the ML phylogeny but no statistically significant change based on the PHYLOMATIC phylogeny.

Based on the island species pool, one of the driest forests (Guánica, which is located at low elevation and on limestone soils) exhibited tree-wide phylogenetic clustering. Across all 15 forests, values of NRI_{ISLAND} tended to decline with mean annual precipitation, suggesting that drier forests generally comprise more phylogenetically clustered subsets of the island species pool than wetter forests. When evaluated with the reduced soil species pool, however, phylogenetic clustering of Guánica became random and only one forest in the moist life zone (Cambalache; located at low elevation and on limestone soils) had significantly clustered NTI with respect to the ML phylogeny only. The two wettest forests (Toro Negro and El Yunque, which are located on higher elevation volcanic soils) exhibited significant phylogenetic evenness in the NRI metric, although the value for Toro Negro was only significant with the ML phylogeny. One interpretation of these patterns is that water limitation represents a strong environmental filter in the dry forests and constrains the composition of local communities to the lineages that are able to persist under these harsh conditions. The issue of water stress in Puerto Rico may be exacerbated by the somewhat confounded nature of underlying geology and precipitation [85]. Specifically, limestone soils tend to occur at lower elevations and receive less precipitation than volcanic soils. The combined influence of these variables likely compounds the effects of limited water availability for plants. In contrast, niche partitioning with respect to other factors (*e.g.*, light use, vulnerability to pathogens) may play a stronger role in the wetter forests on volcanic soils, leading to a phylogenetically more diverse set of co-occurring species. One alternative explanation for this pattern is if *in situ* lineage diversification in Puerto Rico is a more important determinant of local species composition for higher elevation forests. For example, two closely related species of *Tabebuia*, *T. rigida* and *T. schumanniana* (Bignoniaceae) are endemic to El Yunque and Carite mountains, respectively [44].

We acknowledge three main limitations in our ability to interpret these patterns. First, we did not include information on species traits, which are relevant to their occurrence across environmental gradients. Our interpretations depend, in part, on the degree to which functional traits relevant to species occurrence along a gradient of water availability are phylogenetically conserved. Linking key functional traits with phylogenetic relatedness would help to more strongly identify the processes that underlie compositional variation among these forests [70,86]. Second, the occurrence data we used in this analysis lacks information on species abundances. Our analysis may not detect community assembly processes that are more strongly driven by species relative abundances (i.e., dominance) than the simple presence or absence. Finally, although our null model controls for species richness within each plot, statistical power for detecting nonrandom patterns is low for forests with low species richness [87]. Nonetheless, the observed patterns provide a valuable starting point for future work aimed at addressing these limitations and providing additional insight on tree community variation across broad environmental gradients in Puerto Rico.

We found that values of NRI for each forest based on the different phylogenies were highly correlated whereas NTI values for each forest calculated with the two phylogenies were not correlated. These results reinforce the idea that low resolution among terminal tips (congeneric and confamilial taxa) may be especially problematic for recovering consistent patterns with NTI. In general, previous work has suggested that NRI may have greater power to detect nonrandom patterns of community phylogenetic structure than NTI [2,88,89].

In conclusion, our study provides a highly resolved community phylogeny for tropical trees at a regional scale: the island of Puerto Rico. We hope this regional perspective facilitates additional work to better understand the processes governing composition of local tree communities. Our case study confirms the value of a highly resolved phylogeny for detecting nonrandom patterns of phylogenetic community composition. Together with the extensive amount of existing data available in Puerto Rico on environmental conditions, land use history, species distributions and functional traits, we anticipate that the regional phylogeny provided here will help strengthen our historical perspective on the forces generating and structuring the diversity of Puerto Rican forests.

Supporting Information

Table S1   Table of molecular sequences from Puerto Rican tree species included in this analysis, including taxonomic information, collection details, GenBank accessions and voucher specimen details.

Appendix S1   Input file for PATHd8 [61] used to date the constrained ML tree.

Appendix S2   Input file for PATHd8 [61] used to date the unconstrained ML tree.

Appendix S3   A Newick format file (.new) containing five phylogenies for Puerto Rican trees: (1) the dated ultrametric 50% consensus ML phylogeny using ordinal-level (APG-III) constraints, (2) the dated ultrametric 50% consensus ML phylogeny without topological constraints, (3) the undated non-ultrametric 50% consensus ML phylogeny using ordinal-level (APG-III) constraints, (4) the undated non-ultrametric 50% consensus ML phylogeny without topological constraints, (5) a corresponding Phylomatic phylogeny. Bootstrap support is provided for ML trees (i.e., trees 1–4; NA's in trees 1 and 3 refer to APG-III constrained nodes).

Appendix S4   Detailed results from 'nodesig' analysis [67] for each forest using the dated and constrained 50% consensus ML phylogeny and Phylomatic phylogeny.

Acknowledgments

We extend our gratitude to the many people who helped collect and identify specimens including especially Fabiola Areces, Frank Axelrod, Alejandro Cubiña, Freddie Perez, José Sustache, Papo Vives, Alberto Areces, Pedro Acevedo, Marcos Caraballo, Víctor José Vega López, Silvia Bibbo, Christopher J. Nytch, Marcos Rodriguez, and T.J. Agin. Danilo Chinea and provided digitized occurrence data. All necessary permits were obtained for the described study, which complied with all relevant regulations. Specifically, we thank the Departmento de Recursos Naturales y Ambientales (DRNA) of Puerto Rico for permission to collect plant specimens in the forests of Puerto Rico under DRNA permit #2011-IC-046. We also thank the herbaria and staff at the University of Puerto Rico, Rio Piedras and the US National Herbarium for permission to sample tissue from their collections; the assistance of Pedro Acevedo was invaluable for this project. This manuscript benefitted from comments by Damon Little, Robert P. Anderson, Robin Chazdon, Joel Cracraft, Dustin Rubenstein, Camilo Sanin, Brian Weeks, and two anonymous reviewers.

Author Contributions

Conceived and designed the experiments: RM MU DLE NGS JKZ WJK. Performed the experiments: RM DLE. Analyzed the data: RM DLE. Contributed reagents/materials/analysis tools: RM MU DLE WJK. Wrote the paper: RM MU DLE NGS JKZ WJK.

References

1. Losos JB (1996) Phylogenetic perspectives on community ecology. Ecology 77: 1344–1344.

2. Vellend M, Cornwell WK, Magnuson-Ford K, Mooers AØ (2011) Measuring phylogenetic biodiversity. In: Magurran AE, McGill BJ, editors. Biological

diversity: frontiers in measurement and assessment. Oxford: Oxford University Press. 193–206.

3. Cavender-Bares J, Ackerly DD, H KK (2012) Integrating ecology and phylogenetics: the footprint of history in modern-day communities. Ecology 93: S1–S3.

4. Webb CO, Ackerly DD, McPeek MA, Donoghue MJ (2002) Phylogenies and community ecology. Annual Review of Ecology and Systematics 33: 475–505.

5. Cavender-Bares J, Kozak KH, Fine PVA, Kembel SW (2009) The merging of community ecology and phylogenetic biology. Ecology Letters 12: 693–715.

6. Vamosi SM, Heard SB, Vamosi JC, Webb CO (2009) Emerging patterns in the comparative analysis of phylogenetic community structure. Molecular Ecology 18: 572–592.

7. Vane-Wright RI, Humphries CJ, Williams PH (1991) What to protect? Systematics and the agony of choice. Biological Conservation 55: 235–254.

8. Williams PH, Humphries CJ, Vane-Wright RI (1991) Measuring biodiversity: taxonomic relatedness for conservation priorities. Australian Systematic Botany 4: 665–679.

9. Faith DP (1992) Conservation evaluation and phylogenetic diversity. Biological Conservation 61: 1–10.

10. Crozier RH (1997) Preserving the information content of species: genetic diversity, phylogeny, and conservation worth. Annual Review of Ecology and Systematics 28: 243–268.

11. Devictor V, Mouillot D, Meynard D, Jiguet F, Thuiller W, et al. (2010) Spatial mismatch and congruence between taxonomic, phylogenetic and functional diversity: the need for integrative conservation strategies in a changing world. Ecology Letters 13: 1030–1040.

12. Ricklefs RE (1987) Community diversity: relative roles of local and regional processes. Science 235: 167–171.

13. Webb CO, Donoghue MJ (2005) Phylomatic: tree assembly for applied phylogenetics. Molecular Ecology Notes 5: 181–183.

14. Kembel SW, Hubbell SP (2006) The phylogenetic structure of a neotropical forest tree community. Ecology 87: 86–99.

15. Willis CG, Ruhfel B, Primack RB, Miller-Rushing AJ, Davis CC (2008) Phylogenetic patterns of species loss in Thoreau's woods are driven by climate change. Proceedings of the National Academy of Sciences 105: 17029–17033.

16. Kraft NJB, Ackerly DD (2010) Functional trait and phylogenetic tests of community assembly across spatial scales in an Amazonian forest. Ecological Monographs 80: 401–422.

17. Beaulieu JM, Ree RH, Cavender-Bares J, Weiblen GD, Donoghue MJ (2012) Synthesizing phylogenetic knowledge for ecological research. Ecology 93: S4-S13.

18. Swenson NG (2009) Phylogenetic resolution and quantifying the phylogenetic diversity and dispersion of communities. PLoS ONE 4: e4390-e4390.

19. Kress WJ, Erickson DL, Jones FA, Swenson NG, Perez R, et al. (2009) Plant DNA barcodes and a community phylogeny of a tropical forest dynamics plot in Panama. Proceedings of the National Academy of Sciences 106: 18621–18626.

20. Davies JT, Kraft NJB, Salamin N, Wolkovich EM (2012) Incompletely resolved phylogenetic trees inflate estimates of phylogenetic conservatism. Ecology 93: 242–247.

21. Cavender-Bares J, Ackerly DD, Baum DA, Bazzaz FA (2004) Phylogenetic overdispersion in Floridian Oak communities. The American Naturalist 163: 823–843.

22. Sedio BE, Wright SJ, Dick CW (2012) Trait evolution and the coexistence of a species swarm in the tropical forest understory. Journal of Ecology 100: 1183–1193.

23. Ricklefs R, Jenkins DG (2011) Biogeography and ecology: towards the integration of two disciplines. Philosophical Transactions of the Royal Society B-Biological Sciences 366: 2438–2448.

24. Kress WJ, Erickson DL, Swenson NG, Thompson J, Uriarte M, et al. (2010) Advances in the use of DNA barcodes to build a community phylogeny for tropical trees in a Puerto Rican forest dynamics plot. PLoS ONE 5: e15409.

25. Uriarte M, Swenson NG, Chazdon RL, Comita LS, Kress WJ, et al. (2010) Trait similarity, shared ancestry and the structure of neighbourhood interactions in a subtropical wet forest: implications for community assembly. Ecology Letters 13: 1503–1514.

26. Hardy OJ (2008) Testing the spatial phylogenetic structure of local communities: statistical performances of different null models and test statistics on a locally neutral community. Journal of Ecology 96: 914–926.

27. de Bello F, Price JN, Münkemüller T, Liira J, Zobel M, et al. (2012) Functional species pool framework to test for biotic effects on community assembly. Ecology 93: 2263–2273.

28. Lessard J-P, Belmaker J, Myers JA, Chase JM, Rahbek C (2012) Inferring local ecological processes amid species pool influences. Trends in Ecology & Evolution 27: 600–607.

29. Eiserhardt WL, Svenning JC, Borchsenius F, Kristiansen T, Balslev H (2013) Separating environmental and geographical determinants of phylogenetic community structure in Amazonian palms (Arecaceae). Botanical Journal of the Linnean Society 171: 244–259.

30. Brunbjerg AK, Cavender-Bares J, Eiserhardt WL, Ejrnæs R, Aarssen LW, et al. (2014) Multi-scale phylogenetic structure in coastal dune plant communities across the globe. Journal of Plant Ecology 7: 101–114.

31. Münkemüller T, Gallien L, Lavergne S, Renaud J, Roquet C, et al. (2014) Scale decisions can reverse conclusions on community assembly processes. Global Ecology and Biogeography 23: 620–632.

32. Parmentier I, Réjou-Méchain M, Chave J, Vleminckx J, Thomas DW, et al. (2014) Prevalence of phylogenetic clustering at multiple scales in an African rain forest tree community. Journal of Ecology: n/a-n/a.

33. Pickett STA, Bazzaz FA (1978) Organization of an assemblage of early successional species on a soil moisture gradient. Ecology 59: 1248–1255.

34. Silvertown J, Dodd M, Gowing D, Lawson C, McConway K (2006) Phylogeny and the hierarchical organization of plant diversity. Ecology 87: 39–49.

35. Ackerly DD, Cornwell W (2007) A trait-based approach to community assembly: partitioning of species trait values into within- and among-community components. Ecology Letters 10: 135–145.

36. Cavender-Bares J, Keen A, Miles B (2006) Phylogenetic structure of Floridian plant communities depends on taxonomic and spatial scale. Ecology 87: S109-S122-S109-S122.

37. Swenson NG, Enquist BJ, Thompson J, Zimmerman JK (2007) The influence of spatial and size scale on phylogenetic relatedness in tropical forest communities. Ecology 88: 1770–1780.

38. Lessard J-P, Borregaard MK, Fordyce JA, Rahbek C, Weiser MD, et al. (2012) Strong influence of regional species pools on continent-wide structuring of local communities. Proceedings of Royal Society B 279: 266–274.

39. Swenson NG, Umaña MN (2014) Phylofloristics: an example from the Lesser Antilles. Journal of Plant Ecology 7: 166–175.

40. Holdridge LR (1947) Determination of world plant formations from simple climatic data. Science 105: 367–368.

41. Ewel JJ, Whitmore JL (1973) The ecological life zones of Puerto Rico and the U.S. Virgin Islands. USDA Forest Service, Institute of Tropical Forestry.

42. Daly C, Helmer EH, Quiñones M (2003) Mapping the climate of Puerto Rico, Vieques and Culebra. International Journal of Climatology 23: 1359–1381.

43. Bawiec WJ (1998) Geology, geochemistry, geophysics, mineral occurrences, and mineral resource assessment for the commonwealth of Puerto Rico: Open-File Report 98–38, Geological Survey (U.S.).

44. Axelrod FS (2011) A systematic vademecum to the vascular plants of Puerto Rico. Río Piedras, PR: University of Puerto Rico.

45. Acevedo-Rodríguez P, Strong MT (2011) Flora of the West Indies: catalogue of the seed plants of the West Indies. Washington, D.C.: Smithsonian Institution, National Museum of Natural History.

46. USFS (2006) Forest inventory and analysis national core field guide: Caribbean version 3.0. United States Forest Service.

47. Mi X, Swenson NG, Valencia R, Kress WJ, Erickson DL, et al. (2012) The contribution of rare species to community phylogenetic diversity across a global network of forest plots. American Naturalist 180: E17–30.

48. Kress WJ, Erickson DL (2007) A two-locus global DNA barcode for land plants: the coding rbcL gene complements the non-coding trnH-psbA spacer region. PLoS ONE 2: e508. doi:510.1371/journal.pone.0000508.

49. Fazekas AJ, Burgess KS, Kesanakurti PR, Graham SW, Newmaster SG, et al. (2008) Multiple multilocus DNA barcodes from the plastid genome discriminate plant species equally well. PlosONE 3: e2802. (doi:2810.1371/journal.pone. 0002802).

50. Benson DA, Karsch-Mizrachi I, Lipman DJ, Ostell J, Sayers EW (2010) GenBank. Nucleic Acids Research 38: D46-D51.

51. Liu K, Warnow TJ, Holder MT, Nelesen SM, Yu J, et al. (2012) SATé-II: very fast and accurate simultaneous estimation of multiple sequence alignments and phylogenetic trees. Systematic Biology 61: 90–106.

52. Katoh K, Kuma K, Toh H, Miyata T (2005) MAFFT version 5: improvement in accuracy of multiple sequence alignment. Nucleic Acids Research 33.

53. Edgar R (2004) MUSCLE: multiple sequence alignment with high accuracy and high throughput. Nucleic Acids Research 32: 1792–1797.

54. Schoch CL, Crous PW, Groenewald JZ, Boehm EWA, Burgess TI, et al. (2009) A class-wide phylogenetic assessment of Dothideomycetes. Studies in Mycology 64: 1–15-S10.

55. Kivlin SN, Hawkes CV, Treseder KK (2011) Global diversity and distribution of arbuscular mycorrhizal fungi. Soil Biology and Biochemistry 43: 2294–2303.

56. Miller MA, Pfeiffer W, Schwartz T. Creating the CIPRES Science Gateway for inference of large phylogenetic trees; 2010 14 Nov. 2010; New Orleans, LA. 1–8.

57. Darriba D, Taboada GL, Doallo R, Posada D (2012) jModelTest 2: more models, new heuristics and parallel computing. Nature Methods 9: 772.

58. Stevens PF (2001 onwards) Angiosperm Phylogeny Website. Version 12, July 2012 [and more or less continuously updated since]. Available: www.mobot.org/MOBOT/research/APweb.

59. Smith SA, Beaulieu J, Donoghue MJ (2009) Mega-phylogeny approach for comparative biology: an alternative to supertree and supermatrix approaches. BMC Evolutionary Biology 9: 37.

60. APG III (2009) An update of the Angiosperm Phylogeny Group classification for the orders and families of flowering plants: APG III. Botanical Journal of the Linnean Society 161: 105–121.

61. Britton T, Anderson CL, Jacquet D, Lundqvist S, Bremer K (2007) Estimating divergence times in large phylogenetic trees. Systematic Biology 56: 741–752.

62. Magallón S, Castillo A (2009) Angiosperm diversification through time. American Journal of Botany 96: 349–365.

63. Little EL, Wadsworth FH (1964) Common trees of Puerto Rico and the Virgin Islands. Washington, DC: USDA Forest Service.

64. Little EL, Woodbury RO, Wadsworth FH (1974) Trees of Puerto Rico and the Virgin Islands. Washington, DC: USDA Forest Service.

65. Kembel SW, Cowan PD, Helmus MR, Cornwell WK, Morlon H, et al. (2010) Picante: R tools for integrating phylogenies and ecology. Bioinformatics 26: 1463–1464.

66. R Development Core Team (2012) R: A language and environment for statistical computing. v2.9.2 ed. Vienna, Austria: R Foundation for Statistical Computing.

67. Webb CO, Ackerly DD, Kembel SW (2008) Phylocom: software for the analysis of phylogenetic community structure and trait evolution. Bioinformatics 24: 2098–2100.

68. Swenson NG, Erickson DL, Mi XC, Bourg NA, Forero-Montaña J, et al. (2012) Phylogenetic and functional alpha and beta diversity in temperate and tropical tree communities.

69. Swenson NG (2013) The assembly of tropical tree communities – the advances and shortcomings of phylogenetic and functional trait analyses. Ecography 36.

70. Mayfield MM, Levine JM (2010) Opposing effects of competitive exclusion on the phylogenetic structure of communities. Ecology Letters 13: 1085–1093.

71. Rannala B, Huelsenbeck JP, Yang Z, Nielsen R (1998) Taxon sampling and the accuracy of large phylogenies. Systematic Biology 47: 702–710.

72. Poe S, Swofford DL (1999) Taxon sampling revisited. Nature 398: 299–300.

73. Felsenstein J (2004) Inferring phylogenies. Sunderland, Massachusetts: Sinauer Associates.

74. Heath TA, Hedtke SM, Hillis DM (2008) Taxon sampling and the accuracy of phylogenetic analyses. Journal of Systematics and Evolution 46: 239–257.

75. Nabhan AR, Sarkar IN (2012) The impact of taxon sampling on phylogenetic inference: a review of two decades of controversy. Briefings in Bioinformatics 13: 122–134.

76. Stefanović S, Rice DW, Palmer JD (2004) Long branch attraction, taxon sampling, and the earliest angiosperms: Amborella or monocots? BMC Evolutionary Biology 4: 35.

77. CBOL Plant Working Group (2009) A DNA barcode for land plants. Proceedings of the National Academy of Sciences 106: 12794–12797.

78. McMahon MM, Sanderson MJ (2006) Phylogenetic supermatrix analysis of GenBank sequences from 2228 Papilionoid Legumes. Systematic Biology 55: 818–836.

79. Edwards E, Smith SA (2010) Phylogenetic analyses reveal the shady history of C4 grasses. Proceedings of the National Academy of Sciences 107: 2532–2538.

80. Wiens JJ (2006) Missing data and the design of phylogenetic analyses. Journal of Biomedical Informatics 39: 34–42.

81. Phillipe H, Snell EA, Bapteste E, Lopez P, Holland PWH, et al. (2004) Phylogenomics of eukaryotes: impact of missing data on large alignments. Molecular Biology and Evolution 21: 1740–1752.

82. Wiens JJ, Morrill MC (2011) Missing data in phylogenetic analysis: reconciling results from simulations and empirical data. Systematic Biology 60: 719–731.

83. Wiens JJ (1998) Does adding characters with missing data increase or decrease phylogenetic accuracy? Systematic Biology 47: 625–640.

84. Poe S (2003) Evaluation of the strategy of long-branch subdivision to improve the accuracy of phylogenetic methods. Systematic Biology 52: 423–428.

85. Miller GL, Lugo AE (2009) Guide to the ecological systems of Puerto Rico. Gen. Tech. Rep. IITF-GTR-35. In: United States Department of Agriculture, Forest Service, International Institute of Tropical Forestry, editors. San Juan, PR. 437.

86. Adler PB, Fajardo A, Kleinhesselink AR, Kraft NJB (2013) Trait-based tests of coexistence mechanisms. Ecology Letters 16: 1294–1306.

87. Gotelli NJ, Ulrich W (2012) Statistical challenges in null model analysis. Oikos 121: 171–180.

88. Letcher SG (2010) Phylogenetic structure of angiosperm communities during tropical forest succession. Proceedings of the Royal Society B: Biological Sciences 277: 97–104.

89. Letcher SG, Chazdon RL, Andrade ACS, Bongers F, van Breugel M, et al. (2012) Phylogenetic community structure during succession: evidence from three neotropical forest sites. Perspectives in Plant Ecology, Evolution and Systematics 14: 79–87.

90. Gould WA, Alarcon C, Fevold B, Jimenez ME, Martinuzzi S, et al. (2008) The Puerto Rico Gap Analysis Project volume 1: land cover, vertebrate species distributions, and land stewardship. Gen. Tech. Rep. IITF-39.

91. Gesch DB (2007) The National Elevation Dataset. In: Maune D, editor. Digital Elevation Model Technologies and Applications: The DEM Users Manual. 2nd Edition ed. Bethesda, Maryland: American Society for Photogrammetry and Remote Sensing. 99–118.

19

Evaluating Darwin's Naturalization Hypothesis in Experimental Plant Assemblages: Phylogenetic Relationships Do Not Determine Colonization Success

Sergio A. Castro[1]*, **Victor M. Escobedo**[1], **Jorge Aranda**[1], **Gastón O. Carvallo**[2]

1 Laboratorio de Ecología y Biodiversidad, Departamento de Biología, and Center for the Development of Nanoscience and Nanotechnology (CEDENNA), Universidad de Santiago de Chile, Santiago, Chile, **2** Instituto de Biología, Facultad de Ciencias, Pontificia Universidad Católica de Valparaíso, Valparaíso, Chile

Abstract

Darwin's naturalization hypothesis (DNH) proposes that colonization is less likely when the colonizing species is related to members of the invaded community, because evolutionary closeness intensifies competition among species that share similar resources. Studies that have evaluated DNH from correlational evidence have yielded controversial results with respect to its occurrence and generality. In the present study we carried out a set of manipulative experiments in which we controlled the phylogenetic relatedness of one colonizing species (*Lactuca sativa*) with five assemblages of plants (the recipient communities), and evaluated the colonizing success using five indicators (germination, growth, flowering, survival, and recruitment). The evolutionary relatedness was calculated as the mean phylogenetic distance between *Lactuca* and the members of each assemblage (MPD) and by the mean phylogenetic distance to the nearest neighbor (MNND). The results showed that the colonization success of *Lactuca* was not affected by MPD or MNND values, findings that do not support DNH. These results disagree with experimental studies made with communities of microorganisms, which show an inverse relation between colonization success and phylogenetic distances. We suggest that these discrepancies may be due to the high phylogenetic distance used, since in our experiments the colonizing species (*Lactuca*) was a distant relative of the assemblage members, while in the other studies the colonizing taxa have been related at the congeneric and conspecific levels. We suggest that under field conditions the phylogenetic distance is a weak predictor of competition, and it has a limited role in determining colonization success, contrary to prediction of the DNH. More experimental studies are needed to establish the importance of phylogenetic distance between colonizing species and invaded community on colonization success.

Editor: Diego Fontaneto, Consiglio Nazionale delle Ricerche (CNR), Italy

Funding: This research has been financially supported by Fondecyt 11085013, Fondecyt PD 3130399, and Línea 6 of CEDENNA. The funders had no role in study design, data collection and analysis, decision to publish, or preparation of the manuscript.

Competing Interests: The authors have declared that no competing interests exist.

* Email: sergio.castro@usach.cl

Introduction

Biological invasions have attracted the attention of modern ecologists and biogeographers [1] because of their leading role as components of global change [2]. At present, organisms belonging to diverse taxonomic groups are being translocated from one region to another with which they do not share an evolutionary history [3,4]. Although it is estimated that most of the organisms that start this dispersion do not get to become established successfully, sometimes they can constitute a founding colony and become naturalized [5,6]. One of the central challenges in the study of biological invasions has been to understand what factors determine this naturalization process [3–4], [7].

Various hypotheses have been proposed to explain why some species are capable of colonizing successfully (i.e., become naturalized) while others are not [6], [8–9]. A particularly intriguing and controversial role is that played by Darwin's naturalization hypothesis (DNH), which states that naturalization success depends on the phylogenetic relatedness between the colonizer and the members of the recipient community [10]. If this relatedness is close, then the colonization process will be inhibited as a result of the greater competitive intensity that there is – supposedly– between closely related species [11]. Conversely, if the phylogenetic relationship is distant, the establishment would be favored as a result of a lower competitive intensity. Underlying this relationship between phylogenetic distance and invasion success, it is assumed that closely related species shares similar resources and natural enemies [11].

Following the influential paper by Daehler [12], DNH has received renewed interest, generating controversy with respect to its explanatory value on the invasion process [10]. In fact, while some studies have supported the hypothesis [13–19], others have dismissed it [20–29]. Although most of the evidence relies on compositional pattern analysis at a regional scale [21], [30], recently some authors have implemented experimental approaches in communities of microorganisms, providing support to DNH [15], [17]. Even though the ecology of microorganisms is governed by processes equivalent to those that occur in multicellular organisms [31–32], the specific mechanisms that promote successful invasion can differ considerably [32].

In the present study we evaluated DNH in experimental assemblages constituted by vascular plants. For this purpose, plants belonging to a wide taxonomic spectrum were used to establish recipient experimental assemblages, which were then inoculated with seeds of a colonizing plant (*Lactuca sativa*, hereafter *Lactuca*). The experiments were composed of species that differed in their degree of phylogenetic relatedness with respect to *Lactuca*, a fact that allowed us to assess the effect of that factor on the colonizing success of the inoculated species. As far as we can tell, this is the first time that DNH is evaluated experimentally in multicellular communities, specifically in plants.

Methodology

Experimental design

Our experiment involved a total of 15 plant species (Table 1), 14 of which were used to establish receiver assemblages and one (*Lactuca*) was used as colonizing or invading species of those assemblages. Five assemblages were formed and each of them was made up of a subset of five species of the 14 that were available (Table 1). Initially, these assemblages were organized in a taxonomic gradient from strong to weak relatedness with respect to *Lactuca*, and this was later confirmed by means of evolutionary distance metrics (see below). The experimental assemblages were designated A1, A2, A3, A4 and A5 (Table 1), and each of them was replicated eight times. For a control treatment (C), eight plots were set up to evaluate the colonization of *Lactuca* in a monoculture regimen.

The assemblages were set up in small wooden plots (1×1×0.3 m) located in a greenhouse. A homogeneous mixture of sand and compost (1:1) was used as substrate, and evenly distributed in the plots. The main herbivores above ground (i.e., insects, birds and mammals) were completely excluded from the greenhouse. The plots were arranged in a pattern of equidistant rows and columns 2 m apart numbered sequentially. A random procedure allowed assigning each plot to a given kind of assemblage. In turn, each plot was subdivided into a grid of 5×5 cm cells which were numbered consecutively from 1 to 400, and the species that would go in each cell in particular were assigned randomly. Seeds were planted in excess in each cell in order to ensure the future presence of plants, and once the seedlings had become established in each cell, the excess was removed by careful cutting with shears, so that all the species had the same abundance in each plot (i.e., 66 cells per plot, occupying a total of 330 cells per plot), leaving 70 cells empty for the later planting of the colonizing species.

The plots were watered periodically by means of a semi-automated system, in this way ensuring a homogeneous availability of water to the plants. This system consisted of an arrangement of rotary Micro-Jet sprinklers placed at a height of 1.5 m; they were arranged in rows and columns equidistant from each plot to ensure uniform watering. Watering was performed at field capacity every three days.

Three months after starting the planting in the assemblages (the seedlings had reached a height ≥10 cm), the *Lactuca* seeds were sown emulating the invasion of an already established receiver community. This planting process was carried out simultaneously in all the experimental assemblages and in the control treatment (from C, A1–A5) at a density of 70 seeds per plot.

Lactuca colonization success

The colonization success of *Lactuca* was measured by means of five indicators. First, germination was evaluated by quantifying the number and percentage of cells with germinating seeds in each plot. Due to the fast germination of *Lactuca*, this indicator was measured three weeks after sowing the seeds (May 2010), which were considered to have germinated when their epicotyl had grown ≥2 cm. Second, the indicator of colonization success was the survival of *Lactuca* plants, and for that purpose, toward the end of the experiment (January 2011) the number and percentage of surviving plants of the cohort existing at the beginning of the experiment were recorded. The third indicator considered the ratio of plants that reached the flowering stage, counting the number of plants that had one or more flowers in relation to the total number of *Lactuca* plants present in the plot. The fourth indicator evaluated the growth (height above the ground; cm) achieved by a random sample of 20 plants in each plot in January 2011. The fifth indicator was recruiting, counting the number of new *Lactuca* seedlings that appeared spontaneously in the plots (number per plot).

In order to control the concomitant effect of growth of the different species present on the studied assemblages, the height achieved by the plants in each assemblage in each plot was measured. For that purpose a random sample of 20 plants was taken, measuring twice their height (cm) above the ground: at the beginning of the experiment (May 2010) to evaluate their effect on the germination of *Lactuca* seeds, and at the end of the experiment (January 2011) to evaluate their effect on the growth of *Lactuca* plants.

Phylogenetic relatedness among species

A metaphylogeny was reconstructed for the 15 species of angiosperms included in our study using Phylomatic [33] version R20031202, which is based on the APG III phylogeny [34]. Within the family Apiaceae, the phylogenetic relations between *Anethum*, *Coriandrun* and *Petroselinun* were resolved using [35]. The topology of the resulting tree was calibrated by age, based on the divergence times documented by [36], using the BLADJ algorithm implemented in Phylocom [37].

Based on this calibrated tree, two phylogenetic diversity indices were calculated to characterize each assemblage. On the one hand, the average length of the branches was calculated for all the pairs of species of the assemblage (mean phylogenetic distance, MPD; [38–40]) both before (MPD$_{pre}$) and after (MPD$_{post}$) planting *Lactuca*. Furthermore, the average length of the branches between every species and its nearest neighbor in the assemblages was calculated (mean nearest neighbor distance, MNND; [38], [40]) both before (MNND$_{pre}$) and after (MNND$_{post}$) planting *Lactuca*. The values of MPD and MNND are expressed in millions of years (my). Because there was no loss of species over the course of the experiments, the post-invasion phylogenetic diversity shows the effect of adding *Lactuca* to the plots.

As indicators of phylogenetic relatedness between *Lactuca* and the remaining species in each assemblage, we used the average of MPD between *Lactuca* and each member of the assemblages (named MPD$_{Lactuca}$), and the distance between *Lactuca* and the nearest neighbor within assemblages (named MNND$_{Lactuca}$). The two metrics provide different ways in which phylogenetic relatedness can be conceived in the particular DNH context, because MPD$_{Lactuca}$ considers values at the community level, while MNND$_{Lactuca}$ includes only the nearest neighbor.

Table 1. Composition of the experimental assemblages and characterization of their phylogenetic diversity based on evolutionary distances.

Plant species	Assemblages (Treatments)				
	A1	A2	A3	A4	A5
Asteraceae					
Lactuca sativa L.	1	1	1	1	1
Matricaria chamomilla L.	1	0	0	0	0
Apiaceae					
Anethum graveolens L.	1	1	0	0	0
Petroselinum crispum (Mill.) Fuss	1	1	0	0	0
Coriandrum sativum L.	1	1	0	0	0
Solanaceae					
Capsicum baccatum Jacq.	0	1	1	0	0
Solanum melongena L.	0	1	1	0	0
Lamiaceae					
Ocinum basilicum L.	0	0	1	0	0
Brassicaceae					
Brassica oleraceae L.	0	0	1	1	0
Eruca sativa Mill.	0	0	1	1	0
Fabaceae					
Pisum sativum L.	0	0	0	1	1
Trifolium repens L.	1	0	0	0	1
Vicia atropurpurea Desf.	0	0	0	1	1
Vicia faba L.	0	0	0	1	1
Poaceae					
Zea mays L.	0	0	0	0	1
MPD_{pre} (my)	182.3	165.1	208.2	142.6	184.8
MPD_{post} (my)	187.1	184.0	219.4	179.7	212.2
$MNND_{pre}$ (my)	120.0	59.2	94.8	64.0	109.2
$MNND_{post}$ (my)	141.8	142.6	143.8	143.0	143.3
$MPD_{Lactuca}$ (my)	196.8	222.0	242.0	254.0	267.6
$MNND_{Lactuca}$ (my)	88.0	214.0	234.0	254.0	254.0

MPD_{pre} and MPD_{post} represent the average distance of the branch lengths between the pairs of species of the assemblage before and after the invasion by Lactuca, respectively; $MNND_{pre}$ and $MNND_{post}$ represent the average distance of the branch lengths between each species and its nearest neighbor in the assemblage; $MPD_{Lactuca}$ corresponds to the average branch length between Lactuca and each of the members of the assemblage, respectively; and $MNND_{Lactuca}$ was calculated as the branch length between Lactuca and its nearest neighbor. my: million of years.

Statistical analyses

In a first group of analyses we compared the colonization success indicators of *Lactuca* in monocultures (control treatment, C) versus those recorded in the assemblages (treatments A1–A5), using one-way analysis of variance (ANOVA) for germination, survival, flowering, growth, and colonization of *Lactuca* in the different treatments. Prior to each analysis, all the dependent variables were transformed into logarithmic functions (ln (x)). The Tukey test was applied to recognize those treatments that showed statistical significance.

With the purpose of assessing the relation between phylogenetic distance and the colonization success of *Lactuca*, we performed covariance analyses (ANCOVA) for the different assemblages. Here the $MPD_{Lactuca}$ and $MNND_{Lactuca}$ distances were concomitant factors, while the colonization indicators were considered dependent variables. In these analyses, the average as well as the variance of the height reached by the resident plants in the assemblages were included as covariables. All these factors were normalized by means of the logarithmic function (ln (x)). In all our analyses we used the Type III sum of squares.

Results

Pre- and post-invasion phylogenetic diversity

Before the inoculation with *Lactuca*, the assemblages showed MPD_{pre} values varying between 142.6 and 208.2 my, and $MNND_{pre}$ values between 59.2 and 120.0 my (Table 1). After the inoculation with *Lactuca*, the MPD_{post} values increased to between 179.7 and 212.2 my, while those of $MNND_{post}$ were between 141.8 and 143.8 my (Table 1). Therefore, the inoculation of *Lactuca* increased significantly the phylogenetic diversity values measured as MPD and MNND (Table 1; Wilcoxon Signed Ranks test; in both cases $T = 7.5$; $P < 0.05$).

Colonization success

The average germination rate of *Lactuca* among the treatments varied between 55.8 and 98.3% (Figure 1A), with statistical differences between them ($F = 54.2$; $d.f. = 5$; $P<0.05$). These differences were determined by the greater germination in the control treatment with respect to the experimental assemblages, which did not show differences between them (Figure 1A).

The average growth of *Lactuca* in the different treatments varied between 24 and 60 cm (Figure 1B). These values showed statistical differences among treatments ($F = 5.8$; $d.f. = 5$; $P<0.05$), determined by the control treatment and the A2 assemblage with respect to the other assemblages (Figure 1B). On the other hand, the average survival of *Lactuca* varied between 56.8 and 93.8% (Figure 1C), values that also showed significant differences among compared treatments ($F = 5.7$; $d.f. = 5$; $P<0.05$); again these differences were determined by the control treatment with respect to the other assemblages (Figure 1C).

With respect to flowering, this measurement varied between 9.0 and 33.6% (Figure 1D), with statistical differences among the treatments ($F = 10.4$; $d.f. = 5$; $P<0.05$), which were established by the control treatment with respect to the remaining species in the assemblages (Figure 1D). Finally, the average recruitment varied between 1.3 and 8.5 seedlings per plot (Figure 1E), showing statistical differences among the treatments ($F = 10.4$; $d.f. = 5$, $P<0.05$), which were determined by a higher seedling density in the control treatment with respect to that of the experimental assemblages (Figure 1E).

DNH

The phylogenetic relatedness between *Lactuca* and the resident members in each of the assemblages varied between 196.8 and 267.6 my for $MPD_{Lactuca}$ and between 88.0 and 254 my for $MNND_{Lactuca}$ (Table 1). The ANCOVA results showed that neither $MPD_{Lactuca}$ nor $MNND_{Lactuca}$ had significant effect on the germination, growth, survival, flowering, or recruitment of *Lactuca* (see Table 2), showing that the five indicators of *Lactuca* colonization success were not affected by the phylogenetic relatedness between this species and the experimental assemblages. The $MNND_{Lactuca}$ distance showed a marginally significant effect on the growth and recruitment of *Lactuca* (p = 0,052 and 0.085, respectively; see Table 2). However, these results were due to effects recorded in A2 and A4 treatments, no clear trend in support (or refutation) of DNH.

The average height of the assemblages had a significant effect on three of the colonization success indicators, specifically on germination, growth, and flowering of *Lactuca* (Table 2). This effect was determined by a decrease in the colonization success on plots with taller plants. Finally, the variance of the height of the assemblages did not have significant effects on any indicator of the establishment of *Lactuca* (Table 2).

In brief, the colonization success of *Lactuca*, measured by means of five different indicators (germination, growth, survival, flowering, and recruitment) was significantly greater in the control treatments (i.e., *Lactuca* monocultures) than in the experimental assemblages, showing that in the presence of other species the colonization success of *Lactuca* is reduced. However, *Lactuca* colonization did not show significant differences between the experimental assemblages, regardless of the phylogenetic relatedness with the receiver assemblage (measured as $MPD_{Lactuca}$ and $MNND_{Lactuca}$). Finally, the average height achieved by the plants of each assemblage reduced three of the five indicators of colonization success by *Lactuca*.

Discussion

DNH states that if the competitive interactions between phylogenetically close species are more intense, so the colonization success will be reduced when a given taxon colonizes communities that contain related species. Conversely, the colonization success will increase if the invasion occurs in communities consisting of phylogenetically distant taxa. Using an experimental gradient of phylogenetic relatedness between five plant species (receiver assemblages) and a colonizing species (*Lactuca sativa*), we found that phylogenetic relatedness did not influence the colonization success of the inoculated species. Therefore, these findings do not support DNH, and this is valid for a combination of five measures of colonizing success (germination, growth, survival, flowering, and recruitment) and two of phylogenetic relatedness ($MPD_{Lactuca}$ and $MNND_{Lactuca}$).

From the perspective of DNH, our experiment shows two important limitations. First, the experimental plots did not cover the (continuous) spectrum of phylogenetic relatedness between colonizer and recipient assemblage. However, this does not seem to have affected our analyses because using *a posteriori* tests we found no differences between *Lactuca* colonization success recorded in treatment A1 ($MPD_{Lactuca} = 196$ my; $MNND_{Lactuca} = 88$ my) and the other treatments ($MPD_{Lactuca} = 222–267$ my; $MNND_{Lactuca} = 214–254$ my). Second, our experimental design implied that not only the phylogenetic relatedness means greater competitive intensity, but this can also be due to the transfer of specialist natural enemies (e.g., herbivores and parasites) from the receiver assemblage to colonizing species [41,42]. This also reduces the establishment success of colonizing species, a factor that was not assessed in our experiments because we exclude the effect of herbivore interactions. Moreover, the five indicators of successful colonization of *Lactuca* showed inhibitory effects in the experimental assemblages compared to the control treatments (*Lactuca* monocultures). In mechanistic terms, these results probably reflect inhibitory interactions that restrict access to light and nutrients, exercised by species from recipient assemblages upon *Lactuca*.

Our results differ with respect to two studies that so far have evaluated DNH experimentally. Among previous experimental work, one study [15] used a *Serratia marcescens* strain as colonizing taxon, while another strain of the same species was part of the receiver communities; and another, [17] used colonizing species of *Metschnikowia* and *Candida*, which had congeneric representatives in the recipient assemblages. In contrast to these studies, in our experiments we worked with more distant levels of relatedness, since between *Lactuca* and *Matricaria chamomilla*, both species belonging to the same taxonomic family (Asteraceae), we found an MNND distance of 88.0 my. Therefore, the use of the congeneric or same strain (conspecific) taxa may magnify the effect of competition between colonizer and receiver assemblages [41], explaining –at least partially– our discrepant results.

A hypothesis opposed to DNH is related with preadaptation [29], [42–43], which proposes that phylogenetic closeness promotes the colonization process through facilitation among related taxa [41–42], [44–45]. Our results do not support this hypothesis, because colonization by *Lactuca* did not show a positive association with the phylogenetic relatedness of the assemblages. Moreover, the most important mechanisms of *Lactuca* colonization were related to the average height of the plants of the receiver assemblages. This trait, which is an indicator of the availability of light to the colonizer affected the germination, growth, and flowering response of *Lactuca*.

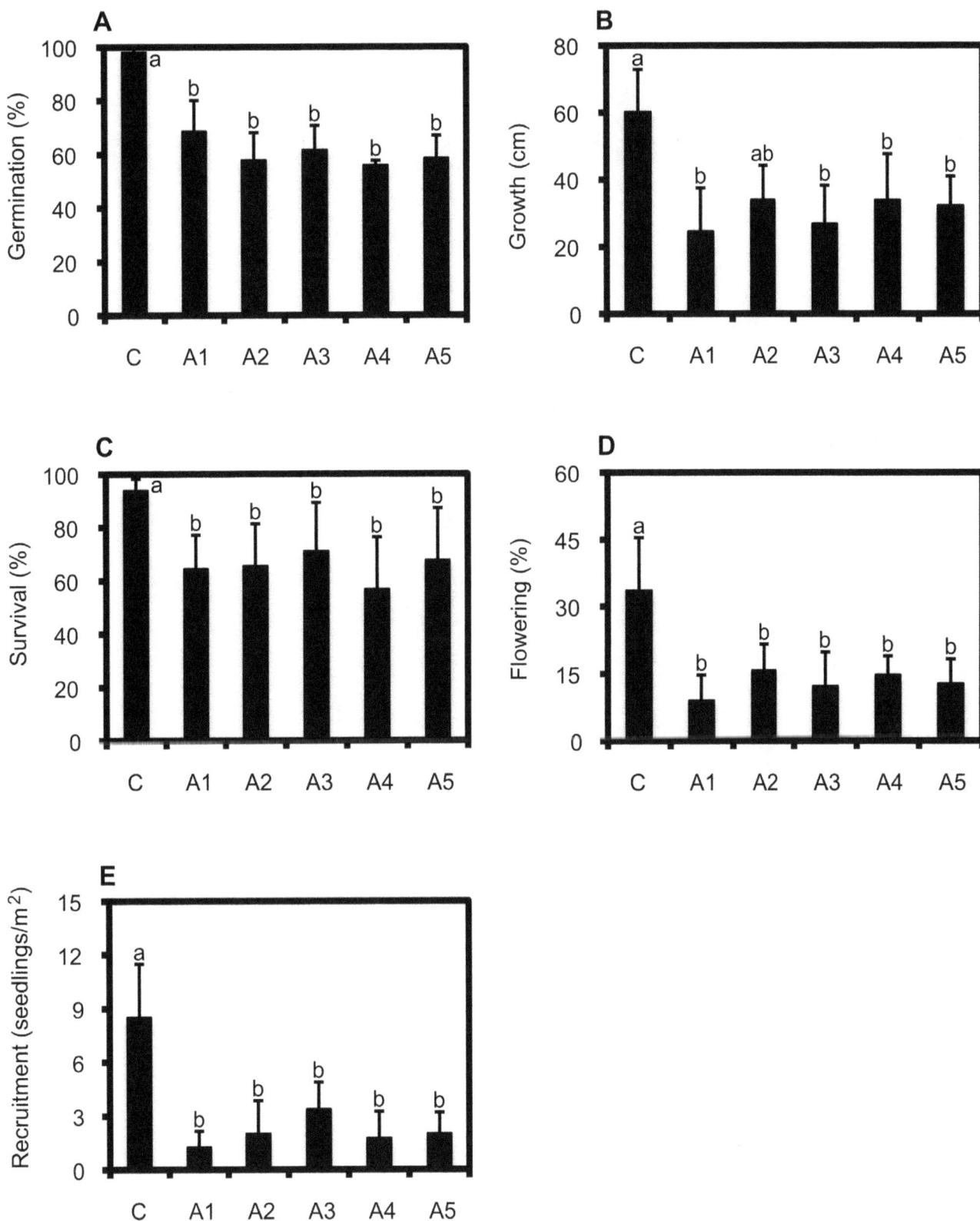

Figure 1. Colonization success (average ± S.D.) of *Lactuca* invading experimental plant communities. Colonization success was assessed as germination (%) (panel A), growth (cm) (panel B), survival (%) (panel C), flowering (%) (panel D), and recruitment (seedlings per m²) (panel E) of inoculation of *Lactuca* recorded in monoculture (control, C) and five assemblages (A1–A5) with which it has different degrees of phylogenetic relatedness.

Table 2. Summary of ANCOVAs that assessed the role of the phylogenetic distance (PD) between *Lactuca* and assemblage members, the average, and the variance of the height of the resident plants on five indicators of colonization success (germination, growth, survival, flowering, and recruiting) of *Lactuca*.

Source	MPD$_{Lactuca}$				MNND$_{Lactuca}$			
	d.f.	SS	F	P	d.f.	SS	F	P
Germination								
PD	4	0.051	0.5	0.708	3	0.051	0.7	0.572
Average	1	0.021	6.7	0.014	1	0.025	6.9	**0.013**
Variance	1	0.016	2.8	0.103	1	0.015	3.4	0.075
Growth								
PD	4	1.636	2.1	0.106	3	1.637	2.8	0.052
Average	1	0.028	10.9	0.002	1	0.033	11.3	**0.002**
Variance	1	1.243	0.2	0.670	1	1.241	0.2	0.639
Survival								
PD	4	0.050	1.2	0.342	3	0.003	0.8	0.476
Average	1	0.100	1.1	0.297	1	0.045	0.1	0.786
Variance	1	0.207	2.3	0.143	1	0.116	1.0	0.329
Flowering								
PD	4	0.037	0.3	0.843	3	0.037	0.4	0.779
Average	1	0.016	4.8	0.035	1	0.020	4.9	**0.033**
Variance	1	0.011	2.1	0.158	1	0.008	2.6	0.114
Recruitment								
PD	4	0.700	1.9	0.138	3	0.695	2.4	0.085
Average	1	0.198	1.7	0.205	1	0.133	1.7	0.203
Variance	1	3.142	0.5	0.496	1	2.970	0.3	0.574

Our results suggest that under the spatial scale used there was a null effect of the phylogenetic distance on colonization success and the invasion's contingent conditions such as the current properties of resident assemblage, and the intrinsic properties of invaders would be more relevant in determining colonization success than factors associated with phylogenetic relatedness, as has been discussed by other authors [10], [41], [43], [46], [47]. This does not mean that phylogenetic distance between invader and resident assemblage does not affect the colonization process. Indeed, the phylogenetic distance would have a non-linear effect on colonization process [43], which would require increasing the spatial scale of the studies, increasing the number of species in resident assemblages, or performing comparative studies that consider a larger number of invader species in several recipient communities [41]. In the absence of more evidence and experimental contrasts, it seems premature to accept the null effect as a definitive answer. However, the available evidence at least allows questioning the generality of DNH, since it has been verified experimentally only when the colonizing species are closely related to the members of the receiving community [15,17], while if the colonizer is phylogenetically more distant (this study), the colonization success becomes independent of the evolutionary relatedness. A recent study [21] evaluated the change in the intensity of the interactions between vascular plant species along a phylogenetic gradient (with MNND distances ranging between 0 my and 81 my). Although these experiments were made between pairs of species and not in communities, the support of DNH was only partial because the evolutionary closeness not only allowed greater inhibitory interaction intensity (such as competition), but it also increased the intensity of facilitating interactions, which have an effect that is the opposite of that expected by DNH [10,43]. From the community viewpoint it would be extremely important to quantify the combined effect of the inhibitory and facilitative interactions on the success of colonization, and how this balance is expressed along the colonizer-community relatedness gradient. Along this line, the evolutionary distance metrics should include most of the components of the community (i.e., herbivores, parasites, facilitators) with which a colonizer can interact, and not only the taxonomic composition of assemblages. These kinds of efforts can be difficult to implement under field conditions, but experimental or modeling approximations can assist in disentangling this complexity [15]. We agree with two previous studies [10,43], which proposed deeper studies that allow establishing the role of the phylogenetic structure of the communities in their susceptibility to being invaded.

An important challenge that still needs to be elucidated is how these experimental conditions reflect the heterogeneity and complexity of the invasive processes under field conditions. For example, it is feasible that invasion by congeneric or conspecific species occurs when a taxon expands its range by means of a reaction-diffusion process [48], colonizing adjacent communities, which likely contain closely related taxa. In contrast, invasion that involves spread to new continents or distant regions probably represents better the case in which the colonizing species has a rather distant relationship with the members of the receiving community [48].

In summary, our results do not show an association (positive or negative) between phylogenetic distance and the colonizing success of an inoculated species, so they do not support DNH. In view of the small number of studies that have evaluated DNH, particularly from the experimental standpoint, it may be premature to generalize on the role of phylogenetic relatedness in determining the result of the invasive process. However, our results and the available background information suggest that DNH may have an explanatory domain restricted to only one part of the phylogenetic spectrum.

Supporting Information

Table S1 Colonization success of *Lactuca* recorded in each experimental plot. The colonization indicators were: Germination (%), Growth (cm), Survival (%), Flowering (%) and Recruitment (No./m2). A1–A5 represent different treatments of recipient plant assemblages and C the control treatment (i.e., *Lactuca* in monoculture). $MPD_{Lactuca}$ are $MNND_{Lactuca}$ are two metrics of phylogenetic distances for *Lactuca* and the recipient assemblages. $MPD_{Lactuca}$ is the average distance between *Lactuca* and each member of the assemblages; $MNND_{Lactuca}$ is the distance between *Lactuca* and its nearest neighbor in each assemblage. As a concomitant factor, the Height (average and variance; both in cm) achieved by the plants in each plot at the end of the experiment was considered.

Author Contributions

Conceived and designed the experiments: SAC. Performed the experiments: SAC VE JA. Analyzed the data: SAC VE GOC. Contributed reagents/materials/analysis tools: SAC GOC JA. Contributed to the writing of the manuscript: SAC GOC.

References

1. Pysek P, Richardson DM (2006) The biogeography of naturalization in alien plants. J Biogeogr 33: 2040–2050.
2. Vitousek PM, D'Antonio CM, Loope LL, Rejmánek M, Westbrooks R (1997) Introduced species: A significant component of human-caused global change. New Zeal J Ecol 21: 1–16.
3. Williamson M (1996) Biological invasions. Oxford: Chapman & Hall, London 256 p.
4. Davis MA (2009) Invasion Biology. New York: Oxford University Press. 244 p.
5. Kolar CS, Lodge DM (2001) Progress in invasions biology: predicting invaders. Trends Ecol Evol 16: 199–204.
6. Sakai AK, Allendorf FW, Holt JS, Lodge DM, Molofsky J, et al. (2001) The population biology of invasive species. Annu Rev Ecol Syst 32: 305–332.
7. Lockwood JL, Hoopes MF, Marchetti MP (2007) Invasion Ecology. Oxford: Blackwell Publishing, Oxford. 304 p.
8. Chesson P (2000) Mechanisms of maintenance of species diversity. Annu Rev Ecol Syst 31: 343–366.
9. Mack RN, Simberloff D, Lonsdale WM, Evans H, Clout M, et al. (2000) Biotic invasions, causes, epidemiology, global consequences, and control. Ecol Appl 10: 689–710.
10. Proches S, Wilson JRU, Richardson DM, Rejmanek M (2008) Searching for phylogenetic pattern in biological invasions. Global Ecol Biogeogr 17: 5–10.
11. Darwin C (1859) On the origin of species. London: J. Murray. 502.

12. Daehler CC (2001) Darwin's naturalization hypothesis revisited. Am Nat 158: 324–330.
13. Strauss SY, Webb CO, Salamin N (2006) Exotic taxa less related to native species are more invasive. Proc Natl Acad Sci USA 103: 5841–5845.
14. Hill SB, Kotanen PM (2009) Evidence that phylogenetically novel non-indigenous plants experience less herbivory. Oecologia 161: 581–590.
15. Jiang L, Tan J, Pu Z (2010) An experimental test of Darwin's naturalization hypothesis. Am Nat 175: 415–423.
16. Gerhold P, Pärtel M, Tackenberg O, Hennekens SM, Bartish I, et al. (2011) Phylogenetically poor plant communities receive more alien species, which more easily coexist with natives. Am Nat 177: 668–680.
17. Peay KG, Belisle M, Fukami T (2012) Phylogenetic relatedness predicts priority effects in nectar yeast communities. Proc R Soc B 279: 749–758.
18. Schaefer H, Hardy O, Silva L, Barraclough T, Savolainen V (2011) Testing Darwin's naturalization hypothesis in the Azores. Ecol Lett 14: 389–396.
19. Pearson DE, Ortega YK, Sears SJ (2012) Darwin's naturalization hypothesis up-close: Intermountain grassland invaders differ morphologically and phenologically from native community dominants. Biol Invasions 14: 901–913.
20. Lambdon PW, Hulme PE (2006) How strongly do interactions with closely-related native species influence plant invasions? Darwin's naturalization hypothesis assessed on Mediterranean islands. J Biogeogr 33: 1116–1125.

21. Burns JH, Strauss SY (2012) More closely related species are more ecologically similar in an experimental test. Proc Natl Acad Sci USA 108: 5302–5307.

22. Duncan RP, Williams PA (2002) Darwin's naturalization hypothesis challenged. Nature 417: 608–609.

23. Ricciardi A, Atkinson SK (2004) Distinctiveness magnifies the impact of biological invaders in aquatic ecosystems. Ecol Lett 7: 781–784.

24. Ricciardi A, Mottiar M (2006) Does Darwin's naturalization hypothesis explain fish invasions? Biol Invasions 8: 1403–1407.

25. Cahill JF, Kembel SW, Lamb EG, Keddy P (2008) Does phylogenetic relatedness influence the strength of competition among vascular plants? Perspect Plant Ecol 10: 41–50.

26. Diez JM, Williams PA, Randall RP, Sullivan JJ, Hulme PE, et al. (2009) Learning from failures: Testing broad taxonomic hypotheses about plant naturalization. Ecol Lett 12: 1174–1183.

27. Escobedo VM, Aranda JE, Castro SA (2011) Hipótesis de Naturalización de Darwin evaluada en la flora exótica de Chile continental. Rev Chil Hist Nat 84: 543–552.

28. Tingley R, Phillips BL, Shine R (2011) Establishment success of introduced amphibians increases in the presence of congeneric species. Am Nat 177: 382–388.

29. Park DS, Potter D (2013) A test of Darwin's naturalization hypothesis in the thistle tribe shows that close relatives make bad neighbors. Proc Natl Acad Sci USA 110: 17915–17920.

30. Cavender-Bares J, Kozak K, Fine P, Kembel S (2009) The merging of community ecology and phylogenetic biology. Ecol Lett 12: 693–715.

31. Green J, Bohannan BJM (2006) Spatial scaling of microbial biodiversity. Trends Ecol Evol 21: 501–507.

32. Martiny JBH, Bohannan BJM, Brown JH, Colwell RK, Fuhrman JA, et al. (2006) Microbial biogeography: Putting microorganisms on the map. Nature Rev Microbiol 4: 102–112.

33. Webb CO, Donoghue MJ (2005) Phylomatic: tree assembly for applied phylogenetics. Mol Ecol Notes 5: 181–183.

34. The Angiosperm Phylogeny Group (2009) An update of the Angiosperm Phylogeny Group classification for the orders and families of flowering plants: APG III. Bot J Linn Soc 161: 105–121.

35. Downie SR, Katz-Downie DS, Watson MF (2000) A phylogeny of the flowering plant family Apiaceae base don chloroplast DNA *RPL16* and *RPOC1* intron sequences: towards a suprageneric classification of subfamily Apioideae. Am J Bot 87: 273–292.

36. Wikstrom N, Savolainen V, Chase MW (2001) Evolution of angiosperms: Calibrating the family tree. Proc R Soc B 268: 2211–2220.

37. Webb CO, Ackerly DD, Kembel SW (2008) Phylocom: Software for the analysis of phylogenetic community structure and trait evolution. Bioinformatics 24: 2098–2100.

38. Webb CO (2000) Exploring the phylogenetic structure of ecological communities: An example for rain forest trees. Am Nat 156: 145–155.

39. Fine PVA, Kembel SW (2011) Phylogenetic community structure and phylogenetic turnover across space and edaphic gradients in western Amazonian tree communities. Ecography 34: 553–656.

40. Webb CO, Ackerly DD, McPeek MA, Donoghue MJ (2002) Phylogenies and community ecology. Annu Rev Ecol Syst 33: 475–505.

41. Strong DR, Lawton JH, Southwood TRE (1984) Insects on plants: Community patterns and mechanisms. Oxford: Blackwell. 313 p.

42. Lewinsohn TM, Novotny V, Basset Y (2005) Insects on plants: diversity of herbivore assemblages revisited. Annu Rev Ecol Evol Syst 36: 597–620.

43. Jones EI, Nuismer SL, Gomulkiewicz R (2013) Revisiting Darwin's conundrum reveals a twist on the relationship between phylogenetic distance and invasibility. Proc Natl Acad Sci USA 110: 20627–20632.

44. MacDougall AS, Gilbert B, Levine JM (2009) Plant invasions and the niche. J Ecol 97: 609–615.

45. Thuiller W, Gallien L, Boulangeat I, De Bello F, Münkemüller T, et al. (2010) Resolving Darwin's naturalization conundrum: A quest for evidence. Divers Distrib 16: 1–15.

46. Bruno JF, Stachowicz JJ, Bertness MD (2003) Inclusion of facilitation into ecological theory. Trends Ecol Evol 18: 119–125.

47. Diez JM, Sullivan JJ, Hulme PE, Edwards G, Duncan RP (2008) Darwin's naturalization conundrum: Dissecting taxonomic patterns of species invasions. Ecol Lett 11: 674–681.

48. Shigesada N, Kawasaki K (2001) Biological invasions: Theory and practice. Oxford: Oxford University Press. 218 p.

20

Phylogenetic Meta-Analysis of the Functional Traits of Clonal Plants Foraging in Changing Environments

Xiu-Fang Xie[1,2], **Yao-Bin Song**[1]*, **Ya-Lin Zhang**[2], **Xu Pan**[2], **Ming Dong**[1,2]*

1 Key Laboratory of Hangzhou City for Ecosystem Protection and Restoration, College of Life and Environmental Sciences, Hangzhou Normal University, Hangzhou, China,
2 State Key Laboratory of Vegetation and Environmental Change, Institute of Botany, Chinese Academy of Sciences, Beijing, China

Abstract

Foraging behavior, one of the adaptive strategies of clonal plants, has stimulated a tremendous amount of research. However, it is a matter of debate whether there is any general pattern in the foraging traits (functional traits related to foraging behavior) of clonal plants in response to diverse environments. We collected data from 97 published papers concerning the relationships between foraging traits (e.g., spacer length, specific spacer length, branch intensity and branch angle) of clonal plants and essential resources (e.g., light, nutrients and water) for plant growth and reproduction. We incorporated the phylogenetic information of 85 plant species to examine the universality of foraging hypotheses using phylogenetic meta-analysis. The trends toward forming longer spacers and fewer branches in shaded environments were detected in clonal plants, but no evidence for a relation between foraging traits and nutrient availability was detected, except that there was a positive correlation between branch intensity and nutrient availability in stoloniferous plants. The response of the foraging traits of clonal plants to water availability was also not obvious. Additionally, our results indicated that the foraging traits of stoloniferous plants were more sensitive to resource availability than those of rhizomatous plants. In consideration of plant phylogeny, these results implied that the foraging traits of clonal plants (notably stoloniferous plants) only responded to light intensity in a general pattern but did not respond to nutrient or water availability. In conclusion, our findings on the effects of the environment on the foraging traits of clonal plants avoided the confounding effects of phylogeny because we incorporated phylogeny into the meta-analysis.

Editor: Alexandra Weigelt, University of Leipzig, Germany

Funding: This research was supported by NFSC grants (31261120580, http://www.nsfc.gov.cn), the Innovative R & D grant (201203) from Hangzhou Normal University, and the Project of Zhejiang Key Scientific and Technological Innovation Team (2010R50039). The funders had no role in study design, data collection and analysis, decision to publish, or preparation of the manuscript.

Competing Interests: The authors have declared that no competing interests exist.

* Email: ybsong@hznu.edu.cn (YBS); dongmingchina@126.com (MD)

Introduction

Essential resources such as light, water and soil nutrients often have heterogeneous distributions in natural habitats [1-4]. Phenotypic plasticity is an adaptive strategy through which plants can cope with environmental variation in space and time [5,6]. Clonal plants occur in many different taxonomic groups [7] and are dominant in many natural and man-made ecosystems [7,8]. The success of clonal plants may occur because their distinctive life-history strategies [9,10] allow them to cope with the heterogeneity of essential resources. One of these strategies is plastic foraging, i.e., the processes whereby an organism searches or ramifies within its habitat to enhance its acquisition of essential resources [6,11,12]. Foraging strategies help clonal plants escape from unfavorable patches and/or exploit favorable ones by altering the clonal morphology of spacers and branching in patchy environments [6,13,14].

The adaptive evolution of clonal life history traits may be limited by physiological or physical constraints [10,15]. Such constraints may limit the plastic foraging [16] responses of species to environmental change [17]. For example, because of their different functions, stolons might be more plastic than rhizomes in

response to a light resource, whereas rhizomes might be more plastic than stolons in response to a nutrient resource [18]. The results of a garden experiment showed that plants with contrasting branching patterns (monopodial versus sympodial) exhibited different foraging traits [17]. Therefore, we may expect the process of plastic foraging to differ among species with different clonal-organ types (i.e., stoloniferous and rhizomatous) and branching-form types (i.e., monopodial and sympodial). The process of plastic foraging of clonal plants may also depend on the response of plants to resources (i.e., water, nutrients or carbohydrates). However, this hypothesis has not been tested because multiple species are involved.

To achieve efficient plastic foraging, clonal plants usually adopt two tactics: branching and spacing. Ramets can regulate the plasticity of branching or spacing in patches with high- or low-level resources [19]. Branching traits include the plasticity of branching intensity (i.e., the number of branches per ramet or node of the rhizome or stolon [13]) and branching angle (i.e., the angle between adjacent branches in the horizontal plane [14]). Spacing traits include the plasticity of spacer length (i.e., the distance between adjacent ramets, which may contain only one internode [13]) and spacer thickness (i.e., specific spacer length, which is the ratio of the spacer length to the dry mass of the spacer). The trade-off between branching and spacing as plastic foraging strategies may be determined by environmental factors [19]. Thus far, whether branching traits are more plastic than spacing traits has not been tested.

A meta-analytic approach is optimal to analyze the responses of clonal plants to various environments (i.e., [20,21]). However, Adams [22] stated that traditional meta-analytic approaches lacked independence because of species with shared evolutionary histories in biology and ecology. Adams [22] proposed that the phylogenetic meta-analysis (PMA) model that incorporates phylogenetic information into the traditional meta-analysis be used. Furthermore, a recent meta-analysis demonstrated that incorporating phylogenetic information significantly changed the pooled effect sizes of traditional meta-analysis [23].

We adopted the phylogenetic meta-analytic method proposed by Lajeunesse [24] to answer the following three questions for clonal plants: 1) Are branching traits more plastic than spacing traits in response to resource availability? 2) Does the foraging behavior of clonal plants differ with different clonal architectures? 3) Do the foraging-related traits depend on the type of resource?

Materials and Methods

Literature survey and data selection criteria

To be comprehensive and to avoid bias in the literature survey, we conducted an exhaustive, strategic search of all literature concerning our scientific questions, relying primarily on the internet search engine Google Scholar, which covers most peer-reviewed papers, theses and dissertations (or degree papers), books, and other published or unpublished academic literature from broad areas of research [25]. To identify studies specific to our questions, the survey was supplemented by additional searches of a number of main databases (e.g., ISI Web of Knowledge). We restricted the search keywords to papers whose topic referred to "forag*", "spacer length", "rhizome length", "stolon length" or "internode length" in combination with "clonal plant". We obtained 715 published papers from which we selected 449 studies as suitable for the meta-analysis (Figure 1; [26]).

For each publication, we recorded the title, author(s), year, location, and other information (see Appendix S1) and examined its potential for meeting the selection criteria for inclusion in the review. Furthermore, only the publications that reported values of plant clonal traits for both the treatment and control groups in (greenhouse, garden or field) experiments were considered, whereas reviews, publications on models and other studies were excluded. In our meta-analysis, we only included studies that reported traits related to foraging strategy (e.g., spacing traits, branching traits) in response to resource availability (Table 1). We classified the resources into three categories (Table 2)—light intensity, nutrient level and water availability—which are usually heterogeneously distributed in nature. Thus, other resources, such as CO_2 [26], were not considered for this review. Furthermore, we excluded the studies in which the means for the treatment and control groups were not reported with the sample size and/or the standard deviation or in which we were unable to infer (or calculate) the sample size or standard deviation from other information provided in the study [27]. Our final data set contained 97 papers published from 1965 to 2013 in 37 journals and provided the data for the meta-analyses (Appendix S2).

Data assembly

From each study, we extracted the mean, a measure of statistical variation (usually standard error or standard deviation) and the sample size for the treatment and control groups for each response variable. We regarded multiple results within a single paper as different results from independent studies when they involved different species and/or treatments [27-31]. We only extracted data once when the same experimental results were published in different papers [32]. To collect original data, the graphs in articles were digitized with GetData (Graph Digitizer v2.22 Datatrend Software, Raleigh, North Carolina, USA) [33], and digitized values were accurate to ±1% of the actual value [27,29].

A total of 1,370 comparisons encompassing 85 clonal plant species from 64 genera belonging to 28 families met our criteria. For each comparison, we calculated Hedges' d as a measure of the effect magnitude because it is the preferred measure of effect size for meta-analysis and because it has a lower Type I error rate than other measures, such as the log-response ratio [34,35]. The absolute value shows the magnitude of the treatment impact. Positive and negative d values signify an increase and a decrease in the effect of the treatment, respectively. A value of zero indicates no difference between the treatment and control groups.

For studies that described experiments with several treatment levels or times, we pooled the effect sizes and variances for each response variable of each species in a study, and we conducted a separate meta-analysis on all traits and treatments of the respective trait category to avoid pseudo-replication (see also [34,35]). The estimated pooled mean effect size and the mean variance were used in the final data sets containing 107 pairwise comparisons with light treatments, 179 pairwise comparisons with nutrient treatments and 26 pairwise comparisons with water treatments (Appendix S1). To pool the effect sizes and all analyses, we chose the random-model approach because we assumed that the differences among comparisons and among studies were not only due to sampling error but also to true random variation, as is the rule for ecological data [36]. All the effect size calculations and pooling were performed with Metawin software, version 2.1 [37].

To apply PMA, we first created a phylogeny including all the plant species using the Phylomatic software online (http://phylodiversity.net/phylomatic/; with option Phylomatic tree R20120829 for plants). The branch length for the phylogeny was estimated using the 'bladj' function in the 'Phylocom' software [38,39] (the constructed phylogeny with branch length is provided in Appendix S3). Using the same procedure, we generated a subset phylogeny for each trait category including the corresponding subset species pool. The branch length was again estimated using the age file in Phylocom software. Because of the restriction of input files executed on phyloMeta v1.3 software [40], we again pooled those multiple effect sizes for the same species. The result was one accumulated weighted effect size and variance for each species within a given trait category. However, this approach inevitably resulted in smaller sample sizes for each trait category ($N_{effect-size} = N_{species}$) [41,42]. The pooling was also conducted with a random-effects model on Metawin, version 2.1 [37].

Data analysis

An inherent problem with meta-analysis is the potential for publication bias, which has been termed the "file-drawer problem" [43]. Therefore, before all analyses, we explored the possibility of publication bias graphically (using a funnel plot and normal quantile plot) [44,45], statistically (using the Spearman rank correlation test) [46] and by calculating a fail-safe number [47,48], which is the number of studies that would have to be added to change the results of the meta-analysis from significant to nonsignificant [49].

For each trait category (Table 1), we calculated the overall effect sizes ($d+$) of every resource category (Table 2) separately across the

RISMA 2009 Flow Diagram

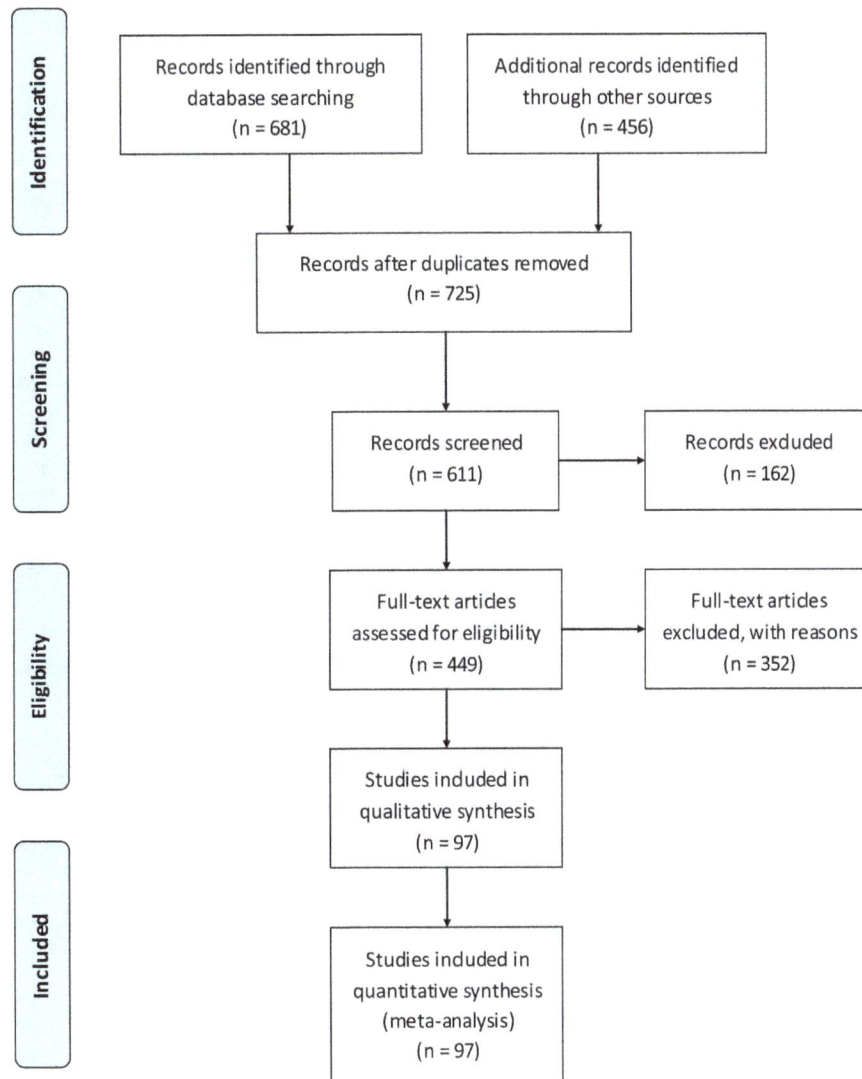

Figure 1. The flow diagram.

Table 1. Foraging traits and their subcategories and categories (foraging tactics).

Foraging tactics	Foraging trait subcategory	Foraging trait
Spacing	Spacer length (SL)	Rhizome length, spacer length, stolon length.
	Specific spacer length (SSL)	Specific rhizome length, specific spacer length, specific stolon length.
Branching	Branching intensity (BI)	Branching index, branching intensity, number of branches, number of rhizomes, number of stolons.
	Branching angle (BA)	Branching angle.

Table 2. Resource treatments used in the studies reported in the literature.

Resource category	Treatments
Light	Darkening (-), light, light increased, low-light (-), partial shading (-), shade (-), shading (-).
Nutrients	Fertilization, litter, low-N (-), N, nutrient, P, sediment type, soil, soil nutrient, soil resources.
Water	Drought (-), low-water (-), soil moisture, soil water, water, water amount, water reduced (-), wet treatment.

(-): Opposite direction of the treatment effect.

samples of case studies with information on the relevant response variables. The overall effect sizes were cumulative effect sizes per species [24,50]. For the interspecific differences in spacer length, we conducted a supplementary analysis on its two determinants—internode length and node number—because a spacer could have more internodes, in addition to a single-internode spacer. To detect the differences between stoloniferous and rhizomatous plants and between monopodial and sympodial plants, we considered clonal organ type (i.e., stoloniferous versus rhizomatous) and branching form type (i.e., monopodial versus sympodial) as moderator variables [41]. The analyses were performed with the software phyloMeta v1.3 [41].

Results

Generalities of sampled studies

The overall data exploration found no evidence of publication bias. The funnel plot of effect size versus sample size showed no skewness (Appendix S3). A plot of the standardized effect sizes against the normal quantiles revealed a straight line (Appendix S3). These two graphical approaches suggest that there was no bias in the results from this meta-analysis. This result was further emphasized by a nonsignificant result of the Spearman rank-order correlation test (R = -0.091, $p > 0.05$). Finally, the weighted fail-safe number 10,139 was much greater than expected (5n + 10 = 1570) without publication bias, which supports the robustness of our results. Thus, we are confident that our results provide reliable estimates of the true effects.

PMA on foraging responses of clonal plants to resource heterogeneity

In the PMA results, the grand mean effect sizes of light on spacing traits (spacer length and specific spacer length) and branching traits (branching intensity) were not significant except for that on branching angle ($d+$ = 0.61, N = 2, 95% CI = 0.01 to 1.20; Figure 2A). The grand mean effect sizes of nutrients on spacing traits and branching traits were not significant except for that on branching intensity ($d+$ = 0.59, N = 26, 95% CI = 0.01 to 1.18; Figure 2B). The grand mean effect sizes of water on spacing traits and branching traits were not significant.

For the clonal organ type, the spacing traits were not significantly different from zero when responding to light in either stoloniferous plants or rhizomatous plants (Figure 2A). The branching intensity response to light was significant for the stoloniferous plants ($d+$ = 1.97, N = 11, 95% CI = 1.06 to 2.89), but it was not significant for the rhizomatous plants (Figure 2A). The branching intensity of the stoloniferous plants was significantly greater than zero when responding to nutrient level ($d+$ = 2.31, N = 13, 95% CI = 1.59 to 3.03), whereas the branching intensity of rhizomatous plants was not significant (Figure 2B). The effect sizes of nutrients on the branching intensity of stoloniferous and rhizomatous were significantly different (Q_b =

11.93, $P < 0.05$). The effect sizes of nutrients on the branching intensity of monopodial and sympodial plants were also significantly different (Q_b = 69.66, $P < 0.05$). However, nutrients had no effects on the spacing traits or the branching angle for either the stoloniferous or the rhizomatous plants (Figure 2B). Moreover, no significant effects of water on spacing traits or branching traits were detected (Figure 2C).

For the branching forms, spacing traits were not significantly different from zero when responding to light either in monopodial or sympodial plants (Figure 2A). The branching traits of both the monopodial and sympodial plants were significantly different from zero when responding to light, except for the branching angle of the monopodial plants for which data were lacking (Figure 2A). Similarly, spacing traits were not significantly different from zero when responding to nutrient level in either the monopodial or the sympodial plants (Figure 2B). The branching intensity of the monopodial and sympodial plants was significantly different from zero when responding to nutrient level (Figure 2B). Neither spacing traits nor branching traits were significantly different from zero when the monopodial and sympodial plants responded to water.

PMA on the responses of internodes to resource heterogeneity

According to the supplementary analyses, the internode length was significantly less than zero when responding to light but was not significantly different from zero when responding to nutrients or water, regardless of the clonal organ type or branching form (Figure 3A). Neither light nor nutrients exerted any significant effects on the node number, but water did have positive effects on the node number (Figure 3B).

Additionally, the results of all effects on the specific space length, branching angle and node number should be interpreted with caution because of the limited number of species and should therefore only be treated as a reference point (Figure 2).

Discussion

Responses of foraging traits to resource heterogeneity

Clonal plants adapt to changing environments by developing different adaptive strategies, mainly in the plasticity of plant traits. Our analyses provided powerful evidence that clonal plants indeed adopted foraging strategies in response to diverse environments [6,51]. Under shaded conditions, clonal plants decreased their branching intensity and tended to increase their spacer length to seek light resources, especially stoloniferous plants. This result was consistent with previous findings from empirical, experimental [52-55] and model [12,56] studies. However, we found that internode length was more flexible in response to light intensity, whereas spacer length had no significant response to environmental heterogeneity. One possible explanation for these findings is that clonal plants may produce spacers with shorter internodes

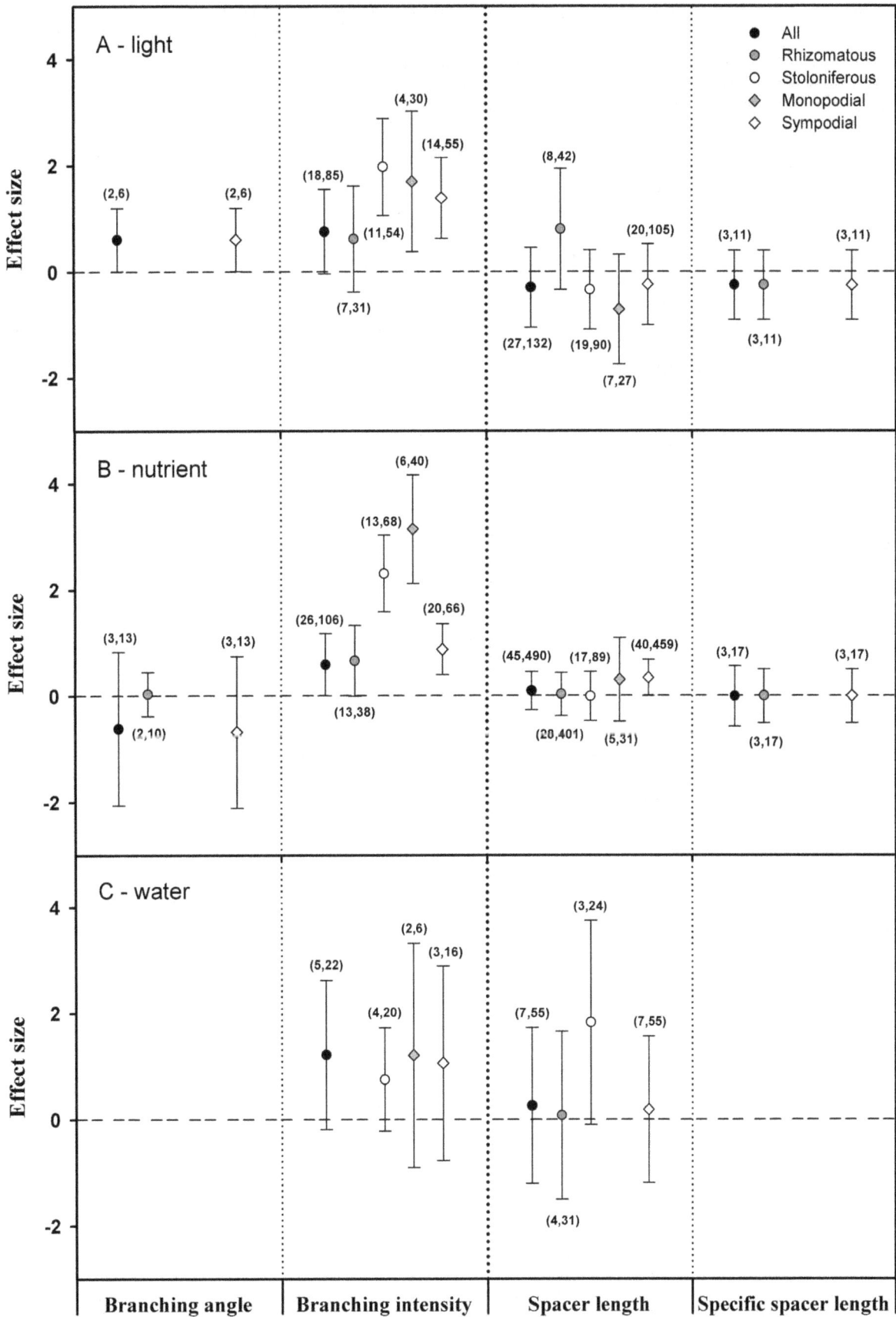

Figure 2. The results from PMA using phyloMeta v1.3. Black circle: the mean effect size of the three types of resources (A: light; B: nutrients; C: water) on all plants. Gray circle: the mean effect size of the three types of resources on rhizomatous plants. White circle: the mean effect size of the three types of resources on stoloniferous plants. Gray diamond: the mean effect size of the three types of resources on monopodial plants. White diamond: the mean effect size of the three types of resources on sympodial plants. Dotted line: the reference line of an effect size equal to zero. The numbers in parentheses: the first is the number of species contained, and the second is the number of cases combined. *BA: branching angle; BI: branching intensity; SL: spacer length; SSL: specific spacer length.

and more nodes under high light conditions, which is supported by the results of our analyses. Thus, the spacer length did not show any significant elongation under shaded conditions. Based on empirical and theoretical studies, another spacing variable (specific spacer length) should increase in response to shading because of limited biomass production in shaded environments [10,57].

However, few studies have focused on the plasticity of specific spacer length, which should be examined in the future.

Our analyses showed that nutrient availability had little effect on spacing traits or branching angle, but it did have a positive effect on branching intensity, even though this effect varied with different clonal architectures. Our findings support previous

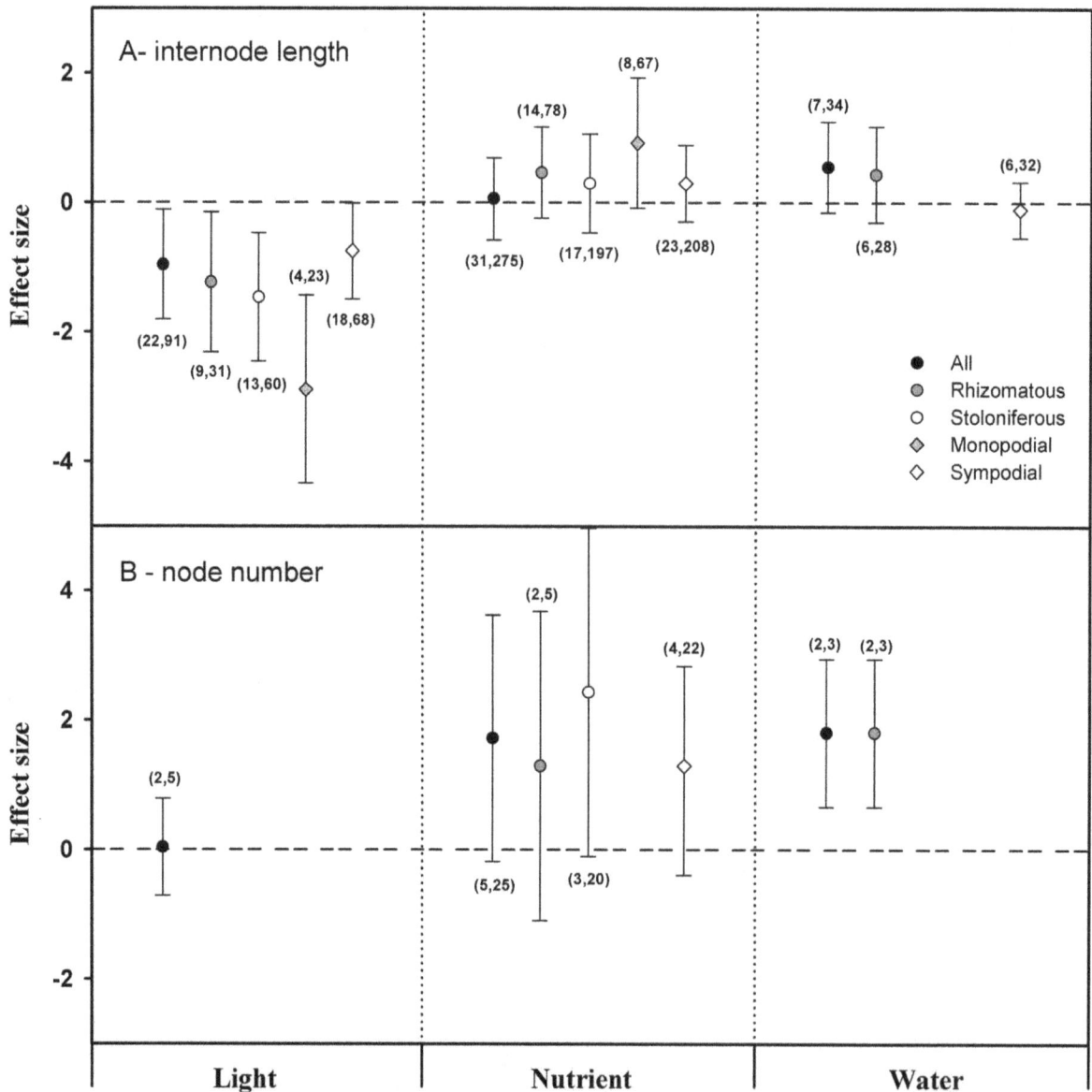

Figure 3. The results from the supplementary analyses (A: internode length; B: node number) with the random-model in PMA using phyloMeta v1.3. Black circle: the mean effect size of the three types of resources on all plants. Gray circle: the mean effect size of the three types of resources on rhizomatous plants. White circle: the mean effect size of the three types of resources on stoloniferous plants. Gray diamond: the mean effect size of the three types of resources on monopodial plants. White diamond: the mean effect size of the three types of resources on sympodial plants. Dotted line: the reference line of an effect size equal to zero. The numbers in parentheses: the first is the number of species contained, and the second is the number of cases combined. *IL: internode length; NN: node number.

research that found that with increasing nutrient availability, branching intensity increased. The branch angle did not exhibit any significant plasticity, and the response of spacer length was species specific [58]. One interpretation may be that the elongation and maintenance of spacer length comes with a cost, but the plant cannot pay this cost when the nutrient availability is too low [59-61], implying that there might be a trade-off between spacing and branching strategies. A significant impact of water on spacer length was not detected in our study. Unfortunately, the relationship between water availability and the other two traits was not clear and the results were not definitive because of limited data. At present, experimental research on the morphological plasticity of clonal plants in response to water availability is relatively lacking, and there is a particular need for such studies (see the example in [62]).

Differences between stoloniferous and rhizomatous plants in foraging tactics

Our results suggested that the tactics used for foraging for light resources were distinct for stoloniferous and rhizomatous species [63]. For stoloniferous plants, light had an appreciably negative effect on internode length and a positive effect on branching intensity. These observations were consistent with foraging theory. However, for rhizomatous plants, there were no significant relationships for light and foraging traits except for the internode length. Therefore, as generally accepted [18,64,65], stoloniferous plants were more sensitive to light than rhizomatous plants. An unexpected result was that we did not find a greater impact of nutrients on the rhizomatous plants compared with the stoloniferous plants, as recognized previously by Dong & de Kroon [18]. However, nutrients had no significant impacts on foraging traits except for the branching intensity of stoloniferous plants. For the rhizomatous plants, increasing nutrient availability had no obvious impact on any foraging trait. This variation in response might be due to the different functions of the different organs; rhizomes serve as organs primarily for the storage of resources and meristems, whereas stolons serve as organs primarily for foraging [18,66]. The data on the effects of water on foraging traits were limited and did not allow us to determine how water influenced the foraging behaviors of clonal plants.

Additionally, our results indicated that the foraging tactics used by monopodial and sympodial plants were similar, with both decreasing their branching intensity under resource-poor conditions and lengthening their internodes under shaded conditions. Thus, the foraging behaviors of clonal plants do not vary with branching form.

Meta-analysis has become a common method in ecological studies, and phylogenetic information is incorporated into meta-analyses with increasing regularity. However, phylogenetic meta-analysis is still in its infancy for use in clonal plant research. In our study, we conducted a parallel analysis for comparison with the traditional meta-analysis method. The results indicated that the incorporation of phylogenetic information into the analyses across our data sets slightly altered the significance of some effect sizes (Appendix S4). Moreover, the phenotype (trait) of a species is derived from the combined effects of local environments and genetic factors, and the genetic factor reflects the evolutionary history (phylogenetic information) [17]. PMA provided us with more convincing results than the traditional meta-analysis because we avoided the confounding influence of genetics and eliminated the possibility that closely related species might have similar responses to a changing environment.

In conclusion, our study is the first to use PMA to analyze the responses of the foraging traits of clonal plants to heterogeneous environments. We summarize the general patterns of our PMA analysis as follows: 1) clonal plants exhibit a higher plasticity of foraging traits in response to light intensity than to nutrient level or water availability; 2) spacer length, and sometimes internode length, is more flexible in response to light heterogeneity, and branch intensity is more sensitive to nutrient heterogeneity; and 3) stoloniferous plants show much stronger morphological plasticity in terms of foraging traits than rhizomatous plants. In this paper, we only tested the foraging hypotheses and aimed to clarify the general patterns of foraging behavior-related traits in clonal plants, but the mechanisms underlying these patterns must be explored in future research.

Supporting Information

Appendix S1 General information and data table with the details from all the single studies.

Appendix S2 List of the studies used for meta-analysis in this article.

Appendix S3 Phylogenetic tree, funnel plot and normal quantile plot.

Appendix S4 Results with the random-model in traditional meta-analysis versus PMA by phyloMeta v1.3.

Acknowledgments

We are very grateful to Dr. Alexandra Weigelt and two anonymous reviewers for their valuable comments on an early version of this paper.

Author Contributions

Conceived and designed the experiments: MD. Performed the experiments: XFX YLZ. Analyzed the data: XFX YBS. Contributed reagents/materials/analysis tools: XFX YBS MD. Contributed to the writing of the manuscript: XFX YBS YLZ XP MD.

References

1. Kolasa J, Pickett STA (1991) Ecological Heterogeneity. Springer, New York, USA.
2. Hutchings MJ, Wijesinghe DK (1997) Patchy habitats, division of labour and growth dividends in clonal plants. Trends in Ecology & Evolution 12: 390-394.
3. Hutchings MJ, John EA (2004) The effects of environmental heterogeneity on root growth and root/shoot partitioning. Annals of Botany 94: 1-8.
4. He WM, Alpert P, Yu FH, Zhang LL, Dong M (2011) Reciprocal and coincident patchiness of multiple resources differentially affects benefits of clonal integration in two perennial plants. Journal of Ecology 99: 1202-1210.
5. Sultan SE (1987) Evolutionary implications of phenotypic plasticity in plants. Evolutionary Biology 21: 127-178.
6. Hutchings MJ, de Kroon H (1994) Foraging in plants: the role of morphological plasticity in resource acquisition. In: Begon M, Fitter AH, editors. Advances in Ecological Research: Academic Press. pp. 159-238.
7. Klimeš L, Klimešová J, Hendriks R, van Groenendael J (1997) Clonal plant architecture: a comparative analysis of form and function. In: de Kroon H, van Groenendael J, editors. The Ecology and Evolution of Clonal Plants. Leiden, The Netherlands: Backhuys Publishers. pp. 1-29.
8. Prach K, Pyšek P (1994) Clonal plants - what is their role in succession? Folia Geobotanica 29: 307-320.
9. Tamm A, Kull K, Sammul M (2002) Classifying clonal growth forms based on vegetative mobility and ramet longevity: a whole community analysis. Evolutionary Ecology 15: 383-401.

10. Fischer M, van Kleunen M (2002) On the evolution of clonal plant life histories. Evolutionary Ecology 15: 565-582.

11. Slade AJ, Hutchings MJ (1987) The effects of nutrient availability on foraging in the clonal herb *Glechoma hederacea*. Journal of Ecology 75: 95-112.

12. Sutherland WJ, Stillman RA (1988) The foraging tactics of plants. Oikos 52: 239-244.

13. Dong M, During HJ, Werger MJA (1997) Clonal plasticity in response to nutrient availability in the pseudoannual herb, *Trientalis europaea* L. Plant Ecology 131: 233-239.

14. Stoll P, Egli P, Schmid B (1998) Plant foraging and rhizome growth patterns of *Solidago altissima* in response to mowing and fertilizer application. Journal of Ecology 86: 341-354.

15. Sachs T (2002) Developmental processes and the evolution of plant clonality. Evolutionary Ecology 15: 485-500.

16. de Kroon H, van Groenendael J (1997) The Ecology and Evolution of Clonal Plants. Leiden: Backhuys Publishers.

17. Dong M, During HJ, Werger MJA (1996) Morphological responses to nutrient availability in four clonal herbs. Plant Ecology 123: 183-192.

18. Dong M, de Kroon H (1994) Plasticity in morphology and biomass allocation in *Cynodon dactylon*, a grass species forming stolons and rhizomes. Oikos 70: 99-106.

19. de Kroon H, Hutchings MJ (1995) Morphological plasticity in clonal plants: the foraging concept reconsidered. Journal of Ecology 83: 143-152.

20. Adams DC (2008) Phylogenetic meta-analysis. Evolution 62: 567-572.

21. Song YB, Yu FH, Keser L, Dawson W, Fischer M, Dong M, van Kleunen M (2013) United we stand, divided we fall: a meta-analysis of experiments on clonal integration and its relationship to invasiveness. Oecologia 171: 317-327.

22. Honnay O, Jacquemyn H (2008) A meta-analysis of the relation between mating system, growth form and genotypic diversity in clonal plant species. Evolutionary Ecology 22: 299-312.

23. Chamberlain SA, Hovick SM, Dibble CJ, Rasmussen NL, Van Allen BG, et al. (2012) Does phylogeny matter? Assessing the impact of phylogenetic information in ecological meta-analysis. Ecology Letters 15: 627-636.

24. Lajeunesse MJ (2009) Meta-analysis and the comparative phylogenetic method. The American Naturalist 174: 369-381.

25. Beckmann M, von Wehrden H (2012) Where you search is what you get: literature mining – Google Scholar versus Web of Science using a data set from a literature search in vegetation science. Journal of Vegetation Science 23: 1197-1199.

26. Moher D, Liberati A, Telzlaff J, Altman DF, The PRISMAGroup (2009) Preferred reporting items for systematic reviews and meta-analyses: The PRISMA statement. PLoS Medicine 6: e1000097.

27. Sullivan L, Wildova R, Goldberg D, Vogel C (2010) Growth of three cattail (*Typha*) taxa in response to elevated CO_2. Plant Ecology 207: 121-129.

28. Gurevitch J, Morrow LL, Alison W, Walsh JS (1992) A meta-analysis of competition in field experiments. American Naturalist 140: 539-572.

29. Wolf FM (1986) Meta-analysis: quantitative methods for research synthesis. Sage Publications, London, UK.

30. Bolnick DI, Preisser EL (2005) Resource competition modifies the strength of trait-mediated predator-prey interactions: A meta-analysis. Ecology 86: 2771-2779.

31. Marczak LB, Thompson RM, Richardson JS (2007) Meta-analysis: trophic level, habitat, and productivity shape the food web effects of resource subsidies. Ecology 88: 140-148.

32. Gurevitch J, Curtis PS, Jones MH (2001) Meta-analysis in ecology. Advances in Ecological Research 32: 199-247.

33. Fedorov S (2008) GetData graph digitizer. Available: http://www.getdata-graph-digitizer.com. Accessed 2014 August 24.

34. Lajeunesse MJ, Forbes MR (2003) Variable reporting and quantitative reviews: a comparison of three meta-analytical techniques. Ecology Letters 6: 448-454.

35. van Kleunen M, Weber E, Fischer M (2010) A meta-analysis of trait differences between invasive and non-invasive plant species. Ecology Letters 13: 235-245.

36. Leimu R, Mutikainen PIA, Koricheva J, Fischer M (2006) How general are positive relationships between plant population size, fitness and genetic variation? Journal of Ecology 94: 942-952.

37. Gurevitch J, Hedges LV (2001) Meta-analysis: combining the results of independent experiments. In: Scheiner SM, Gurevitch J, editors. Design and analysis of ecological experiments. New York: Oxford University Press. pp. 347-369.

38. Rosenberg MS, Adams DC, Gurevitch J (2000) MetaWin: statistical software for meta-analysis: version 2.0. Sunderland: Sinauer Associates.

39. Webb CO, Donoghue MJ (2005) Phylomatic: tree assembly for applied phylogenetics. Molecular Ecology Notes 5: 181-183.

40. Webb CO, Ackerly DD, Kembel SW (2008) Phylocom: software for the analysis of phylogenetic community structure and trait evolution. Bioinformatics 24: 2098.

41. Lajeunesse MJ (2011) phyloMeta: a program for phylogenetic comparative analyses with meta-analysis. Bioinformatics 27: 2603-2604.

42. Nakagawa S, Santos ESA (2012) Methodological issues and advances in biological meta-analysis. Evolutionary Ecology: 1-22.

43. Carmona D, Lajeunesse MJ, Johnson MTJ (2011) Plant traits that predict resistance to herbivores. Functional Ecology 25: 358-367.

44. Rosenthal R (1979) The file drawer problem and tolerance for null results. Psychological Bulletin 86: 638-641.

45. Wang MC, Bushman BJ (1998) Using the normal quantile plot to explore meta-analytic data sets. Psychological Methods 3: 46-54.

46. Gates S (2002) Review of methodology of quantitative reviews using meta-analysis in ecology. Journal of Animal Ecology 71: 547-557.

47. Begg CB (1994) Publication bias. In: Cooper H, Hedges LV, editors. The Handbook of Research Synthesis: New York, NY, US: Russell Sage Foundation. pp. 399-409.

48. Rosenthal R (1991) Meta-analytic procedures for social research (rev. ed.): Thousand Oaks, CA, US: Sage Publications, Inc.

49. Rosenberg MS (2005) The file-drawer problem revisited: a general weighted method for calculating fail-safe numbers in meta-analysis. Evolution 59: 464-468.

50. Aguilar R, Ashworth L, Galetto L, Aizen MA (2006) Plant reproductive susceptibility to habitat fragmentation: review and synthesis through a meta-analysis. Ecology Letters 9: 968-980.

51. Dong M (1996) Clonal growth in plants in relation to resource heterogeneity: foraging behavior. Acta Botanica Sinica 10: 828-835.

52. Slade AJ, Hutchings MJ (1987) The effects of light intensity on foraging in the clonal herb *Glechoma hederacea*. Journal of Ecology 75: 639-650.

53. Hedges LV, Olkin I (1985) Statistical methods for meta-analysis: Academic Press New York.

54. Bell DL, Galloway LF (2008) Population differentiation for plasticity to light in an annual herb: Adaptation and cost. American Journal of Botany 95: 59-65.

55. Evans JP, Cain ML (1995) A spatially explicit test of foraging behavior in a clonal plant. Ecology 76: 1147-1155.

56. Dong M (1993) Morphological plasticity of the clonal herb *Lamiastrum galeobdolon* (L.) Ehrend. & Polatschek in response to partial shading. New Phytologist 124: 291-300.

57. Song YB, Chen LY, Xiong W, Dai WH, Dong M (2014) Variation of functional clonal traits along altitude in two fern species. Pakistan Journal of Botany, accepted.

58. van der Hoeven EC, de Kroon H, During HJ (1990) Fine-scale spatial distribution of leaves and shoots of two chalk grassland perennials. Plant Ecology 86: 151-160.

59. Oborny B, Englert P (2012) Plant growth and foraging for a patchy resource: A credit model. Ecological Modelling 234: 20-30.

60. de Vries F, Liiri M, Bjørnlund L, Setälä H, Christensen S, et al. (2012) Legacy effects of drought on plant growth and the soil food web. Oecologia 170: 821-833.

61. van Kleunen M, Fischer M, Schmid B (2000) Costs of plasticity in foraging characteristics of the clonal plant *Ranunculus reptans*. Evolution 54: 1947-1955.

62. Thompson L, Harper JL (1988) The effect of grasses on the quality of transmitted radiation and its influence on the growth of white clover *Trifolium repens*. Oecologia 75: 343-347.

63. Dong M, Pierdominici MG (1995) Morphology and growth of stolons and rhizomes in three clonal grasses, as affected by different light supply. Plant Ecology 116: 25-32.

64. Weigelt A, Steinlein T, Beyschlag W (2005) Competition among three dune species: the impact of water availability on below–ground processes. Plant Ecology 176: 57-68.

65. Lovell P, Lovell P (1985) The importance of plant form as a determining factor in competition and habitat exploitation. In: White J, editors; Studies on Plant Demography: a Festschrift for John L Harper Academic Press, New York. pp. 209-221.

66. Hutchings MJ (1988) Differential foraging for resources, and structural plasticity in plants. Trends in Ecology & Evolution 3: 200-204.

PERMISSIONS

LIST OF CONTRIBUTORS

Jing Xu, Chunhui Zhang, Wei Liu and Guozhen Du
State Key Laboratory of Grassland and Agroecosystems, School of Life Science, Lanzhou University, Lanzhou, P.R. China

Wenlong Li
State Key Laboratory of Grassland and Agroecosystems, School of Pastoral Agriculture Science and Technology, Lanzhou University, Lanzhou, P.R. China

John K. Senior
School of Plant Science, University of Tasmania, Hobart, TAS, Australia

Jennifer A. Schweitzer and Joseph K. Bailey
School of Plant Science, University of Tasmania, Hobart, TAS, Australia
Department of Ecology and Evolutionary Biology, University of Tennessee, Knoxville, Tennessee, United States of America

Julianne ÓReilly-Wapstra
School of Plant Science and National Centre for Future Forest Industries, University of Tasmania, Hobart, TAS, Australia

Samantha K. Chapman and Adam Langley
Department of Biology, Villanova University, Villanova, Pennsylvania, United States of America

Dorothy Steane
School of Plant Science, University of Tasmania, Hobart, TAS, Australia
University of the Sunshine Coast, Sippy Downs, Queensland, Australia

Hafiz Maherali
Department of Integrative Biology, University of Guelph, Guelph, Ontario, Canada

John N. Klironomos
Department of Biology, I.K. Barber School of Arts and Sciences, University of British Columbia – Okanagan, Kelowna, British Columbia, Canada

María Natalia Umaña, Ángela Cano and Pablo R. Stevenson
Universidad de Los Andes, Laboratorio de Ecología de Bosques Tropicales y de Primatología, Centro de Investigaciones Ecológicas La Macarena, Bogotá, Colombia

Natalia Norden
Universidad de Los Andes, Laboratorio de Ecología de Bosques Tropicales y de Primatología, Centro de Investigaciones Ecológicas La Macarena, Bogotá, Colombia
Pontificia Universidad Javeriana, Departamento de Ecología y Territorio, Bogotá, Colombia

Elizabeth R. Ellwood and Richard B. Primack
Department of Biology, Boston University, Boston, Massachusetts, United States of America

Stanley A. Temple
Department of Forest and Wildlife Ecology, University of Wisconsin, Madison, Wisconsin, United States of America
Aldo Leopold Foundation, Baraboo, Wisconsin, United States of America

Nina L. Bradley
Aldo Leopold Foundation, Baraboo, Wisconsin, United States of America

Charles C. Davis
Department of Organismic and Evolutionary Biology, Harvard University Herbaria, Cambridge, Massachusetts, United States of America

Natsumi Kanzaki
Department of Forest Microbiology, Forestry and Forest Products Research Institute, Tsukuba, Ibaraki, Japan

Ryusei Tanaka
Division of Parasitology, Faculty of Medicine, University of Miyazaki, Miyazaki, Miyazaki, Japan

Robin M. Giblin-Davis
Fort Lauderdale Research and Education Center, University of Florida/IFAS, Davie, Florida, United States of America

Kerrie A. Davies
Centre for Evolutionary Biology and Biodiversity, School of Agriculture, Food and Wine, The University of Adelaide, Waite Campus, Glen Osmond, South Australia, Australia

José R. Ferrer-Paris and Ada Sánchez-Mercado
Kirstenbosch Research Centre, South African National Biodiversity Institute, Cape Town, Western Cape, Republic of South Africa

Botany Department, University of Cape Town, Cape Town, Western Cape, Republic of South Africa
Centro de Estudios Botánicos y Agroforestales, Instituto Venezolano de Investigaciones Científicas, Maracaibo, Estado Zulia, Venezuela

Ángel L. Viloria
Centro de Ecología, Instituto Venezolano de Investigaciones Científicas, Caracas, Distrito Capital, Venezuela

John Donaldson
Kirstenbosch Research Centre, South African National Biodiversity Institute, Cape Town, Western Cape, Republic of South Africa
Botany Department, University of Cape Town, Cape Town, Western Cape, Republic of South Africa

Yingmei Peng
State Key Laboratory of Genetic Resources and Evolution, Kunming Institute of Zoology, Chinese Academy of Sciences, Kunming, PR China

Jing Cai
Shenzhen Key Laboratory for Orchid Conservation and Utilization, National Orchid Conservation Center of China and Orchid Conservation and Research Center of Shenzhen, Shenzhen, China
Center for Biotechnology and BioMedicine, Graduate School at Shenzhen, Tsinghua University, Shenzhen, China

Wen Wang and Bing Su
State Key Laboratory of Genetic Resources and Evolution, Kunming Institute of Zoology, Chinese Academy of Sciences, Kunming, PR China

Fritz Hans Schweingruber
Swiss Federal Research Institute WSL, Birmensdorf, Switzerland

Pavel Říha and Jiří Doležal
Section of Plant Ecology, Institute of Botany, Academy of Sciences of the Czech Republic, Třeboň, Czech Republic
Department of Botany, Faculty of Science, University of South Bohemia, České Budějovice, Czech Republic

Shi-Bao Zhang
Key Laboratory of Economic Plants and Biotechnology, Kunming Institute of Botany, Chinese Academy of Sciences, Kunming, Yunnan, China
Key Laboratory of Tropical Plant Ecology, Xishuangbanna Tropical Botanical Garden, Chinese Academy of Sciences, Kunming, Yunnan, China

Zhi-Jie Guan
Key Laboratory of Economic Plants and Biotechnology, Kunming Institute of Botany, Chinese Academy of Sciences, Kunming, Yunnan, China
State Key Laboratory of Plant Physiology and Biochemistry and College of Agronomy and Biotechnology, China Agricultural University, Beijing, China

Mei Sun and Kun-Fang Cao
Key Laboratory of Tropical Plant Ecology, Xishuangbanna Tropical Botanical Garden, Chinese Academy of Sciences, Kunming, Yunnan, China

Juan-Juan Zhang and Hong Hu
Key Laboratory of Economic Plants and Biotechnology, Kunming Institute of Botany, Chinese Academy of Sciences, Kunming, Yunnan, China

Cody E. Hinchliff and Stephen Andrew Smith
Department of Ecology and Evolutionary Biology, University of Michigan. Ann Arbor, Michigan, United States of America

Suzana Alcantara
Departamento de Botânica, Instituto de Biociências, Universidade de São Paulo, São Paulo, SP, Brazil
Department of Botany, Field Museum of Natural History, Chicago, Illinois, United States of America

Richard H. Ree
Department of Botany, Field Museum of Natural History, Chicago, Illinois, United States of America

Fernando R. Martins
Departamento de Biologia Vegetal, Instituto de Biologia, Universidade Estadual de Campinas – UNICAMP, Campinas, SP, Brazil

Lúcia G. Lohmann
Departamento de Botânica, Instituto de Biociências, Universidade de São Paulo, São Paulo, SP, Brazil

Fabian Runge
University of Hohenheim, Institute of Botany, Stuttgart, Germany

Beninweck Ndambi
University of Hohenheim, Institute of Botany, Stuttgart, Germany
University of Hohenheim, Institute of Plant Production and Agroecology in the Tropics and Subtropics, Stuttgart, Germany

Marco Thines
Biodiversity and Climate Research Centre (BiK-F), Frankfurt (Main), Germany

Senckenberg Gesellschaft für Naturforschung, Frankfurt (Main), Germany
Johann Wolfgang Goethe University, Department of Biological Sciences, Institute of Ecology, Evolution and Diversity, Frankfurt (Main), Germany

Rodrigo S. Rios and Cristian Salgado-Luarte
Departamento de Biología, Universidad de La Serena, La Serena, Chile

Ernesto Gianoli
Departamento de Biología, Universidad de La Serena, La Serena, Chile
Departamento de Botánica, Universidad de Concepción, Concepción, Chile

Jae-Cheon Sohn, Jerome C. Regier and Charles Mitter
Department of Entomology, University of Maryland, College Park, Maryland, United States of America

Donald Davis
Department of Entomology, National Museum of Natural History, Smithsonian Institution, Washington DC, United States of America

Jean-François Landry
Agriculture and Agri-Food Canada, Eastern Cereal and Oilseed Research Centre, C.E.F., Ottawa, Canada

Andreas Zwick
Department of Entomology, State Museum of Natural History, Stuttgart, Germany

Michael P. Cummings
Laboratory of Molecular Evolution, Center for Bioinformatics and Computational Biology, University of Maryland, College Park, Maryland, United States of America

Kristýna Vazačová and Zuzana Münzbergová
Department of Botany, Faculty of Science, Charles University, Prague, Czech Republic
Institute of Botany, Academy of Sciences of the Czech Republic, Průhonice, Czech Republic

Eugen Egorov and Martin Brändle
Department of Ecology, Faculty of Biology, Philipps Universität, Marburg, Germany

Daniel Prati, Markus Fischer, Barbara Schmitt and Stefan Blaser
Institute of Plant Sciences, University of Bern, Bern, Switzerland

Walter Durka and Stefan Michalski
Department Community Ecology, Helmholtz Centre for Environmental Research, Halle, Germany

Robert Muscarella and María Uriarte
Department of Ecology, Evolution and Environmental Biology, Columbia University, New York, New York 10027, United States of America

David L. Erickson and W. John Kress
Department of Botany, MRC-166, National Museum of Natural History Smithsonian Institution, P.O. Box 37012, Washington, D. C., 20013, United States of America

Nathan G. Swenson
Department of Plant Biology, Michigan State University, East Lansing, Michigan 48824, United States of America

Jess K. Zimmerman
Department of Environmental Science, University of Puerto Rico, San Juan, Puerto Rico 00925, United States of America

Sergio A. Castro, Victor M. Escobedo and Jorge Aranda
Laboratorio de Ecología y Biodiversidad, Departamento de Biología, and Center for the Development of Nanoscience and Nanotechnology (CEDENNA), Universidad de Santiago de Chile, Santiago, Chile

Gastón O. Carvallo
Instituto de Biología, Facultad de Ciencias, Pontificia Universidad Católica de Valparaíso, Valparaíso, Chile

Xiu-Fang Xie and Ming Dong
Key Laboratory of Hangzhou City for Ecosystem Protection and Restoration, College of Life and Environmental Sciences, Hangzhou Normal University, Hangzhou, China
State Key Laboratory of Vegetation and Environmental Change, Institute of Botany, Chinese Academy of Sciences, Beijing, China

Yao-Bin Song
Key Laboratory of Hangzhou City for Ecosystem Protection and Restoration, College of Life and Environmental Sciences, Hangzhou Normal University, Hangzhou, China

Ya-Lin Zhang and Xu Pan
State Key Laboratory of Vegetation and Environmental Change, Institute of Botany, Chinese Academy of Sciences, Beijing, China

Index